# Parallel Solution Methods in Computational Mechanics

## Wiley Series in Solving Large-scale Problems in Mechanics

The objective of this series is to publish timely monographs, state-of-the art accounts, and special events proceedings on advanced topics related to the solution of large-scale and/or computationally intensive problems in mechanics.

Titles published in this series encompass recent progress and achievements in computational solid and structural mechanics, and computational fluid dynamics, including parallel processing, linear-nonlinear eigenvalue solution methods, mesh generation, coupled-field problems, adaptive methods, metal forming, optimum design, stochastic methods, and knowledge-based methods.

### *Series Editor*

**Professor M Papadrakakis**
Institute of Structural Analysis and Seismic Research
National Technical University of Athens
Zografou Campus
CR – 15773 Athens
Greece

# Parallel Solution Methods in Computational Mechanics

*Edited by*

**M. Papadrakakis**
*National Technical University of Athens, Greece*

John Wiley & Sons
Chichester • New York • Weinheim • Brisbane • Singapore • Toronto

National          01243 779777
International     (+44) 1243 779777
e-mail (for orders and customer service enquiries): cs-books@wiley.co.uk
Visit our Home Page on http://www.wiley.co.uk
or http://www.wiley.com

*Other Wiley Editorial Offices*

John Wiley & Sons, Inc., 605 Third Avenue,
New York, NY 10158-0012, USA

VCH Verlagsgesellschaft mbH, Pappelallee 3,
D–69469 Weinheim, Germany

Jacaranda Wiley Ltd, 33 Park Road, Milton,
Queensland 4064, Australia

John Wiley & Sons (Asia) Pte Ltd, 2 Clementi Loop #02-01,
Jin Xing Distripark, Singapore 129809

John Wiley & Sons (Canada) Ltd, 22 Worcester Road,
Rexdale, Ontario M9W 1L1, Canada

***British Library Cataloguing in Publication Data***

A catalogue record for this book is available from the British Library

ISBN  0 471 95696 1

Reproduced from camera-ready copy supplied by the editor.
Printed and bound in Great Britain by Bookcraft, Midsomer-Norton, Somerset
This book is printed on acid-free paper responsibly manufactured from sustainable forestation,
for which at least two trees are planted for each one used for paper production.

# Contents

# Preface

This is the second volume in the series Solving Large-Scale Problems in Mechanics devoted to high-performance computing using the new generation of computers with parallel and distributed processing capabilities. The last two decades have been characterized by two important developments in scientific computing. First, the increasing speed and expanded storage capacity of modern high-performance computers with new hardware components and parallel architectures. Second, the new advanced numerical methods and programming techniques for handling and solving large-scale problems. The bigger and faster computers because, the larger are the problems scientists and engineers want to solve. The mathematical models become ever more sophisticated and the computer simulations even more complex and extensive. However, the use of computers in engineering applications has traditionally lagged hardware advances. New algorithms and codes are required in order to exploit effectively these modern computer architectures, as programs suitable for conventional computers achieve very modest improvement performance on these new machines. There is therefore an urgent need to develop and test powerful solution and data handling techniques capable of exploiting the potential of modern computers and of accomplishing solution of complex engineering problems in acceptable computing time.

This second volume is born of that need. It comprises a number of self-contained chapters contributed by leading researchers in the field dealing with recent advances in the solution of large-scale problems in mechanics. The newcorner in the field of parallel and distributed computing will receive sufficient introduction to comprehend the pertinent issues and to apply the technologies presented. The seasoned researcher, on the other hand, will benefit from presentation of the latest work of the authors both in theory and algorithmic-computational analysis as well as in applications to real problems.

**Chapter 1** illustrates how the emergence of high-performance computer architectures, visualization, virtual reality and advanced telecommunication in the mid 1990s and beyond promises to dramatically alter the way designers, manufactures, engineers and scientists will do their work. Examples are presented to demonstrate how the application of the latest developments in computer science can lead to heightened productivity of the design engineering team. Chapters 2 and 3 review parallel techniques using domain decomposition for solving large-scale structural mechanics problems. In **Chapter 2**, the basic model elasticity problem is described together with the general domain decomposition technique, the selection of the proper preconditioner, the global coarse solver on the interface, an abstract convergence theory, different partitioning techniques and implementation details. It is shown that the additive Schwarz algorithms, provide an efficient abstract iterative procedure for solving linearly elliptic problems in parallel and different practical implementations are introduced for solving the interface problem. **Chapter 3** discusses three domain decomposition formulations combined with the preconditioned conjugate gradient method using properly selected preconditioners. The global subdomain implementation in which the iterative algorithm is applied on the global matrix, the Schur complement implementation in which the preconditioned conjugate gradient method is applied on the interface problem after eliminating the internal degrees of freedom and the dual subdomain implementation which operates on overlapping subdomains on the global level after partitioning the domain into a set of totally disconnected subdomains. Furthermore, the solution of linear systems with multiple right-hand sides is addressed together with topology and shape optimization problems as well as stochastic finite element analysis problems with Monte Carlo simulation.

**Chapter 4** discusses the application of parallel adaptive multigrid methods for elasticity, plasticity and eigenvalue problems. The emergence of multigrid methods represents one of the more interesting recent developments in numerical analysis. This chapter presents some methods the author has used to solve a variety of solid mechanics problems using multigrid methods. A basic algorithm is first discussed to solve linear matrix equations which is then embedded within Newton-Raphson iteration for history-dependent nonlinear problems. Several methods for computing eigensolutions are then presented, followed by a discussion of multigrid methods incorporating adaptive mesh refinement. **Chapter 5** presents the accuracy aspects of multi-time step explicit integration schemes in structural dynamic analysis. The results on numerical experimentation provided suggest that multi-time step integration should not be used indiscriminately. It is observed that a subcycling solution can offset the accuracy achieved by the introduction of small elements into the mesh. Thus, while selected finer spatial discretization is used to further reduce the errors in the solution, selected coarser temporal discretization should be used very carefully so as not to induce errors back into the solution.

**Chapter 6** reviews Krylov subspace methods for the solution of general sparse matrix problems in parallel environments. Three different approaches are discussed. The first extracts parallelism from a standard iterative solver implementing the standard operation such as triangular solves, inner products, and other vector operation in parallel. In the second class, either an intrinsically parallel preconditioner is employed or a strategy, such as reordering, is used to enhance parallelism of the preconditioner. These strategies include multicoloring techniques and polynomial preconditioners among others. The application framework is that of large sparse linear systems that are not necessarily well structured. **Chapter 7** highlights some important aspects of coupled nonlinear dynamic aeroelastic response problems. The solution of the governing three field equations with mixed implicit-implicit and explicit-implicit staggered procedures are analyzed with particular reference to accuracy, stability, subcycling, distributed computing, I/O transfers, and parallel processing. A general and flexible framework for implementing the partitioned analysis with non-matching fluid-structure interfaces on heterogeneous and/or parallel computational platforms is also described. **Chapter 8** discusses finite element analysis and active element placement techniques for control-structure interaction applications. The dynamic response analyses and design optimization methods related to the placement of active elements for shape and vibration control of active and/or passive dumping enhanced systems are extremely computationally intensive problems. This chapter presents finite element based massively parallel processing computational algorithm-solution procedures for control-structure interaction systems. Emphasis is placed on distributed piezoelectrically actuator-sensor controlled systems of the multiaxially active, collocated piezoelectric sensor-actuator structural element type.

In **Chapter 9**, the theory, parallelization and application of the stochastic search method of evolution strategies for continuous and mixed-discrete structural optimization problems are introduced. The evolution strategies imitate biological evolution in nature and have a characteristic that render themselves amenable to natural parallelization. Instead of a single point, they work simultaneously with a population of designs points in the space variables. Compared to conventional optimization methods, the stochastic search methods are relative simple and easy to adjust to almost any optimization problem, they are well suited for parallelization and can locate a global optimum with a certain probability. **Chapter 10** presents high performance computing techniques for flow simulations. All simulations are carried out in three dimensions and are based on finite element formulations applicable to unstructured meshes. The latter are either semi-discrete, for solving problems with fixed spatial domains, or space-time formulations, for solving flow problems with moving boundaries and interfaces. The large coupled, nonlinear equation systems that need to be solved at every time step are solved using iterative strategies. Diagonal and nodal-block-diagonal preconditioners are used together with more sophisticated ones such as the clustered element-by-element and the mixed element-by-element and cluster companion preconditioners. The solution vector is updated using the GMRES technique.

**Chapter 11** discusses several examples of the application of computational fluid dynamics and computational electromagnetics technology to aerospace and other commercial products and outlines the role of computer simulation and high performance computing in reducing the design-cycle cost and time to market. The development of modern aerospace vehicles increasingly requires a computational environment referred to as Computational Sciences that encompasses many disciplines. The mission of Computational Science is to intergrate such multidisciplinary technology to achieve synergism in the design process. Several examples of computational aerospace applications are described together with computational commercial applications such as automobile simulation, thermal modeling, image enhancement, micro-electromechanical systems, computational manufacturing, bioelectromagnetics. In **Chapter 12** a multi-color neural network application with feed back mechanism is discussed for parallel finite element fluid analysis. This chapter focuses on the artificial neural network and its applications to the field of computational mechanics. Recent research activities on neuro-computations and the feasibilities of the neural network as a computational mechanics tool, especially in the area of computational fluid dynamics, are discussed. The neuro-solver is utilized for the Poissons equation, which has to be solved for the pressure at every time step in the incompressible viscous flow analysis. The feasibilities of the neuro-solver are investigated with respect to the convergence behaviour and the parallel efficiency. **Chapter 13** discusses parallel automatic mesh generation and adaptive mesh control techniques. Automated and adaptive techniques provide the promise of reliably solving complex problems to the derived level of accuracy, while relieving the user from the time consuming and error prone tasks associated with mesh generation. This chapter is focused on the techniques required to support automated adaptive analysis on distributed memory MIMD parallel computers. As the mesh is adapted partition work load becomes unbalanced, therefore procedures must be available to effectively modify the mesh partitions to regain load balance for the next computational step. For this purpose a set of data structures and algorithms are discussed for the effective parallel control of evolving meshes together with an efficient three-dimensional automatic mesh generation and a parallel adaptive solver.

I would finally like to express my appreciation to the contributing authors for entering into this venture and for their spirit of collaboration during the preparation of this book. I would also like to express my gratitude to C. Apostolopoulou for her patience and perseverance in collecting the camera-ready chapters and ensuring the unformity in style and layout of this volume.

**M. Papadrakakis**

# List of Contributors

**R. K. Agarwal**          *Aerospace Engineering Department*
                           *Wichita State University*
                           *Wichita*
                           *KS 67260, USA*

**C. Farhat**              *Department of Aerospace Engineering Sciences*
                           *Centre for Aerospace Structures*
                           *University of Colorado at Boulder*
                           *Boulder*
                           *CO 80309 - 0429, USA*

**S. Gupta**               *Engineering Analysis Services Inc*
                           *Bingham Center*
                           *30800 Telegraph Road*
                           *Suite 3700*
                           *Bingham Farms, MI 48025, USA*

**P. Le Tallec**           *Université de Paris - Dauphine &*
                           *INRIA - Rocquencourt*
                           *BP 105*
                           *F - 78153 Le Chesnay Cedex*
                           *France*

**H. Okuda**               *Department of Quantum Engineering and Systems*
**G. Yagawa**              *Science*
                           *University of Tokyo*
                           *7 - 3 - 1 Hongo*
                           *Bunkyo-ku*
                           *Tokyo 113, Japan*

**M. Papadrakakis**        *Institute of Structural Analysis and Seismic Research*
                           *National Technical University of Athens*
                           *Zografou Campus*
                           *15773 Athens*
                           *Greece*

**I. D. Parsons**         *Department of Civil Engineering*
                         *University of Illinois*
                         *Urbana*
                         *IL 61801, USA*

**E. J. Plaskacz**        *Argonne National Laboratory*
                         *9700 South Cass Avenue*
                         *Argonne*
                         *IL 60439, USA*

**M. Ramirez**            *Engineering System Research Center*
                         *3172 Etcheverry Hall*
                         *University of California at Berkeley*
                         *Berkeley*
                         *CA 94720, USA*

**Y. Saad**               *Department of Computer Science*
                         *University of Minnesota*
                         *Minneapolis*
                         *MN 55455, USA*

**V. Shankar**            *Rockwell International Science Center*
                         *Thousand Oaks*
                         *CA 91360, USA*

**M. S. Shephard**        *Scientific Computation Research Center*
**J. E. Flaherty**        *Rensselaer Polytechnic Institute*
**H. L. de Cougny**       *Troy*
**C. L. Bottasso**        *NY 12180 - 3590*
**C. Ozturan**            *USA*

**R. C. Shieh**           *MRJ Inc*
                         *10560 Arrowhead Drive*
                         *Fairfax*
                         *VA 22030, USA*

**G. Thierauf**           *Department of Civil Engineering*
**J. Cai**                *University of Essen*
                         *D - 45117 Essen*
                         *Germany*

**M. Vidrascu**           *INRIA - Rocquencourt*
                         *BP 105*
                         *F - 78153 Le Chesnay Cedex*
                         *France*

Οὐκ ἦν τὸ φοβούμενον λύειν ὑπὲρ τῶν κυριωτάτων μὴ κατειδύτα τίς ἤ τοῦ σύμπαντος φύσις, ἀλλ᾽ ὑποπτεύοντά τι τῶν κατὰ τοὺς μύθους.
Ἐπικούρου, Κύριαι Δόξαι 12


There is no way to dispel the fear about matters of supreme importance, for someone who does not know what the nature of the universe is but retains some of the fears based on mythology.
*Epicurus, Ratae Sententiae 12*
*(transl. A. Long- D. Sedley)*

# 1

# Overcoming Engineering Design Time and Distance Hurdles Via High-Performance Computing

E. J. Plaskacz [1]

## 1.1 INTRODUCTION

Finite element analysis is the dominant tool in the numerical solution of the partial differential equations. Finite element analysis readily accomodates nonlinearities and complex geometries. This versatility has resulted in its extensive use. Akin (1982) identifies the following three major categories of finite element applications.

**Equilibrium problems** include the stress analysis of static structures, electrostatics, magnetostatics, steady state thermal conduction and fluid flow in porous media. There are no time-dependent terms in their solutions.

**Eigenvalue problems** include the computation of the critical frequencies of a bridge, the critical load of a structural column and energy levels in quantum mechanics. If a system of equations possesses solutions satisfying the given boundary conditions for only certain distinct values of a parameter $\lambda$, such a value of $\lambda$ is called an eigenvalue and the corresponding solution is an eigenfunction.

**Propagation problems** have solutions with time-dependent terms. These systems include earthquake response of multistory buildings, fluid-structure interaction problems, pollutant dispersion, vehicle aerodynamics, contact-impact simulations and automobile crashworthiness.

Widespread application of the technique as a design tool for linear static analysis has been prevalent for thirty years. The treatment of dynamic and nonlinear problems entails an increase in the number of numerical operations. The widespread availability of high-speed digital computers during the 1970s extended the range of applications to dynamic and nonlinear analyses. The 1980s witnessed the introduction of vector and coarse grained parallel supercomputers. The speed of the computational hardware has increased dramatically over the past thirty years. However, it has failed to keep

---

[1]  Argonne National Laboratoty, 9700 South Cass Avenue, Argonne, Illinois 60439

*Parallel Solution Methods in Computational Mechanics* edited by M. Papadrakakis
© 1997 John Wiley & Sons Ltd

pace with the burgeoning resource requirements of nonlinear and multidisciplinary simulations.

The aim of this chapter is to illustrate how the emergence of high-performance computer architectures, visualization, virtual reality and advanced telecommunications in the mid 1990s promises to dramatically alter the way designers, manufacturers, engineers and scientists will do their work. Examples are presented to demonstrate how the application of the latest developments in computer science can lead to heightened productivity of the design engineering team. Through the use of parallel-distributed computing, engineers are able to perform more simulations in the same amount of time and hence find better designs. State-of-the-art telecommunications and virtual reality can be exploited to alleviate the distance constraint faced by geographically distributed collaborators. These developments promise the ability to bring researchers together through telepresence, enabling them to simultaneously study a simulation and exchange ideas.

Accordingly, this chapter revolves around the themes of time and distance as major obstacles hindering the engineering design process. Overcoming these hurdles through the application of the latest advances in computer science is the focus of the two major sections of the chapter. The goal of this chapter is to accelerate the application of these technologies in the production engineering environment. Therefore, the chapter extends beyond the usual compendium of state-of-the-art techniques with the inclusion of tutorial sections on parallel computing, message passing and computational steering. The newcomer to the field of parallel distributed computing will receive sufficient introduction to comprehend the pertinant issues and to apply the technologies presented in this as well as subsequent chapters. The seasoned researcher, on the other hand, will benefit from presentations of our latest work in parallel distributed algorithms, computational steering and collaborative engineering.

## 1.2 OVERCOMING TIME HURDLES

The iterative nature of engineering design is governed in time by the cycle of design and analysis. Design engineers are often faced with the prospect of changing a design to simultaneously satisfy constraints imposed by manufacturability, serviceability, and safety. Traditionally, the large blocks of time required for testing each combination of parameters has restricted the number of simulations which could be performed before the product deadlines were impacted. This, in turn, restricts productivity, optimality, and in most cases, creativity. For example, in studies of crashworthiness, impact and penetration, it is not unusual for an analysis to require 100 hours of CPU time on current supercomputers, despite the simplicity of the models being studied. The computational requirements for a single Belytschko-Tsay shell element (used extensively in finite element analysis) with five integration points through the thickness are on the order of 3500 floating point operations per time-step and 600 words of memory. Crashworthiness models currently consist of approximately 5000 to 100,000 shell elements and run for about 100,000 time steps. Incorporation of detailed passenger compartment, air-bag, and occupant models can easily double or triple these requirements. The difficulty in obtaining blocks of time this large in a production environment severely impairs the role of such simulations in engineering decisions.

High-performance computer architectures have proven themselves to be an effective avenue for bridging the gap between computational needs and the power of computational hardware in many disciplines. Our experience with message-passing in the context of nonlinear explicit transient finite element simulations on the 512 processor Intel Delta resulted in a speedup of 333 relative to a uniprocessor run. The practical significance of a two orders-of-magnitude increase in computational speed is that it enables designers and engineers to perform more simulations and hence find better designs. In addition, large-scale detailed simulations become manageable and the need to break up the problem and to model each region separately is reduced.

This section describes the evolution of a finite element program with explicit time integration from a sequential form to a parallel message-passing incarnation. The resulting parallel code has been used on a wide range of parallel platforms. The finite elements are allocated among multiprocessors through various domain decomposition strategies. At each time step, only internal forces for nodes at subdomain interfaces are exchanged. The Argonne-developed **p4** package is used to manage interprocessor communication. The code is portable among platforms supporting Berkeley Unix 4.3BSD interprocess communication sockets. Results are presented for problems exhibiting large-deformation, elastic-plastic dynamic transient response. Speedup factors are presented for shared memory MIMD multiprocessors, networks of workstations, and distributed memory MIMD multiprocessors.

## BACKGROUND

The performance of the currently dominant sequential Von Neumann computer architecture (SISD) is asymptotically approaching a peak speed prescribed on the underlying hardware by manufacturability and by fundamental physical limits. Two important physical limits which have received wide-spread attention are the speed of light and the effectiveness of heat dissipation. But perhaps the most compelling reason for considering parallel architectures is their cost-effectiveness. For a single processor, the relationship between incremental cost and processor performance follows a law of diminishing return. As the underlying physical limit is approached, the cost for a given incremental increase in processor computational power rises dramatically. Also, the market demand for these highly specialized and proprietary supercomputer chips is very small relative to the demand for common workstation and personal computer chips. In many instances, a parallel machine comprised of many such commodity processors not only has more speed than a uniprocessor or coarse-grained vector-parallel supercomputer, but is also less expensive. Computational scientists are in almost universal agreement that dramatic increases in supercomputer performance will be obtained through efficient parallel computers. Alternate multiprocessor computer architectures have been devised. These architectures fall under three broad categories (Flynn (1972) Desrouchers (1987)):

**SIMD** Single Instruction Multiple Data. All processors execute the same instruction; however, each processor uses its own data.

**MISD** Multiple Instruction Single Data. Each processor has a unique instruction stream which operates on the same data stream.

**MIMD** Multiple Instruction Multiple Data. Each processor has its own independent instruction and data streams. In general, processors are operating asynchronously and communication between processors is minimal.

Each of these alternate architectures requires changes in the algorithms and the underlying code. The programmer must consider details of the computer architecture to a degree unprecedented in the SISD era. The computer architecture also dictates the type of algorithm which is most appropriate.

Parallel implementations of finite elements have been considered by several investigators. Lyzenga, Raefsky, and Hager (1988) have implemented conjugate gradient solvers on moderately massive hypercube computers. They used an iterative solver based on conjugate gradients. In contrast with this work, the element nodal forces were computed by taking the product of the element stiffnesses and the displacement vector. Nour-Omid and Park (1987) have reported on the implementation of implicit solvers on Hypercube MIMD machines with large partitioned memories. Hajjar and Abel (1988) have described the implementation of an iterative solver based on conjugate gradients on a cluster of Digital Vax workstations. Their network topology was restricted to ethernet connected workstations communicating via the DECnet protocol. Their application was the seismic analysis of framed structures. Farhat and Crivelli (1989) have proposed general methods for multiprocessor finite element formulations based on stiffnesses. Johnson and Mathur (1990) have described the implementation on the CONNECTION Machine of a finite element program which also uses an iterative solver with the internal nodal forces computed via the element stiffness. Flanagan and Taylor (1987) have described the "vectorization" of a transient dynamics finite element code. Belytschko and Gilbertsen (1987) have described the implementation of explicit time integration with subcycling in a MIMD computer with shared memory. Malone (1990) describes an implementation of an explicit finite element analysis code featuring a shell element based on degenerated three-dimensional continuum elements on a distributed memory hypercube computer. Belytschko, Plaskacz, Kennedy, and Greenwell (1990) and Belytschko and Plaskacz (1992) describe the adaptation of a finite element program with explicit time integration to a massively parallel SIMD supercomputer - the CONNECTION Machine. An order-of-magnitude increase in the computational speed over a conventional supercomputer was reported. This speedup was difficult to obtain, and substantial redesign of the algorithm and underlying data structure was necessary to adapt them for use on the Connection Machine. The salient feature of the "Exchange" algorithm is the minimization of interprocessor communication at the expense of redundant computations and storage. Because the computational capabilities of the Connection Machine far exceed its communication abilities, the Exchange algorithm is much faster on the Connection Machine-2 than alternate schemes in spite of the redundancies. In all cases cited above, the resulting codes were highly platform specific and, in general offered limited portability. For example, codes developed and optimized for shared memory vector supercomputers could not easily be adapted for efficient execution on the Connection Machine.

Code portability has been the biggest obstacle for the migration of parallel computing technology beyond the realm of large universities and national laboratories. Even large corporations which have the financial resources necessary for the purchase of the latest and most powerful high-performance computers could not justify the cost of totally rewriting and reverifying legacy applications for execution on the latest parallel platform. In many cases, this would involve hundreds of thousands of lines of code. Even worse was the choice of either making a commitment to a single vendor or having to repeat the painful software porting process with the next hardware purchase.

One software mechanism for expressing parallel algorithms which has appeared in various incarnations over the past decade is message-passing. The message-passing model views the processors in a parallel computer as having local memory. However, each processor is able to communicate with other processors by sending and receiving messages. In many of its early forms, message-passing was implemented through hardware vendor proprietary libraries. The research community addressed portability constraints through the contribution of a group of public-domain portable message-passing libraries such as PARMACS (Bomans, Roose and Hempel (1990)), PICL (Geist, Heath, Peyton and Worley (1990)), TCGMSG (Harrison (1991)), p4 (Boyle, Butler, Disz, Glickfeld, Lusk, Overbeek, Patterson and Stevens (1987)) and (Butler and Lusk (1992)), PVM (Beguelin, Dongarra, Geist, Manchek, Sunderam (1991)), Chameleon (Gropp and Smith (1993)), and Zipcode (Skjellum, Smith, Doss, Leung and Morari (1994)).

When used in conjunction with one of these libraries, an algorithm formulated in terms of message passing is substantially more portable to a wider class of machines. Once the algorithm is restructured as a set of processes communicating through messages, the program can run on systems as diverse as a uniprocessor workstation, multiprocessors with and without shared memory, a group of workstations that communicate over a local area network, or any combination of the above.

## FINITE ELEMENT FORMULATION

After spatial semi-discretization is performed, the governing equations of motion for the finite element model are:

$$\mathbf{Ma}(t) = \mathbf{f}(t) \tag{1.1}$$

$$\mathbf{f}(t) = \mathbf{f}_{ext}(t) - \mathbf{f}_{int}(t) \tag{1.2}$$

Since the equation of motion must be satisfied at each time step, we have

$$\mathbf{M\ddot{u}}^n = \mathbf{f}^n \tag{1.3}$$

$$\mathbf{f}^n = \mathbf{f}_{ext}^n - \mathbf{f}_{int}^n \tag{1.4}$$

for time step n. Eq. (1.3) is integrated in time. The central difference algorithm implemented in the program proceeds through six steps.

*Step 1. Initial conditions*

The solution of a second-order ordinary differential equation requires two initial conditions   i.e. $\mathbf{u}^0$ and $\mathbf{\dot{u}}^0$. Define

$$\mathbf{u}(0) = \mathbf{u}^0 \tag{1.5}$$

$$\mathbf{\dot{u}}\left(-\tfrac{\Delta t}{2}\right) = \mathbf{\dot{u}}^0 \tag{1.6}$$

$n = t = 0$ initialize elements

*Step 2. Loop over elements:* $e = 1$ to $n_e$

(a) Gather displacements

$$\mathbf{u}_e = \mathbf{L}_e \mathbf{u} \tag{1.7}$$

(b) Evaluate strains

$$\epsilon_e = \mathbf{B}_e \mathbf{u}_e^n \tag{1.8}$$

(c) Evaluate stress

$$\sigma_e^{n+1} = \sigma_e^n + \Delta\sigma = \mathbf{S}\left(\epsilon_e^n, \epsilon_e^{n+1}\right) \tag{1.9}$$

(d) Compute element internal forces

$$\mathbf{f}_{int}^e = \int_{\Omega_e} \mathbf{B}^T \sigma_e d\Omega \tag{1.10}$$

(e) Assembly

$$\mathbf{f}_{int} = \sum_e \mathbf{L}_e^T \mathbf{f}_{int}^e \tag{1.11}$$

$$\mathbf{f}_{ext} = \sum_e \mathbf{L}_e^T \mathbf{f}_{ext}^e \tag{1.12}$$

end loop over elements

*Step 3. Loop over nodes:* $I = 1$ to $n_n$

(a) Update accelerations

$$\ddot{\mathbf{u}}_I^n = \mathbf{M}_I^{-1} \mathbf{f}_I^n \tag{1.13}$$

(b) Update velocities

$$\dot{\mathbf{u}}_I^{n+\frac{1}{2}} = \dot{\mathbf{u}}_I^{n-\frac{1}{2}} + \Delta t \ddot{\mathbf{u}}_I^n \tag{1.14}$$

(c) Update displacements

$$\mathbf{u}_I^{n+1} = \mathbf{u}_I^n + \Delta t \dot{\mathbf{u}}_I^{n+\frac{1}{2}} \tag{1.15}$$

end loop over nodes

*Step 4.* $t \leftarrow t + \Delta t$ ; $n \leftarrow n + 1$ ; go to Step 2

In the above

$$\mathbf{M} = \text{global mass matrix;}$$
$$\mathbf{a} = \text{nodal acceleration vector;}$$
$$n = \text{number of time step;}$$
$$t = \text{time;}$$
$$\mathbf{u}^n, \dot{\mathbf{u}}^n, \ddot{\mathbf{u}}^n = \text{nodal displacements, velocities and accelerations, at } t = n\Delta t;$$
$$\mathbf{f}^n, \mathbf{f}_{ext}^n, \mathbf{f}_{int}^n = \text{resultant, external and internal nodal forces, at } t = n\Delta t;$$
$$\epsilon^n, \sigma^n = \text{strain and stress tensors of element } e \text{ at } t = n\Delta t;$$
$$\mathbf{B}_e = \text{strain} - \text{displacement matrix of element } e;$$
$$\mathbf{L}_e = \text{Boolean connectivity matrix of element } e;$$
$$\Omega_e = \text{element domain.}$$

If **M** is diagonal and lumped then there is no solution of simultaneous equations and the method is explicit. The Boolean connectivity array $L_e$ is never computed; instead the operations indicated by Eqs. (1.11) and (1.12) are implemented by simply adding the entries of the element array into the appropriate locations in the global array.

## FINITE ELEMENT ANALYSIS, THE EXCHANGE ALGORITHM AND MESSAGE-PASSING

The first task a prospective user of any parallel computer platform faces is finding the parallelism in the physics underlying his application code. All finite element codes exhibit large degrees of parallelism. For finite element codes using explicit time integration, a large part of the time is consumed by element internal force computations. Instead of performing the element internal force calculations sequentially, they may be parallelized through a spatial domain decomposition. That is, the computational work may be distributed through the assignment of individual elements or groups of elements comprising a portion of the finite element mesh to each processor. When finite element analysis is performed on either traditional sequential computers or shared-memory vector computers, one finite element mesh is developed to represent the physical model under study. In contrast, since each processor in a parallel computing platform only operates on part of the full problem, it only needs the portion of the data on which it operates.

### Genesis of the Exchange Algorithm

The next task is the restructuring of the application code through the incorporation of algorithms and data structures which best express that parallelism. For explicit finite element analysis on the Connection Machine, the three data structures and algorithms which were tested are depicted in Fig.1.1. The Exchange algorithm emerged as the best option from this field of three algorithms. In the Exchange algorithm, a unit of parallel work was defined as a single finite element and its incident nodes. Thus, the accelerations of most of the nodes in the finite element mesh are integrated redundantly. However, the only interprocessor communication required is the passing of internal forces from each element to its immediate neighbors. The salient feature of the Exchange algorithm is the minimization of interprocessor communication at the expense of redundant computations and storage.

The SIMD computational model, while providing a trivial mechanism for the allocation of work among processors, was too restricted for use on general nonlinear finite element applications with unstructured meshes and heavy interprocessor communication, such as impact-contact simulations. In addition, it required a substantial part of the application code to be rewritten. However, we were able to exploit the computational power of high-performance computer architectures to test postulated hypotheses, using high resolution meshes to further our understanding of failure mechanisms. Belytschko, Chiang, and Plaskacz (1994) describe the application of the Connection Machine for large-scale finite element simulations with meshes containing 64000 elements. The objective of their work was to examine shear band formation in the dynamic response of viscoplastic plane-strain specimens under tension.

1.            Calculation of Internal Force on the Connection Machine

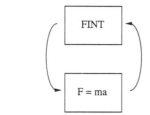

Solution of Equation of Motion on Front End

2.        Establishing a Finite Element Data Structure with Two Domain Types

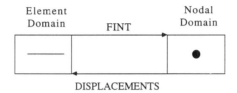

DISPLACEMENTS

3.      Establish a Finite Element Data Structure with a Single Domain Type

**Figure 1.1** Alternate Algorithms For Finite Element Analysis on the Connection Machine

## Evolution of the Exchange Algorithm from SIMD to MIMD form

Single-Instruction Multiple-Data fine-grained massively-parallel computers such as the Connection Machine-2 favor an element-per-processor data structure; interelement communication and interprocessor communication become equivalent. On any parallel platform with coarser granularity, larger computational tasks are necessary and intertask communication becomes more complicated. A process called domain decomposition is used to subdivide the finite element grid into $N_p$ grids, where $N_p$ is the number of processors that will be used for the computation. Domain decomposition has received a great deal of attention from researchers in many disciplines. Accordingly, it has spawned a large amount of literature. A set of references is the series of conference proceedings published by SIAM (Glowinski (1988))(Chan (1989)) (Chan (1990))(Glowinski (1991))(Keyes (1992)).

Domain decomposition is illustrated in Fig.1.2. The two halves of the figure represent the two parts of the preprocessor stage of our finite element procedure. The upper half of the figure shows a complete finite element mesh (i.e., the computational domain) representative of the mesh generation stage. For parallel computing, the first step is to generate input data using either in-house or commercial mesh generation software. This is also the first step when doing sequential or vector computing. The second step is to process the original input data set through the domain decomposer and produce $N_p$ input data sets, one for each processor. The lower half of Fig.1.2 shows how the original mesh could be subdivided (i.e. decomposed) into three submeshes for parallel computing using three processors. Each subdomain is in itself a complete finite element mesh. Therefore analysts have the option of using the same postprocessing software for each subdomain as they would normally do for a job executed on a single processor machine.

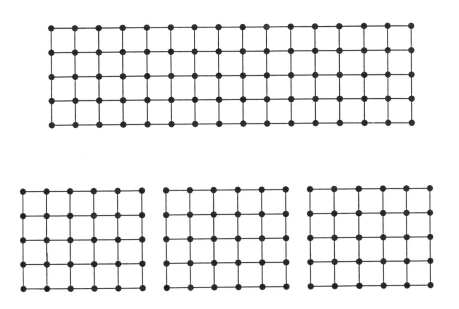

**Figure 1.2** An Example of Domain Decomposition

It should be noted that elements can belong only to one subdomain, but nodes may belong to multiple subdomains. The nodes that lie completely within a subdomain have the forces available for the solution of the equations of motion. However, the forces at the interfaces of contiguous domains must be exchanged before the equations can be solved. Note: this leads to redundant calculations of the equations of motion for the interface nodes; however, this is not significant.

Message passing was selected here to exchange information among processors because of portability concerns. The Argonne-developed **p4** message-passing system provides a powerful, portable tool for the adaptation of finite element codes to a wide range of parallel platforms.

## THE P4 SYSTEM FOR DISTRIBUTED PARALLEL COMPUTING

**P4** is a subroutine library developed at Argonne National Laboratory for programming a variety of parallel machines. Its predecessor was the M4-based "Argonne macros" system described in Boyle, Butler, Disz, Glickfeld, Lusk, Overbeek, Patterson and Stevens (1987) from which **p4** takes its name. The current **p4** system maintains the same basic computational models described there (monitors for the shared-memory model, message-passing for the distributed-memory model, and support for combining the two models) while significantly increasing ease and flexibility of use. **P4** addresses constraints and problems with the macro package and has been implemented through a much cleaner methodology. The macro package was implemented with an underlying C base, using the M4 macro processor to provide an interface to the Fortran layer. These three layers made it difficult to incorporate changes, and for certain types of operating systems, did not lend themselves to adaptation. The **p4** package is completely written in C, without any macros, and provides an interface for Fortran. The **p4** package operates within the environment of the user program, rather than building code around the user's application.

## Establishing and Managing Distributed Processes Using p4

A complete description of the **p4** package is found in Butler and Lusk (1992). A summary of the **p4** capabilities used to implement message passing is presented below.

The "`#include "p4.h"`" statement must appear in all C programs that use any **p4** features. For Fortran programs, the statement "`#include "p4f.h"`" is used. The header files p4.h and p4f.h contain function declarations and symbolic constant definitions. Since the example code presented in the next section is in Fortran we will restrict our discussion in this section to the Fortran component of the **p4** library.

**P4** always assumes that a master process exists. In addition to performing its own computations, the master is responsible for initiating the parallel environment. The master process next calls the function `p4crpg()` which reads a user-defined "process group" or "procgroup" file. **P4** assumes that the slave processes (process groups) are described in the procgroup file. The name of the procgroup file is arbitrary, but procgroup is the default name. The process group file is a flat text file, accessed by the master process, that describes the slaves to be created and where all slave processes are to be executed. From this information, a procgroup table that describes all created processes is built and given to each process. Using the procgroup table, the processes can communicate with one another. The format of a procgroup file is as follows:

| | | | |
|---|---|---|---|
| local | num_slaves | [full_slave_path_name] | [userid] |
| remote_machine | num_slaves | full_slave_path_name | [userid] |

The first line of a procgroup file must be "`local n`" where n is the number of slave processes that are running on the host machine, i.e., the machine running the master process. The subsequent lines contain either three or four fields:

| | | |
|---|---|---|
| *remote_machine* | = | Remote machine on which slave processes will be created. |
| *num_slaves* | = | Number of slaves to be created on the machine. |
| *full_slave_path_name* | = | Full pathname of the slave process executable. |
| *userid* | = | Optional field which is used if the user's login name or userid on the remote machine is different from that running the master on the host machine. |

Two options exist for the initiation of slave processes. Slave processes running on the same machine communicate through shared memory and are started by a UNIX *fork*. Slave processes running on remote machines are started by the UNIX *rsh* (remote shell) command and pass messages through sockets. The flexibility to run the same code on shared-memory machines via fork or on remote machines via rsh is implemented through code segments called wrappers. **P4** places different wrappers around the entire program run by a slave process depending on how these slaves are specified in the procgroup file. **p4** looks for a function or subroutine called slave around which these wrappers are to be placed. Thus, the **p4** slave process must contain a subroutine/function called slave in order to link properly. In the current study, the master and slave processes ran on different machines, hence they were started by rsh and passed messages through sockets.

## Interprocess Message Communication

To send messages among the slave and master processes, one of the following subroutines must be called:

   p4send   *(type, process_id, data, data_len)*
   p4sendr  *(type, process_id, data, data_len)*
   p4sendx  *(type, process_id, data, data_len, data_type)*
   p4sendrx *(type, process_id, data, data_len, data_type)*
where:

| | | |
|---|---|---|
| *type* | = | User definable message tag. (type integer) |
| *process_id* | = | Process id of the destination. (type integer) |
| *data* | = | Variable or structure containing the message. (any type) |
| *data_len* | = | Length of the message in bytes. (type integer) |
| *data_type* | = | Conversion code for data if required. (type integer) |

All of these subroutines will send the *data* message to the designated *process_id*. We will first define the function parameters, followed by a discussion highlighting similarities and differences in their capabilities. It is essential that these distinctions be taken into account when choosing one form of the **p4send()** function over another.

The *type* field is a user-defined message type tag. The *data* message can be of any type: real, integer, character, Fortran arrays and common blocks. The *data* parameter must be the starting address of the data to be sent. Fortran by convention passes values by address, therefore the name of the variable is sufficient for the data parameter. Arrays can be sent by using the name of the array for the *data* parameter and common blocks are sent by using the name of the first element within the common block. The *data_len* field represents the length of the data to be sent in bytes. This requires knowing the word size of the machine sending the message. The correct number of bytes of data to be sent must be specified with this parameter. The *process_id* parameter designates the process number the data will be sent to. The *data_type* parameter specifies the conversion to be performed on the data. This is required when passing data to machines that have different internal data representations. If conversion is not required, a "0" must be specified for the *data_type* parameter.

The variations of **p4send** provide added functionality. Procedures having an "r" in their name do not return from the send call to execute subsequent instructions until an acknowledgement is received from the process *process_id* (the "r" stands for rendezvous). Those procedures having an "x" in their name take an extra argument that specifies the type of data in the message. These procedures will use that information to call XDR (External Data Representation) for data conversion if the message is being passed to a machine with a different architecture where the internal representation may be different. To receive messages, use the following subroutine:

p4recv (*type, process_id, data, data_len, recv_len, return_code*)
where:

| | | |
|---|---|---|
| *type* | = | User definable message tag. (type integer) |
| *process_id* | = | Select only messages from process_id. (type integer) |
| *data* | = | Variable or structure to receive message. |
| *data_len* | = | Length of buffer in bytes. |
| *recv_len* | = | Length of received message. (type integer) |
| *return_code* | = | Completion code for receive. (type integer) |

The *type* and *process_id* parameters can be used to select certain types of messages. In that case, the receive subroutine will cause the invoking process to block until the arrival of a message satisfying the conditions specified by those two parameters. The user may relax unnecessary restrictions on the order incoming messages are processed by assigning the value -1 to either one or both of these parameters. The data parameter designates the address of where to place the incoming data. The *data_len* parameter must be set to designate the number of bytes to receive. The *recv_len* parameter will be set to the number of bytes actually received. Finally, at the end of the master process, a call is made to p4cleanup(). All global memory and structures that were created by **p4** are released and all slaves launched by the master process are killed.

## P4 *IN THE CONTEXT OF EXPLICIT TRANSIENT FINITE ELEMENT ANALYSIS*

In explicit finite element algorithms, the most computationally intensive portion of the program is the calculation of element internal forces. In studies of the application of fine-grain parallelism to explicit finite element codes on the CONNECTION Machine, it was observed that the computation speed greatly exceeded the interprocessor communication speed. An exchange procedure was developed which minimized interprocessor communication at the expense of redundant computations and storage. The disparity between computation speed and communication speed is even greater on a network of workstations. To reduce the communication overhead, the exchange algorithm was incorporated into the message-passing code.

The transition from the CONNECTION Machine architecture to parallel finite element analysis via message-passing also entailed a change in programming paradigm. The SIMD programming paradigm is discussed in great detail by Hillis (1987). Here we will present a brief summary of that discussion. In SIMD machines, such as the CONNECTION Machine, all processors were controlled from a single instruction stream that is broadcast to all the processing elements simultaneously. Each processor has the option of executing an instruction or ignoring it, depending on the processor's internal state. Every processor does not necessarily execute the same set of instructions. However, each processor is presented with the same sequence of instructions. Processors not executing an instruction must "wait out" while the active processors execute. Although we were aware of research underway at that time aimed at building SIMD compilers for MIMD architectures, we chose not to pursue that route based on the following observations.

The success of the SIMD architecture was strongly dependent on the SIMD character of the underlying application. Many applications such as ray tracing and Monte Carlo simulations induce little to no "waiting out" as different processors execute different branches of conditional statements. In nonlinear finite element analysis, large amounts of processor idle time appear when different groups of elements execute linear and nonlinear segments of the constitutive model subroutines. Accordingly, we chose to pursue a computational paradigm which is more suited to the application and the underlying hardware capability.

Many researchers have successfully used the single-program multiple-data (SPMD) mode of execution to port legacy code originally written using the shared memory programming model on the distributed memory machines. Typically, in the SPMD model, each processor runs the same executable on only part of the data. If an operand required for the execution of a statement resides at another processor, then the value of that operand must be obtained through some sort of message-passing.

For our implementation of parallel finite element analysis via message passing we chose the combination of domain decomposition, a portable message-passing library and the SPMD programming paradigm. This combination can be characterized as loosely-coupled communicating sequential processes. Each processor performs computations for the mesh data residing locally, i.e., its subdomain. Thus, the tightly synchronous execution order characterized by the SIMD paradigm is absent. Also absent are the data and control dependencies characteristic of shared-memory MIMD programs. Since each subdomain is in itself a complete finite element mesh, subdomain computations for each time step are highly independent. The interprocessor dependencies are focussed at the interface nodes. Here message passing must be used to ensure that the sum of the parts is equal to the whole.

## A Simple Example With Benchmarks on A Network of Workstations

A two-node rod element for one-dimensional waves is presented to illustrate the incorporation of message passing subroutines in a finite element program. The rod element was chosen for brevity and clarity. The partial differential equations for an elastic rod are identical to those given by Eqs. (1.1) through (1.4). The displacement, strain, and stress simplify to:

$$u = u_x \tag{1.16}$$

$$\epsilon = \epsilon_x \tag{1.17}$$

$$\sigma = \sigma_x \tag{1.18}$$

For a one-dimensional element, the shape functions and gradient matrices are given by:

$$\mathbf{N} = \tfrac{1}{L}\begin{bmatrix} 1 - x & x \end{bmatrix} \tag{1.19}$$

$$\mathbf{B} = \tfrac{1}{L}\begin{bmatrix} -1 & +1 \end{bmatrix} \tag{1.20}$$

where L is the length of the element. The strain-displacement equation simplifies to:

$$\epsilon = \mathbf{B}u_e = \tfrac{1}{L}\begin{bmatrix} u_2^e - u_1^e \end{bmatrix} \tag{1.21}$$

The constitutive law simplifies to:

$$\sigma = E\epsilon \tag{1.22}$$

The internal force and mass matrices are computed on the element level by:

$$\left\{ \begin{matrix} f_1^{int} \\ f_2^{int} \end{matrix} \right\}^e = \int_{\Omega^e} \mathbf{B}^T \sigma d\Omega = A\sigma \left\{ \begin{matrix} -1 \\ +1 \end{matrix} \right\} \tag{1.23}$$

$$\left\{ \begin{matrix} M^1 \\ M^2 \end{matrix} \right\}^e = \tfrac{1}{2}\rho AL \left\{ \begin{matrix} +1 \\ +1 \end{matrix} \right\} \tag{1.24}$$

Our early work in parallel distributed finite element analysis via message-passing was centered around determining its feasibility. Accordingly, we worked with a very simple code and a parallel extension which sacrificed efficiency for ease of implementation. The minimal extensions to the basic 1D explicit finite element code for execution in a parallel distributed environment are briefly summarized below.

These extensions fall under three categories: a program for a master process, a program or programs for the slave processes, and a table providing a mapping between slave processes and processors. By convention, the master process is process number zero and the slave processes are numbered sequentially starting with one. The inter-relationship between the master and slave processes is depicted in Fig.1.3.

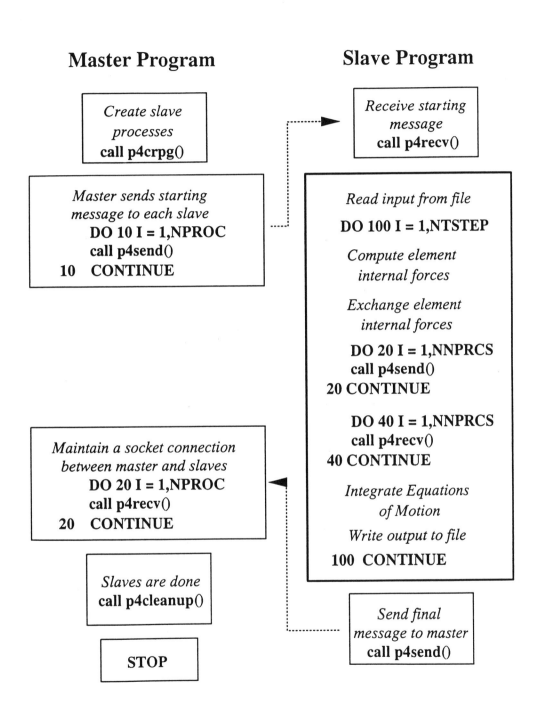

## Master Program

**Create slave processes**
**call p4crpg()**

**Master sends starting message to each slave**
DO 10 I = 1,NPROC
call p4send()
10   CONTINUE

**Maintain a socket connection between master and slaves**
DO 20 I = 1,NPROC
call p4recv()
20   CONTINUE

**Slaves are done**
**call p4cleanup()**

**STOP**

## Slave Program

**Receive starting message**
**call p4recv()**

**Read input from file**
DO 100 I = 1,NTSTEP

*Compute element internal forces*

*Exchange element internal forces*

DO 20 I = 1,NNPRCS
call p4send()
20 CONTINUE

DO 40 I = 1,NNPRCS
call p4recv()
40 CONTINUE

*Integrate Equations of Motion*

*Write output to file*
100  CONTINUE

**Send final message to master**
**call p4send()**

**Figure 1.3** Transformation of a Sequential Code to a Parallel Message-Passing Based Code

Execution begins and ends with the master program. In our example code we designate the master with being responsible for initializing the parallel environment, sending a starting message to each of the slaves, and pausing until it receives a message indicating process termination from each of the slaves. Also, we depict only the slaves as being responsible for finite element computations. On some platforms it is advantageous to make the master process a special case of the slave process. That is, the master process would perform finite element computations in the time interval between creating and destroying the slave processes. However, since this option was not available on the platforms benchmarked here, we will not consider it any further.

After calling p4init(), the master process calls subroutine p4crpg() to create the slave processes. When subroutine p4crpg() is called, the slave processes specified in the procgroup file are created. For example, the five line procgroup file:

```
local       0
brutus      1    /home/ejp/P4/1D/slave
theseus     1    /home/ejp/P4/1D/slave
achilles    1    /home/ejp/P4/1D/slave
caesar      1    /home/ejp/P4/1D/slave
```

specifies a total of four slave processes, each process being on a different computer (brutus, theseus, achilles, caesar). Each slave process will run the executable in the file /home/ejp/P4/1D/slave. The master process then executes the section of code listed below.

```
C*
C*    Maintain socket connections between master and slave processes
C*
      DO 10 I = 1,NPROC
         call p4recv(tagend,I,msg,length,recvln,retcde)
   10 CONTINUE
```

This loop causes the master process to wait until it receives a message from each slave process indicating that the slave computation is complete.

Upon its creation, each slave process read its own input file. To minimize code changes for our preliminary investigations, these input files were identical. Thus each slave process generated and stored all of the data for the entire mesh. The redundant storage of mesh data simplified the incorporation of send/receive into our explicit transient finite element codes at the cost of restricting the model size which could be analyzed before workstation memory was exceeded. The slave processes determined their own process id through a call to the function p4myid(); this value was stored in the variable NID. The call to the function p4ntotids() returned the number of processors that are executing; subtracting one from this value will yield the number of slaves that are executing. The number of slaves is stored in the variable NPROC. The section of code illustrating these calls is given below.

```
            NPROC = p4ntotids()-1
            NID = p4myid()
```

Each slave process then determined the groups of nodes and elements it will process through the computation of four integer quantities: LSTRTE (start element), LENDE

(end element), LSTRTN (start node), and LENDN (end node). The section of code
for the computation of these quantities is given below.

```
C*
C*   Determine the range of nodes and elements
C*   to be processed by each slave process
C*
     LSTRTE = (NID - 1) * NELE / NPROC + 1
     LENDE = (NID) * NELE / NPROC
```

In the above equations, NELE is the total number of elements in the finite element
mesh. Farhat (1988) coined the term "maladroit" to characterize this type of domain
decomposition. Basically, it is only effective for structured meshs, i.e. meshes with a
regular element numbering. For the 1D case, this implies a consecutive numbering of
elements from end to end. We will also (initially) restrict ourselves to meshes with
regular node numbering as well. The following two equations are used to determine
the nodes in the subdomain.

```
     LSTRTN = (NID - 1) * NNODE / NPROC + 1
     LENDN = (NID) * NNODE / NPROC + 1
```

In the above equations, NNODE is the total number of nodes in the finite element
mesh.

Each processor executes the serial finite element algorithm outlined in Section Finite
Element Formulation. First, the element internal forces are computed, followed by
nodal force-vector assembly, and for parallel computations, an exchange of element
internal forces for nodes at subdomain interfaces. However, before the nodal equations
of motion can be integrated, messages must be exchanged to update the nodal internal
force vector. The exchange of internal forces pertaining to interface nodes is required
because, in Eq. (1.11), contributions from elements in neighboring subdomains are not
added in fint.

For a one-dimensional elastic wave problem, each element has two nodes, each
subdomain has one or two adjacent subdomains with which it share one common
node. For our initial studies, we assumed a regular numbering of nodes, elements, and
subdomains depicted in Fig.1.4.

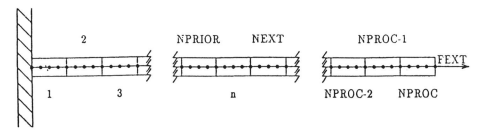

**Figure 1.4** Numbering Convention Adopted for One Dimensional Elastic Wave Analysis

Thus, for any subdomain NID, the slave process ids of the two adjacent subdomains
are given by:

```
         NPRIOR = NID - 1
         NEXT = NID + 1
```

If NEXT is greater than NPROC, then the right end of subdomain NID is a boundary, and if NPRIOR is zero, then the left end of the subdomain is a boundary.

Knowing where the data must be sent and from where to expect data to be received is the first requirement for the successful implementation of a parallel message-passing based algorithm. The second requirement is determining which data must be exchanged and forming the messages. ¿From Eq. (1.23) we observe that the fint vector for the one dimensional rod element consists of two scalar values, one for each node. Thus, only a single real scalar value needs to be exchanged between any two adjacent subdomains. The section of code for the construction of messages for the exchange of internal forces is given below.

```
C*
C*   Exchange internal forces among subdomain interface nodes
C*
     length = 4
     IF (NPRIOR .NE. 0) THEN
         FINC = FINT(1,LSTRTN)
         call p4send(tagnde2,NPRIOR,FINC,length,retcde)
     ENDIF
     IF (NEXT .LE. NPROC) THEN
         FINC = FINT(1,LENDN)
         call p4send(tagnde1,NEXT,FINC,length,retcde)
     ENDIF
     IF (NEXT .LE. NPROC) THEN
         call p4recv(tagnde2,NEXT,FINC,length,recvln,retcde)
         FINT(1,LENDN) = FINT(1,LENDN) + FINC
     ENDIF
     IF (NPRIOR .NE. 0) THEN
         call p4recv(tagnde1,NPRIOR,FINC,length,recvln,retcde)
         FINT(1,LSTRTN) = FINT(1,LSTRTN) + FINC
     ENDIF
```

This exchange forms a barrier synchronization between a subdomain and its neighbors. A subdomain cannot proceed beyond this barrier synchronization point until the exchange of messages is completed. Different subdomains can reach this barrier synchronization at different points in time. A further discussion of the cause and effect of this phenomenon is presented in Plaskacz, Ramirez, and Gupta (1994). Finally, the equations of motion are integrated (Step 3). No kinematic quantities, i.e., displacements, velocities or accelerations, are exchanged. The equations of motion for the interface nodes are integrated redundantly.

This process continues until the final time step. As each slave completes the final time step, it sends a message to the master signalling that it has completed its work using the section of code given below.

```
C*
C*   Inform master process that computations are complete
C*
     call p4send(tagend,0,msg,length,retcde)
```

Upon receipt of the completion message from all of the slave processes, the master process kills all the slave processes by calling subroutine p4cleanup() and shut down.

The numerical studies reported next are intended to illustrate the variation in attainable speedups as a function of problem size for explicit dynamic transient finite element analysis over a network of Sun workstations and the cost of interprocess communication. The times recorded in the tables below are wall-clock times. Timings for the one-dimensional elastic wave problem are given in Tables 1.1 and 1.2. From these two tables, we can observe that the communications overhead associated with distributed computing becomes prohibitive with small models. Distributed computing results in a performance degradation with a 200 element mesh. A 1500 time step simulation utilizing two SPARCstations took over twice as long as the same computation performed on one SPARCstation. Scanning across a row of these tables, one observes the effect of increasing the number of SPARCstations for a fixed problem size. Scanning down a column of these tables illustrates the effect of increasing problem size for a fixed number of processors. In general, for a fixed number of processors, speedup increases with problem size. However, for a given problem size, speedup does not always increase with the number of processors. Each one-dimensional elastic wave element requires 7 multiplications and 7 additions per time step for internal force computation. Speedups diminish as communication overhead catches up with gains in computational speed. In Table 1.2, the overhead attributed to redundant generation of input data is minimized by considering longer analyses. That is, the redundant generation of input data is amortized over a larger number of timesteps which in turn lowers its influence on speedup. While speedups are slightly greater, the trends are identical to those described for Table 1.1.

**Table 1.1**  Performance of a one-dimensional elastic wave analysis on a Sun network (1500 time steps): execution time (seconds) and speedup relative to a single workstation

| Problem Size (elements) | Number of SPARCstations | | | |
|---|---|---|---|---|
|  | 1 | 2 | 4 | 6 |
| 200 | 16.68 | 41.57 | 157.55 | 180.45 |
|  | (1.0) | (0.40) | (0.10) | (0.10) |
| 2000 | 86.57 | 64.08 | 106.91 | 173.49 |
|  | (1.0) | (1.35) | (0.81) | (0.50) |
| 4000 | 149.80 | 102.52 | 8.216 | 114.68 |
|  | (1.0) | (1.45) | (1.81) | (1.30) |
| 8000 | 284.82 | 171.47 | 118.67 | 154.96 |
|  | (1.0) | (1.66) | (2.40) | (1.84) |
| 16000 | 551.32 | 312.36 | 281.94 | 360.00 |
|  | (1.0) | (1.77) | (1.96) | (1.84) |
| 32000 | 1100.54 | 606.59 | 398.89 | 340.80 |
|  | (1.0) | (1.81) | (2.76) | (3.23) |
| 64000 | 2200.25 | 1192.64 | 712.32 | 734.88 |
|  | (1.0) | (1.84) | (3.09) | (3.00) |

**Table 1.2**   Performance of a one-dimensional elastic wave analysis on a Sun network (9000 time steps): execution time (seconds) and speedup relative to a single workstation

| Problem Size | Number of SPARCstations | | | |
| --- | --- | --- | --- | --- |
| (elements) | 1 | 2 | 4 | 6 |
| 200 | 71.52 | 120.52 | 880.32 | 792.86 |
|  | (1.0) | (0.59) | (0.10) | (0.10) |
| 2000 | 444.10 | 435.65 | 842.77 | 948.93 |
|  | (1.0) | (1.02) | (0.53) | (0.47) |
| 4000 | 884.95 | 701.56 | 829.80 | 906.04 |
|  | (1.0) | (1.26) | (1.94) | (0.98) |
| 8000 | 1673.28 | 1004.93 | 1031.20 | 932.62 |
|  | (1.0) | (1.66) | (1.62) | (1.79) |
| 16000 | 3236.73 | 1760.54 | 1604.60 | 1559.73 |
|  | (1.0) | (1.84) | (2.01) | (2.08) |
| 32000 | 6220.05 | 3364.04 | 1890.24 | 1864.95 |
|  | (1.0) | (1.85) | (3.29) | (3.33) |
| 64000 | 12776.7 | 6619.83 | 3534.92 | 3059.07 |
|  | (1.0) | (1.93) | (3.61) | (4.18) |

## A More Sophisticated Example

Upon obtaining favorable results with the computationally minimal test code described previously, the performance of parallel finite element analysis via message passing was studied in a more computationally demanding setting. For this purpose, the shell element described in Belytschko, Lin, and Tsay (1984) was used. It is a four-node element using the standard bilinear isoparametric shape functions. This element is widely used in explicit programs for nonlinear transient analysis, primarily because of its speed. The speed results mainly from the use of only a single stack of quadrature points in each element; for an elastic element, only one quadrature point is used. Thus, the element is typically four times as fast as fully integrated elements (2x2 Gaussian quadrature) in explicit programs. Since the one-point quadrature element is rank deficient, it gives rise to spurious singular modes. Suppression of these spurious singular modes is given in Belytschko, Ong, Liu and Kennedy (1984) and Flanagan and Belytschko(1981)(1984). The element is shown in Fig.1.5.

Another important characteristic of this element is the use of a corotational formulation. A triad of unit vectors $e_1$, $e_2$, and $e_3$ is established at the element quadrature points where $e_1$ and $e_2$ are tangent to the midsurface, and $e_3$ is perpendicular to the plane of the plate. This triad is embedded within the element and rotates with the material; however the vectors $e_1$ and $e_2$ remain tangent to the midsurface. The element strains are evaluated in terms of the corotational basis vectors. As a result, the rate form of a constitutive equation does not require a correction for frame invariance.

A set of body coordinates is defined to coincide with the principal directions of the moment of inertia of each node. The orientation of the nodal body coordinate system is described by the unit vectors $\mathbf{b}_i$.

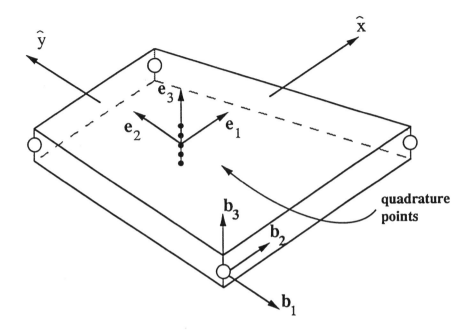

**Figure 1.5** Belytschko - Tsay Shell Element

Since closed form solutions are not available for nonlinear transient programs, solutions obtained by finite elements are typically compared to experimental results. The cylindrical panel problem shown in Fig.1.6 has been used as a benchmark for many nonlinear transient programs. Experimental results have been obtained for this shell by Balmer and Witner (1964). The material properties of the panel are summarized in Table 1.3. An initial velocity of 5650 in/sec is applied to the 3.08 x 10.205 in area indicated in Fig.1.6. The panel is simply supported at its ends and clamped at the sides.

The analytical model takes advantage of symmetry, so only half the panel is modelled; 1089 nodes and 1024 elements were used for the half panel. A von Mises elastic-plastic material model with five integration points through the thickness was used. The timings include complete output of nodal coordinates, element connectivities, initial conditions and boundary conditions for each subdomain.

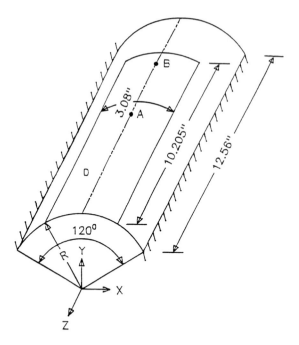

**Figure 1.6** Impulsively Loaded Cylindrical Panel

**Table 1.3**   Material Properties and Parameters for Cylindrical Panel

| | |
|---|---|
| Radius | $R = 2.9375$ in. |
| Length | $l = 12.560$ in. |
| Thickness | $t = 0.1250$ in. |
| Angle | $\alpha = 120°$ |
| Density | $\rho = 2.5 \times 10^{-4}$ lb.sec$^2$/in.$^4$ |
| Young's Modulus | $E = 1.05 \times 10^7$ psi |
| Poisson's Ratio | $\nu = 0.33$ |
| Yield Stress | $\sigma_Y = 4.4 \times 10^4$ psi |
| Plastic Modulus | $E_P = 0.0$ psi |
| Initial Velocity | $v_o = 5650$ in./sec |

Timings for the impulsively loaded cylindrical panel are given in Table 1.4. This dataset was chosen because the regular mesh structure lent itself well to a simple domain decomposition technique. The mesh was described by the number of element rows and columns. Due to the regular numbering of this mesh, the appropriate elements in each subdomain and interface nodes could be described by simple algebraic expressions. Speedups were limited by the overhead entailed by generating the entire mesh in each processor and interprocessor communication.

**Table 1.4**  Performance on a Sun network for nonlinear shell problems: execution time (seconds) and speedup relative to a single workstation

|  | Number of SPARCstations | | |
|---|---|---|---|
|  | 1 | 2 | 4 |
| Cylindrical Panel | 3227.28 | 1874.87 | 1223.14 |
| (1024 elements) | (1.0) | (1.72) | (2.64) |

Finite elements meshes typically describe irregularly shaped objects. These "real world" meshes are typically unstructured. By unstructured, we mean meshes whose element and node numbering is irregular. These meshes do not lend themselves to the simple domain decomposition techniques outlined above. To effectively bring the power of parallel distributed computing to bear on unstructured meshes demands a more sophisticated domain decomposition algorithm.

A utility was developed which first reads nodal coordinate, element connectivity, boundary condition, and loading data from an input deck. Next, it decomposes the mesh into subdomains and records subdomain adjacencies. Finally, it generates an input deck for each slave process describing its subdomain. This preprocessing program is based on algorithms and code described in Farhat (1988) and Al-Nasra and Nguyen (1991). Separate input and output files are established for each subdomain. There is a one-to-one correspondance between processors and subdomains. Accordingly, each processor only reads the input file corresponding to its subdomain. Similarly, each processor only writes to the output file corresponding to its subdomain. This removes the restriction of generating and storing the entire mesh on each processor and will facilitate the analysis of larger models and unstructured meshes. The numerical studies in Table 1.5 reflect the use of domain decomposition as a preprocessing step prior to parallel finite element analysis.

**Table 1.5**  Performance on a Sun network for nonlinear shell problems: execution time (seconds) and speedup relative to a single workstation

|  | Number of SPARCstations | | |
|---|---|---|---|
|  | 1 | 2 | 4 |
| Cylindrical Panel | 3042.8 | 1576.8 | 846.7 |
| (1024 elements) | (1.0) | (1.93) | (3.59) |

*Shared Memory MIMD Multiprocessor (Encore Multimax)*

The Encore Multimax incorporates 20 (0.75 MIPS) processors, with 64Mbytes of shared memory. The processors are arranged on 10 Dual Processor Cards. Each Dual Processor Card provides two independent 10 MHz National Semiconductor 32032 processors and one shared cache memory. On each Dual Processor Card are two National Semiconductor 32081 floating point units with a 2.5 MFLOPS rating. Timings for the impulsively loaded cylindrical panel running on the Encore Multimax are given in Table 1.6.

**Table 1.6**  Performance on the Encore Multimax for nonlinear shell problems: execution time (seconds) and speedup relative to a single processor

|  | Number of Processors |  |  |  |  |
|---|---|---|---|---|---|
|  | 1 | 2 | 4 | 8 | 16 |
| Cylindrical Panel | 27661.2 | 14258.6 | 7131.0 | 3789.0 | 2370.2 |
| (1024 elements) | (1.0) | (1.94) | (3.88) | (7.30) | (11.67) |

*Distributed Memory MIMD Multiprocessor (Intel Hypercube)*

The Intel iPSC/860 hypercube has 8 nodes with one processor and 16 Mbytes of memory per node. In the Boolean n-cube, each successive n-cube is established by linking the corresponding vertices of the previous cube. A single processor can be thought of as a "zero-cube". Connecting two zero-cubes with a single wire yields a "one-cube". Connecting two "one-cubes" at their corresponding vertices yields a "two-cube". Repeating this process yields a "three-cube" with $2^3$ (8) vertices. A three-cube can arrange 8 processors with no processor more than 3 wires away. Performance benchmarks of the Intel hypercube are presented in Table 1.7.

**Table 1.7**  Performance on an Intel hypercube for nonlinear shell problems: execution time (seconds) and speedup relative to a single workstation (processor)

|  | Number of Intel i860 Processors |  |  |  |
|---|---|---|---|---|
|  | 1 | 2 | 4 | 8 |
| Cylindrical Panel | 1401.1 | 726.5 | 378.8 | 218.8· |
| (1024 elements) | (1.0) | (1.93) | (3.70) | (6.40) |

*Distributed Memory MIMD Multiprocessor (Intel Delta)*

The Intel Touchstone Delta system is jointly owned by Argonne National Laboratory and 12 other members of the Concurrent Supercomputing Consortium. The system supports various types of processing nodes (numeric, mass storage, gateway and service). Numeric nodes form the computational core of the system. Mass storage nodes enable the numeric nodes to access the tape and disk drives in the mass storage system. Gateway nodes provide access to the system over LAN lines. The service nodes handle resource management, such as enabling the user to log into the system from the network and to obtain and release a collection of numeric nodes. The Delta consists of 528 processing nodes, each with 16 megabytes of local memory and a RISC processor, the i860, giving total peak theoretical performance of 32 billion floating point operations per second. The nodes are interconnected as a two-dimensional mesh. Interprocessor communication is handled by mesh router chips. One mesh router chip is installed to the routing plane for each node in the mesh. The Delta has an aggregate memory capacity of more than 8 gigabytes, and secondary (disk) storage of 90 gigabytes.

Three levels of discretization are presented for the cylpanel dataset: 1024 nonlinear shell elements, 8192 elements, and 16384 elements. A Von mises elastic-plastic material model with five integration points through the thickness was used. For all three cases, the timings include complete output of nodal coordinates, element connectivities, initial conditions and boundary conditions for each subdomain. In order to compute the Delta megaflop rate, the problem was also run on a CRAY X-MP/18. The timings recorded in Table 1.8 are wall-clock times for the Delta. Cray timings are CPU times measured from the vectorized code described in Gallopoulos (1990).

For 1024 elements, substantial speedups were attained even though the data set was small. The problem was run for 1600 time steps. Nodal deflections for 31 nodes were output every eight time steps. Each processor read and wrote data to a unique pair of files. Thus, for interface nodes, the deflections were output redundantly.

The 67.3 CPU seconds measured on the Cray X-MP/18 corresponds to 68.28 MFLOPS. Hence the execution time of 21.2 seconds on 128 Delta nodes is about 3.17 times faster than Cray X-MP/18 and corresponds to 217 MFLOPS. It may be noted here that since Delta timings are wall-clock times, this rating of Delta is a lower-bound on performance. Further, timings reported here are for a very small data set. For the 128 node Delta run, there are only 8 elements per processor. If we use 512 nodes of the Delta, there will be only 2 elements per processor. This reduces the computation to communication time ratio considerably. Better performance was attained on the Delta with the larger data sets.

The 8192 element mesh was run for 5000 time steps with nodal deflections for 2 nodes being output every hundred time steps. The 1470 CPU seconds measured on the Cray X-MP/18 corresponds to 77.2 MFLOPS. Hence execution time of 116.6 seconds on 512 Delta nodes is about 12.61 times faster than Cray X-MP/18 and corresponds to 973 MFLOPS.

The 16384 element mesh, was also run for 5000 time steps with results for 2 nodes being output every hundred time steps. The 3010 CPU seconds measured on the Cray X-MP/18 corresponds to 77.7 MFLOPS. Hence execution time of 206.9 seconds on 512 Delta nodes is about 14.55 times faster than Cray X-MP/18 and corresponds to 1130 MFLOPS or 1.13 GFLOPS.

**Table 1.8** Performance of cylindrical panel nonlinear shell problem on Delta: execution time (seconds) and speedup relative to a single processor

| Elements | Number of Intel i860 Processors | | | | | | |
|---|---|---|---|---|---|---|---|
| | 1 | 16 | 32 | 64 | 128 | 256 | 512 |
| 1024 | 1354.3 | 98.4 | 55.1 | 30.3 | 21.2 | 22.1 | |
| | (1.0) | (13.77) | (24.56) | (44.67) | (63.88) | (61.26) | |
| 8192 | 33445.7 | 2300.7 | 1176.5 | 622.9 | 340.4 | 192.1 | 116.6 |
| | (1.0) | (14.54) | (28.43) | (53.69) | (98.25) | (174.11) | (286.84) |
| 16384 | 68927.7 | 4510.8 | 2317.4 | 1208.8 | 624.2 | 362.6 | 206.9 |
| | (1.0) | (15.28) | (29.74) | (57.02) | (110.43) | (190.09) | (333.15) |

*Hewlett Packard Workstation Network*

The Hewlett Packard workstation cluster consisted of nine machines. The file server was an HP model 715 with 16MB of memory and 525 MB of disk space. The other eight machines were HP model 735's with 32 MB of memory and 1 GB of disk space each. The workstation cluster was placed on its own ethernet subnetwork to reduce outside network traffic. Timings for the impulsively loaded cylindrical panel running on the Hewlett Packard workstation cluster are given in Table 1.9.

**Table 1.9** Performance of cylindrical panel nonlinear shell problem on Hewlett Packard Workstation Network: execution time (seconds) and speedup relative to a single processor

| | Number of Hewlett Packard Workstations | | | | |
|---|---|---|---|---|---|
| | 1 | 2 | 4 | 6 | 8 |
| Cylindrical Panel | 922.432 | 574.893 | 302.952 | 208.263 | 163.229 |
| (1024 elements) | (1.0) | (1.60) | (3.04) | (4.43) | (5.65) |

*IBM SP1 Supercomputer*

The IBM SP1 Supercomputer consists of 128 compute nodes. Each compute node is an IBM RS/6000 Model 370 processor with 128MB of memory and 1 GB of disk space. The nodes are connected by a high-speed Omega switch. Timings for the impulsively loaded cylindrical panel running on the IBM SP1 are given in Table 1.10.

**Table 1.10** Performance of cylindrical panel nonlinear shell problem on the IBM SP1 Supercomputer: execution time (seconds) and speedup relative to a single processor

|  | Number of RS/6000 Processors | | | | |
|---|---|---|---|---|---|
| Elements | 1 | 2 | 4 | 6 | 8 |
| 1024 | 714.610 | 388.715 | 262.310 | 383.484 | 950.250 |
|  | (1.0) | (1.84) | (2.72) | (1.86) | (0.75) |
| 16384 | 35652.8 | 18190.2 | 9688.3 | 6936.3 | 5904.9 |
|  | (1.0) | (1.96) | (3.68) | (5.14) | (6.04) |

## *REMOTE INTERACTIVE COMPUTING FOR DESIGN AND OPTIMIZATION*

The previous section focused on using message-passing to bring the power of parallel distributed computing to bear on finite element simulations, with the sole objective of accelerating those computations. Accordingly, the computations were performed on networks of homogeneous processors. In this section we will demonstrate the effectiveness of executing different components of an interactive simulation on heterogeneous distributed platforms that are best suited to their individual computational needs. The objective is the acceleration of the overall design process through the introduction of interactivity.

Traditionally, engineering design has involved running finite element codes to simulate the response of a structure under applied loads, writing the output data to a file, and "post-processing" the results at a later time. In general, the term "post-processing" refers to the reduction of finite element results to a manageable level of information. Prior to the availability of computer graphics, much of this data reduction was done manually. This process was tedious, error-prone, and only feasible with the crude models being studied at the time. Initial computer graphics devices were specialized as well as centralized. Thus, the role of computer graphics was restricted to being passive; that is, the desired drawings were specified before the computing was done, and executed some time after the information was produced. The analyst needed to draw on his intuition to apriori choose the necessary drawings. To remove this uncertainty, the next generation analysis codes stored all the information in data files and introduced interactive computer graphics to display the results. However, the graphics were still being executed some time after the information was produced. An example of the early role interactive computer graphics played in input and output processing is presented in Biffle (1977).

As computational capabilities continued to increase and hardware costs drop, computer graphics became a necessity; the execution times for many models dropped to a point where submitting them as batch jobs became unnecessary. Engineers were able to observe finite element simulations on a graphics terminal as the computation proceeded, select the desired drawings and obtain hardcopy for future reference and documentation. This trend will continue; the emergence of high-performance computer architectures in the mid 90's promises to dramatically alter the way scientists and engineers will do their work. Recently, workstations have attained graphics capabilities worthy of consideration in large supercomputing problems. Grimsrud and Loring (1989) have reported on the use of workstations in a distributed environment. In their work, they used the Cray Research MPGS (Multipurpose Graphics System) package where the supercomputer handles cpu and memory intensive tasks, while downloading visual information to a graphics workstation.

Somewhere, in the future along this evolutionary path, will be the numerical laboratory. Specifically, engineers will be able to observe finite element simulations on a graphics workstation as the computation proceeds on a compute server. Furthermore, they will be able to intervene and dynamically alter the parameters of the model in response to their observations. Some of the resulting possibilities – checking the progress of a simulation, terminating unfruitful design directions midstream, and altering parameters to attain better designs – are presented in this section.

## Distributed Computing

An environment is provided where the analyst can interactively redesign an s-rail (a front-end automobile structural component), by introducing lighter materials at noncritical sections without drastically reducing the collapse load. The analyst's goal is the optimal design from the weight-cost-manufacturability perspective. This demonstration features a distributed interactive environment with computation being performed on a remote supercomputer, and visualization on a local Silicon Graphics workstation. Communication between the two computers is managed by message-passing.

The graphical user interface code was chosen to be the master process for our application. In addition to performing its own computations, the master is responsible for initiating the parallel environment. Therefore, after initializing the interactive environment, the graphics interface code calls the function p4crpg( ) which reads a user-defined "process group" or "procgroup" file and launches the slave code on the compute server. Our procgroup file on the Silicon Graphics Crimson VGXT workstation is as follows:

```
local          0
xmp.ctd.anl.gov 1  /usr/plaskacz/SIGGRAPH/simulation/frame  EdP
```

When the master process is invoked on the Silicon Graphics workstation, one slave process is started on the Cray X-MP which runs the executable "frame" in the directory "/usr/plaskacz/SIGGRAPH/simulation". The "frame" executable is a compilation of the explicit transient finite element algorithm outlined in Section Finite Element Formulation coded in the FORTRAN programming language. The finite element mesh used in these simulations is shown in Fig.1.7. Hallquist, Benson, and Goudreau (1986) have documented the geometry definitions of this data set, which was obtained from Suzuki Motor Company of Japan. A total of 1122 nodes and 1100 elements were used for this mesh. Impact was modelled by prescribing a uniform velocity of 800 in./sec. (45 miles/hour) across the mesh with a clamped row of nodes at one end.

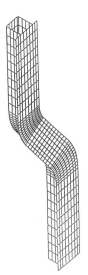

**Figure 1.7** Finite Element Discretization for Structural Rail

Fig.1.8 depicts a typical display from the interactive numerical crashworthiness simulation. Three modes of interaction are supported: interaction with the visual display (e.g. changing object positions and color contour maps), interaction with the simulation (e.g. altering material properties), and finally, interactive control over the amount of communication between the graphical user interface and the computational kernal (e.g. changing the interval between successive updates to the graphics display window).

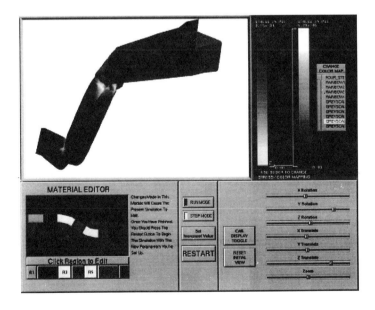

**Figure 1.8** Interactive Numerical Crashworthiness Simulation on Distributed Platforms

The following sections describe the construction of the graphical user interface, a detailed description of its features, as well as the interaction between the graphical user interface and the finite element simulation.

A summary of the **p4** capabilites used to implement message passing between the graphical user interface and the computational kernel is presented below. The accessibility of **p4** functions from both applications written in the C and Fortran programming languages proved invaluable for our work. This allowed us to write our graphical user interface code in C, which in turn enabled us to easily incorporate functions generated by the Forms Designer. Furthermore, we were able to use the optimized Fortran code described in Gallopoulos 1990 for the finite element computations being performed on the Cray XMP.

## Graphical User Interface Toolkit

The construction of graphical user interfaces for programs is normally a time-consuming process. Recently, software developers have focussed on producing packages which accelerate the construction of graphical user interfaces. The Forms Library (Overmars 1991) is a graphical user interface toolkit for Silicon Graphics workstations.

A form is the fundamental concept behind the Forms Library. Hardcopy forms such as employment applications are comprised of labels and boxes where information is to be provided (e.g. name, address, education, previous employment history, etc.). Analogously, the Forms Library constructs a window in which interaction objects such as buttons, sliders, and dials may be placed. The user can interact with the different objects (widgets) on the form to indicate his choices. The Forms Library supports two basic object types: static objects and interactive objects. Static Objects include a box, text, bitmap, clock or chart. These objects never change state and therefore permit no interaction. Interactive objects include a button, slider, dial, positioner, counter, input, menu, choice, browser or timer. Interactive Objects respond to the users' mouse actions. In addition, the Forms library is extensible, allowing the user to create a custom object and link it into the library.

The Form Designer was used to do the brunt of the layout for most of the graphical user interface (GUI). The Forms Designer is a menu-driven utility program which facilitates the construction of forms by allowing a graphical user interface developer to interactively place objects on forms, move them around, scale them, and modify their attributes. It is important to note that the Forms Designer only provides C code for the form layout. The programmer must go back into his code and insert the code delineating the actions to be taken when the user interacts with a particular object on a form, e.g. pushes a button with the mouse. The C source code generated by the Designer was extended in order to integrate it correctly into the program. This code defines all the interfaces, their types, their pixel locations in that particular form, color, etc. Also present in this code are routines that identify callback functions associated with particular object actions (eg. slider motion).

## Graphical User Interface Components

The graphical user interface consists of six windows. The first window contains a rendered image of the model with either stress or strain contours. In the next window is the color contour legend. The options for rendering the stress or strain contours include several rainbow and grey scale color maps. The third window contains sliders controlling image transformations, i.e. translations and rotations as well as a button to reset the transformation matrix. The ability to translate and rotate the image of the structural rail allows the analyst to view the rail from an alternate angle as well as to focus in on areas of interest. The next window is a button for the toggling between displaying the rail alone or in the context of the automobile body. The fifth window consists of buttons to control frequency of image updating, which in turn determines the amount of message passing between the two codes. Finally, the last window consists of the material editor. The modes of interaction supported by the forms is summarized below:

### Computational Results Display Window

The rendered image displays an effective plastic strain contour or stress contour superimposed on the deformed rail. The initial rail geometry is constructed from coordinate and connectivity files read in by the visualization code at the start of the simulation. Updated nodal coordinates and stresses and/or effective plastic strains are obtained from the finite element code through message-passing.

### Color Map Form

The color map form controls the mapping between the stress and/or effective plastic strain and color. Depressing the "Change Color Map" button causes a browser form to appear listing each available palette. A user chooses the desired palette by positioning the cursor on the appropriate entry and clicking the mouse button.

### Transformation Form

During the course of a simulation, the designer may choose to focus in on an area of interest. The transformation form consists of six sliders which support three-dimensional image translation and rotation. The three slides for rotation serve to input the Euler angles describing rotations about the x, y and z coordinate axes. The transformation matrix is dependent on the order of application of the Euler angles as well as their values. To return to the initial view, using the sliders, the designer would have to "undo" the three rotations in the reverse order they were originally applied. To remove the necessity for "bookkeeping", a "reset transformation" button is provided to reinitialize the transformation matrix.

### Automobile Display Form

The rail may be displayed alone or within the context of the entire vehicle. An automobile body, rendered as a transparent solid, may be superimposed to effectively represent the structural rail computation in an automotive context while retaining rail visibility.

*Run Step Stop Restart Form*

During the course of the simulation, the designer may wish to control the amount of communication overhead by adjusting the frequency by which the rendered image is updated. Depressing the "Increment" button causes a slider to appear, allowing the designer to choose values in the range between 0 and 100 for the number of simulation time steps between two successive image updates. The designer can thereby progress faster through uninteresting parts of the simulation by setting the value of increment to a relatively large number, or obtain a time step by time step update by setting the value of increment to 1. The "Step" button allows the designer to halt the update of the rendered rail and step through and study an interesting part of a simulation frame-by-frame. The "Stop" button allows the analyst to stop the simulation midstream, and dynamically alter parameters using the material editor in response to their observations.

*Material Editor Menu*

The material editor form enables an analyst to interactively redesign certain properties of the structural rail. Current capabilities facilitate the lengthwise subdivision of the structural frame into sections to which different material types may be assigned. One objective the analyst may seek to attain is the introduction of lighter materials at noncritical sections without drastically reducing the collapse load. The analyst may make initial approximations of noncritical sections apriori using engineering judgement and the results of the first single-material simulation. These approximations of section length and appropriate material composition are further refined using the results of later computations.

The focus of this form is a schematic of the structural rail, color-coded by region. Seven regions may be defined through the seven corresponding menu buttons displayed below the schematic. The menu buttons are defined to display a drop-down menu when pressed. This menu gives the user two choices: change the region size, or change the material property of the region.

A material properties database consisting of the density, Poisson's ratio, and piecewise linear approximation for the stress-strain curve for every material under consideration is included in the analysis code's input data. This data is read in prior to the initial single-material simulation. The graphical user interface maintains an array containing the region sizes and material identification number for each region. When a user chooses to change the material of a region through the material editor, the callback routine updates the corresponding array entry.

When the user resizes a region, the schematic image changes to reflect the changes in size on the regions making up the structural rail. Resizing is simply a change in the allocation of element rows between regions. When a user changes region sizes, the callback routine typically must update several array entries to reflect the new allocation.

When the user has finished interacting with the material editor form, a click of the restart button sends the program to a callback routine, which resets some global variables. These in turn transfer program control to the outermost loop, causing a restart message to be sent to the analysis program via **p4**, along with the array containing the new sizes and materials for the seven regions.

The algorithm for the graphical user interface code is given in Table 1.11. After reading in the initial rail geometry, initializing forms and launching the finite element code on the compute server, the graphical user interface code enters a rendering loop where a Forms function gets called once per iteration. The Forms library handles the polling and queueing of Forms events. All of the objects were structured with callback functions, so the program goes straight to the callback. Any information needed is obtained by querying the status of objects using Forms functions, and then making appropriate function calls to carry out the interaction with the user.

**Table 1.11**   Flowchart for Graphical User Interface

*Step 1* Spawn the finite element process on the compute server.

*Step 2* Allocate memory.

*Step 3* Read initial material properties for structural rail from file.

*Step 4* Initialize graphical user interface.

*Step 5* Rendering loop

> Send "Restart" message to process on compute server.
> Send material property selections to finite element code.
> Graphic user interface active loop.
>> Query the event queue.
>>> If an event has been queued
>>> (for example, a button has been depressed),
>>> enter call-back function and execute appropriate steps.
>>> Color Map Callback
>>>> • Modify software color look-up tables.
>>>> • Update frame and colormap windows.
>>> Image Transformation Callback
>>>> • Modify transformation matrix.
>>>> • Apply transformation and update frame window.
>>> Auto Display Toggle Callback
>>>> • Switch between automobile and structural frame displays.
>>>> • Update frame display.
>>> Run Step Stop Restart Callback
>>>> • Halt update of rendered rail when analyst selects "Stop".
>>>> • Halt update of rendered rail and step through simulation frame-by-frame when user selects "Step" button.
>>>> • Send material property selections and go to [5] when the analyst selects "Restart".
>>> Material Editor Callback
>>>> • Set conditionals that control data update of frame object.
>>>> • Execute appropriate form display updates and data updates.
>> Check for **p4** messages from compute server process.
>>> If new time step data has arrived, update nodal displacements and stress/strain vector list.
> Update main window display.
> End GUI active loop.
> End rendering loop.

*Step 6* Kill remote compute process and local GUI process.

**Encapsulation Of A Finite Element Analysis Code As A Slave Process To The GUI Code**

The algorithm for explicit time integration is given in Section Finite Element Formulation. The extension of this algorithm for remote interactive computing for design and optimization is presented in Table 1.12. First element internal forces are computed; then the nodal kinematic variables are updated. The updated displacements along with stresses and/or strains are then stored into an array and sent upon request to the graphics (master) code. The request for additional time steps as well as the updated information are exchanged between the two platforms via the **p4** message-passing library. An outline of the analysis code is presented as pseudo-FORTRAN in Table 1.12. The two main loops in a finite element code, calculation of element internal forces and updating nodal kinematic quantities, appear as statement number 1 (Compute a time step) in Table 1.12. All time steps must be computed regardless of whether or not they will ever be sent. However, only time steps which possibly will be needed by the rendering code at some future point are stored. Prior to computing the next time step, the conditional statements presented in Table 1.12 must be evaluated.

The outcome of the first conditional determines whether the displacements and stress/strains computed will be saved. The current time step is compared to the value of a variable NXTSTP. If the current time step is less than but not equal to NXTSTP, then the results are not saved. If the current time step is greater than or equal to NXTSTP, then it is saved. Time steps greater than NXTSTP must be saved since it is impossible to know apriori how the designer may alter the value of "Increment".

The next conditional tests whether the user interface is waiting for a message from the analysis code. It is possible that the analysis code will either be ahead of the graphical user interface code, i.e. computing time steps before they are requested or behind the graphical user interface code, i.e. receiving requests for time steps before they have been computed. If the analysis code has computed the time step which has been requested, it sends it. Otherwise, the analysis code updates a variable "ktchup" (catch-up) which keeps track of how many more time steps must be computed before the analysis code has the information being requested. When "ktchup" is equal to zero, the analysis code has caught up and has the information being requested by the graphical user interface.

If the graphical user interface code is not waiting for a message from the analysis code, then the analysis code executes a third conditional to test for the arrival of a new message. If no messages have arrived, the next time step will be computed. If new messages have arrived, then the analysis code must test whether a time step is being requested or the simulation is being restarted. A "Restart" message consists of an array of 14 integers describing the 7 rail segment lengths and the material assigned to each segment. The nodal masses and critical time step are recomputed on the basis of this information, the time step, element stresses and strains; and nodal kinematic variables are reinitialized and the simulation is restarted from time zero.

**Table 1.12**   Communication Details Within The Analysis Code

> Do 100 I = 1, NSTEPS
>   1 *Compute a time step*
>       *Test Whether This Time Step Will Ever Be Sent*
>           **TRUE**
>               *Save it*
>           **FALSE**
>               *Go to 1*
>       *ENDIF*
>       *Test For Unrequited Messages*
>           **TRUE**
>               *Test Whether Requested Time Step Has Been Computed*
>                   **TRUE**
>                       *Send it*
>                   **FALSE**
>                       *Decrement ktchup*
>                       *Go to 1*
>                   *ENDIF*
>           **FALSE**
>               *Continue*
>       *ENDIF*
>       *Test for message arrival*
>           **TRUE**
>               *Test for message type*
>                   *Type = "Send time step"*
>                       *Test whether requested time step has been computed*
>                           **TRUE**
>                               *Send it*
>                           **FALSE**
>                               *Compute ktchup*
>                               *Go to 1*
>                           *ENDIF*
>                   *Type = "Restart"*
>                   *Reinitialize*
>                   *Go to 1*
>           **FALSE**
>               *Go to 1*
>       *ENDIF*
> 100 CONTINUE

*SUMMARY*

A highly-portable methodology has been presented for exploiting parallelism in computer codes originally written for execution on sequential computers. The foundations of our approach are domain decomposition, the exchange algorithm, the SPMD (single-program multiple-data) mode of program execution, and the **p4** message-passing package. The code is portable among platforms supporting Berkeley Unix 4.3 BSD interprocess communication sockets. Our experiences with this formulation have been extremely favorable. In our early work we demonstrated that by exploiting existing network connections and implementing message-passing through **p4**, a group of workstations is able to perform finite element analyses in parallel. The elastic rod element was benchmarked at $1.7 \times 10^{-6}$ sec./ element-cycle on the CRAY X-MP/14 (FORTRAN cft 1.5 compiler) (Plaskacz 1990). A 64000 element problem was run for 9000 time steps on a network of six SUN SPARCstations. The wall-clock time was 3059.07 sec. This corresponds to a speed of $5.3 \times 10^{-6}$ sec./ element-cycle or approximately 1/3 the speed of the CRAY. These encouraging results motivated us to pursue additional studies in a computationally more demanding setting, the nonlinear explicit transient analysis of shell structures. The portability of the message passing package enables the use of one parallel finite element code across multiple platforms. The spectrum of computer architectures studied and benchmarked included a single workstation, a cluster of workstations, and many high performance scalable parallel platforms. The software is usable by researchers in R&D institutions who have access to the latest high-performance parallel platforms, as well as engineers in small and large companies who only have access to underutilized workstations.

One fringe benefit of a parallel message-passing formation is that it has enabled us to formulate parallel algorithms at a high-level. Working at a higher-level provided an insulating layer which allowed us to maintain application code stability in spite of extreme hardware volatility. The code's robustness has allowed us to maintain a sustained effort in applying parallel computer architectures towards the solution of engineering problems. This sustained effort has come without the necessity of a total code rewrite at each successive generation of parallel computing hardware. We were able to add more features to the code as opposed to repeatedly rewriting the same capabilities. Accordingly, we are actively working to explore advanced capabilities such as mesh rezoning (Plaskacz (1995a)) and impact-contact simulations (Plaskacz (1995b)). Dynamic relaxation is a straightforward extension of dynamic transient finite element analysis for the static solution of structural mechanics problems. A parallel dynamic relaxation algorithm is described in Kulak, Plaskacz and Pfeiffer (1995).

A personal workstation on every engineer's desk is becoming commonplace in modern corporate engineering departments. As workstation processing speeds and network transfer rates increase, computing across a distributed network of workstations will become an attractive option. This is especially true for many small to midsize companies with limited access to supercomputers or companies whose day-to-day production work does not justify the acquisition of supercomputers. It is our hope that by following the methodology presented in this paper, many of these firms can use the benefit of parallel-distributed finite element analysis to reduce the turn-around time of their computer simulations.

We have demonstrated the feasibility of running a simulation on network-connected heterogeneous computing platforms. Our example featured the synergistic combination of an analysis code running on a supercomputer coupled with a graphical user interface running locally on a workstation. Distributed interactive simulation promises to radically alter the way scientists and engineers do their work by coupling a visual processing supercomputer (the human brain) with electronic computers excelling in data manipulation. Of the brain's ten-billion neurons, almost half are devoted to the analysis of visually gathered information. As the interface takes over the burdensome task of encoding and the user is free to perform higher cognitive tasks in which he excels over the computer, each component of this thinking team is utilized in its most efficient and natural mode. The advent of computers having more CPU cycles per second will make it increasingly more critical to be able to use those cycles to shield the user from the immense quantities of data that will bombard him. Incorporating greater degrees of interactivity through virtual reality will continue until visualization becomes a seamless extension of reality. Future engineering applications will be developed which will in effect allow the user to create a visual model that can be studied in a manner analogous to today's experimental models.

We have explored the use of a numerical laboratory where an engineer can design an automobile structural component. The capabilities supported include monitoring the progress of a simulation, terminating unfruitful design directions midstream and altering design parameters in response to what he sees. Thus, he may use engineering judgement to adjust the material parameters towards an optimal design. Alternatively, he may use the numerical laboratory to build intuition previously attainable only through repetitive costly physical testing.

## 1.3 OVERCOMING DISTANCE HURDLES

High-performance computers are shifting engineering design bottlenecks away from simulation turn-around time and towards information sharing between the experts. True collaborations involve ongoing communication. Traditionally, for geographically distributed participants, this has involved traveling. Teleconferencing has reduced this burden somewhat but nevertheless remains inadequate due to its lack of ubiquity. State-of-the-art telecommunications and virtual reality promise the ability to bring researchers together through telepresence, to take tours through the facilities of collaborators and to take virtual tours through a simulation with ideas being exchanged. Among the possible applications for these new technologies is the paradigm of concurrent engineering wherein the design, analysis, redesign, manufacturing feasibility, and overall optimization is addressed concurrently. Concurrent, interactive engineering design and analysis has the potential for substantially reducing product development time.

This section presents an overview of our work with the CAVE Virtual Reality Environment, which facilitates its use as a mechanism for effective collaborations between participants who are geographically distributed.

### BACKGROUND

The CAVE (Cave Automatic Virtual Environment) is a room used for the three-dimensional visualization of images. The CAVE has several advantages over other virtual reality systems. Typically, in most other virtual reality systems, a user abandons reality to enter into virtual reality. That is, he must don special headgear which totally blocks out reality and diverts his total attention to the virtual world.

He cannot readily interact with anything not explicitly programmed into the virtual world. In contrast, virtual reality in the CAVE is targeted to *supplement* reality as opposed to *replacing* reality. One area where this capability is invaluable is vehicle prototyping and driver training applications. Actual vehicle passenger compartments can be placed into a CAVE. The remainder of the vehicle and its surroundings may be created in the virtual environment. The transition between reality and virtual reality can be made to be almost seamless.

The CAVE's utility is not restricted to providing lone researchers with a three-dimensional visualization capability. The CAVE's strongest feature is that it provides and promotes collaborative engineering. A single CAVE can be used by a group of experts to take a virtual tour through a simulation. Each expert can participate in the redesign of a component in the presence of other experts who can assess the impact of the proposed change on other aspects of component behavior. Engineered structures as commonplace as buildings and bridges as well as more esoteric structures such as aerospace planes can be more optimally designed given a more collaborative environment. For example, a design change made to improve vehicle crashworthiness can be immediately assessed by a noise and vibration specialist without any intermediaries, paperwork, or time delay. By linking multiple CAVEs together by high-speed networks, collaboration among geographically distributed partners is possible.

## CAVE OVERVIEW

The CAVE Automatic Virtual Environment is a 10 x 10 x 10 ft, multi-person, high-resolution, 3-dimensional video and audio environment (see Fig.1.9). A CAVE environment consists of a projectable floor and three rear projection screen walls. A projector is set up overhead and the image is bounced off a mirror to reflect onto the floor. Three projectors and mirrors are also set up behind the three projection screen walls. Images are projected in stereo into the CAVE with a resolution of 1024 x 760 pixels. The purpose of the CAVE is to achieve a real life perspective of a 3-dimensional process that could otherwise only be viewed in two dimensions. Unlike a representation on a two-dimensional screen, the CAVE does not allow for misrepresentation of proximities.

**Figure 1.9** CAVE Hardware (http://www.ncsa.uiuc.edu/evl/html/CAVE.html)

The same reference, right-handed, coordinate system is shared by all walls in the CAVE (see Fig.1.10).

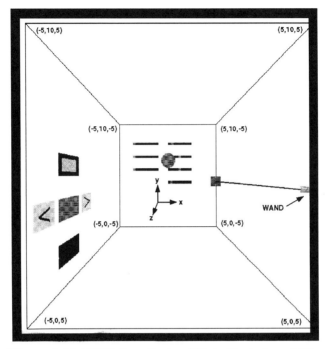

**Figure 1.10** CAVE Coordinates, VRviz Controls and Wand

Stereo glasses are used while in the CAVE in order to view the virtual environment in 3D. Two views are produced for every image, one for the left eye and one for the right eye. In order to see in stereo, a small button on the right side of the glasses is pressed. Stereo emitters around the edges of the CAVE synchronize the stereo glasses and the screen update. The glasses are synchronized to the screen at an update rate of 96 Hz. The two halves of the stereo image are seen 48 times per second by each eye separately. The brain then combines these two views into one 3-dimensional image.

Graphics in the CAVE are produced by a Silicon Graphics Onyx which houses three Reality Engine graphics CPUs. These Reality Engines have direct RGB outputs to Electrohome video projectors. In order to provide audio as well as visual effect, the CAVE utilizes a surround-sound audio system. This system is driven by a SGI Indy running audio server software.

One viewer in the CAVE wears a special pair of stereographic glasses with a six-degree-of-freedom head-tracking device attached. A hand held device called a wand (which is basically a 3-D mouse) is also tracked by a tracking system which is mounted on top of the CAVE. The tracking system records the current position of the wand and the headset, both of which act as visualization tools to the user. The head tracker controls the perspective of the image based on its location and elevation in the CAVE. As the viewer moves inside the CAVE, the correct stereoscopic perspective projections are automatically produced. The CAVE application developer can assign different functions to each of the three buttons on the wand.

## VRVIZ OVERVIEW

The focus of the post-processing software VRviz is the three dimensional visualization of finite element data in the CAVE. VRviz is a program written in ANSI C with the use of Silicon Graphics GL (Graphics Library). This program is written to complement finite element analysis programs. A brief description of the functions of VRviz, the virtual reality finite element post-processing program follows below.

VRviz has several visualization options. These options are available through two visual control panels on the CAVE walls. These panels consist of buttons and sliders which are manipulated by using the wand as a "laser pointer". To manipulate a control, the user presses a button on the wand to get a red beam of light. The beam which emanates from the wand can be projected on any CAVE wall. A point of intersection with the CAVE wall is calculated based on the direction of the wand and the position of the user. The algorithm for this computation is presented in Dwyer 1995. A blue box is drawn at the point of intersection to give the user a better visualization of where the beam is hitting. By adjusting the wand's position and orientation, the user is able to position the blue box inside a control to either push a button or move a slider.

Fig.1.11 depicts the buttons on the left-hand wall of the CAVE. This group of buttons is analogous in function to the controls on a VCR. The simulation results may be replayed in slow motion, viewed frame-by-frame or frozen at any time step. The "GO" button prescribes replay at the fastest speed. "SLOW" replays the simulation results in slow motion. This control in VRviz allows the evolution of deformations and stresses to be studied, thus allowing designers to gather ideas for changes and alternate designs. The ">" and "<" buttons allow the user to go forward or backward, respectively, by one time step. The image will then remain on that time step until another button is activated. The "STOP" button stops the animatation.

**Figure 1.11** Replay Control Panel

The seven slider controls on the back wall are depicted in Fig.1.12. These sliders enable the user to translate the image, rotate it about any axis, or rescale it at any time. These capabilities allow the user to navigate around, or even inside, the image in order to effectively analyze possible problem areas and redesign as necessary. The control bars labeled LEFT/RIGHT, DOWN/UP, and BACK/FORWARD control the image location in the CAVE. The LEFT/RIGHT control will bring the center of the image from x = -20 in the far left corner to x = 20 in the right corner. The DOWN/UP bar will translate the center of the image from y = -20 to y = 20. The BACK/FORWARD button will bring the center of the object from z = -20 to z = 10. The ROTX, ROTY, and ROTZ sliders all rotate the entire image about the respective axis from 0 to 360 degrees. The ZOOM control will zoom in or out.

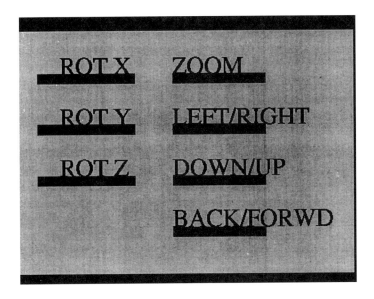

Figure 1.12 View Control Panel

The simulation shown in Fig.1.13 is that of a box beam impacting a rigid wall. An attached mass of 1400 kg at the free end is prescribed an initial velocity of 15.64 m/s. The beam measures 0.15 m in length, 0.03 m in depth and width and has a thickness of 0.0015 m. Due to symmetry, only one-quarter of the boxbeam is modeled by a mesh of 336 shell elements and 385 nodes. The two frames in Fig.1.13 are snapshots of the simulation which reveal shell elements coming into contact with one another as the box beam buckles. A single-surface slideline is defined over the entire mesh. Contact between any two elements of the mesh is possible.

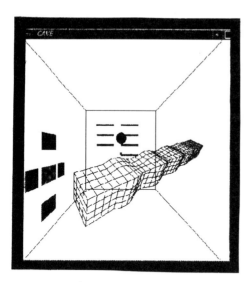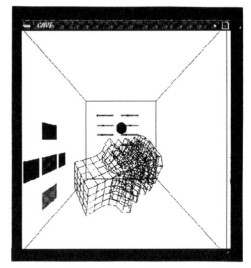

**Figure 1.13** Box Beam Simulation Displayed in CAVE

## COLLABORATIVE ENGINEERING WITH THE CAVE

The CAVE-to-CAVE software library is a collection of functions designed to facilitate shared visualization experiences among remote virtual environments anywhere on the internet. The CAVE-to-CAVE library establishes three processes. The broker process constructs and maintains a database containing information describing the session participants. The server process first informs the broker of its existence by registering with the broker. Next, the server process registers the information stream available for the client processes to subscribe to. To minimize the amount of information transmitted over the internet (for both speed and security purposes), the finite element data is assumed to reside locally at each participant's CAVE. Only nine parameters need to be broadcast to synchronize the displays in all CAVEs participating in a session. These nine parameters are: FrameId, RunStatus, ZoomStatus, TransXStatus, TransYStatus, TransZStatus, RotXStatus, RotYStatus, and RotZStatus. Client processes register with the broker, subscribe to an information stream, and execute a callback function each time a message is received. In the case of VRviz, a function named DrawData uses the parameters broadcast from the server process and the locally stored finite element data to display the same image in each of the CAVEs controlled by server processes. Fig.1.14 is a schematic representation of a shared visualization of an impact-contact calculation between two remote CAVEs. To present the "illusion" that the users are all in the same room, each remote user is represented as a stick figure. In the future, these crude stick figure images will be replaced with more true-to-life representations.

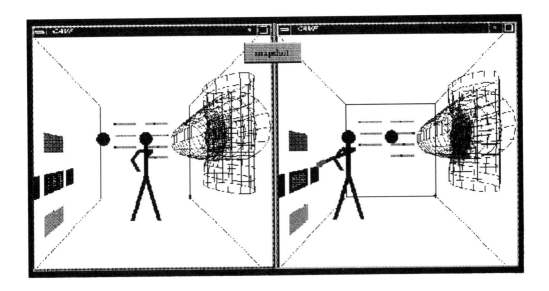

**Figure 1.14** VRviz Application Being Displayed in Multiple CAVEs

*SUMMARY*

In this section we have presented an overview of the basic capabilities of the CAVE Virtual Reality Environment. We have described our use of the CAVE as a "control room" for the interactive visualization of finite element results. Our application allows the user to navigate around, or even inside the image in order to effectively analyze possible problem areas and redesign as necessary. This is accomplished through sliders and buttons controlling image orientation and size as well as a group of buttons analogous in function to the controls on a VCR. These allow the user to step through the simulation at various speeds and even stop and pause at frames of particular interest.

## ACKNOWLEDGMENTS

This work was performed at Argonne National Laboratory, a contract laboratory of the United States Department of Energy under contract W-31-109-Eng-38.

This research was performed in part using the Intel Touchstone Delta System operated by Caltech on behalf of the Concurrent Supercomputing Consortium. Access to this facility was provided by Argonne National Laboratory. Use of the IBM SP1 system, operated by the Mathematics and Computer Science (MCS) Division at Argonne National Laboratory, is gratefully acknowledged. Performance results presented for a network of Hewlett Packard workstations were obtained on a workstation cluster at the National Center for Supercomputing Applications (NCSA) at the University of Illinois at Urbana-Champaign.

The programming assistance of Fred Dech, Mary Kuhn and Steve Karlovsky is gratefully acknowledged. The author is especially indebted to his wife Liz who painstakingly proofread this manuscript, correcting many errors in grammar, and offering numerous suggestions for its improvement.

## REFERENCES

Akin J.E. 1982. *Application and Implementation of Finite Element Methods*, Academic Press, New York.

Balmer H.A. and Witmer E.A. 1964. *Theoretical-Experimental Correlation of Large Dynamic and Permanent Deformation of Impulsively Loaded Simple Structures*, Air Force Flight Dynamics Laboratory, Report FDP-TDR-64-108, Wright-Patterson AFB, Ohio.

Beguelin A., Dongarra J., Geist G.A., Manchek R. and Sunderam V. 1991. *A user's guide to PVM: Parallel virtual machine*, Technical Report TM-11826, Oak Ridge National Laboratory.

Belytschko T. 1983, "An Overview of Semidiscretization and Time Integration Procedures," in *Computational Methods for Transient Analysis*, ed. by T. Belytschko and T. J. R. Hughes, North Holland, Amsterdam, pp. 1–66.

Belytschko T., Chiang H.Y. and Plaskacz E.J. 1994. "High resolution two- dimensional shear band computations: imperfections and mesh dependence," *Computer Methods in Applied Mechanics and Engineering*, 119: 1–15.

Belytschko T. and Chiapetta R.L., 1973. *A Computer Code for Dynamic Stress Analysis of Media-Structure Problems with Nonlinearities (SAMSON)*, Air Force Weapons Laboratory, KAFB. Report AFWL-TR-72-104.

Belytschko T., Chiapetta R.L. and Bartel H. 1976. "Efficient Large Nonlinear Transient Analysis by Finite Elements," *International Journal for Numerical Methods in Engineering*, 10: 579–596.

Belytschko T. and Gilbertsen N. 1987. "Concurrent and Vectorized Mixed Time Explicit Nonlinear Structural Dynamics Algorithms" in *Parallel Computations and Their Impact on Mechanics*, ed. by A. K . Noor, ASME, New York, pp 279–290.

Belytschko T., Lin J.I. and Tsay C.S. 1984. "Explicit Algorithms for the Nonlinear Dynamics of Shells," *Computer Methods in Applied Mechanics and Engineering*, 42: 225–251.

Belytschko T., Ong J. S-J, Liu W.K. and Kennedy J.M. 1984. "Hourglass Control in Linear and Nonlinear Problems," *Computer Methods in Applied Mechanics and Engineering*, 43: 251–276.

Belytschko T. and Plaskacz E.J. 1992. "SIMD Implementation of a Nonlinear Transient Shell Program with Partially Structured Meshes," *International Journal for Numerical Methods in Engineering*, 33: 997–1026.

Belytschko T., Plaskacz E.J., Kennedy J.M. and Greenwell D.L. 1990. "Finite Element Analysis on the CONNECTION Machine," *Computer Methods in Applied Mechanics and Engineering*, 81: 229–254.

Biffle J.H. 1977. "Use of Graphics in Three-Dimensional Structural Analysis Of Solids," *Interactive Computer Graphics In Engineering*, L.E. Hulbert, ed. New York: American Society of Mechanical Engineers, pp. 41–57.

Bomans L., Roose D. and Hempel R. 1990. "The Argonne/GMD macros in FORTRAN for portable parallel programming and their implementation on the Intel iPSC/2," *Parallel Computing*, 15: 119–132.

Boyle J., Butler R., Disz T., Glickfeld B., Lusk E., Overbeek R., Patterson J. and Stevens R. 1987. *Portable Programs for Parallel Processors*, Holt, Rinehart and Winston, New York.

Butler R. and Lusk E. 1992. *User's Guide to the p4 Programming System*, Technical Report ANL-92/17, Mathematics and Computer Science Division.

Canfield T., Minkoff M. and Plaskacz E.J. 1991. The Advanced Software Development and Commercialization (ASDAC) Project: Progress Report PR-2, ANL/TM 488 & CSRD Rpt. No. 1129, Argonne, IL 60439 & Urbana, IL 61801.

CAVE User's Guide 1995. University of Illinois at Chicago, Electronic Visualization Laboratory, cavesupport@evl.eecs.uic.edu.

Chan T.F. 1989. Proceedings of the Second International Symposium on Domain Decomposition Methods for Partial Differential Equations, Los Angeles, California, January 14-16, 1988, Philadelphia: SIAM.

Chan T.F. 1990. Proceedings of the Third International Symposium on Domain Decomposition Methods for Partial Differential Equations, Houston, Texas, 1989, Philadelphia: SIAM.

Desrouchers G.R. 1987. *Principles of Parallel and Multiprocessing*, Intertext Publications, Inc., McGraw-Hill Book Company, New York.

Disz T., Papka M., Pellegrino M. and Stevens R. 1995. "Sharing Visualization Experiences Among Remote Virtual Environments," *Proceedings of the International Workshop on High Performance Computing for Computer Graphics and Visualization*, Swansea, Wales, July 1995.

Doty R.J. 1988. "FE Analysis: The Decathlon For Computers," *Mechanical Engineering*, November 1988, 62–68.

Dwyer N. 1995. "Implementing and Using BSP Trees," *Dr. Dobbs Journal*, Vol. 20, Issue 7, pp. 46-49, July 1995.

Farhat C. 1988. "A Simple and Efficient Automatic FEM Domain Decomposer," *Computers & Structures*, 28: 579–602.

Farhat C. and Crivelli L. 1989. "A General Approach to Nonlinear FE Computations on Shared Memory Multiprocessors," *Computer Methods in Applied Mechanics and Engineering*, 72: 153 – 172.

Flanagan D.P. and Belytschko T. 1984. "Eigenvalues and Stable Time Steps for the Uniform Strain Hexahedrom and Quadrilateral," *Journal of Applied Mechanics*, 51: 35–40.

Flanagan D.P. and Belytschko T. 1981. "Simultaneous Relaxation in Structural Dynamics," *Journal of the Engineering Mechanics Division*, American Society of Civil Engineers, 107: 1039–1055.

Flanagan D.P. and Taylor L.M. 1987. "Structuring Data for Concurrent Vectorized Processing in a Transient Dynamics Finite Element Program," in *Parallel Computations and Their Impact on Mechanics*, ed. by A. K. Noor, ASME, New York, pp. 291–299.

Flynn M.J. 1972. "Some Computer Organizations and Their Effectiveness," *IEEE Transactions on Computers*.

Gallopoulos E. 1990. The Advanced Software Development and Commercialization (ASDAC) Project: Progress Report PR-1, ANL/TM 484 & CSRD Rpt. No. 1047, Argonne, IL 60439 & Urbana, IL 61801.

Gallopoulos E. 1992. The Advanced Software Development and Commercialization (ASDAC) Project: Progress Report PR-5, Argonne, IL 60439 & Urbana, IL 61801.

Geist G.A., Heath M.T., Peyton B.W. and Worley P.H. 1990. *PICL: A portable instrumented communication library*, Technical Report TM-11130, Oak Ridge National Laboratory, Oak Ridge, TN.

Glowinski R. 1988. Proceedings of the First International Symposium on Domain Decomposition Methods for Partial Differential Equations, Ecole Nationale des Ponts et Chaussees, Paris, France, January 7-9, 1987, Philadelphia: SIAM.

Glowinski R. 1991. Proceedings of the Fourth International Symposium on Domain Decomposition Methods for Partial Differential Equations, Moscow, F.S.F.S.R., May 21–25, 1990, Philadelphia: SIAM.

Grimsrud A. and Lorig G. 1989. "Implementing a Distributed Process between Workstation and Supercomputer," *Application of Supercomputers in Engineering, Proceedings of the First International Conference*, Southampton, UK, September 1989, C. A. Brebbia and A. Peters, eds., Elsevier, Amsterdam, pp. 133–144.

Gropp W.D. and Smith B. 1993. *Chameleon parallel programming tools users manual*, Technical Report ANL-93/23, Argonne National Laboratory.

Gropp W., Lusk E. and Skjellum A. 1994. *Using MPI: Portable Parallel Programming with the Message-Passing Interface*, The MIT Press, Cambridge, Massachusetts.

Hajjar J.F. and Abel J. 1988. "Parallel Processing For Transient Nonlinear Structural Dynamics Of Three-Dimensional Framed Structures Using Domain Decomposition," *Computers and Structures*, 30: 1237–1254, Pergamon Press.

Hallquist J.O., Benson D.J. and Goudreau G.L. 1986. "Implementation of a Modified Hughes-Liu Shell into a Fully Vectorized Explicit Finite Element Code," *Finite Elements For Nonlinear Problems*, P. G. Bergan, K. J. Bathe, and W. Wunderlich, eds., Springer-Verlag, Berlin, pp. 465–479.

Harrison R.J. 1991. Portable tools and applications for parallel computers, *International Journal of Quantum Chemistry*, 40: 847–863, John Wiley and Sons.

Hillis W. Daniel 1987. *The CONNECTION Machine*, The MIT Press, Cambridge, MA.

Johnson S.L. and Mathur K.K. 1990. "Data Structures and Algorithms for the Finite Element Method on a Data Parallel Computer," *International Journal for Numerical Methods in Engineering*, 29: 881–908.

Kardestuncer H. and Norrie D. H. *Finite Element Handbook*, McGraw Hill, New York, 1987.

Kennedy J.M., Belytschko T. and Lin J.I. 1986. "Recent Developments in Explicit Finite Element Techniques and Their Application to Reactor Structures," *Nuclear Engineering and Design*, 97: 1–24.

Keyes D.E. 1992. Proceedings of the Fifth International Symposium on Domain Decomposition Methods for Partial Differential Equations, May 6-8, 1991, Norfolk, Virginia, Philadelphia: SIAM.

Kulak, R. F., Plaskacz, E. J., Pfeiffer, P. A. 1995. "Structural Mechanics Computations on Parallel Computing Platforms," in *Fluid-Structures Interaction and Structural Mechanics-1995*, eds. C. Y. Wang, et al., 1995 ASME/JSME Pressure Vessel and Piping Conference, Honolulu, Hawaii, ASME Publication PVP-Vol. 310, pp. 129-133, July 23-27, 1995.

Lyzenga G.A., Raefsky A. and Hager B.H. 1988. "Finite Elements and the Method of Conjugate Gradients on a Concurrent Processor" in J. Fox and G. A. Lyzenga (eds.) *Solving Problems on Concurrent Processors*, Vol. 2. Prentice Hall, Englewood Cliffs, New Jersey.

Malone J.G. 1990. "Parallel Nonlinear Dynamic Finite Element Analysis Of Three-Dimensional Shell Structures," *Computers and Structures*, 35: 523–539, Pergamon Press.

Morino L., Leech J.W. and Witmer E.A. 1971. "An Improved Numerical Calculation Technique for Large Elastic-Plastic Transient Deformations of Thin Shells: Part 2 - Evaluation and Applications," *Journal of Applied Mechanics*, 38: 429–436.

Al-Nasra M. and Nguyen D.T. 1991. "An Algorithm For Domain Decomposition In Finite Element Analysis," *Computers and Structures*, 39: 277–289.

Nour-Omid B. and Park K.C. 1987. "Solving Structural Mechanics Problems on the CALTECH Hypercube Machine," *Computer Methods in Applied Mechanics and Engineering*, 61: 161–176, North-Holland.

Overmars M. H. 1991. *Forms Library A Graphical User Interface Toolkit for Silicon Graphics Workstations*, Version 2.0 (Public Domain), Department of Computer Science, Utrecht University. P.O. Box 80.089, 3508 TB Utrecht, the Netherlands.

Plaskacz E.J 1990. *Efficient Allocation of Computational Resources for Finite Element Applications*, PhD Dissertation, Northwestern University, Evanston, IL.

Plaskacz E.J. 1995a. "On Mesh Rezoning Algorithms for Parallel Platforms," in *Fluid-Structure Interaction and Structural Mechanics-1995*, eds. C. Y. Wang, et al., 1995 ASME/JSME Pressure Vessel and Piping Conference, Honolulu, Hawaii, ASME Publication PVP-Vol. 310, pp. 175-179, July 23-27, 1995.

Plaskacz E.J. 1995b. "On Impact-Contact Algorithms for Parallel Distributed Memory Computers," *Computational Mechanics 95 Theory and Applications: Proceedings of the International Conference on Computational Engineering Science*, July 30–August 3, 1995, Mauna Lani, Hawaii, eds. S.N. Atluri, G. Yagawa and T.A. Cruse, pp. 369–374.

Plaskacz E.J. and Belytschko T. 1990. "Measurement and Exploitation of Mesh Structure in a SIMD Environment," *Proceedings of the 1990 ASME International Computers In Engineering Conference and Exposition*, II. ASME.

Plaskacz E.J., Ramirez M.R. and Gupta S. 1992. "On Distributed Processing Applications in Finite Element Analysis," Proceedings of the Engineering Mechanics Division, ASCE, May 1992.

Plaskacz E.J., Ramirez M.R. and Gupta S. 1994, "Non-Linear Explicit Transient Finite Element Analysis On The Intel Delta," *Computing Systems in Engineering*, 5: 1–17, Elsevier.

Skjellum A., Smith S.G., Doss N.E., Leung A.P. and Morari M. 1994. "The design and evolution of zipcode", *Parallel Computing*, 20: 565–596, Elsevier.

Snir S., Otto S.W., Huss-Lederman S., Walker D.W. and Dongarra J. 1996. *MPI: The Complete Reference*, The MIT Press, Cambridge, Massachusetts.

# 2

# Solving Large Scale Structural Problems on Parallel Computers using Domain Decomposition Techniques

P. Le Tallec,[1] M. Vidrascu [2]

## 2.1 INTRODUCTION

The primary objective of domain decomposition methods is the efficient solution on parallel machines of problems in Computational Mechanics set on complex geometries and approximated on very fine grids. This objective is achieved by splitting the original domain of computation in smaller simpler subdomains, computing local simplified solutions, and using efficient algebraic solvers to properly interface these solutions.

The purpose of this review paper is to describe these domain decomposition techniques from the point of view of large scale three-dimensional structural problems. The practical use of such techniques for treating real life problems involves three ingredients :

- automatic techniques to partition complex structures (described by their finite element mesh) into simple compact substructures;

- local domain decomposition algorithms to interface subdomain solutions. Such algorithms must be consistent from the mechanical point of view and usually involve at each step and on each subdomain the solution of local displacement and/or traction problems;

- a coarse global solver to coordinate the rigid body motions of the different subdomains when solving equilibrium problems on a large number of subdomains.

Such techniques turn out to be ideally suited to modern parallel computers, because of the built-in parallelism of the algorithm and of the good localisation of the associated data. The present paper successively describes the basic model elasticity problem, the general domain decomposition technique, the preconditioner and global coarse solver

[1] Université de Paris-Dauphine and INRIA-Rocquencourt
[2] INRIA-Rocquencourt, B.P 105, F78153 Le Chesnay Cedex.

*Parallel Solution Methods in Computational Mechanics* edited by M. Papadrakakis

which is used, an abstract convergence theory, different partitioning techniques and implementation details, and finally presents various numerical results obtained on parallel machines.

## 2.2 A MODEL PROBLEM

The model problem considered hereafter is a three-dimensional linear structural mechanics problem. The solution of such problems arise in one form or another in many calculations of large linear or nonlinear structures.

This problem can be written in the following form :

Find the displacement field $u(x)$, of a three-dimensional structure $\Omega$, subjected to a given external loading.

The external forces acting onto the body can be reduced to surface tractions $f^\Gamma$ acting on the part $\partial\Omega_2$ of the boundary $\partial\Omega$ and body forces $f^\Omega$. The displacement $u(x)$ is imposed to the value $u = u_D$ on the remaining part $\partial\Omega_1 = \partial\Omega - \partial\Omega_2$ of the boundary.

For such problems, the governing equilibrium equations can be written in the classical variational form :

$$\int_\Omega E(x)\, \varepsilon(u) : \varepsilon(v) = \int_\Omega f^\Omega.vdx + \int_{\partial\Omega_2} f^\Gamma.vda, \quad \forall v \in H(\Omega), \qquad (2.1)$$

with $H(\Omega)$ the space of compatible virtual displacements

$$H(\Omega) = \{v \in H^1(\Omega, \mathbb{R}^3),\ v = 0 \quad \text{sur} \quad \partial\Omega_1\},$$

with E(x) the (possibly anisotropic) local elasticity tensor, and $\varepsilon(u)$ the linearized strain tensor

$$\varepsilon(u) = \frac{1}{2}\left(\nabla u + (\nabla u)^t\right).$$

In view of its numerical solution, the above variational formulation must first be approximated by finite element methods, that is by replacing the space $H(\Omega)$ by a finite element space $H_h(\Omega)$. In many complex cases, this space is constructed with quadrilateral (respectively hexahedral if $\Omega \subset \mathbb{R}^3$) isoparametric finite elements $T_l$, defined on a given regular "triangulation" of $\Omega$ (see Fig. 2.1) by

$$H_h(\Omega) = \left\{ v_h : \bar{\Omega} \to \mathbb{R}^3, v_h \text{ continuous}, \ v_h = 0 \text{ on } \partial\Omega_1, \right.$$

$$(2.2)$$

$$\left. v_{h|T_l} = v_l \circ \varphi_l^{-1}, v_l \in [Q_2'(\hat{\Omega})]^3, \forall l = 1,\ldots,N_h \right\}.$$

This finite element is the widely used *serendipity element* ( Zienkiewicz O. 1971, p108) and is a good compromise between accuracy and cost-efficiency. Local values of the displacements are characterized by their values at vertices and at midedges. Other choices of finite element are of course possible and will be also used in this text.

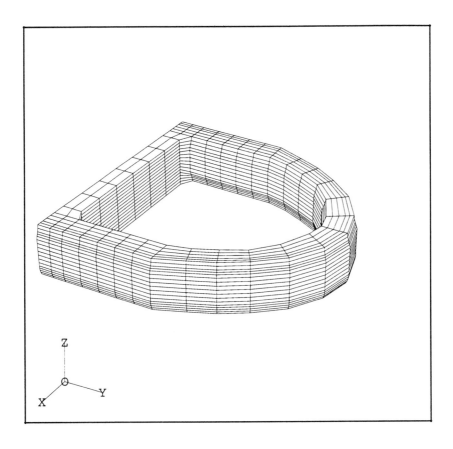

Figure 2.1: *Three-dimensional triangulation.*

On this finite element space, after expansion of the unknown displacement field into the basis $(\phi_l)_l$ of nodal finite element shape functions

$$u(x) = u_D(x) + \sum_l X_l \phi_l(x),$$

the original equilibrium equations reduce to the linear algebraic system

$$\mathbf{K}\, X = F.$$

Above, the stifness matrix $\mathbf{K}$ is given by :

$$\mathbf{K} = \left(\int_\Omega E(x)\, \varepsilon(\phi_l) : \varepsilon(\phi_m)\, dx\right)_{lm}.$$

For large three-dimensional structures, this finite element discretization of Equation (2.1) leads to a large linear system, (50 000 equations or more), with no particular structure. The computation of the stiffness matrix and the solution of the system by a direct method is then extremely time and memory consuming. On the other hand, this system is often very ill conditioned which means that its solution by classical iterative techniques is quite difficult. The development of effective parallel algorithms for solving this problem is thus quite a challenge. Domain decomposition methods, which mix local direct solvers and a global iterative interface solver turn out to be a very good answer to this challenge.

## 2.3 A BASIC DOMAIN DECOMPOSITION ALGORITHM

Domain decomposition methods are now widely studied, and are well understood in the framework of linear elliptic problems. Most methods refer to the so-called iterative substructuring techniques and use non overlaping domain partitions which split the original domain into small disjoint subdomains and reduce the original problem to an interface problem solved by an iterative conjugate gradient method.

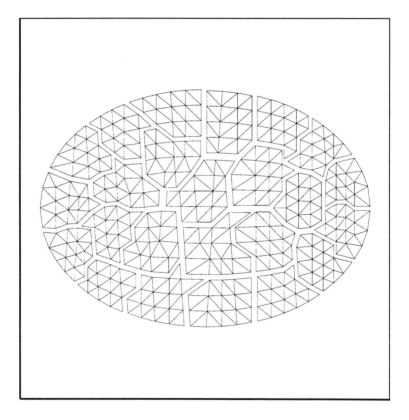

Figure 2.2: *Triangulation and decomposition of the mesh in nonoverlapping subdomains*

The first step is thus to split the domain into small local nonoverlapping subdomains

(Figure 2.2). This can be done by hand or using an automatic tool as described later (see Simon H. 1991, Farhat C. and Lesoinne M. 1993, & 2.6).

The second step is to construct the interface problem. Let $\mathbf{K}^i$ denote the stifness matrix of the subdomain $\Omega^i$

$$(\mathbf{K}_i)_{lm} = \int_{\Omega_i} E\, \varepsilon(\phi_l) : \varepsilon(\phi_m)\, dx$$

and $F^i$ the corresponding right hand side. These matrices and right hand sides can obviously be computed independently on each subdomain. For second or higher order finite elements, this local construction already involves a lot of computations, and is usefully done in parallel on different processors. For each subdomain the total set of degrees of freedom is then split into two subsets, the internal degrees of freedom $\overset{\circ}{X}{}^i$ associated to nodes which are strictly inside the subdomain $\Omega_i$, and the interface degrees of freedom $\bar{X}^i$ associated to nodes lying on the boundary between two or more neigboring subdomains. Under this notation the linear system describing the local equilibrium of domain $\Omega_i$ becomes :

$$\mathbf{K}^i X^i = \begin{bmatrix} \overset{\circ}{\mathbf{K}}{}^i & \mathbf{B}^i \\ \mathbf{B}^{it} & \bar{\mathbf{K}}^i \end{bmatrix} \begin{bmatrix} \overset{\circ}{X}{}^i \\ \bar{X}^i \end{bmatrix} = \begin{bmatrix} \overset{\circ}{F}{}^i \\ \bar{F}^i \end{bmatrix}.$$

It is then possible to eliminate internal degrees of freedom $\overset{\circ}{X}{}^i$ by

$$\overset{\circ}{X}{}^i = (\overset{\circ}{\mathbf{K}}{}^i)^{-1}(\overset{\circ}{F}{}^i - \mathbf{B}^i \bar{X}^i). \tag{2.3}$$

By writing then the equilibrium equations associated to the interface nodal degrees of freedom, if we denote by $\bar{X} = \bigcup_i \bar{X}^i$ the entire set of degrees of freedom on the interfaces, by $\bar{X}^i = \mathbf{R}^i \bar{X}$ the restriction of $\bar{X}$ on the boundary of $\Omega_i$, and if we eliminate the internal degrees of freedom $\overset{\circ}{X}{}^i$ by (2.3), we obtain the reduced interface system

$$\sum_i \mathbf{R}^{it} \left( \bar{\mathbf{K}}^i \mathbf{R}^i \bar{X} + \mathbf{B}^{it}(\overset{\circ}{\mathbf{K}}{}^i)^{-1} \left[ \overset{\circ}{F}{}^i - \mathbf{B}^i \mathbf{R}^i \bar{X} \right] \right) = \sum_i \mathbf{R}^{it} \bar{F}^i.$$

Introducing the so-called local Schur complement matrix

$$\mathbf{S}^i = \bar{\mathbf{K}}^i - \mathbf{B}^{it}(\overset{\circ}{\mathbf{K}}{}^i)^{-1}\mathbf{B}^i, \tag{2.4}$$

this interface system writes :

$$\left(\sum_i \mathbf{R}^{it}\mathbf{S}^i\mathbf{R}^i\right)\bar{X} = \sum_i \mathbf{R}^{it}(\bar{F}^i - \mathbf{B}^{it}(\overset{\circ}{\mathbf{K}}{}^i)^{-1}\overset{\circ}{F}{}^i). \tag{2.5}$$

Problem (2.5) is equivalent to our original equilibrium problem, but is only written in terms of the interface unknowns $\bar{X}$. In many engineering codes, this problem is solved by a direct method. These so-called direct substructuring techniques (Przemieniecki W. (1963), Bjorstad P.E. and Hvidsten A. (1988), Bell K., Hatelstad B., Hansteen O.E. and Araldsen P.O (1973)) first build the local Schur complement matrix $\mathbf{S}^i$ by a frontal elimination technique, stopping the elimination process when all internal nodes are eliminated. This is done in parallel subdomain per subdomain. Then the global Schur complement matrix

$$\mathbf{S} = \sum_i \mathbf{R}^{i^t} \mathbf{S}^i \mathbf{R}^i$$

is assembled and is factored into $\mathbf{S} = \mathbf{L}^t \mathbf{L}$. Afterwards, the solutions of (2.5) are obtained by the usual backward and forward substitution process. This strategy is very efficient for complex structures with small interfaces as those encountered in offshore rigs (Bjorstad P.E. and Hvidsten A. 1988). It is not feasible for large three-dimensional structures cut in many subdomains because the dimension of $\mathbf{S}$ gets rapidly very large (the interface problem has often more than $10^4$ unknowns).

The main idea of modern domain decomposition methods is to solve problem (2.5) by an iterative preconditioned conjugate gradient algorithm such as :

. Let $d\bar{X} = \sum_i \mathbf{R}^{i^t} d\bar{X}^i$ be a known descent direction used for correcting the interface vector of displacement unknowns $\bar{X}$. We then obtain a new descent direction by the following sequence of operations. We first compute on each subdomain the solution $d\mathring{X}^i$ of the local Dirichlet problem :

$$\mathring{\mathbf{K}}^i d\mathring{X}^i = -\mathbf{B}^i d\bar{X}^i. \tag{2.6}$$

Note that problems (2.6) are independent on each subdomain and thus can be solved in parallel.

. We then compute the variation of the residual $d\bar{R}$ equal to the algebraic sum of interface reaction forces

$$d\bar{R} = (\sum_i \mathbf{R}^{i^t} \mathbf{S}^i \mathbf{R}^i) d\bar{X} = \sum_i (\mathbf{R}^{i^t} \bar{\mathbf{K}}^i d\bar{X}^i - \mathbf{B}^{i^t} d\mathring{X}^i). \tag{2.7}$$

. We deduce a preconditioned gradient $d\bar{G}$ by solving an additional interface problem :

$$\mathbf{M} d\bar{G} = d\bar{R}; \tag{2.8}$$

. The updated descent direction is then obtained by the standard conjugate gradient algorithm

$$\bar{X} = \bar{X} - \rho_n d\bar{X},$$
$$\bar{G} = \bar{G} - \rho_n d\bar{G},$$
$$d\bar{X} = d\bar{G} + \beta_n d\bar{X}.$$

Observe that these iterative techniques never require the explicit calculation of matrix $\mathbf{S}$. The main issue conditioning the success and parallel efficiency of such techniques is the choice of the preconditioner $\mathbf{M}$. This preconditionner must be easy to implement in a parallel environment and must lead to a scalable domain decomposition algorithm when the number of processors increase.

## 2.4 CONSTRUCTION OF THE PRECONDITIONER

*IMPORTANCE OF THE PRECONDITIONER*

If we take $\mathbf{M} = \mathbf{Id}$ (no preconditioner) in the above algorithm, the proposed strategy is not mechanically consistent because it tries to update interface displacement unknowns $\bar{X}$ by normal reaction force jumps $\sum_i (\mathbf{R}^{i^t} \bar{\mathbf{K}}^i d\bar{X}^i - \mathbf{B}^{i^t} d\mathring{X}^i)$. It can actually be proved that the condition number of the Schur matrix $\mathbf{S}$, monitoring the efficiency of the above conjugate gradient algorithm used with $\mathbf{M} = \mathbf{Id}$, varies as

$$\text{Cond}(\mathbf{S}) \approx \frac{1}{H^2} (1 + max \frac{H_i}{h_i}),$$

and blows up either for small discretisation steps $h$ or for small subdomain diameters $H$ (i.e. for partitions with many subdomains). A preconditioner is therefore necessary.

In very specific cases such as a Poisson equation solved on a uniform rectangular grid, the exact spectrum of $\mathbf{S}$ can be computed (Bjorstad P.E. and Widlund O. (1986), Chan T. and Hou T.Y (1991), Goovaerts D., Chan T. and Piessens R. (1991)). But using this exact information to build accurate preconditioners is unpractical when dealing with real geometries, elasticity operators and variables coefficients. In another direction, standard algebraic preconditioners such as diagonal scaling turn out to be also unpractical (they require the explicit knowledge of $\mathbf{S}_i$) and inefficient. A possible variant would be the diagonal probing technique of Chan (Chan T. 1989) but such a method is delicate to implement in three-dimensional situations and is not fully efficient (Chan T.and Mathew T. 1992). It turns out in fact that the interface problem (2.5) require specific preconditioners which take advantage of its particular structure. Such preconditioners must have nice parallel properties, must be able to handle arbitrary elliptic operators and discretization grids, and their performance must be insensitive to the discretization step $h$ and to the number of subdomains. Many such preconditioners have appeared in the litterature, following the early work of Bramble J., Pasciak J. and Schatz A. 1986 (Chan T.and Mathew T. 1994, Cowsar L., Mandel J. and Wheeler M. 1993, Dryja M. and Widlund O. 1992, Mandel Jan 1990, Widlund O. 1988). For three dimensional elasticity results, efficient results have been obtained using either the wirebasket algorithm of Smith (Smith B. 1992-1) or the Neumann-Neumann preconditioner (Le Tallec P., De Roeck Y.H. and Vidrascu M. 1991, Le Tallec P. 1994).

**Remark 2.1** *The interface problem can also be written in a dual form, acting on the values of interface reaction forces instead of the values of the interface displacement unknowns. Efficient conjugate gradient solution procedures (the so-called FETI method) have then been introduced in Farhat C. and Roux F.X. 1992 for solving this dual interface problem in the framework of general structural problems (Farhat C. and Roux F.X. 1994).*

*NEUMANN-NEUMANN PRECONDITIONER*

The specific preconditioner detailed hereafter is the so-called Neumann-Neumann algorithm proposed by Aghoskov V. 1988, Bourgat J.F., Glowinski R., Le Tallec P.

and Vidrascu M. 1989, Morice P. 1989, between others, with different choices of the weighting factors $\mathbf{D}^i$. The idea is to use as a preconditioner for the interface operator $\mathbf{S} = \sum_i \mathbf{R}^{i^t} \mathbf{S}^i \mathbf{R}^i$ the following weighted sum of inverses (Le Tallec P. 1994) :

$$\mathbf{M}^{-1} = \sum_i \mathbf{D}^i (\mathbf{S}^i)^{-1} \mathbf{D}^{i^t}. \tag{2.9}$$

Above, $\mathbf{D}^i$ are diagonal ponderation matrices verifying

$$\sum_i \mathbf{D}^i \mathbf{R}^i = \mathrm{Id}.$$

For implementation reasons (flexibility and parallelism), the map $\mathbf{D}^i$ must be as local as possible. The generic choice consists in defining $\mathbf{D}^i$ on each interface degree of freedom $\bar{v}(P_l)$ by :

$$\mathbf{D}^i \bar{v}(P_l) = \frac{\rho_i}{\rho} \bar{v}(P_k)$$

if the $l$ degree of freedom of V corresponds to the $k$ degree of freedom of $V_i$,

$$\mathbf{D}^i \bar{v}(P_l) = 0,$$

if not. Here $\rho_i$ is a local measure of the stiffness of subdomain $\Omega_i$ (for example an average Young modulus on $\Omega_i$) and $\rho = \sum_{P_l \in \Omega_j} \rho_j$ is the sum of $\rho_j$ on all subdomains $\Omega_j$ containing $P_l$.

To compute the inverse $(\mathbf{S}^i)^{-1}$, we also observe that the matricial definition of $\mathbf{S}^i$ implies that we have

$$\begin{bmatrix} 0 \\ \mathbf{I} \end{bmatrix} \mathbf{S}^i = \begin{bmatrix} \overset{\circ}{\mathbf{K}}{}^i & \mathbf{B}^i \\ \mathbf{B}^{i^t} & \bar{\mathbf{K}}_i \end{bmatrix} \begin{bmatrix} -(\overset{\circ}{\mathbf{K}}{}^i)^{-1} \mathbf{B}^i \\ \mathbf{I} \end{bmatrix} = \mathbf{K}^i \begin{bmatrix} -(\overset{\circ}{\mathbf{K}}{}^i)^{-1} \mathbf{B}^i \\ \mathbf{I} \end{bmatrix}.$$

Hence the matrix form of $(\mathbf{S}^i)^{-1}$ is

$$(\mathbf{S}^i)^{-1} = (0, \mathbf{I})(\mathbf{K}^i)^{-1} \begin{pmatrix} 0 \\ \mathbf{I} \end{pmatrix}. \tag{2.10}$$

Thus the preconditioning step used in the Neumann-Neummann algorithm consists finally in solving in parallel the local Neumann problem :

$$\mathbf{K}^i dG^i = \begin{bmatrix} 0 \\ \mathbf{D}^{i^t} d\bar{R} \end{bmatrix}. \tag{2.11}$$

These values, defined up to a rigid body motion, are then added together to obtain the preconditioned gradient :

$$d\bar{G} = \sum_i \mathbf{D}^i d\bar{G}^i. \tag{2.12}$$

The interest of this preconditioner is that it is fully parallel, and that it can handle nonstructured meshes and arbitrary shaped subdomains. Its weakness is its lack of scalability. Indeed, the conditionning of the interface matrix $\mathbf{M}^{-1}\mathbf{S}$ grows when the diameter of the subdomains $H$ and discretisation step $h$ decrease, following the formula De Roeck Y.H. and Le Tallec P. 1991

$$cond(\mathbf{M}^{-1}\mathbf{S}) = [\frac{c}{H}(1 + ln\frac{H}{h})]^2. \tag{2.13}$$

The formula (2.13) means that the number of iterations of the conjugate gradient algorithm grows linearly when the number of subdomains increases. Thus this algorithm is efficient only for a small number of subdomains. An additional coarse grid preconditionner is required in order to obtain a scalable algorithm and to pass rapidly information between subdomains which are far apart in the partition. In fact, all scalable iterative domain decomposition algorithms are proved to require such coarse solvers and this is confirmed by numerical experiments.

*COARSE GRID BALANCING PRECONDITIONER*

The main idea of the coarse grid solver proposed by Mandel Jan 1993 is the following :

Instead of taking arbitrary rigid bodies in the solution of the local Neumann problems (2.11), we choose them in order to minimise the residual of the next iteration.

For this purpose, we first introduce on each subdomain independent degrees of freedom $(r_i(\alpha))_{\alpha=1,6}$ characterizing the different rigid body motions of the subdomain. In three dimensional elasticity, we get six or less rigid body motions $(z_\beta^i)_{\beta=1,6}$ per subdomain, whose matrix representation $U_\beta^i$ is solution of :

$$\begin{aligned}(\mathbf{K}^i U^i)_l &= 0 \ \forall l \neq r_i(\alpha), \\ U^i_{r_i(\alpha)} &= \delta_{\alpha\beta}, \ \forall \alpha = 1 \text{ to } 6.\end{aligned} \tag{2.14}$$

The modified balanced Neumann-Neumann preconditioner is then given by the following operations :

i) compute the local residual

$$F^i = \begin{pmatrix} 0 \\ \mathbf{D}^{i^t} d\bar{R} \end{pmatrix};$$

ii) solve the local Neumann problem with zero rigid body motions

$$(\mathbf{K}^i \Psi^i)_l = F_l^i, \ \forall l \neq r_i(\alpha),$$

$$\Psi^i_{r_i(\alpha)} = 0, \ \forall \alpha = 1 \text{ à } 6; \tag{2.15}$$

iii) set the interface correction to the sum

$$d\bar{G} = \sum_i \mathbf{D}^i d\bar{G}^i = \sum_i \mathbf{D}^i \bar{\Psi}^i + \sum_{i,\alpha} b_\alpha^i \mathbf{D}^i \bar{z}_\alpha^i, \tag{2.16}$$

where the coefficients $b_\alpha^i$ in front of the local rigid body motions $\bar{z}_\alpha^i$ are additional unknowns which we find by minimizing the residual at next step.

From the definition of the Schur complement, this residual $R$ is given by :

$$R = d\bar{R} - \mathbf{S} \, d\bar{G}.$$

Its energy norm is :

$$\|R\|_S^2 = \text{err} = \langle \mathbf{S}^{-1}(d\bar{R} - \mathbf{S}\, d\tilde{G}), d\bar{R} - \mathbf{S}\, d\tilde{G} \rangle,$$

that is

$$\text{err} = \langle \mathbf{S}^{-1}\, d\bar{R} - \sum_i \mathbf{D}^i \bar{\Psi}^i - \sum_{i,\alpha} b_\alpha^i \mathbf{D}^i \bar{z}_\alpha^i \,,$$
$$d\bar{R} - \mathbf{S} \sum_i \mathbf{D}^i \bar{\Psi}^i - \sum_{i,\alpha} b_\alpha^i \mathbf{S}\mathbf{D}^i \bar{z}_\alpha^i \rangle.$$

The minimisation of this residual norm with respect to the coefficients $b_\alpha^i$ amounts to the solution of the following coarse linear problem of order ($6\times$ number of subdomains) :

$$\sum_{i,\alpha} b_\alpha^i \langle \mathbf{S}\mathbf{D}^i \bar{z}_\alpha^i, \mathbf{D}^j \bar{z}_\beta^j \rangle + \langle \sum_i \mathbf{D}^i \bar{\Psi}^i, \mathbf{S}\mathbf{D}^j \bar{z}_\beta^j \rangle$$
$$- \langle d\bar{R}, \mathbf{D}^j \bar{z}_\beta^j \rangle = 0, \forall j, \forall \beta. \qquad (2.17)$$

Altogether, the final preconditioner is given by (2.16), with $b_\alpha^i$ solution of the above coarse system (2.17). This improved preconditioner stays efficient independently of the number of subdomains and it can handle irregular decompositions of arbitrary domains discretised by different types of finite elements. The whole algorithm can thus be used in a very general framework and in particular for solving three dimensional elasticity problems. Its practical implementation was done within the finite element software Modulef and efficiently run on massively parallel systems using 100 subdomains or more.

## 2.5 ABSTRACT CONVERGENCE THEORY

### ABSTRACT ADDITIVE SCHWARZ METHOD

Let us consider the solution of the abstract variational problem

$$a(u,v) = \langle f,v \rangle, \forall v \in V, u \in V, \qquad (2.18)$$

where $V$ is a given Hilbert space with duality product $\langle .,. \rangle$, and $a$ an elliptic continuous symmetric bilinear form defined on $V$.

We suppose that the space $V$ can be decomposed into the sum

$$V = R_0{}^t\, V_0 + R_1^t\, V_1 + R_2^t\, V_2 + \ldots + R_N^t\, V_N. \qquad (2.19)$$

On each subspace $V_i$, we introduce a symmetric elliptic bilinear form $b_i(.,.)$. We denote by $A : V \to V'$ the linear operator associated to the form $a$

$$\langle Au, v \rangle = a(u,v), \forall u, v \in V,$$

by $B_i : V_i \to V_i'$ the linear operator associated to the form $b_i$, and by $R_i^t$ the extension map from $V_i$ into $V$.

With this notation, the additive Schwarz method for solving our original problem

$$Au = f \quad \text{in} \quad V'$$

is defined as the conjugate gradient method operating on $A$ preconditioned by the following sum of local operators

$$M^{-1} = R_0^t B_0^{-1} R_0 + \ldots + R_N^t B_N^{-1} R_N. \tag{2.20}$$

This preconditioner is quite easy to compute since its action on a given element $L \in V'$ is simply equal to the sum

$$M^{-1} L = \sum_{i=0}^{N} R_i^t u_i,$$

where $u_i \in V_i$ is the solution of the local variational problem

$$b_i(u_i, v_i) = \langle L, R_i^t v_i \rangle, \forall v_i \in V_i.$$

## CONVERGENCE RESULTS

The efficiency of the above preconditioned conjugate gradient method is best estimated by the following convergence theorem (see Golub G. and Van Loan C. 1983 for more details) :

**Theorem 2.1** : *The sequence* $(u^n)$ *generated by the additive Schwarz algorithm satisfies*

$$\|u^n - u\|_A \leq \left( \frac{\sqrt{Cond(M^{-1}A)} - 1}{\sqrt{Cond(M^{-1}A)} + 1} \right)^n \|u^0 - u\|_A, \tag{2.21}$$

*where the condition number* $Cond(M^{-1}A)$ *is given by*

$$Cond(M^{-1}A) = \frac{\lambda_{max}(M^{-1}A)}{\lambda_{min}(M^{-1}A)}. \tag{2.22}$$

The above lowest and largest eigenvalues $\lambda_{min}$ and $\lambda_{max}$ of $M^{-1}A$ can then be estimated by two basic lemmas. The first partition lemma supposes that any vector $u$ of $V$ can be decomposed at low energy into a sum of local vectors $z_i$ :

**Lemma 2.1** : *If any* $u \in V$ *can be decomposed into* $u = R_0^t z_0 + R_1^t z_1 + \ldots + R_N^t z_N$, *with*

$$\sum_{i=0}^{N} b_i(z_i, z_i) \leq C_0 a(u, u), \tag{2.23}$$

*then we have*

$$\lambda_{min}^{-1}(M^{-1}A) \leq C_0. \tag{2.24}$$

**Proof :** By construction of the local solution $u_i$ we have

$$
\begin{aligned}
a(u, u) &= a\left(u, \sum_{i=0}^{N} R_i^t z_i\right) \\
&= \sum_{i=0}^{N} a(u, R_i^t z_i) \\
&= \sum_{i=0}^{N} b_i(u_i, z_i) \\
&\leq \left(\sum_{i=0}^{N} b_i(u_i, u_i)\right)^{\frac{1}{2}} \left(\sum_{i=0}^{N} b_i(z_i, z_i)\right)^{\frac{1}{2}} \\
&\leq C_0^{\frac{1}{2}} \left(\sum_{i=0}^{N} a(u, R_i^t u_i)\right)^{\frac{1}{2}} a(u, u)^{\frac{1}{2}} \\
&\leq C_a^{\frac{1}{2}} a(u, M^{-1} A u)^{\frac{1}{2}} a(u, u)^{\frac{1}{2}}.
\end{aligned}
$$

From this, we directly deduce the required estimate

$$
\lambda_{min}^{-1}(M^{-1}A) = \sup_{u \in V} \frac{a(u, u)}{a(u, M^{-1} A u)} \leq C_0.
$$

Using similar techniques, the largest eigenvalue can be estimated by estimating the amount of overlapping present in the decomposition $V = \sum_{i=0}^{N} V_i$ (Dryja M. and Widlund O. 1992) :

**Lemma 2.2** : *Let us suppose that the subspaces $V_i$ satisfy the following strengthened Cauchy-Schwarz inequality*

$$
a(R_i^t u_i, R_j^t u_j) \leq \varepsilon_{ij} a(R_i^t u_i, R_i^t u_i)^{\frac{1}{2}} a(R_j^t u_j, R_j^t u_j)^{\frac{1}{2}}, \forall u_i \in V_i, \forall u_j \in V_j, \forall 1 \leq i, j \leq N. \tag{2.25}
$$

*Suppose also that the bilinear form $a$ is locally bounded on each subspace $V_i$ by*

$$
a(R_i^t u_i, R_i^t u_i) \leq \omega_+ b_i(u_i, u_i), \forall u_i \in V_i, \forall i = 0, N. \tag{2.26}
$$

*Then*

$$
\lambda_{max}(M^{-1}A) \leq \omega_+ (\rho(\varepsilon) + 1), \tag{2.27}
$$

*with $\rho(\varepsilon)$ the spectral radius of the matrix $(\varepsilon_{ij})$.*

Combining the above lemmas, we then obtain the basic estimate

$$
cond(M^{-1}A) \leq C_0 \omega_+ (\rho(\varepsilon) + 1)
$$

which is used in the convergence analysis of most domain decomposition techniques (with or without overlapping) (Dryja M. and Widlund O. 1992), or more generally to study the convergence properties to various multilevel techniques (Bramble J., Pasciak J. and Xu J. 1990, Xu J. 1992).

## THE ABSTRACT NEUMANN-NEUMANN PRECONDITIONER

We have seen earlier that the interface problem (2.5) takes the abstract form

$$(\sum_i \mathbf{R}^{i^t} \mathbf{S}^i \mathbf{R}^i)\bar{X} = \bar{F} \in V, \qquad (2.28)$$

with $\mathbf{R}^{i^t}$ the restriction from the space $V$ of global interface values $\bar{X}$ to the space $V_i$ of local interface values $\bar{X}^i$. Such an abstract problem can be solved in all generality by the Neumann-Neumann algorithm introduced earlier, which preconditions the sum $S = \sum \mathbf{R}^{i^t} \mathbf{S}^i \mathbf{R}^i$ by a two level weighted sum of the inverses $\mathbf{M}^{-1} = \sum \mathbf{D}^i (\mathbf{S}^i)^{-1} \mathbf{D}^{i^t}$. From a theoretical point of view, this algorithm turns out to be a particular case of the additive Schwarz method studied in the previous section.

To recover such a theoretical interpretation which is quite useful for its convergence analysis, we will describe below the Neumman-Neumann algorithm in the abstract general framework of (2.28). For this purpose, we first need to *choose*

1. a partition of unity $\mathbf{D}^i : V_i \to V$ satisfying

$$\sum_{i=1}^{N} \mathbf{D}^i \mathbf{R}^i = \mathbf{Id}|_V;$$

2. a local coarse space $\bar{Z}_i$ containing potential local singularities, such that

$$Ker\mathbf{S}^i \subset \bar{Z}_i \subset V_i;$$

3. a complement $V^0\_i$ of $\bar{Z}_i$

$$V_i = V^0\_i + \bar{Z}_i,$$

on which $\mathbf{S}^i$ is coercive, and on which we define the solution $\mathbf{S}_{io}^{-1} L$ of the associated local elliptic problem

$$\langle \mathbf{S}^i (\mathbf{S}_{io}^{-1} L), v \rangle = \langle L, v \rangle,$$
$$\forall v \in V^0\_i, \mathbf{S}_{io}^{-1} L \in V^0\_i, \forall L \in V_i'.$$

We then introduce

1. the global coarse space

$$V_0 = \sum_{i=1}^{N} \mathbf{D}^i \bar{Z}_i,$$

2. the $\mathbf{S}$ orthogonal projection $\mathbf{P} : V \to V_0$,

3. the local fine space

$$V_i^{\perp} = (\mathbf{I} - \mathbf{P})\mathbf{D}^i V^0\_i \subset V$$

endowed with the scalar product

$$b_i(\bar{u}_i, \bar{v}_i) = \langle \mathbf{S}^i [(\mathbf{I} - \mathbf{P})\mathbf{D}^i]^{-1} \bar{u}_i, [(\mathbf{I} - \mathbf{P})\mathbf{D}^i]^{-1} \bar{v}_i \rangle$$

where $[(\mathbf{I} - \mathbf{P})\mathbf{D}^i]^{-1} : V_i^{\perp} \to V_i^{00}$ is the inverse of $(\mathbf{I} - \mathbf{P})\mathbf{D}^i$ in the space

$$V_i^{00} = \{\bar{v}_i \in V^0\_i, \langle \mathbf{S}^i \bar{v}_i, \bar{z}_i \rangle = 0, \forall \bar{z}_i \in V^0\_i \cap Ker(\mathbf{I} - \mathbf{P})\mathbf{D}^i\}.$$

The Neumann-Neumman algorithm is then a standard additive Schwarz algorithm solving

$$a(u, v) := \langle \mathbf{S}u, v \rangle = \langle f, v \rangle, \forall v \in V,$$

using the interface decomposition

$$V = V_0 + \sum_i V_i^{\perp}, \tag{2.29}$$

where $V_0$ is endowed with the $a$ scalar product $b_0(u, v) = a(u, v)$ and where where $V_i^{\perp}$ is endowed with the above scalar product $b_i$.

To see how this abstract algorithm is indeed related to the original version of the Neumann-Neumann algorithm, and to explain how it can be numerically implemented, we detail below the application of the operator $\mathbf{M}^{-1}\mathbf{S}$ to a given vector $\bar{u}$ of V. By construction of a Schwarz additive method, we have

$$\mathbf{M}^{-1}\mathbf{S}\bar{u} = \bar{u}_0 + \sum_{i=1}^{N} \bar{u}_i,$$

where $\bar{u}_0 = \mathbf{P}\bar{u}$ satisfies

$$\langle \mathbf{S}\mathbf{P}\bar{u}, \bar{w}_0 \rangle = \langle \mathbf{S}\bar{u}, \bar{w}_0 \rangle, \forall \bar{w}_0 \in V_0, \mathbf{P}\bar{u} \in V_0,$$

and where $\bar{u}_i$ is solution of the local problem

$$b_i(\bar{u}_i, \bar{v}_i) = \langle \mathbf{S}_i[(\mathbf{I} - \mathbf{P})\mathbf{D}^i]^{-1}\bar{u}_i, [(\mathbf{I} - \mathbf{P})\mathbf{D}^i]^{-1}\bar{v}_i \rangle = \langle \mathbf{S}\bar{u}, \bar{v}_i \rangle, \tag{2.30}$$
$$\forall \bar{v}_i \in V_i^{\perp}.$$

With respect to the variable $v_i = [(\mathbf{I} - \mathbf{P})\mathbf{D}^i]^{-1}\bar{v}_i \in V_i^{00}$, this writes

$$\langle \mathbf{S}_i u_i, v_i \rangle = \langle \mathbf{S}\bar{u}, (\mathbf{I} - \mathbf{P})\mathbf{D}^i v_i \rangle, \forall v_i \in V_i^{00}, u_i \in V_i^{00}. \tag{2.31}$$

By construction, the right hand side of this equation takes zero values on the $\mathbf{S}_i$ orthogonal complement $Ker(\mathbf{I} - \mathbf{P})\mathbf{D}^i$ of $V_i^{00}$ into $V^0_{\_}i$. Therefore by construction of $V_i^{00}$, this equation is equivalent to

$$\langle \mathbf{S}_i u_i, v_i \rangle = \langle \mathbf{S}\bar{u}, (\mathbf{I} - \mathbf{P})\mathbf{D}^i v_i \rangle = \langle \mathbf{S}(\mathbf{I} - \mathbf{P})\bar{u}, \mathbf{D}^i v_i \rangle, \tag{2.32}$$
$$\forall v_i \in V_i^0, u_i \in V_i^0.$$

By definition of $\mathbf{S}_{i0}^{-1}$, we finally get

$$\bar{u}_i = (\mathbf{I} - \mathbf{P})\mathbf{D}^i \mathbf{S}_{i0}^{-1}\mathbf{D}^{i^t}\mathbf{S}(\mathbf{I} - \mathbf{P})\bar{u} \tag{2.33}$$

and hence our preconditioner writes

$$\mathbf{M}^{-1}\mathbf{S}\bar{u} = \mathbf{P}\bar{u} + (\mathbf{I} - \mathbf{P})(\sum_i \mathbf{D}^i \mathbf{S}_{io}^{-1}\mathbf{D}^{i^t})(\mathbf{I} - \mathbf{P})^t\mathbf{S}\bar{u}, \tag{2.34}$$

which is the basis of all practical calculations of $\mathbf{M}^{-1}\mathbf{S}$, and is the original form of the previously introduced Neumann-Neumman algorithm.

**Remark 2.2** *The factor* $(\mathbf{I} - \mathbf{P})^t$ *was not present in our first description of the Neumann-Neumann algorithm in section 2.4. This is due to the fact that, when one uses the exact Schur interface operator* $\mathbf{S}$ *for defining the projection* $\mathbf{P}$ *and when one starts with an initial guess* $u^0$ *in* $V_0^\perp$, *then all further directions of descent* $\bar{u}$ *also belong to* $V_0^\perp$ *and therefore satisfy*

$$(\mathbf{I} - \mathbf{P})^t \mathbf{S}\bar{u} = \mathbf{S}\bar{u}.$$

*This property is no longer valid if the projection operator is associated to a simplified interface operator as it will be the case for nonlinear or for nonmatching grid problems. There, the application of the Neumann-Neumman algorithm requires the explicit calculation of the projection* $(\mathbf{I} - \mathbf{P})^t \mathbf{S}$.

**Remark 2.3** *The decomposition (2.29) is an admissible* $\mathbf{S}$ *orthogonal decomposition of the space* $V$

$$V = V_0 \oplus_\perp \sum_{i=1}^{N} V_i^\perp.$$

**Remark 2.4** *The space* $V_i^\perp$ *does not depend on the practical construction of* $V^0\_i$, *but the scalar product* $b_i$ *and hence the proposed preconditioner does depend on the construction of this local space* $V_i^o$.

## CONVERGENCE ANALYSIS

Using either the general convergence theory outlined earlier for additive Schwarz methods, or a specific proof, we can prove the following convergence theorem for analyzing the performance of the Neumann-Neumann algorithm (Le Tallec P., Mandel J. and Vidrascu M. 1996).

**Theorem 2.2** *The abstract Neumann-Neumann preconditioner given by (2.34) satisfies*

$$cond(\mathbf{M}^{-1}\mathbf{S}) \leq N \sup_i \sup_{\bar{v}_i \in V_i^{oo}} \frac{\|\mathbf{D}^i \bar{v}_i\|_S^2}{\|\bar{v}_i\|_{S_i}^2},$$

*with* $N$ *the maximum number of neighbors of a given subdomain.*

**Proof.** From lemma 2.1, the lower eigenvalue of $\mathbf{M}^{-1}\mathbf{S}$ is bounded if we can decompose any function of $\bar{v} \in V$ into a sum

$$\bar{v} = \bar{v}_0 + \sum_i \bar{v}_i, \bar{v}_0 \in V_0, \bar{v}_i \in V_i^\perp,$$

satisfying

$$b_0(\bar{v}_0, \bar{v}_0) + \sum_i b_i(\bar{v}_i, \bar{v}_i) \leq C_0^2 a(\bar{v}, \bar{v}). \tag{2.35}$$

For this purpose, using the $\mathbf{S}$ orthogonal projection operator $\mathbf{P}$ and the partition of unity $\mathbf{D}^i$, we decompose $\bar{v}$ into

$$\bar{v} = \mathbf{P}\bar{v} + (\mathbf{I} - \mathbf{P})\bar{v} = \bar{v}_0 + \bar{v}_\perp$$

with
$$\bar{v}_\perp = \sum_i (\mathbf{I} - \mathbf{P})\mathbf{D}^i \mathbf{R}^i \bar{v}_\perp = \sum_i (\mathbf{I} - \mathbf{P})\mathbf{D}^i (\mathbf{R}^i \bar{v}_\perp - \bar{z}_i) = \sum_i \bar{v}_i.$$

Above, $z_i \in \mathrm{Ker}(\mathbf{I} - \mathbf{P})\mathbf{D}^i$ is such that $\mathbf{R}^i \bar{v}_\perp - \bar{z}_i$ belongs to $V_i^{00}$, implying that $(\mathbf{I} - \mathbf{P})\mathbf{D}^i(\mathbf{R}^i \bar{v}_\perp - \bar{z}_i)$ is indeed an element of $V_i^\perp$. From the definition of $b_i(\cdot, \cdot)$ and the $\mathbf{S}^i$ orthogonality between $\bar{z}_i$ and all elements of $V_i^{00}$, we then deduce

$$
\begin{aligned}
\sum_i b_i(\bar{v}_i, \bar{v}_i) &= \sum_i \langle \mathbf{S}^i(\mathbf{R}^i \bar{v}_\perp - \bar{z}_i), \mathbf{R}^i \bar{v}_\perp - \bar{z}_i \rangle \\
&\leq \sum_i \langle \mathbf{S}^i \mathbf{R}^i \bar{v}_\perp, \mathbf{R}^i \bar{v}_\perp \rangle \\
&= \langle \mathbf{S} \bar{v}_\perp, \bar{v}_\perp \rangle = a(\bar{v}_\perp, \bar{v}_\perp).
\end{aligned}
$$

By this and the $\mathbf{S}$ orthogonality of $\bar{v}_0$ and $\bar{v}_\perp$, the estimate (2.35) holds with $C_0 = 1$.

The upper eigenvalue is bounded by lemma 2.2. On one hand, the spectral radius $\rho(\varepsilon_{ij})$ is bounded by the number of neigbors, that is for a given $i$ by the maximum number of spaces $V_j$ for which

$$\exists \bar{v}_i \in V_i, \bar{v}_j \in V_j : \langle \mathbf{S} \bar{v}_i, \bar{v}_j \rangle \neq 0.$$

On the other hand, we need to bound

$$b_i(\bar{v}_i, \bar{v}_i) \leq \omega_+ a(\bar{v}_i, \bar{v}_i), \forall \bar{v}_i \in V_i^\perp, \forall i.$$

But, by definition of $V_i^{00}$, we have $\bar{v}_i = (\mathbf{I} - \mathbf{P})\mathbf{D}^i \bar{v}_i^0$ for some $v_i^0 \in V_i^{00}$, and thus, using the definition of $b_i(\cdot, \cdot)$ and the fact that $\mathbf{P}$ is an $\mathbf{S}$-orthogonal projection, we can finally bound $\omega_+$ by

$$
\begin{aligned}
\omega_+ &= \sup_{\bar{v}_i \in V_i^\perp} \frac{\langle \mathbf{S} \bar{v}_i, \bar{v}_i \rangle}{b_i(\bar{v}_i, \bar{v}_i)} \\
&= \sup_{\bar{v}_i^0 \in V_i^{00}} \frac{\langle \mathbf{S}(\mathbf{I} - \mathbf{P})\mathbf{D}^i \bar{v}_i^0, (\mathbf{I} - \mathbf{P})\mathbf{D}^i \bar{v}_i^0 \rangle}{\langle \mathbf{S}_i v_i^0 \, v_i^0 \rangle} \\
&\leq \sup_{\bar{v}_i^0 \in V_i^{00}} \frac{\langle \mathbf{S} \mathbf{D}^i \bar{v}_i^0, \mathbf{D}^i \bar{v}_i^0 \rangle}{\langle \mathbf{S}_i v_i^0 \, v_i^0 \rangle},
\end{aligned}
$$

concluding the proof.


## 2.6 IMPLEMENTATION ISSUES

The practical implementation of the domain decomposition algorithm of section 2.3 using the preconditioner of section 2.4 is done in three steps :

- **Partition** for generating the different subdomains $\Omega_i$ starting from a given finite element triangulation of $\Omega$.

- **Preparation step** involves the computation of all operators used in the iterative process.

- **Solution step** executes the proposed preconditioned conjugate gradient algorithm.

## AUTOMATIC MESH PARTITIONING

The first step in the solution of a practical problem by domain decomposition methods is to split the domain into nonoverlapping subdomains. When dealing with a significant number of subdomains, an automatic mesh partitioning algorithm is needed (Farhat C. and Roux F.X. 1994, Simon H. 1991). This partitioning procedure must lead to partitions for which

- each subdomain keeps a good aspect ratio,

- the interfaces stay regular,

- the subdomains are *well balanced* meaning that the amount of *work* per subdomain must be uniform. Depending on the solution algorithm this may be equivalent to the equidistribution of the number of nodes or of elements,

- the interface are of minimal length. This is an important feature from the numerical point of view because it determines the number of unknowns of the interface problem (2.5). It is also essential for parallel implementation as it governs the amount of communication between processors.

As described in Farhat C. and Roux F.X. 1994 or in Farhat C. and Lesoinne M. 1993, there is a large number of partitioning techniques. The simplest is probably the greedy algorithm. In this strategy, all subdomains are built by a frontal technique. The front is initialized on each subdomain by taking the remaining most isolated boundary node, and is advanced by adding recursively all neigboring nodes until the subdomain preassigned size has been reached.

Another popular technique is the recursive dissection algorithm in which the original mesh is recursively cut into halves. In the simplest case, at each dissection step, the new interface is perpendicular to the maximal diameter of the considered domain, or to one of its principal axis of inertia. In a more elaborate and expensive strategy proposed by Simon H. 1991, the interface is taken as the surface of zero values of the second eigenvector of a Laplace operator defined on the global mesh. It can be shown that the corresponding solution approximately minimizes the interface size, and leads to rather regular shapes.

A more recent approach is based on K means techniques which were previously developed in automatic data classification. This method, also called dynamic clusters, is initialized by a first partition which is is then iteratively improved by first computing the barycenters $G_s$ of the different subdomains, and by affecting then any element $e$ to the subdomain $s$ with the closest barycenter (measured by an adequate weighted norm).

In any case, the output of the partitioning step is a regular nonoverlapping partition of the finite element mesh of $\Omega$ into subdomains made of elements of $\Omega$.

*PREPARATION STEP*

The algorithm involves at each step the solution of Dirichlet problems (2.6) and Neumann problems (2.11). For each subdomain, these local problems will be solved by a direct method. As the corresponding matrices do not change from one iteration to another it is possible to compute and factorize them once for ever. Then the *solution of the local linear systems* will be a simple forward-backward substitution.

The preparation step executes thus the following tasks :

- (i) For each subdomain compute and factorize the matrices $\mathbf{K}^i$ (respectively $\mathring{\mathbf{K}}^i$) of the Neumann problems (2.11) (respectively the Dirichlet problems (2.6)).

- (ii) Compute the interface and weighting coefficients $\mathbf{D}^i$.

- (iii) For the coarse problem, for each subdomain :

    - Compute the rigid bodies $z_\alpha^i$ by solving (2.11) with at most six right hand sides).

    - Compute the vectors $\mathbf{SD}^i \bar{z}_\alpha^i$ ( solution of a Dirichlet problem (2.6) with multiple right hand sides. If $NEIGH$ is the number of neighbouring subdomains of domain $j$, the number of right hand sides is equal to $6 \times (NEIGH + 1)$).

    - Compute the coarse matrix $\mathbf{C}_{i\alpha,j\beta} = \langle \mathbf{SD}^i \bar{z}_\alpha^i, \mathbf{D}^j \bar{z}_\beta^j \rangle$ ( $6 \times (NEIGH + 1)$ dot products).

Computations (i) and (iii) are obviously independent on each subdomain and thus can be run in parallel.

*SOLUTION STEP*

The practical implementation of our conjugate gradient loop involves the following steps :

- (i) Compute the solution $d\mathring{X}^i$ of the Dirichlet problem (2.6).

- (ii) Compute $d\bar{R}$ by (2.7).

- (iii) Preconditioning step

    - (a) Solve the Neumann problem (2.11).
    - Compute the right hand side of the coarse problem (2.17) and then solve it.
    - Compute $d\bar{G}$ by (2.16).

- Actualisation of the descent direction

Obviously steps (i) and (iii) (a) can be run in parallel and only involve local data. These are the most time consuming parts of the algorithm (about 95 % to 98 %) and this explains the parallel efficiency of the proposed algorithm.

## 2.7 APPLICATIONS AND NUMERICAL RESULTS
### *THREE-DIMENSIONAL LINEAR ELASTICITY*

We now conclude our presentation by presenting several practical applications of Domain Decomposition Methods, starting with our original elasticity problem. After partition, each subdomain $\Omega_i$ has a local interface

$$\Gamma_i = \Gamma \cap \partial\Omega_i = \partial\Omega_i \backslash \partial\Omega,$$

and is characterizd by its aspect ratio $\alpha_i$ which is the ratio between the diameter of the biggest sphere included in $\Omega_i$ and the diameter $H_i$ of $\Omega_i$.

We solve the resulting interface problem (2.5) by the modified Neumann-Neunmman algorithm of Section 2.4, choosing the local coarse spaces $Z_i$ as the space of rigid body motions

$$Z_i = \{v_j \in H_h(\Omega_i), \varepsilon(v_j) = \frac{1}{2}(\nabla v_j + \nabla v_j^t) = 0\}. \tag{2.36}$$

This space reduces to $\{0\}$ for all subdomains $\Omega_i$ which have a nontrivial intersection with the Dirichlet boundary $\partial\Omega_1$. More precisely, defining the degrees of freedom $(r_i(\alpha))_{\alpha=1,6}$ and the subdomain rigid body motions $(z_\beta^i)_{\beta=1,6}$ as in (2.14), the spaces $\bar{Z}_i$ and $V^0{}_i$ are taken as

$$V^0{}_i = \{X^i, X^i_{r_i(\alpha)} = 0, \forall \alpha = 1 \text{ to } 6\},$$
$$\bar{Z}_i = \{X^i = U^i_{r_i(\alpha)}, \alpha = 1 \text{ to } 6\}.$$

For this choice, we have the following quite general convergence theorem (Le Tallec P. 1994) :

**Theorem 2.3** *Using the above Neumann-Neumann preconditioner in the framework of three-dimensional linear elasticity problems, the condition number of the operator* $\mathbf{M}^{-1}\mathbf{S}$ *is bounded by*

$$Cond(\mathbf{M}^{-1}\mathbf{S}) \leq \frac{C}{\alpha_i^2}[1 + ln \max_i \frac{H_i}{h_i}]^2, \tag{2.37}$$

*the constant $C$ being independent of the subdomains diameters $H_i$, discretization steps $h_i$, aspect ratios $\alpha_i$ and averaged stiffness coefficients $\rho_i$.*

The above result guarantees the scalability of the proposed algorithm with respect to the number of subdomains ($H_i$ independance), and its robustness with respect to strongly heterogeneous elasticity coefficients ($\rho_i$ independance). This independence with respect to coefficient jumps is due to our specific choice of weighting factor

$$\mathbf{D}^i\bar{v}(P_l) = \frac{\rho_i}{\rho}\bar{v}(P_k).$$

The following test describes a three-dimensional complex automotive structure made of aluminium. The structure is fixed on three lateral bolts and twisted through a prescribed rotation of its internal axis. It is discretized into first order tetrahedral finite elements. The mesh contains 122 609 elements, 25 058 nodes (75 174 degrees of freedom). The aim of this test is to experimentally compare the original Neumann-Neumann algorithm with the version with coarse grid and, on the other hand, to study the influence of different mesh partitions on the behaviour of the algorithm. Some of these partitions are obviously not optimal for this algorithm. They are nevertheless used for computations just for comparison purpose and to test the reliability of our global domain decomposition algorithm. We will consider six different decompositions :

i) K means : the Modulef implementation as described in & ,

ii) Frontal algorithm,

iii) Greedy algorithm,

iv) Recursive principal inertia algorithm purposely used with a bad choice of inertia axis (Rpi),

v) Greedy algorithm with K means optimisation,

vi) Rpi algorithm with K means optimisation.

The algorithms ii)-vi) are those implemented in TOP/DOMDEC (Sharp M. and Farhat C. 1994). All decompositions use the same mesh and contain 24 subdomains. Table summarizes the convergence behaviour. For each partition the number of iterations for both the original and the modified algorithm to reach a relative residual of $10^{-6}$ are indicated. The same table displays the number of degrees of freedom on the boundaries (LFRON), which indicates the size of the interface problem, and the conditioning of this system.

Table 2.1: *Convergence properties*

| Decomp | Neumann | Cond | Coarse | Cond | LFRON |
|--------|---------|------|--------|------|-------|
| K means | 133 | 0.30066E+11 | 58 | 0.24590E+05 | 9 960 |
| rpi | > 500 | 0.86266E+10 | pl 146 | 0.51910E+07 | 60 051 |
| rpi+K means | 121 | 0.83381E+07 | 41 | 0.87619E+03 | 8 094 |
| greedy | 161 | 0.33483E+07 | 65 | 0.21836E+05 | 13 432 |
| greedy+K means | 116 | 0.11178E+06 | 42 | 0.86884E+03 | 8 205 |
| frontal | 290 | 0.11875E+06 | 152 | 0.81268E+05 | 29 562 |

The irregular shape of the domains obtained by the recursive principal inertia algorithm explains the difficulty of the convergence for this partitioning. For the Neumann-Neumann case a residual of $10^{-2}$ was obtained after 500 iterations. This partitioning was not a good choice! It is interesting to observe the good properties of the K means algorithm. In all situations it gives very good results. It can improve the performance not only of a good decomposition (as the greedy algorithm) but also of a decomposition not adapted for this algorithm (as the rpi). All these experiments were run on 24

Figure 2.3: *K means decomposition*

processors of a KRS1 computer in non dedicated time. The CPU time for the K means method using the Neumann-Neumann preconditioner are reported in table . The time measurements are not accurate as the computer was not used in single user mode. At each iteration there is a synchronisation step, some computations are performed only on one processor. The corresponding time is denoted by critical section.

## NONCONFORMING GRIDS

Nonconforming mesh refinement is quite useful in practice and can easily be handled by most domain decomposition techniques. To illustrate this point, let us start from a global conforming finite element mesh of the whole domain $\Omega$. This mesh is first partitioned as usual into conforming nonoverlapping subdomains $\Omega_i$. In order to improve the local accuracy of the finite element solution, we then refine the finite element mesh

Figure 2.4: *Frontal decomposition*

of several subdomains (but not of all subdomains) by subdividing each original element of these subdomains into $2, 4, 8, 16, \ldots$ subelements of same nature. After such local refinements the global mesh is no longer conforming : on the interface between a refined domain $\Omega_i$ and an unrefined domain $\Omega_j$, several nodes of $\Omega_i$ will have no equivalent on $\Omega_j$ (Figure 2.5).

To handle this lack of conformity, we can use a general mortar element approach as Bernardi C., Maday Y. and Patera A. 1989 or in Le Tallec P., Sassi T. and Vidrascu M. 1995. On the mathematical side, this technique imposes that the solution on two neighboring subdomains $\Omega_i$ and $\Omega_j$ share the same $L^2$ projection $Tr_{ijh}$ on the mortar space which is defined on their common interface. For elliptic problems, it is proved that such discretization strategies lead to optimal approximation errors. On the numerical side, the resulting discrete problem can again be reduced to an interface problem

Table 2.2: *CPU time for a sequential and a parallel run*

| # proc | Initialisation | Iterations | Critical section (included in the iteration step) |
|:------:|:--------------:|:----------:|:-------------------------------------------------:|
| 1      | 116.62s        | 16208.12s  | 226.98s                                           |
| 24     | 8.64s          | 817.28 s   | 179.5s                                            |

associated to the generalized Schur complement matrix,

$$\sum_i \tilde{\mathbf{R}}_i^t \begin{pmatrix} \mathbf{K}^i & Tr_{ijh}^t \\ Tr_{ijh} & 0 \end{pmatrix}^{-1} \tilde{\mathbf{R}}_i \bar{X} = F. \tag{2.38}$$

This mortar element case can still be solved by standard domain decomposition algorithms, within the three major changes :

- the pointwise traces are replaced on each face $F_{ij} = \partial\Omega_i \cap \partial\Omega_j$ by local $L^2$ averaged traces $Tr_{ijh}$;

- the pointwise interface restriction $\mathbf{R}_i$ is replaced by a global restriction operator $\tilde{\mathbf{R}}_i$ which maps the global trace $\bar{X}$ into the local right-hand sides $X^{ij}$ of the subdomain generalised Dirichlet problems

$$\mathbf{K}^i X^i + \sum_{i<j} Tr_{ijh}^t \Lambda_{ij} - \sum_{isllsj} Tr_{ijh}^t \Lambda_{ji} = F^i, \forall i,$$
$$Tr_{ijh} X^i = \bar{X}^{ij}, \forall j \neq i.$$

- the space of global traces $\bar{X}$ is now defined face by face. Edges and vertices do not play any role in this definition. In fact, the definition of the trace on any geometric vertex or edge will no longer be unique and will depend of the particular face $F_{ij}$ on which it is taken.

There is also a simpler solution, based on the so called slave node approach used in Bramble J., Ewing R., Parashkevov R. and Pasciak J. 1992.. In this approach, all finite element displacement fields are imposed to be pointwise continuous at all subdomain interfaces. In other words, the finite element space definition is kept as

$$H_h(\Omega) = \Big\{ v_h : \bar{\Omega} \to \mathbb{R}^3, v_h \text{ continuous, } v_h = 0 \text{ on } \partial\Omega_1,$$

$$v_{h|T_l} = v_l \circ \varphi_l^{-1}, v_l \in [Q_2'(\hat{\Omega})]^3, \text{ for all elements } T_l \text{ of all subdomains } \Omega_i \Big\}.$$

With this choice, on any nonconforming interface, the values of the displacement field at any interface node $P_l^i$ of the refined subdomain $\Omega_i$ which is not shared by the

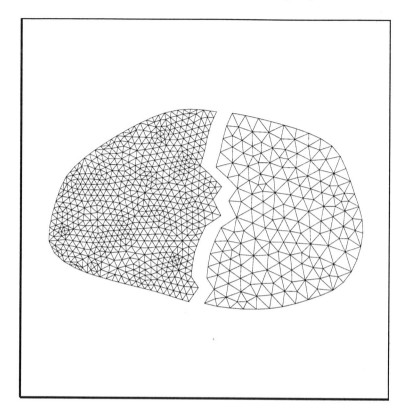

Figure 2.5: *Nonconforming mesh refinement : refined nodes on the left domain have no counterpart on the right domain.*

neighboring subdomain $\Omega_j$ are constrained to be equal to the value at this point of the $\Omega_j$ finite element interpolation of this field

$$u(P_l^i) = \sum_{k \in T \cap \partial \Omega_j} u(P_k^j) \phi_k^T(P_l^i).$$

Here $T$ is the finite element of the subdomain $\Omega_j$ which contains $P_l^i$, $P_k^j$ are the interface nodes of this element, and $\phi_k^T(x)$ is the nodal element shape function associated to the node $P_k^j$. But by construction, the interface nodes $P_k^j$ of $\Omega_j$ are shared by $\Omega_i$, and by the imposed continuity of the displacement field at the interface, we have in addition

$$u(P_k^i) = u(P_k^j).$$

Consequently, the values of the displacement field on the extra nodes of $\partial \Omega_i$ are finally constrained by the relations

$$u(P_l^i) = \sum_{k \in T \cap \partial \Omega_i} u(P_k^i) \phi_k^T(P_l^i). \qquad (2.39)$$

The additional degrees of freedom introduced by refinement on $\partial\Omega_i$ are therefore not directly related to any degree of freedom of $\Omega_j$ but only to degrees of freedom of $\Omega_i$. They must then be considered as internal degrees of freedom of $\Omega_i$ that is as elements of the set $\mathring{X}^i$ and do not participate to the interface problem. In other words, the mesh refinement of the subdomain $\Omega_i$ will modify the local stiffness matrix $\mathbf{K}^i$ (new finite elements are added and the internal kinematic constraint (2.39) must be taken into account), will add elements to the set $\mathring{X}^i$ of internal unknowns, but will not modify the set $\bar{X}^i$ of the interface degrees of freedom. In particular, the interface problem keeps the same structure and dimension, which means that we can still use the (much cheaper) interface preconditioner of the unrefined case. With this choice, the domain decomposition algorithm to be used for solving partially refined problems takes the final form :

. Let $d\bar{X} = \sum_i \mathbf{R}^{i^t} d\bar{X}^i$ be a known descent direction.

. Compute on each subdomain $\Omega_i$ the solution $d\mathring{X}^i$ of the Dirichlet problem

$$\mathring{\mathbf{K}}^i d\mathring{X}^i = -\mathbf{B}^i d\bar{X}^i.$$

For the refined subdomains, this problem must be solved on the refined mesh and must take into account the kinematic constraint (2.39).

. Compute the variation of the residual $d\bar{R} = \sum_i (\mathbf{R}^{i^t} \bar{\mathbf{K}}^i d\bar{X}^i - \mathbf{B}^{i^t} d\mathring{X}^i)$.

. Precondition this gradient by

$$d\bar{G} = \mathbf{M}^{-1} d\bar{R},$$

using the preconditioner $\mathbf{M}^{-1} = \mathbf{P} + (\mathbf{I} - \mathbf{P})(\sum_i \mathbf{D}^i \mathbf{S}_{io}^{-1} \mathbf{D}^{i^t})(\mathbf{I} - \mathbf{P})^t$ of the original unrefined problem.

. Update the direction of descent and reiterate.

We will illustrate this strategy on a simple 3D elasticity problem. The domain considered is an isotropic 3D elastic beam (the Young coefficient is E=70 000 MPa and the Poisson ratio is .3). This problem is ill conditionned because the length of the beam (20m) is large compared to the dimension of its section (a square of size 2m). The beam is discretised by using regular triangulations made of $Q'2$ 20 nodes hexaedric finite elements. The beam is clamped at one extremity and a vertical surface force $F^\Gamma = (0, 0, -1) MPa$ is applied at the opposite end. A reference solution to this problem is computed on a very fine mesh (2048 elements, more than 10 000 nodes) using a standard Cholesky factorisation. The domain is then split into two subdomains of equal size and the resulting problem is solved on three types of meshes :

- the first computation uses two compatible domains with coarse meshes (8 elements and 81 nodes per subdomain)(*Coarse/Coarse* in Table 2.3);

- the second computation uses two compatible domains with fine meshes (64 elements and 425 nodes per subdomain)(*Fine/Fine* in Table 2.3);

- the last computation uses two incompatible meshes (one coarse, one fine) (*Incompatible* in Table 2.3), where the refined mesh corresponds to the clamped extremity.

For all the problems, the domain decomposition algorithm converged in 5 iterations, indicating that the refinement does not increase the number of iterations required for convergence. We then compare in Table 2.3 the accuracy of the different solutions by measuring their difference with the reference solution on the two first sections of the beam. We observe there that the incompatible solution is as accurate as the uniformly refined solution.

Table 2.3: *Error for the incompatible meshes*

| Section | u | Value | Incompatible | Fine/Fine | Coarse/Coarse |
|---------|---|-------|--------------|-----------|---------------|
| z=2.5 | $u_x$ | $.22 * 10^{-4}$ | 5.5 % | 5.5 % | 9 % |
|  | $u_y$ | $.25 * 10^{-3}$ | 4.5 % | 4.5 % | 10 % |
|  | $u_z$ | $.19 * 10^{-3}$ | 2.5 % | 2.5 % | 18 % |
| z=5 | $u_x$ | $.19 * 10^{-4}$ | 0.3 % | 0.3 % | 15 % |
|  | $u_y$ | $.98 * 10^{-3}$ | 2.3 % | 2.3 % | 7 % |
|  | $u_z$ | $.37 * 10^{-3}$ | 1 % | 1 % | 3 % |
| z=10 | $u_x$ | $.13 * 10^{-4}$ | 5.3 % | 0 % | 1.5 % |
|  | $u_y$ | $.35 * 10^{-2}$ | 1.3 % | 1.4 % | 4 % |
|  | $u_z$ | $.64 * 10^{-3}$ | 0.7 % | 0.7 % | 3 % |
| z=12.5 | $u_x$ | $.96 * 10^{-5}$ | 0.5 % | 0 % | 0 % |
|  | $u_y$ | $.53 * 10^{-2}$ | 1.2 % | 1 % | 3.6 % |
|  | $u_z$ | $.73 * 10^{-3}$ | 1.2 % | 0.6 % | 2.4 % |

## NONLINEAR ELASTICITY

We now turn to the numerical solution of nonlinear three-dimensional elasticity problem. For simplicity, we will consider situations where the constitutive laws does not involve any internal variable. By writing the weak form of the equilibrium equations (in a fixed reference configuration $\Omega$, with $\partial\Omega = \partial\Omega_1 \cup \partial\Omega_2$), by eliminating the Piola-Kirchoff stress tensor $T$ by standard hyperelastic constitutive laws

$$T(x) = \frac{\partial\psi}{\partial F}(x, (Id + \nabla u)^t(Id + \nabla u)), \quad F = Id + \nabla u,$$

and by taking into account the imposed boundary condition, we can write the problem of computing the large deformations of a nonlinear elastic structure as the variational problem

$$\int_\Omega \frac{\partial\psi}{\partial F}(x, (Id + \nabla u)^t(Id + \nabla u)) : \nabla v = \int_\Omega f^\Omega \cdot v + \int_{\partial\Omega_2} f^\Gamma \cdot v,$$

$$\forall v \in H(\Omega). \tag{2.40}$$

This total lagrangian formulation is highly nonlinear due to the specific form of the energy density $\psi$ as a function of $\nabla u$. After discretisation, the solution of this nonlinear problem (2.40) is classically obtained by a Newton's algorithm which iteratively solves its linearized version :

$$\int_\Omega a_F(\nabla du_h, \nabla v_h) = L_F(v_h), \forall v_h \in H_h(\Omega), du_h \in H_h(\Omega), \qquad (2.41)$$

under the notation

$$a_F(\nabla w, \nabla v) \;=\; \nabla w^T : \frac{\partial^2 \psi}{\partial F^2} : \nabla v \qquad (2.42)$$

$$L_F(v) \;=\; -\int_\Omega \frac{\partial \psi}{\partial F} : \nabla v + \int_\Omega f^\Omega \cdot v + \int_{\partial\Omega_2} f^\Gamma \cdot v. \qquad (2.43)$$

Each step of this Newton algorithm is thus a linear elasticity problem, on which we apply our standard Neumann-Neumann domain decomposition algorithm. A further simplification can nevertheless be added in this nonlinear case by keeping the same interface preconditioner during several consecutive or during all Newton iterations. The loss in performance observed by not updating the operators $S_{i0}$ and $P$ in the preconditioner $M$ is very small compared to the economy done in the preparation step by not recomputing and refactoring the Neumann and the coarse matrices.

To validate the ability of the algorithm to solve nonlinear highly inhomogeneous problems we will study the deformation of a human trachea subjected to a transmural pressure. This model takes into account the different constitutive tissues of the trachea : the stiff cartilage, assumed to be compressible, and the soft tissue, the intercartilaginous membranae, assumed to be incompressible. The geometry used corresponds to the dimensions of the midtrachea. The trachea was presumed to be sufficiently long that the deformation of the tracheal midportion is not influenced by the conditions at the upper and lower boundaries. A single slice has been modeled, which is composed of one cartilaginous ring and the upper and lower midportion of the adjacent intercartilaginous membranes. For symmetry reasons only half of the trachea is studied. The domain is discretized into 1615 Q2 hexaedra (27 nodes par element). The finite element mesh contains 14 821 nodes (54 463 degress of freedom with 22323 on the interface). This domain was decomposed into 20 subdomains (Figure 2.6).

The pressure considered in this example is $p = 0.01$ imposed into two steps of loading. Each step was solved in 4 Newton iterations. At each Newton step the convergence of the domain decomposition algorithm was obtained in 50 to 70 iterations, using the interface preconditioner computed at the initial step. We can therefore observe in this particular example how straightforward is the extension of Domain Decomposition Techniques to nonlinear problems.

## PLATE AND SHELLS

Domain decomposition algorithms can also be applied to the solution of real life plate and shell problems. To illustrate this last point, let us consider a thin plate occupying a domain $\Omega$ in $\mathbb{R}^2$, which is clamped on the part $\partial\Omega_c$ of its boundary and simply supported along $\partial\Omega_s - \partial\Omega_c$. Then, a Kirchhoff-Love model characterizes the vertical displacement $u$ of the plate as the solution of the variational problem

$$a(u, v) = L(v), \forall v \in H(\Omega), u \in H(\Omega), \qquad (2.44)$$

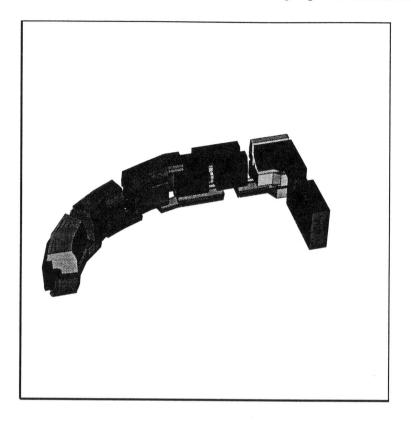

Figure 2.6: *Decomposition of half trachea*

under the notation

$$a(u,v) = \int_\Omega \varepsilon(\vec{\theta}(u)) : K : \varepsilon(\vec{\theta}(v)) \tag{2.45}$$

$$L(v) = \int_\Omega fv + \int_{\partial\Omega - \partial\Omega_c} m_g \partial_n v + \int_{\partial\Omega - \partial\Omega_s} gv, \tag{2.46}$$

$$H(\Omega) = \{v \in H^2(\Omega), v = 0 \quad \text{on} \quad \partial\Omega_s, \partial_n v = 0 \quad \text{on} \quad \partial\Omega_c\}. \tag{2.47}$$

Above, $f$ denotes the density of vertical forces, $m_g$ the density of flexion moments applied on the part $\partial\Omega - \partial\Omega_c$ of the boundary where the plate is free to rotate, $g$ is the density of boundary vertical loading and $H(\Omega)$ denotes the space of kinematically admissible displacement fields. Moreover $\varepsilon(\vec{\theta}) = \frac{1}{2}(\nabla\vec{\theta} + \nabla^T\vec{\theta})$ is the curvature tensor, $\vec{\theta}(u) = \nabla u$ represents the inplane rotation of the plate, $K$ the plate flexural stiffness. For a simple isotropic plate of thickness $t$, made of an homogeneous elastic material with Young modulus $E$ and Poisson coefficient $\nu$, this flexural stiffness is given by

$$\varepsilon(\vec{\theta}(u)) : K : \varepsilon(\vec{\theta}(v)) = \frac{Et^3}{12(1-\nu^2)}((1-\nu)\nabla^2 u : \nabla^2 v + \nu\Delta u\Delta v).$$

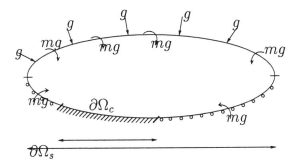

Figure 2.7: *Plate configuration. The plate is clamped on $\partial\Omega_c$ and simply supported on $\partial\Omega_s - \partial\Omega_c$*

We now try to solve such a plate problem using a finite element discretisation of $\Omega$ based on DKT triangles. In a DKT triangle (Discrete Kirchhoff Triangle) (Batoz J.L., Bathe K.J. and Ho W.H. 1980), the vertical displacement $u$ of the plate and its inplane rotation $\vec{\theta}$ are treated as two separate variables. On each triangle $T$ of the triangulation, the displacement $u(M)$ is a third order polynomial of the barycentric coordinates $\lambda_i$ of $M$

$$u(M) = \sum_i a_i\lambda_i + \sum_{i<j} \left(b_{ij}\lambda_i\lambda_j + c_{ij}\lambda_i^2\lambda_j\right), \qquad (2.48)$$

the rotation $\vec{\theta}$ is a reduced second order polynomial

$$\vec{\theta} = \vec{\theta}(a_i)\lambda_i + 4\alpha_{ij}\lambda_i\lambda_j a_i\vec{a}_j \qquad (2.49)$$

and $u$ and $\vec{\theta}$ satisfy the Kirchhoff hypothesis

$$\frac{\partial u}{\partial s} = \vec{\theta}.\frac{a_i\vec{a}_j}{|a_ia_j|} \qquad (2.50)$$

along each side $a_ia_j$ of the triangle. By writing (2.50) at each vertex $a_i$, this implies

$$\vec{\theta}(a_i) = \nabla u(a_i).$$

By integrating (2.50) along $a_ia_j$, we also get

$$u(a_j) - u(a_i) = \frac{1}{2}(\vec{\theta}(a_i) + \vec{\theta}(a_j)).a_i\vec{a}_j + \frac{2}{3}\alpha_{ij}|a_ia_j|^2.$$

Therefore, the discrete Kirchhoff hypothesis (2.50) gives the local rotation $\vec{\theta}|_T$ as a function of the displacement $u$ by

$$\vec{\theta}|_T \;=\; \vec{\theta}_d(u) := \nabla u(a_i)\lambda_i + 6\nabla_{ij}^2(u)\frac{a_i\vec{a}_j}{|a_ia_j|}\lambda_i\lambda_j, \qquad (2.51)$$

$$\nabla_{ij}^2(u) = \frac{1}{|a_ia_j|}[u(a_j) - u(a_i) - \frac{1}{2}(\nabla u(a_i) + \nabla u(a_j)).a_i\vec{a}_j]. \qquad (2.52)$$

This approximation leads to the following definition of our finite element space

$$H_d(\Omega) = \{v \in C^0(\Omega), v|_T \text{ satisfies } (2.48), \vec{\theta}_d(v) \in C^0(\Omega),$$
$$v = 0 \text{ on } \partial\Omega_s, \vec{\theta}_d(v) = 0 \text{ on } \partial\Omega_c\}. \tag{2.53}$$

Elements $v$ of this space are then characterized by the values $v(a_i)$ of the vertical displacement and $\vec{\theta}_d(v(a_i)) = Dv(a_i)$ of the inplane rotation at each vertex $a_i$ of the triangulation. The corresponding discrete variational formulation of the plate problem (2.44) becomes

$$\int_\Omega \varepsilon(\vec{\theta}_d(u)) : K : \varepsilon(\vec{\theta}_d(v)) = L(v), \forall v \in \mathbb{H}_d(\Omega), u \in \mathbb{H}_d(\Omega). \tag{2.54}$$

The extension to shell problems is obtained by considering as usual the shell as an assembly of triangular plates. The nodal degrees of freedom are then the cartesian components of the nodal displacement $\vec{v}(a_i)$ and of the rotation $\vec{\theta}(a_i)$. The stiffness matrix is obtained by adding to the flexion energy $\varepsilon(\vec{\theta}) : K : \varepsilon(\vec{\theta})$ a membrane energy which takes into account the contributions of the inplane displacements and of the drilling degrees of freedom (Bergan P. and Felippa C.A. 1985). Nonlinear curved shells can also be considered as in Carrive-Bédouani M., Le Tallec P. and Mouro J. 1995.

In any case, these plate or shell problems have exactly the same structure as the three-dimensional elasticity problems of Section 2.2. Therefore they still can be reduced to interface problems, the interface unknowns being now the interface values of the displacements and of the rotations. To apply our Neumann-Neumann algorithm to these problems, we then need to choose the local spaces $\bar{Z}_i$ and $V^0_{-i}$. The convergence theorem of Section 2.5 guarantees then that this algorithm is efficient for plate problems if all elements of $V^{oo}_i$ have zero normal displacements $v^n(P_c)$ at all corners $P_c$ of subdomain $\Omega_i$. The first choice of local spaces is then to take

$$V^0_{-i} = \{\bar{v}_i = Trv_i|_{\Gamma_i} \in V_i, v_i^n(P_c) = 0, \text{ for all corners } (P_c) \text{ of } \Omega_i\},$$

$$\bar{Z}_i = \{\bar{z}_i \in V_i, \langle S_i\bar{z}_i, \bar{w}_i\rangle = 0, \forall \bar{w}_i \in V^0_{-i}\}.$$

The next choice, originally proposed in Le Tallec P., Mandel J. and Vidrascu M. 1995 first decomposes each local space of finite elements $H(\Omega_i)$ into

$$H(\Omega_i) = H^0(\Omega_i) \oplus P_1$$

and set

$$V^0_{-i} = TrH^0(\Omega_i)|_{\Gamma_i}$$

$$\bar{Z}_i = TrP_1 + \sum_{\text{corner}} S_{io}^{-1}L_c$$

with $L_c$ a point normal load applied to $P_c$.

**Remark 2.5** *By definition, a corner is a point of the interface where at least three subdomains meet. We will suppose that each subdomain has at least three nonaligned corners. If not, and if $\partial\Omega_i \cap \partial\Omega_s$ is empty, we will add a fictitious corner to the corresponding subdomain so that the Schur operator is indeed coercive on $V^0_{-i}$.*

Figure 2.8: *Deap half sphere*

The first test problem considers a deep spherical shell of radius 5 subjected to a gravitational loading (Figure 2.8). The original mesh $h$ contains 2650 planar DKT elements, and is partitioned into 24 subdomains.

Each element is then cut into 4 ( mesh $h/2$), and 16 ( mesh $h/4$). We compare again in this test case the performances of the different choices of coarse spaces for different thickness and different mesh refinements. In Table 2.4 *Coarse* is the standard preconditionner, $Corner_d$ corresponds to the first choice of local space and $Corner_f$ to the second choice. For different thicknesses $t$ the number of iterations, *nbiter* and an estimation of the smallest and largest eigenvalues $\lambda_{min}$ and $\lambda_{max}$ are indicated. As predicted by the theory, we verify that we indeed have $\lambda_{min} = 1$. The results (Table 2.4) are quite typical and prove that the domain decomposition strategy is not very sensitive to mesh refinement. On the other hand, due to membrane coupling, the proposed method is less efficient when the thickness of the shell decreases.

The aim of the next test is to show the ability of the method to handle large real-life shell problems. The simulation computes the deformations of a curved aortic blood

Table 2.4: *Condition Number for the spherical shell*

| Coarse | | | | | |
|---|---|---|---|---|---|
| t | nbiter | $\lambda_{min}$ | $\lambda_{max}$ | condit | mesh |
| .3 | 55 | 0.10141E+01 | 0.34626E+03 | 0.34145E+03 | h |
| | 73 | 0.10124E+01 | 0.12476E+04 | 0.12323E+04 | h/2 |
| | 95 | 0.10111E+01 | 0.52231E+04 | 0.51657E+04 | h/4 |
| .03 | 92 | 0.10118E+01 | 0.78075E+03 | 0.77161E+03 | h |
| | 106 | 0.10123E+01 | 0.11416E+04 | 0.11277E+04 | h/2 |
| | 128 | 0.10112E+01 | 0.23326E+04 | 0.23068E+04 | h/4 |
| Corner-f | | | | | |
| t | nbiter | $\lambda_{min}$ | $\lambda_{max}$ | condit | mesh |
| .3 | 26 | 0.10146E+01 | 0.18167E+02 | 0.17906E+02 | h |
| | 36 | 0.10119E+01 | 0.51838E+02 | 0.51230E+02 | h/2 |
| | 53 | 0.10102E+01 | 0.16376E+03 | 0.16211E+03 | h/4 |
| .03 | 48 | 0.10117E+01 | 0.75847E+02 | 0.74974E+02 | h |
| | 59 | 0.10118E+01 | 0.11888E+03 | 0.11749E+03 | h/2 |
| | 73 | 0.10111E+01 | 0.20446E+03 | 0.20221E+03 | h/4 |
| Corner-d | | | | | |
| t | nbiter | $\lambda_{min}$ | $\lambda_{max}$ | condit | mesh |
| .3 | 18 | 0.10172E+01 | 0.92599E+01 | 0.91031E+01 | h |
| | 27 | 0.10144E+01 | 0.26452E+02 | 0.26077E+02 | h/2 |
| | 40 | 0.10153E+01 | 0.62246E+02 | 0.61306E+02 | h/4 |
| .03 | 24 | 0.10169E+01 | 0.18162E+02 | 0.17859E+02 | h |
| | 28 | 0.10177E+01 | 0.38631E+02 | 0.37958E+02 | h/2 |
| | 37 | 0.10156E+01 | 0.86683E+02 | 0.85347E+02 | h/4 |

vessel subjected to an internal blood pressure as computed by a CFD code.

The vessel wall was 0.125 thick (one tenth of the pipe radius), its Young modulus was $E = 3 * 10^7$, its Poisson ratio of $\nu = .3$. The mesh contains 19707 planar DKT elements (elements SHE3 of the University of Colorado) and 10814 nodes. It was automatically partitioned into 24 subdomains (Figure 2.9). The calculation was run on a HP735 workstation. The deformed configuration is shown in Figure 2.10.

The number of iterations and CPU time for the various preconditionners are reported in Table 2.5 :*neum* is the original preconditioner without coarse grid solver, *coarse* is the preconditioner where the coarse space is limited to local rigid motions and *corner* uses the corner modes within the coarse.

## 2.8 CONCLUDING REMARKS

Domain decomposition methods are a natural candidate for solving large scale partial differential equations on coarse grain, distributed memory parallel computers. Their efficient application to the numerical solution of linearly elliptic problems of the type

Figure 2.9: *Mesh of the curved pipe*

Figure 2.10: *Curved pipe : deformed configuration*

Table 2.5: *Convergence of the biomedical problem.*

| precond | # iter | condition number | CPU |
|---------|--------|------------------|-----|
| neum | 336 | .13866E+08 | 346.33+6984.4=7330.8 |
| coars | 104 | .10931E+04 | 699.05+2238.97=2938.02 |
| corner | 58 | .14906E+03 | 1085.71+1289.49=2375.2 |

$Au = L$ in $V'$ is not completely straightforward. Nevertheless, we have seen in this chapter that the additive Schwarz algorithms provide an efficient abstract iterative procedure for solving such problems in parallel and we have introduced different practical implementations for solving interface problems in the field of structural mechanics.

The challenge is now to solve vibroacoustic problems or to identify the natural frequencies and modes of complex elastic structures. The underlying operator (of Helmholtz type) is no longer positive, its natural setting uses complex arithmetics and its discretisation at high frequencies requires very fine grids.

## REFERENCES

Aghoskov V. 1988. Poincaré-Steklov's operators and domain decom position methods in finite dimensional spaces, *R. Glowinski, G. Golub and J. Periaux (eds) 1988.*

Batoz J.L., Bathe K.J. and Ho W.H. 1980. A study of three-node triangular bending element, *Int. J. Numer. Engrg.*, **15**, 1771–1812.

Bell K., Hatelstad B., Hansteen O.E. and Araldsen P.O 1973. NORSAM, a programming system for the finite element method, users manual, part 1 ( NTH, Trondheim).

Bergan P. and Felippa C.A. 1985. A triangular membrane element with rotational degrees of feeedom. *Comp. Meth. Appl. Mech. Eng.* 50, pp. 25-69.

Bernardi C., Maday Y. and Patera A. 1989. A new nonconforming approach to domain decomposition: the mortar element method, in H. Brezis and J.L. Lions eds *Nonlinear Partial Differential Equations and their Applications* (Pitman).

Bjorstad P.E. and Hvidsten A. 1988. Iterative Methods for Substructured Elasticity Problems in Structural Analysis, in R. Glowinski, G. Golub and J. Periaux (eds) 1988.

Bjorstad P.E. and Widlund O. 1986. Iterative methods for the solution of elliptic problems on regions partitioned in substructures, *SIAM J. Numer. Anal* **23**, 1097–1120.

Bourgat J.F., Glowinski R., Le Tallec P. and Vidrascu M. 1989. Variational formulation and algorithm for trace operator in domain decomposition calculations, in R. Glowinski,G. Golub and J. Periaux (eds) 1988.

Bramble J., Ewing R., Parashkevov R. and Pasciak J. 1992. Domain decomposition methods for problems with partial refinement *SIAM J. Sci. Stat. Comp.* **13** 397–410.

Bramble J., Pasciak J. and Schatz A. 1986. The construction of preconditioners for elliptic problems by substructuring, *Math. Comp.* **47** 103–134.

Bramble J., Pasciak J. and Schatz A. 1989. The construction of preconditioners for elliptic problems by substructuring, IV, *Math. Comp.* **53** 1–24.

Bramble J., Pasciak J. and Xu J. 1990. Parallel Multilevel Preconditioners, *Math. Comp.* **55** 1–22.

Carrive-Bédouani M., Le Tallec P. and Mouro J. 1995. Approximations par Eléments Finis d'un modèle de coques minces géométriquement exact, *Revue Européenne des Eléments Finis*, to appear.

Chan T. 1989. Boundary Probe Preconditioners for Fourth Order Elliptic Problems, in T. Chan, R. Glowinski, J. Periaux and O. Widlund (eds) 1989.

Chan T., Glowinski R., Periaux J. and Widlund O. (eds) 1989. *Proceedings of the Second International Symposium on Domain Decomposition Methods for Partial Differential Equations, Los Angeles, California, January 14-16, 1988*, (SIAM, Philadelphia).

Chan T., Glowinski R. (eds) 1990. *Proceedings of the Third International Symposium on Domain Decomposition Methods for Partial Differential Equations, Houston, Texas, March 20-22, 1989*, (SIAM, Philadelphia).

Chan T. and Hou T.Y 1991. Eigendecomposition of domain decomposition interface operators for constant coefficient elliptic problems, *SIAM J. Sci. Stat. Comp.* **12**, 1471–1479.

Chan T., Keyes D., Meurant G., Scroggs S. and Voigt R. (eds) 1992. *Proceedings of the fifth international symposium on Domain Decomposition Methods for Partial Differential Equations, Norfolk, May 1991*, (SIAM, Philadelphia).

Chan T.and Mathew T. 1992. The interface probing technique in domain decomposition, *SIAM J. Matrix Anal. and Applications* **13**, 212–238.

Chan T.and Mathew T. 1994. Domain Decomposition Algorithms, *Acta Numerica* .

Cowsar L., Mandel J. and Wheeler M. 1993. Balancing Domain Decomposition for Mixed Finite Elements, *Technical Report 93-08*, (Department of Mathematical Sciences, Rice University).

De Roeck Y.H. and Le Tallec P. 1991. Analysis and test of a local domain decomposition preconditioner, in R. Glowinski, Y. Kuznetsov, G. Meurant, J. Periaux and O. Widlund (eds) 1991.

De Roeck Y.H., Le Tallec P. and Vidrascu M. 1992. A Domain Decomposed Solver for Nonlinear Elasticity, *Comp. Meth. Appl. Mech. Eng.* **99** 187–207.

Dryja M. and Widlund O. 1990. Towards an unified theory of domain decomposition algorithm for elliptic problems, in T. Chan, R. Glowinski (eds) 1990.

Dryja M. and Widlund O. 1992. Additive Schwarz Methods for Elliptic Finite Element Problems in Three-Dimensions, in T. Chan, D. Keyes, G. Meurant, S. Scroggs and R. Voigt (eds) 1992.

Dryja M. and Widlund O. 1993. Some recent results on Schwarz type domain decomposition algorithms, in A. Quarteroni (ed) 1993.

Dryja M. and Widlund O. 1995. Schwarz Methods of Neumann-Neumann type for Three-Dimensional Elliptic Finite Element Problems. *Comm. Pure Appl. Maths* Vol XLVIII 121-155 .

Farhat C. and Lesoinne M. 1993. Automatic Partitioning of Unstructured Meshes for the Parallel Solution of Problems in Computational Mechanics *Int. J. Num. Meth. in Eng.* **36**, 745–764.

Farhat C. and Roux F.X. 1994. Implicit parallel processing in structural mechanics, *Computational Mechanics Advances,* Vol 2, **1** North-Holland, 1-124.

Farhat C. and Roux F.X. 1992. An unconventional Domain Decomposition Method for an Efficient Parallel Solution of Large Scale Finite Element Systems, *SIAM J. Sci. Stat. Comput.* **13**, 379–396.

Glowinski R., Golub G. and Periaux J. 1988. *Proceedings of the First International Symposium on Domain Decomposition Methods for Partial Differential Equations, Paris, France, January 7-9, 1987,* (SIAM, Philadelphia).

Glowinski R., Kuznetsov Y., Meurant G., Periaux J. and Widlund O. (eds) 1991. Proceedings of the fourth international symposium on Domain Decomposition Methods for Partial Differential Equations, Moscou, June 1990, (SIAM, Philadelphia).

Golub G. and Van Loan C. 1983. Matrix computations, (North Oxford Academic, Oxford).

Goovaerts D., Chan T. and Piessens R. 1991. The eigenvalue spectrum of domain decomposed preconditioners, *Appl. Numer. Math.* **8**, 389–410.

Keyes D. 1992. Domain Decomposition: a bridge between nature and Parallel Computers *Symposium on Adaptive, Multilevel and Hierachical Computational Strategies, ASME Winter Annual Meeting, November 8-13* (Anaheim, California).

Keyes D. and Gropp W. 1987. A comparison of domain decomposition techniques for elliptic partial differential equations and their parallel implementation *SIAM J. Sci. Stat. Comp.* **8**, 166–202.

Keyes D. and Xu J.(eds) 1995. Proceedings of the seventh international symposium on Domain Decomposition Methods for Partial Differential Equations, Penn State, October 93, (AMS, Providence).

Le Tallec P. 1994. Domain Decomposition Methods in Computational Mechanics. *Computational Mechanics Advances,* Vol 1, **2** North-Holland, 121-220.

Le Tallec P., De Roeck Y.H. and Vidrascu M. 1991. Domain decomposition methods for large linearly elliptic three dimensional problems,*J. Comp. Appl. Math.* **34**, 93–117.

Le Tallec P., Mandel J. and Vidrascu M. 1995. Balancing Domain Decomposition for Plates, in D. Keyes and J. Xu (eds) 1995.

Le Tallec P., Mandel J. and Vidrascu M. 1996. A Neumann-Neumann Domain Decomposition Algorithm for Solving Plate and Shell Problems, *SIAM J. Numer. Anal.,* submitted.

Le Tallec P., Sassi T. and Vidrascu M. 1995. Three-dimensional Domain Decomposition Methods with Nonmatching Grids and Unstructured Coarse Solvers, in D. Keyes and J. Xu (eds) 1995.

Mandel Jan 1990. Two-level Domain Decomposition Preconditioning for the p-version Finite Element Method in Three Dimensions, *Int. J. Num. Meth. Eng.* **29**, 1095-1108.

Mandel Jan 1993. Balancing Domain Decomposition, *Communications in Numerical Methods in Engineering* **9**, 233–241.

Mandel Jan and Brezina M. 1993. Balancing Domain Decomposition: theory and performances in two and three dimensions, *Technical report* (Computational Mathematics Group, University of Colorado at Denver, March 1993).

Marini L.D. and Quarteroni A. 1989. A relaxation procedure for domain decomposition methods using Finite Elements, *Numer. Math.* **55**, 575–598.

Morice P. 1989. Transonic computations by a multidomain technique with potential and euler solvers. J Zienep and H Oertel Editors, *Symposium Transsonicum III, IUTAM Symposium GOTTINGEN 24-27 may 1988*, (Springer Verlag Berlin).

Przemieniecki W. 1963. Matrix structural analysis of substructures. *Am. Inst. Aero. Astro. J.* **1**, 138–147.

Quarteroni A. 1990. Domain Decomposition Method for the Numerical Solution of Partial Differential Equations, *Technical report UMSI90/246* (Supercomputer Institute, University of Minnesota).

Quarteroni A. (ed) 1993. Proceedings of the sixth internationalsymposium on Domain Decomposition Methods for Partial Differential Equations, Como, June 1992, (AMS, Providence).

Schwarz H.A. 1869. Uber einige Abbildungensaufgaben, *J. fur die Reine und Angewandte Mathematik* **70** 105–120.

Sharp M. and Farhat C. 1994. *TOP/DOMDEC A Totally Object Oriented Program for Visualisation, Domain Decomposition and Parallel Processing User's Manual*, PGSoft and The University of Colorado, Boulder.

Simon H. 1991. Partitioning of unstructured problems for parallel processing, *Comp. Systems in Eng.*,**2** 135–148.

Smith B. 1991. A Domain Decomposition Algorithm for Elliptic Problems in Three Dimensions, *Num. Math..* **60**, 219–234.

Smith B. 1992-1. An optimal domain decomposition preconditioner for the finite element solution of linear elasticity problems, *SIAM J. Sci Stat. Comp.* **13**, 364–378.

Smith B. 1992-2. A Parallel Implementation of an Iterative Substructuring Algorithms for Problems in Three Dimensions, *SIAM J. Sci. Comp.* **14**, 406–423.

Widlund O. 1983. Iterative methods for elliptic problems on regions partitioned into substructures and the biharmonic Dirichlet problem, *Proc. 6th International Conference in Applied Science and Engineering, Versailles*, (North-Holland, Amsterdam, New-York).

Widlund O. 1988. Iterative substructuring methods: algorithms and theory for elliptic problems in the plane, in R. Glowinski, G. Golub and J. Periaux (eds) 1988.

Xu J. 1992. Iterative Methods by space decomposition and subspace correction, *SIAM Review* **34**, 581–613.

Zienkiewicz O. 1971. The Finite Element Method in Engineering Science (Mc Graw-Hill, New York, Toronto, London).

# 3

# Domain Decomposition Techniques for Computational Structural Mechanics

M.Papadrakakis [1]

## 3.1 INTRODUCTION

In structural mechanics the partition of complex structural models into smaller and more manageable pieces, termed as substructures, has primarily served since the early sixties (Przemieniecki 1963) as a technique to facilitate solving large-scale problems and to provide some managerial advantages in the design and manufacturing process. A revived interest in this approach has been observed with the increasing availability of multiprocessor computers. It is due to the strong emergence of this new generation of computers that the method of substructures has attracted growing popularity over the last few years. The numerical analysis community used the substructuring concept as a way of breaking up the solution of elliptic problems into problems on smaller domains and provided a new understanding and a theoretical formulation for the use of iterative solution methods in this process. Thus, a class of innovative solution methods, known as domain decomposition has emerged. The domain decomposition methods have evolved into computational efficient algorithms both in terms of computing time and memory requirements. These methods are iterative in nature as the solutions on the individual subdomains are adapted by an iterative process in order to obtain the overall solution. They could also be described as hybrid methods since direct solvers are usually implemented as well on the subdomain level.

This chapter presents three domain decomposition formulations combined with the Preconditioned Conjugate Gradient (PCG) method for solving large-scale linear problems in mechanics. The first approach is the Global Subdomain Implementation (GSI) in which a subdomain-by-subdomain PCG algorithm is implemented on the global stiffness matrix. The dominant matrix-vector operations of the stiffness and the preconditioning matrices are performed locally on the basis of a multi-element group partitioning of the entire domain which is obtained with a mesh partitioning algorithm. The approximate inverse of the global stiffness matrix, which acts as the preconditioning matrix, is expressed by a truncated Neumann series and is formed by an additive composition of local - subdomain contributions. In order to improve

[1]   Institute of Structural Analysis and Seismic Research,
      National Technical University of Athens, Zografou Campus, 15773 Athens, GREECE

*Parallel Solution Methods in Computational Mechanics* edited by M. Papadrakakis
© 1997 John Wiley & Sons Ltd

the efficiency of the preconditioner, in terms of number of operations, a rejection by magnitude criterion is adopted combined with a compact storage scheme. Thus, the off-diagonal terms of the preconditioning matrix that influence to a lesser extend the outcome of its multiplication with the residual vector are rejected. An important feature of this global subdomain preconditioner is its local additive nature which is inherently parallelizable. Such implementation may be envisaged as mapping one or more subdomains onto each processor, while independent preconditioning matrices could be formed and operated for each subdomain.

In the second approach the PCG algorithm is applied on the interface problem after eliminating the internal degrees of freedom. For this Schur complement implementation the preconditioner is formulated from local Schur complement contributions of each subdomain. The approximate inverse of the global Schur complement is the summation of local Schur complement contributions and is expressed, as in the case of the GSI approach, by truncated Neumann series. An improvement to the efficiency of the preconditioner, is also achieved in this case by adopting a rejection by magnitude criterion and a compact storage scheme. This approach is called the Primal Subdomain Implementation (PSI) on the interface to distinguish from the third approach which is called the Dual Subdomain Implementation (DSI) on the interface.

The DSI approach operates on overlapping subdomains on the global level after partitioning the domain into a set of totally disconnected subdomains. The governing equilibrium equations are derived by invoking stationarity of the energy functional subject to displacement constraints which enforce the compatibility conditions on the subdomain interface. The augmented equations are solved for the Lagrange multipliers after eliminating the unknown displacements. The resulting interface problem is in general indefinite, due to the presence of floating subdomains which do not have enough prescribed displacements to eliminate the local rigid body modes. The solution of the indefinite problem is performed by a conjugate projected gradient algorithm using a lumped type preconditioner. This approach is introduced by Farhat and Roux (1991) as the method of Finite Element Tearing and Interconnecting (FETI). In the present study an improved version of the FETI method is proposed which results in a cost-effective version with enhanced convergence properties and reduced storage requirements.

In addition to the previously mentioned solution approaches the solution of linear systems with multiple right-hand sides is addressed on the basis of either PSI or DSI approaches in connection with a reorthogonalization procedure. Topology and shape optimization problems are also discussed by studying the impact of efficient hybrid solution schemes on the overall performance of the optimization procedure. The topology optimization problem is considered as a reanalysis problem, while solution techniques are also applied in sensitivity analysis which is the most important and time-consuming task of shape optimization problems. Finally, the solution of stochastic finite element analysis problems is addressed. The weighted integral method and the Monte Carlo simulation are implemented in conjunction with innovative solution schemes to produce robust and efficient solutions for the stochastic finite element analysis of large-scale structural problems.

## 3.2 THE GLOBAL SUBDOMAIN IMPLEMENTATION(GSI)

The assembling of the stiffness matrix $K$ in the finite element equation $Ku = f$ can be expressed as

$$K = \sum_{j=1}^{s} \tilde{K}^{(j)} \tag{3.1}$$

in which $\tilde{K}^{(j)}$ is the so called macro-element matrix of the subdomain given by $B^{(j)^t} K^{(j)} B^{(j)}$, where $B^{(j)}$ is the global connectivity matrix of subdomain $j$ which maps the global to local d.o.f., $K^{(j)}$ is the subdomain stiffness matrix and $s$ is the number of subdomains. The global unknown variables $u$ can be related to the unassembled ones by the transformation $u^e = Bu$, where $u^e$ is defined from the subdomain unknown variables in global d.o.f. as $u^e = [u^{(1)} u^{(2)} \ldots u^{(s)}]^t$ and $B = [B^{(1)} B^{(2)} \ldots B^{(s)}]^t$. Similarly, the unassembled global right-hand side vector can be defined as $f^e = [f^{(1)} f^{(2)} \ldots f^{(s)}]^t$ and $f = B^t f^e$. As can be seen by these formulas matrix $B$ acts as a scatter operator, while matrix $B^t$ acts as a gather operator.

### THE GSI-PCG ALGORITHM

Based on the previous definitions, the global subdomain implementation of the preconditioned conjugate gradient algorithm (GSI-PCG) for solving $Ku = f$ with preconditioning matrix $C_k$ may be stated, using the shared memory programming model, as follows:

- initialise:
$$u_o = 0, \quad r_o = Ku_o - f = f, \quad z_o = C_k^{-1} r_o, \quad d_o = z_o$$

- for $m = 0, 1, \ldots$

  - coefficient matrix-vector operation:

  $$d^e{}_m = Bd_m = \left[ d^{(1)} d^{(2)} \ldots d^{(s)} \right]^t_m$$

  $$v^{(j)}{}_m = K^{(j)} d^{(j)}{}_m \qquad\qquad j = 1, 2, \ldots, s$$

  $$v^e{}_m = \left[ v^{(1)} v^{(2)} \ldots v^{(s)} \right]^t_m$$

  $$v_m = B^t v^e{}_m$$

  - vector operation:

  $$\alpha_m = \frac{r_m^t z_m}{d_m^t v_m} \tag{3.2a}$$

  $$u_{m+1} = u_m + \alpha_m d_m$$

$$r_{m+1} = r_m - \alpha_m v_m$$

- convergence check:

$$\text{if } \|r_{m+1}\| \leq \varepsilon \|r_0\| \text{ stop, else continue}$$

- preconditioning step:

$$z_{m+1} = C_k^{-1} r_{m+1}$$

- vector operation:

$$\beta_m = \frac{r_{m+1}^t z_{m+1}}{r_m^t z_m}$$

$$d_{m+1} = z_{m+1} + \beta_m d_m$$

• end for

The single-program multiple-data (SPMD) model of code execution is considered the most successful way to port a code originally written using the shared memory programming model on distributed memory machines. Typically in the SPMD model each processor runs the same executable on the data corresponding to its subdomain. The SPMD version of algorithm (3.2a) with a local preconditioner for each subdomain may be stated as follows:

• initialise:

$$u_o^{(j)} = 0, \quad r_o^{(j)} = f_g^{(j)}, \quad z_o^{(j)} = C_k^{(j)^{-1}} r_o^{(j)}$$

$$\alpha^{(j)} = r_o^{(j)^t} z_o^{(j)} \qquad\qquad j = 1, 2, \ldots, s$$

- communications:

$$\alpha' = \sum_{j=1}^{s} \alpha^{(j)}$$

$$z_o = B^t z_o^e$$

$$z_o^e = B z_o$$

- vector operation:

$$d_o^{(j)} = z_o^{(j)} \qquad\qquad j = 1, 2, \ldots, s$$

• for $m = 0, 1, \ldots$

- coefficient matrix-vector operation:

$$v_m^{(j)} = K^{(j)} d_m^{(j)} \qquad\qquad j = 1, 2, \ldots, s$$

- vector operation:

$$\alpha^{(j)} = d_m^{(j)^t} v_m^{(j)} \qquad\qquad j = 1, 2, \ldots, s$$

- communications:

$$\alpha'' = \sum_{j=1}^{s} \alpha^{(j)} \qquad\qquad (3.2b)$$

$$v_m = B^t v_m^e$$

$$v_m^e = Bv_m$$

– scalar operation:

$$\alpha_m = \frac{\alpha'}{\alpha''}$$

– vector operation:

$$u_{m+1}^{(j)} = u_m^{(j)} + \alpha_m d_m^{(j)}$$

$$r_{m+1}^{(j)} = r_m^{(j)} - \alpha_m v_m^{(j)} \qquad\qquad j = 1, 2, \ldots, s$$

– preconditioning step:

$$z_{m+1}^{(j)} = C_k^{(j)}{}^{-1} r_{m+1}^{(j)} \qquad\qquad j = 1, 2, \ldots, s$$

– vector operation:

$$\gamma^{(j)} = (r_i)_{m+1}^{(j)}{}^t (r_i)_{m+1}^{(j)} + (r_{bk})_{m+1}^{(j)}{}^t (r_b)_{m+1}^{(j)}$$

$$\beta^{(j)} = r_{m+1}^{(j)}{}^t z_{m+1}^{(j)} \qquad\qquad j = 1, 2, \ldots, s$$

– communications:

$$\beta' = \sum_{j=1}^{s} \beta^{(j)}$$

$$\gamma = \sum_{j=1}^{s} \gamma^{(j)}$$

– convergence check:

$$\text{if } \sqrt{\gamma} \leq \epsilon \|r_0\| \text{ stop, else continue}$$

– communications:

$$z_{m+1} = B^t z_{m+1}^e$$

$$z_{m+1}^e = B z_{m+1}$$

– scalar operation:

$$\beta_m = \frac{\beta'}{\alpha'}$$

$$\alpha' = \beta'$$

– vector operation:

$$d_{m+1}^{(j)} = z_{m+1}^{(j)} + \beta_m d_m^{(j)} \qquad\qquad j = 1, 2, \ldots, s$$

• end for

In this algorithm $f_g^{(j)}$ satisfy $f = B^t f_g^e$ with $f_g^e = [f_g^{(1)} f_g^{(2)} \ldots f_g^{(s)}]^t$, $r_i^{(j)}$, $r_b^{(j)}$ correspond to the residual vector of the internal and interface d.o.f., while $r_{bk}^{(j)}$ consists of the residuals of the interface d.o.f. divided by the number of neighbouring subdomains of each interface d.o.f..

## AN ADDITIVE GSI PRECONDITIONER

The implementation of algorithms (3.2a) or (3.2b) requires the solution at each iteration of either the global system $C_k z_m = r_m$ or the local systems $C_k^{(j)} z_m^{(j)} = r_m^{(j)}$ respectively. For the two algorithms to be equivalent, the relationship

$$C_k^{-1} = \sum_{j=1}^{s} \tilde{C}_k^{(j)-1}$$

should be satisfied. It is apparent that the more $C_k$ resembles $K$, the more the ellipticity of the transformed stiffness matrix is reduced and the convergence rate of PCG is improved.

A GSI preconditioner based on polynomial preconditioning techniques is proposed by Papadrakakis and Bitzarakis (1996). In order to establish approximate expressions for $K^{-1}$, an auxiliary matrix $\bar{K} = L^{-1} K L^{-t}$ is defined where $LL^t$ is a block diagonal matrix of $K$. Then, the inverse of $\bar{K}$ is expressed as

$$\bar{K}^{-1} = (I - V)^{-1} \tag{3.3}$$

with

$$V = L^{-1} \left( LL^t - K \right) L^{-t} \tag{3.4}$$

For $\|V\| < 1$ the inverse of $(I - V)$ can be written by the polynomial expansion

$$(I - V)^{-1} \approx \eta_0 I + \eta_1 V + \eta_2 V^2 + \ldots + \eta_k V^k \tag{3.5}$$

where $\eta_k$ are properly selected coefficients for accelerated convergence.

Following the same definition for the subdomain quantities of $V$ as for $K$ with

$$\tilde{V}^{(j)} = L^{-1} \left( \tilde{L}^{(j)} \tilde{L}^{(j)^t} - \tilde{K}^{(j)} \right) L^{-t}$$

the inverse expression (3.5) may be written as

$$(I - V)^{-1} \approx \eta_0 I + \eta_1 \sum_{j=1}^{s} \tilde{V}^{(j)} + \eta_2 \left( \sum_{j=1}^{s} \tilde{V}^{(j)} \right)^2 + \ldots + \eta_k \left( \sum_{j=1}^{s} \tilde{V}^{(j)} \right)^k \tag{3.6}$$

An approximation to the inverse can be constructed by approximating the power of the sum by the sum of the power as follows:

$$(I - V)^{-1} \approx \eta_0 I + \eta_1 \sum_{j=1}^{s} \tilde{V}^{(j)} + \eta_2 \sum_{j=1}^{s} \tilde{V}^{(j)2} + \ldots + \eta_k \sum_{j=1}^{s} \tilde{V}^{(j)k} \tag{3.7}$$

Thus, the inverse preconditioning matrix

$$C_k^{-1} = L^{-t} (I - V)^{-1} L^{-1} \tag{3.8}$$

may be written

$$C_k^{-1} = \eta_0 (LL^t)^{-1} + \sum_{j=1}^{s} \tilde{R}^{(j)} \qquad (3.9)$$

Based on the above formulation the operation $z_{m+1} = C_k^{-1} r_{m+1}$ can be performed in parallel and the preconditioning step in the shared memory programming model of algorithm (3.2a) becomes:

– matrix-vector operations:

$$z_{m+1} = \eta_0 (LL^t)^{-1} r_{m+1}$$

$$r_{m+1}^e = B r_{m+1}$$

$$\bar{z}_{m+1}^{(j)} = R^{(j)} r_{m+1}^{(j)} \qquad\qquad j = 1, 2, \ldots, s$$

$$\bar{z}_{m+1}^e = [\bar{z}^{(1)} \ldots \bar{z}^{(s)}]_{m+1}^t \qquad (3.10a)$$

$$\bar{z}_{m+1} = B^t \bar{z}_{m+1}^e$$

$$z_{m+1} = z_{m+1} + \bar{z}_{m+1}$$

For the distributed memory model of algorithm (3.2b) the preconditioning step becomes:

– matrix-vector operations:

$$z_{m+1}^{(j)} = \eta_0 (LL^t)^{(j)^{-1}} r_{m+1}^{(j)}$$

$$\bar{z}_{m+1}^{(j)} = R^{(j)} r_{m+1}^{(j)} \qquad\qquad j = 1, 2, \ldots, s$$

– vector operation:

$$z_{m+1}^{(j)} = z_{m+1}^{(j)} + \bar{z}_{m+1}^{(j)} \qquad (3.10b)$$

– communications:

$$z_{m+1} = B^t z_{m+1}^e$$

$$z_{m+1}^e = B z_{m+1}$$

### The Polynomial Coefficients $\eta_i$

Polynomial preconditioning consists of choosing a polynomial $p(\bar{K})$ to replace $\bar{K}^{-1}$. The simplest polynomial expansion is given by the Neumann series of Eq.(3.5) with $\eta_0 = \ldots = \eta_k = 1$ : $p(\bar{K}) = I + V + V^2 + \ldots$ . More efficient polynomials can be obtained by bringing the spectrum of the preconditioned matrix as close as possible to that of the identity. One way to achieve this is to minimize the residuals $\left\| I - p(\bar{K})\bar{K} \right\|$, where $\|\cdot\|$ represents the $L_2$-norm.

According to Johnson et al. (1983) and Saad (1985) this can be achieved by defining the best polynomial which minimizes

$$\max_{\mu \epsilon [a,b]} |1 - \mu p\left(\mu\right)| \tag{3.11}$$

where $[a, b]$ is some interval containing the eigenvalues of $\bar{K}$, with $0 < a < b$. The Chebychev polynomials or alternatively a least squares minimization procedure may be implemented to solve this problem. The latter approach takes the form: Find $p$ that minimizes $\|I - p(\mu)\mu\|_w$ where $\|\cdot\|_w$ is the $L_2$-norm on $[a, b]$ with respect to some weight function $w(\mu)$. It has been observed by Saad (1985) that least squares polynomials tend to perform better than those based on the Chebychev ones, since they lead to a better overall clustering of the spectrum.

Following the derivation of Saad (1985) the parameters $a$ and $b$ are computed as the Gershgorin estimates of the extreme eigenvalues $\mu_{min}$ and $\mu_{max}$ of the matrix $\bar{K}$. However, since the Gershgorin lower bound "a" may be non positive even when $\bar{K}$ is positive definite and the upper bound "b" is usually overestimated, the least squares polynomials are thus defined in the interval $[0, \bar{b}]$, where $\bar{b} = \rho b$ and $\rho$ is a user defined reduction factor based on the experience of the user in different applications. From our experience we have found that $\rho$ should take values in the range [0.35-0.45] for plane stress problems, in the range [0.40-0.50] for 3D problems discretized with solid elements and in the range [0.30-0.55] for shell problems. The computed first two polynomials take the form

$$p_1\left(\mu\right) = -\frac{5}{4}\bar{b} + \mu \tag{3.12}$$

$$p_2\left(\mu\right) = \frac{7}{8}\bar{b}^2 - \frac{7}{4}\bar{b}\mu + \mu^2 \tag{3.13}$$

and the coefficients in Eq.(3.5) correspond to the coefficients of $\mu$ in Eqs.(3.12) and (3.13), respectively. Thus, for the first order approximation $\eta_0 = -5\bar{b}/4$, $\eta_1 = 1$ and for the second order approximation $\eta_0 = 7\bar{b}^2/8$, $\eta_1 = -7\bar{b}/4$, $\eta_2 = 1$.

## Computational Aspects

The preconditioning approach based on a first or second order polynomial expansion of the inverse $\bar{K}$ has the following features: The multiplication $R^{(j)}r_m^{(j)}$ is performed explicitly rather than implicitly, since implicit multiplication would involve extended computations in each iteration. When this multiplication is performed explicitly $R^{(j)}$ is formed only once and stored in single precision arithmetic before the iterative procedure begins.

In order to further reduce the operations involved in this multiplication a rejection by magnitude criterion is adopted for the off-diagonal terms of $R^{(j)}$ by comparing them with the corresponding assembled diagonal terms:

$$(R_{kl})^2 \leq \psi D_k^{-1} D_l^{-1} \tag{3.14}$$

$D_k$ is the global component of the diagonal entries of $K$ at the k-th d.o.f.. The control parameter $\psi$ takes values $0 < \psi < 1$. This criterion controls the terms to be retained in $R^{(j)}$ according to their magnitude, resulting in a sparse matrix $R^{(j)}$ without significantly impairing the improvement of the ellipticity of the preconditioned matrix. Due to the presence of many zero terms, a compact storage scheme is implemented for $R^{(j)}$ in order to reduce both storage requirements and the computational cost of the preconditioning matrix-vector multiplication. In this compact storage scheme only the non-zero terms of $R^{(j)}$ are stored together with their addresses and an auxiliary integer vector of length equal to the number of d.o.f. of subdomain $j$.

In the GSI-PCG algorithm the subdomain contributions are considered to be contributions by groups of elements depending on the partition of the domain into a number of subdomains. This subdivision on the domain level is equivalent with an overlapping block partitioning on the algebraic level, in which each block corresponds to the assembled stiffness matrix of each subdomain. Thus, operations are performed at the level of each subdomain $j$ $(j = 1, \ldots, s)$ and the formation of the global stiffness matrix is avoided.

The replacement of $\left(\sum \tilde{V}^{(j)}\right)^k$ by $\sum \left(\tilde{V}^{(j)}\right)^k$ in Eq.(3.7) aims in reducing the multiplications between subdomains in order to keep operations and communication overheads to a minimum and facilitate the parallel implementation of the pre-conditioning step.

In order to investigate the effect of the block size of $L$, defining $C_k$ in Eq.(3.8), on the quality of the preconditioning matrix, the efficiency of block diagonal preconditioning with respect to the block size is then examined.

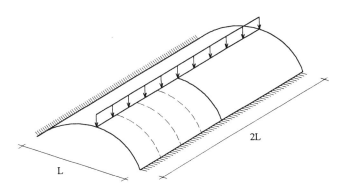

**Figure 3.1** Geometric representation of the shell example

## BLOCK DIAGONAL PRECONDITIONING

A subdivision of the basic global equations $Ku = f$ on the algebraic level, corresponding to the subdivision of the variables in $s$ subsets $u = [u_1 u_2 \ldots u_s]^t$, gives the diagonal submatrices $K_{11}, K_{22}, \ldots K_{ss}$ which may be decomposed by Cholesky factorization according to $K_{ii} = L_{ii}L_{ii}^t$. Defining $\bar{L}$ to be of the block diagonal form $\bar{L} = \lceil L_{11}L_{22}\ldots L_{ss} \rfloor$, the preconditioning matrix is then given by $C_k = \bar{L}\bar{L}^t$ which corresponds to an incomplete factorization of the stiffness matrix. Then, the transformed stiffness matrix $\bar{K} = \bar{L}^{-1}K\bar{L}^{-t}$ has unit block diagonal submatrices and $\bar{L}_{ii}^{-1}K_{ij}\bar{L}_{jj}^{-t}$ are the off-diagonal submatrices.

Based on the above definitions, the CG algorithm, with a block diagonal preconditioner on the algebraic level, performs iterations on the transformed space $\bar{r}_j = L_{jj}^{-1}f$, $\bar{u} = L_{jj}^t u$, $j = 1,\ldots,s$. After the convergence check $\|\bar{r}_{m+1}\| \leq \varepsilon\|\bar{r}_o\|$ is satisfied the required subvectors of $u$ are computed by $u_j = L_{jj}^t\bar{u}_j$. For this implementation vectors $\{v_j\}_m$ and $\{z_j\}_m$ of the PCG algorithm are computed as follows :

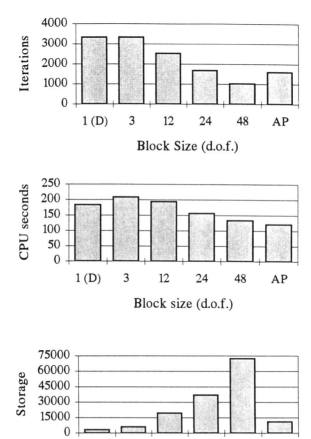

**Figure 3.2** Example 1: Performance of GSI-PCG with Diagonal (D), Block Diagonal and Additive Polynomial (AP) ($\psi = 0.01$) preconditioners. Division into 8 subdomains (Storage refers to non-zero terms of the preconditioner. Stiffness matrix non-zero terms: 28,996)

$${\{v_j'\}_m = L_{jj}^{-t}\{d_j\}_m}$$

$${\{v_j\}_m = \{d_j\}_m + L_{jj}^{-t}\left(\sum_{j\neq k} K_{jk}\{v_k'\}_m\right)} \qquad\qquad j = 1,\ldots,s$$

$$v_m = [v_1 v_2 \ldots v_s]_m^t \qquad\qquad\qquad (3.15)$$

$$\{z_j\}_m = L_{jj}^{-t} L_{jj}^{-1}\{r_j\}_m$$

The two extreme cases of the block-type preconditioning correspond to the diagonal scaling, when the order of the blocks is 1×1, and to the complete Cholesky factorization when there is one block of order $n \times n$ (where $n$ is the total number of global d.o.f.). As the block size increases, the convergence rate is improving since $C_k$ is approaching $K$, but as more fill-in is occurring the computation time per iteration is increasing. Block-type factorization preconditioners have been proposed by Jennings and Malik (1978) and Concus et al. (1985) among others. In this implementation the block-type preconditioning is combined with the domain decomposition concept where the multiplication $K_{jk}\{v_k'\}_m$ is performed on the subdomain level and not on the global level.

## NUMERICAL TESTS

The GSI-PCG algorithm is applied to three structural analysis test problems. Example 1 is a clamped orthogonal plate with a central point load. Due to symmetry the quarter of the plate is discretized with a mesh of 32 × 32 quadrilateral elements (3,000 d.o.f.).

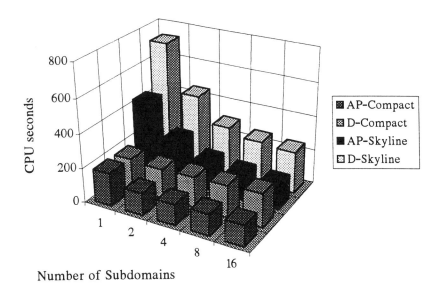

**Figure 3.3** Example 1: Influence of the stiffness matrix storage handling on the performance of GSI-PCG with Diagonal (D) and Additive Polynomial (AP) preconditioners for different number of subdomains

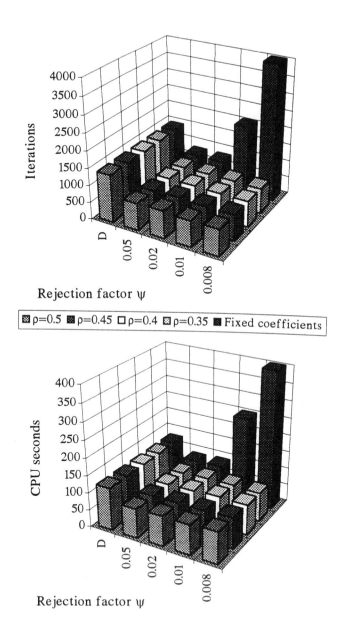

**Figure 3.4** Example 2: Performance of GSI-PCG with Diagonal (D) and Additive Polynomial (AP) preconditioners for different rejection factors $\psi$ and reduction factors $\rho$ of the Gershgorin estimate $b$. Division into 8 subdomains

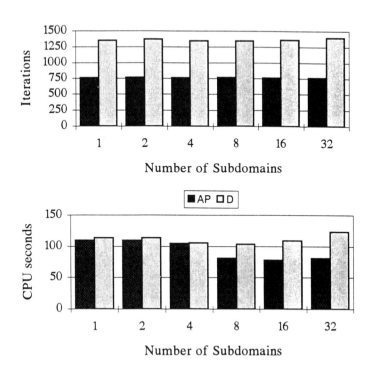

**Figure 3.5** Example 2: Performance of GSI-PCG with Diagonal (D) and Additive Polynomial (AP) preconditioners for different number of subdomains

The element aspect ratio (ratio of the largest dimension to the smallest) is taken 4 in order to introduce ill-conditioning to the stiffness matrix. Example 2 is a cantilever beam subjected to transverse end loading. The beam is discretized with a mesh of $96 \times 24$ quadrilateral plane stress elements giving 4,800 d.o.f.. The element aspect ratio is also taken to be 4. Example 3 is a parabolic cylindrical shell, shown in Fig.3.1, clamped in its straight edges with a loading along its longitudinal axis of symmetry. Two discretizations are tested for the one quarter of the shell: In Test 1 a $12 \times 24$ mesh of 9-node Lagrangian elements results in 5,734 d.o.f., while in Test 2 the corresponding discretization is $24 \times 48$ elements resulting in 22,990 d.o.f.. The element aspect ratio is approximately one and the width over the radius of curvature ratio is 0.0125. When not explicitly stated a second order expansion is used for the polynomial preconditioning. The numerical tests for Examples 1, 2 and Test 1 of Example 3 are performed on a SG Indigo R4000 workstation, while the numerical tests for Test 2 of Example 3 are performed on a SG Challenge with 16 R4400 processors. The termination tolerance criterion was taken as $\varepsilon = 10^{-6}$.

Fig.3.2 gives a comparison between the GSI-block diagonal preconditioning and the GSI-additive preconditioning for Example 1 which is divided into 8 subdomains in one-way dissection subdomains. For the block diagonal preconditioning, the size of the block is either 1 (which corresponds to diagonal or Jacobi preconditioning) or a multiple of the number of d.o.f. per structural node (which for the plate example is 3). For the additive preconditioning case, the factor $L$ of Eq.(3.8) is taken $D^{1/2}$, where $D$ is the diagonal matrix with the diagonal entries of $K$. For the block diagonal case, it can be observed that as the block size increases the convergence rate is improving, but as more fill-in is occurring the computation time is not substantially decreasing.

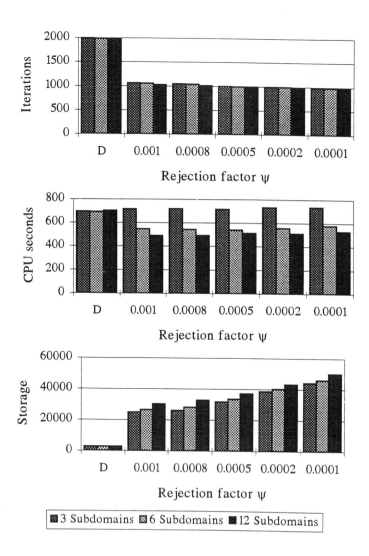

**Figure 3.6** Example 3 - Test 1: Performance of GSI-PCG with Diagonal (D) and Additive Polynomial (AP) preconditioners for different rejection factors $\psi$ ($\rho=0.50$). Divisions into 3, 6 and 12 subdomains (Storage refers to non-zero terms of the preconditioner. Stiffness matrix non-zero terms: 208,045 - 213,468 and 223,468 respectively)

As can be seen in Fig.3.2 the additive preconditioning is competitive to block diagonal preconditioning in terms of CPU time and particularly in terms of computing storage. In order for the block type preconditioner to be effective the size of the block has to be relatively large but this produces excessive storage requirements due to extensive fill-in. It is thus anticipated that an additive preconditioning matrix $C_k$ with an $L$ factor of a block size bigger than 1 would not be effective and for this reason the point diagonal form for $L$ is adopted in all subsequent tests of the GSI-PCG with additive preconditioning.

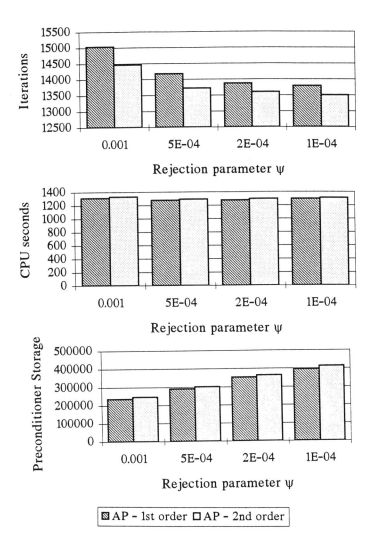

**Figure 3.7** Example 3 - Test 2: Performance of GSI-PCG for different rejection factors $\psi$ of the Additive Polynomial preconditioner in 16 processors (storage in 4-byte words)

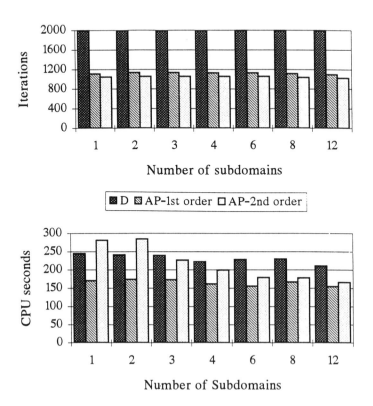

**Figure 3.8** Example 3 - Test 1: Performance of GSI-PCG with Diagonal (D) and Additive Polynomial (AP) preconditioners for different number of subdomains (sequential mode)

Fig.3.3 shows the influence of the stiffness matrix handling on the total CPU time of the GSI-PCG for different numbers of subdomains for the plate example. It can be observed that the skyline storage of the matrices has a substantial influence on the computing time due to the increased operations with zero terms on the matrix-vector multiplications. The compact storage, on the other hand, seems not to be affected by the number of subdivisions of the entire domain. The remaining tests have been performed with a compact storage handling of the stiffness matrix.

Fig.3.4 depicts the performance of the additive polynomial preconditioner with respect to the rejection factor $\psi$ which controls the non-zero terms to be retained in matrix $R^{(j)}$ of Eq.(3.9) and the reduction factor $\rho$ of the Gershgorin estimate $b$ for the evaluation of polynomial coefficients $\eta_i$. The fixed coefficients correspond to the values $\eta_0 = 1, \eta_1 = -1, \eta_2 = 1/4$. Fig.3.5 shows the influence of the number of subdomains on the performance of the GSI-PCG method for the cantilever beam.

Fig.3.6 depicts the performance of diagonal and additive preconditioners for different values of the rejection factor $\psi(\rho = 0.50)$ for Test 1 of the shell example divided into 3, 6 and 12 subdomains, while Fig.3.7 depicts the performance of the additive preconditioner with respect to the rejection factor $\psi$ for Test 2 of the shell example divided into 16 subdomains. Test 1 is performed sequentially and Test 2 in parallel. It should be mentioned that the fixed coefficients ($\eta_0 = 1, \eta_1 = -1, \eta_2 = 1/4$) in the Neumann expansion did not achieve convergence for this example. The subdivision adopted is the one-way dissection shown in Fig.3.1. Figs.3.8 and 3.9 show the performance of GSI-PCG algorithm for different number of subdomains of Tests 1 and 2, respectively. Fig.3.10 shows the speed up factors achieved when dedicated or multi-user computing modes are employed and the number of subdomains are kept fixed independently of the number of processors. Table 3.1 depicts the storage requirements of GSI-PCG algorithm for Test 2 using a compact storage scheme for the subdomain stiffness matrices.

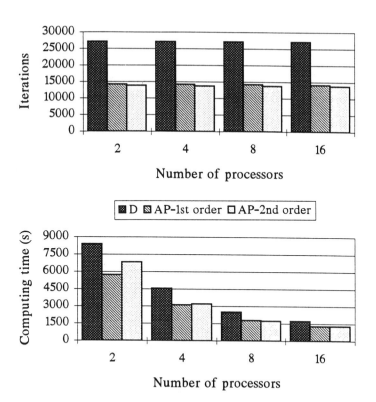

**Figure 3.9** Example 3 - Test 2: Performance of GSI-PCG with Diagonal (D) and Additive Polynomial (AP) preconditioners for different number of processors (parallel mode)

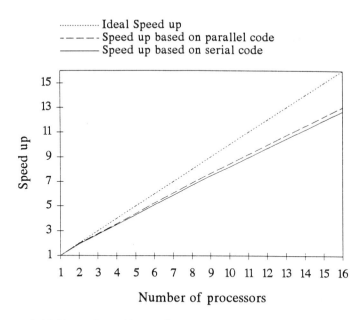

Figure 3.10 Example 3 - Test 2: Speed-up factors for the GSI-PCG algorithm

Table 3.1   Example 3 - Test 2: Storage requirements for the GSI-PCG algorithm in 4-byte words

| number of subdomains | K-storage (compact scheme) | D | AP-1st order | AP-2nd order |
|---|---|---|---|---|
| 2 | 1,710,808 | 45,980 | 260,436 | 264,151 |
| 4 | 1,722,440 | 45,980 | 264,744 | 269,397 |
| 8 | 1,750,392 | 45,980 | 273,360 | 279,889 |
| 16 | 1,806,248 | 45,980 | 290,592 | 300,873 |

## 3.3 THE PRIMAL SUBDOMAIN IMPLEMENTATION (PSI) ON THE INTERFACE

Let us consider a finite element domain $\Omega$ subdivided into $s$ non-overlapping subdomains $(\Omega_j , j = 1, \ldots, s)$. The mesh nodes which are common to the subdomain interfaces define a global interface noted by $N_b$. Numbering first the nodal point unknowns within all $\Omega_j$ not belonging to $N_b$ and last the interface nodes belonging to $N_b$, results in an arrow pattern for the global stiffness matrix. Thus, the equation of equilibrium $Ku = f$ for the displacement based finite element method can be written as

$$
\begin{bmatrix}
K_{ii}^{(1)} & 0 & & 0 & \tilde{K}_{ib}^{(1)} \\
0 & K_{ii}^{(2)} & & 0 & \tilde{K}_{ib}^{(2)} \\
& & \ddots & & \vdots \\
0 & 0 & & K_{ii}^{(s)} & \tilde{K}_{ib}^{(s)} \\
\tilde{K}_{bi}^{(1)} & \tilde{K}_{bi}^{(2)} & \cdots & \tilde{K}_{bi}^{(s)} & K_{bb}
\end{bmatrix}
\begin{bmatrix}
u_i^{(1)} \\
u_i^{(2)} \\
\vdots \\
u_i^{(s)} \\
u_b
\end{bmatrix}
=
\begin{bmatrix}
f_i^{(1)} \\
f_i^{(2)} \\
\vdots \\
f_i^{(s)} \\
f_b
\end{bmatrix}
\tag{3.16}
$$

with

$$
K_{bb} = \sum_{j=1}^{s} \tilde{K}_{bb}^{(j)}
\tag{3.17a}
$$

and

$$
f_b = \sum_{j=1}^{s} \tilde{f}_b^{(j)}
\tag{3.17b}
$$

where

$$
\tilde{K}_{bb}^{(j)} = B_b^{(j)^t} K_{bb}^{(j)} B_b^{(j)}, \quad \tilde{K}_{ib}^{(j)} = K_{ib}^{(j)} B_b^{(j)}, \quad \tilde{K}_{bi}^{(j)} = B_b^{(j)^t} K_{bi}^{(j)}, \quad \tilde{f}_b^{(j)} = B_b^{(j)^t} f_b^{(j)} \tag{3.18}
$$

$K_{ii}^{(j)}, K_{bb}^{(j)}$ correspond to the internal and the interface d.o.f. of the subdomain $\Omega_j$ respectively, while $K_{ib}^{(j)}, K_{bi}^{(j)}$ correspond to the interaction of the internal and the interface d.o.f.. $\tilde{K}_{bb}^{(j)}, \tilde{K}_{ib}^{(j)}, \tilde{K}_{bi}^{(j)}$ and $\tilde{f}_b^{(j)}$ are the macro-element matrices and vector respectively and $B_b^{(j)}$ is the interface connectivity matrix of subdomain $j$ which maps the local interface to global interface d.o.f..

The concept behind this interface PSI is to reduce the initial system of equations to a system comprising only the interface d.o.f.. Applying a static condensation to the subdomain internal d.o.f., Eq.(3.16) can be transformed into the following problem for the interface unknowns:

$$
\left( K_{bb} - \sum_{j=1}^{s} \tilde{K}_{bi}^{(j)} K_{ii}^{(j)^{-1}} \tilde{K}_{ib}^{(j)} \right) u_b = f_b - \sum_{j=1}^{s} \tilde{K}_{bi}^{(j)} K_{ii}^{(j)^{-1}} f_i^{(j)}
\tag{3.19}
$$

or

$$
S u_b = \hat{f}_b
\tag{3.20}
$$

where

$$
S = \sum_{j=1}^{s} \tilde{S}^{(j)} = \sum_{j=1}^{s} B_b^{(j)^t} S^{(j)} B_b^{(j)}
\tag{3.21}
$$

$$
S^{(j)} = K_{bb}^{(j)} - K_{bi}^{(j)} K_{ii}^{(j)^{-1}} K_{ib}^{(j)}
\tag{3.22}
$$

$$
\hat{f}_b = f_b - \sum_{j=1}^{s} B_b^{(j)^t} K_{bi}^{(j)} K_{ii}^{(j)^{-1}} f_i^{(j)}
\tag{3.23}
$$

Matrix $S$ is the Schur complement of $K_{bb}$ in $K$ and each subdomain matrix $S^{(j)}$ corresponds to a local static condensation operator or a local Schur complement.

## THE INTERFACE PSI-PCG ALGORITHM

For slender structures (i.e. space trusses) and mesh partitions with "narrow" interfaces, it may be feasible to assemble the Schur complement matrix $S$. However, for large-scale three-dimensional arbitrary structures where $N_b$ is large, assembling the interface problem can place extreme demands on both storage and computational resources, and therefore a direct method for solving system (3.20) could be prohibitively expensive. The usual alternative implementation for solving the Schur complement matrix equations is to apply PCG methods. In applying the PCG method one needs to evaluate the matrix vector product $S\{d_b\}_m$ for the direction vector $\{d_b\}_m$ at iteration $m$. In this case, $S$ does not need to be explicitly assembled. Instead, each matrix-vector product can be efficiently performed using subdomain-by-subdomain matrix-vector products and local forward and backward substitutions.

The interface unknown variables can, in this case, be related to the unassembled interface unknown variables by the transformation $u_b^e = B_b u_b$, where $u_b^e$ is defined from the subdomain unknown variables in global d.o.f. as $u_b^e = \left[ u_b^{(1)} \ldots u_b^{(s)} \right]^t$, and $B_b = \left[ B_b^{(1)} \ldots B_b^{(s)} \right]^t$. Similarly, the unassembled interface right-hand side vector can be defined as $\hat{f}_b^e = \left[ \hat{f}_b^{(1)} \ldots \hat{f}_b^{(s)} \right]^t$ and $\hat{f}_b = B_b^t \hat{f}_b^e$.

The implementation of the PCG algorithm in the domain decomposition formulation of Eq.(3.16) based on the Schur complement equation (3.20), using the shared memory programming model, may be stated as follows:

- initialise:

$$\{u_b\}_0 = 0, \quad \{r_b\}_0 = \hat{f}_b, \quad \{z_b\}_0 = C_s^{-1}\{r_b\}_0, \quad \{d_b\}_0 = \{z_b\}_0$$

- for $m = 0, 1, \ldots$

    - coefficient matrix-vector operation:

$$\{d_b^e\}_m = B_b \{d_b\}_m = \left[ d_b^{(1)} d_b^{(2)} \ldots d_b^{(s)} \right]_m^t$$

$$\left\{ v_b^j \right\}_m = S^{(j)} \left\{ d_b^{(j)} \right\}_m \qquad\qquad j = 1, 2, \ldots, s$$

$$\{v_b^e\}_m = \left[ v_b^{(1)} v_b^{(2)} \ldots v_b^{(s)} \right]_m^t$$

$$\{v_b\}_m = B_b^t \{v_b^e\}_m$$

    - vector operation:

$$\alpha_m = \frac{\{r_b\}_m^t \{z_b\}_m}{\{d_b\}_m^t \{v_b\}_m} \qquad\qquad (3.24)$$

$$\{u_b\}_{m+1} = \{u_b\}_m + \alpha_m \{d_b\}_m$$

$$\{r_b\}_{m+1} = \{r_b\}_m - \alpha_m\{u_b\}_m$$

- convergence check:

$$\text{if } \|\{r_b\}_{m+1}\| \leq \varepsilon \|\{r_b\}_0\| \text{ stop, else continue}$$

- preconditioning step:

$$\{z_b\}_{m+1} = C_s^{-1}\{r_b\}_{m+1}$$

- vector operation:

$$\beta_m = \frac{\{r_b\}_{m+1}^t\{z_b\}_{m+1}}{\{r_b\}_m^t\{z_b\}_m}$$

$$\{d_b\}_{m+1} = \{z_b\}_{m+1} - \beta_m\{d_b\}_m$$

- end for

Once the interface variables $u_b$ are determined, the internal variables $u_i^{(j)}$ can be solved from

$$u_i^{(j)} = -K_{ii}^{(j)-1}\left(\tilde{K}_{ib}^{(j)}u_b - f_i\right) \tag{3.25}$$

The matrix-vector product $S^{(j)}\left\{d_b^{(j)}\right\}_m$ can be reformulated as a subdomain problem with non-zero Dirichlet boundary conditions $d_b^{(j)}$:

$$\begin{bmatrix} K_{ii}^{(j)} & K_{ib}^{(j)} \\ K_{bi}^{(j)} & K_{bb}^{(j)} \end{bmatrix}\begin{bmatrix} d_i^{(j)} \\ d_b^{(j)} \end{bmatrix} = \begin{bmatrix} 0 \\ v_b^{(j)} \end{bmatrix} \tag{3.26}$$

where $v_b^{(j)}$ represents the interface tractions that correspond to the displacements $d_b^{(j)}$. After eliminating $d_i^{(j)}$ the second equation becomes

$$v_b^{(j)} = S^{(j)}d_b^{(j)} \tag{3.27}$$

and consequently

$$v_b = B_b^t v_b^e \tag{3.28}$$

## THE ADDITIVE INTERFACE PSI PRECONDITIONER

The preconditioning matrix associated with PSI on the interface is taken as an approximation to the inverse Schur complement matrix which results from a Neumann series expansion following the same approach as for the GSI additive preconditioner. The inverse of $\bar{S} = D_s^{-1/2}SD_s^{-1/2}$ is expressed as

$$\bar{S}^{-1} = (I - V)^{-1} \tag{3.29}$$

with

$$V = D_s^{-1/2}\left(D_s - S\right)D_s^{-1/2} \tag{3.30}$$

where

$$D_s = \sum_{j=1}^{s} B_b^{(j)^t} D_s^{(j)} B_b^{(j)} \tag{3.31}$$

is the assembled global diagonal matrix with the diagonal entries of $S$ and $D_s^{(j)}$ is the diagonal matrix with the diagonal entries of $S^{(j)}$.

Following the derivation in Section 3.2 with $S^{(j)}$, $D_s^{(j)}$ instead of $K^{(j)}$, $L^{(j)}L^{(j)^t}$ respectively, the preconditioning matrix may now be written as

$$C_s^{-1} = D_s^{-1/2}(I - V)^{-1}D_s^{-1/2} \tag{3.32}$$

or

$$C_s^{-1} = \eta_0 D_s^{-1} + \sum_{j=1}^{s} \tilde{R}_s^{(j)} \tag{3.33}$$

with

$$\tilde{R}_s^{(j)} = D_s^{-1/2}\left(\eta_1 \tilde{V}^{(j)} + \eta_2 \tilde{V}^{(j)^2} + \ldots + \eta_k \tilde{V}^{(j)^k}\right)D_s^{-1/2}$$

As in the GSI additive preconditioner case, the polynomial coefficients $\eta_i$ are derived by a least squares minimization procedure in order to bring the spectrum of the preconditioned matrix as close as possible to that of unity. Furthermore, a rejection by magnitude criterion is adopted and the resulting sparse matrix $R_s^{(j)}$ is stored using a compact storage scheme so that both storage requirements and the computational cost of the preconditioning matrix-vector multiplication are reduced.

Based on the above formulation the global interface operation $\{z_b\}_{m+1} = C_s^{-1}\{r_b\}_{m+1}$ can be performed in parallel and the preconditioning step of algorithm (3.24) becomes:

- vector operation:

$$\{z_b\}_{m+1} = \eta_0 D_s^{-1}\{r_b\}_{m+1}$$

$$\{r_b^e\}_{m+1} = B_b\{r_b\}_{m+1}$$

- preconditioning step

$$\{\bar{z}_b^{(j)}\}_{m+1} = R_s^{(j)}\{r_b^{(j)}\}_{m+1} \qquad\qquad j = 1, 2, \ldots, s \tag{3.34}$$

$$\{\bar{z}_b^e\}_{m+1} = [\bar{z}_b^{(1)} \ldots \bar{z}_b^{(s)}]_{m+1}^t$$

$$\{\bar{z}_b\}_{m+1} = B_b^t\{\bar{z}_b^e\}_{m+1}$$

$$\{z_b\}_{m+1} = \{z_b\}_{m+1} + \{\bar{z}_b\}_{m+1}$$

In the interface PSI-PCG algorithm the Schur complement of each subdomain can also be formed with a modified decomposition algorithm proposed by Han and Abel (1984). This matrix decomposition generates a rectangular matrix $W$ by performing forward and backward substitutions according to

$$K_{ii}^{(j)}W^{(j)} = K_{ib}^{(j)} \tag{3.35}$$

Then the subdomain Schur complement follows from

$$S^{(j)} = K_{ii}^{(j)} - W^{(j)}W^{(j)^t} \qquad (3.36)$$

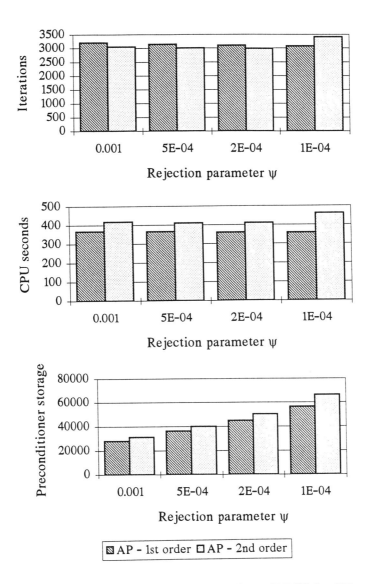

**Figure 3.11** Example 3 - Test 2: Performance of interface PSI-PCG for different rejection factors $\psi$ of the Additive Polynomial (AP) preconditioner in 16 processors (storage in 4-byte words)

This approach requires less than half the computations of the conventional approach described by Eq.(3.22) but it needs to store $W^{(j)}$ in full. When memory limitations exist, either the $S^{(j)}$ is not formed and the Schur complement matrix-vector multiplications are performed implicitly, or $W^{(j)}$ is not formed completely but only as many columns of it as the available storage (George and Liu 1981).

## NUMERICAL TESTS

The interface PSI-PCG algorithm is applied to the clamped shell Example 3 of Section 3.2. The numerical results of Example 3-Test 1 are performed on a SG Indigo R4000 Workstation, while those of Example 3-Test 2 are performed on a SG Power Challenge with 16 R4400 processors. The results presented in this section are performed with the explicitly formulated Schur complement as described in Eq.(3.36) with tolerance criterion $\varepsilon = 10^{-6}$ applied on the interface d.o.f..

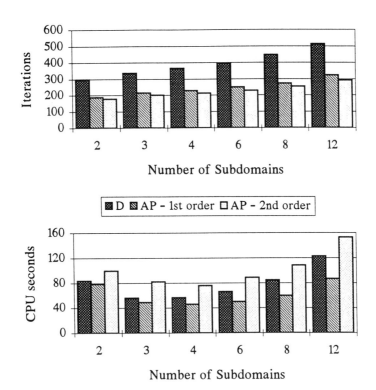

**Figure 3.12** Example 3 - Test 1: Performance of interface PSI-PCG with Diagonal (D) and Additive Polynomial (AP) preconditioners for different number of subdomains (sequential mode)

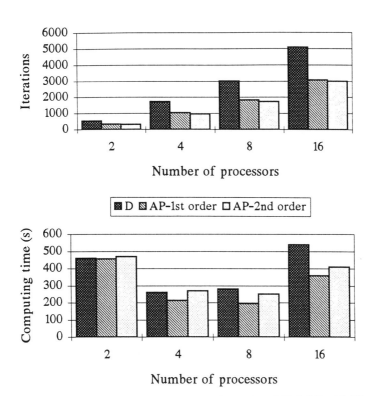

**Figure 3.13** Example 3 - Test 2: Performance of interface PSI-PCG with Diagonal (D) and Additive Polynomial (AP) preconditioners for different number of processors (parallel mode)

**Table 3.2** Example 3 - Test 2: Computing time allocation in seconds for the interface PSI-PCG algorithm

| number of processors | | 2 | 4 | 8 | 16 |
|---|---|---|---|---|---|
| | condensation | 450.00 | 135.00 | 43.00 | 16.00 |
| D | $C_s$ formulation | – | – | – | – |
| | PCG iterations | 8.90 | 124.80 | 240.00 | 525.00 |
| | condensation | 450.00 | 135.00 | 43.00 | 16.00 |
| AP-1st order | $C_s$ formulation | 0.10 | 0.50 | 0.55 | 0.65 |
| | PCG iterations | 5.90 | 78.30 | 151.50 | 340.50 |
| | condensation | 450.00 | 135.00 | 43.00 | 16.00 |
| AP-2nd order | $C_s$ formulation | 6.45 | 62.45 | 62.50 | 66.00 |
| | PCG iterations | 5.70 | 72.50 | 145.00 | 328.90 |

Fig.3.11 depicts the performance of the interface PSI-PCG additive preconditioner with respect to the rejection factor $\psi$ for Test 2 divided into 16 subdomains. Figs.3.12 and 3.13 show the performance of interface PSI-PCG algorithm for Tests 1 and 2

respectively, where Test 1 is performed sequentially and Test 2 in parallel. Fig.3.14 shows the speed up factors achieved when dedicated or multi-user computing modes are employed and the number of subdomains is kept fixed independently of the number of processors. Finally, Tables 3.2 and 3.3 depict the computing time and the storage requirements respectively for Test 2 of the PSI-PCG algorithm. A skyline storage scheme is adopted for the Schur complement in this case.

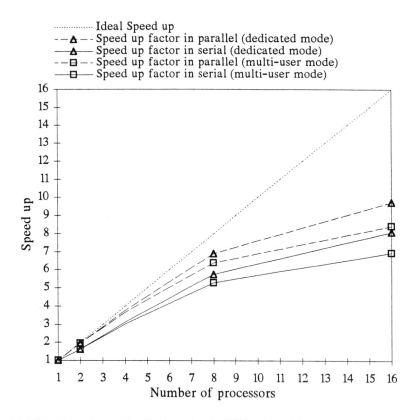

**Figure 3.14** Speed-up factors for the interface PSI-PCG algorithm

**Table 3.3**  Example 3 - Test 2: Storage requirements for the interface PSI-PCG algorithm in 4-byte words

| number of processors | storage for S Eq.(3.36) | D | AP-1st order | AP-2nd order |
|---|---|---|---|---|
| 2 | 218,500,000 | 476 | 2,920 | 4,682 |
| 4 | 169,750,000 | 1,428 | 11,138 | 9,826 |
| 8 | 147,250,000 | 3,332 | 22,668 | 25,685 |
| 16 | 142,250,000 | 7,140 | 56,346 | 40,091 |

## 3.4 THE DUAL SUBDOMAIN IMPLEMENTATION (DSI) ON THE INTERFACE

The overlapping subdomain implementation on the global level for the case of two subdomains, may be partitioned into a set of totally non-overlapping subdomains

$$
\begin{bmatrix}
K_{ii}^{(1)} & K_{ib}^{(1)} & & \\
K_{bi}^{(1)} & K_{bb}^{(1)} & & \\
& & K_{bb}^{(2)} & K_{bi}^{(2)} \\
& & K_{ib}^{(2)} & K_{ii}^{(2)}
\end{bmatrix}
\begin{bmatrix}
u_i^{(1)} \\
u_b^{(1)} \\
u_b^{(2)} \\
u_i^{(2)}
\end{bmatrix}
=
\begin{bmatrix}
f_i^{(1)} \\
f_b^{(1)} \\
f_b^{(2)} \\
f_i^{(2)}
\end{bmatrix}
\tag{3.37a}
$$

subject to the constraint equation

$$
u_b^{(1)} = u_b^{(2)} \tag{3.37b}
$$

where $u_i^{(j)}, u_b^{(j)}$ are the displacements of the internal and interface d.o.f. of subdomain $j$ respectively. In the case of $s$ subdomains Eq.(3.37a) is written as

$$
K^e u^e = f^e \tag{3.38a}
$$

where $K^e = \lceil K^{(1)} K^{(2)} \dots K^{(s)} \rfloor$ is the block diagonal unassembled global matrix and the compatibility conditions between the common d.o.f. of the subdomains are now enforced by

$$
\sum_{j=1}^{s} B^{(j)} u^{(j)} = 0 \tag{3.38b}
$$

where $B^{(j)}$ is a signed Boolean matrix which localizes a subdomain quantity to the subdomain interface. This matrix has to be distinguished, however, from the $B^{(j)}$ matrix used in Section 3.2.

The governing equilibrium equations of the constrained problem, defined by Eqs.(3.38), are derived by invoking stationarity of the energy functional $\Pi = \frac{1}{2}(u^{et} K^e u^e) - f^{et} u^e$ subject to the constraints (3.38b). This problem is transformed into an unconstrained optimization problem of the Lagrangian functional

The governing equilibrium equations of the constrained problem, defined by Eqs.(3.38), are derived by invoking stationarity of the energy functional $\Pi = \frac{1}{2}(u^{et} K^e u^e) - f^{et} u^e$ subject to the constraints (3.38b). This problem is transformed into an unconstrained optimization problem of the Lagrangian functional

$$
L(u, \lambda) = \frac{1}{2} u^{et} K^e u^e - f^{et} u^e - \lambda^t \sum_{j=1}^{s} B^{(j)} u^{(j)} \tag{3.39}
$$

with the necessary conditions

$$
\nabla_u L(u, \lambda) = K^e u^e - f^e - \sum_{j=1}^{s} {B^{(j)}}^t \lambda = 0 \tag{3.40}
$$

$$\nabla_\lambda L(u, \lambda) = \sum_{j=1}^{s} B^{(j)} u^{(j)} = 0 \tag{3.41}$$

or

$$\begin{bmatrix} K^e & B^{e\,t} \\ B^e & 0 \end{bmatrix} \begin{bmatrix} u^e \\ \lambda \end{bmatrix} = \begin{bmatrix} f^e \\ 0 \end{bmatrix} \tag{3.42}$$

where $B^e = [B^{(1)} B^{(2)} \ldots B^{(s)}]$. The vector of Lagrange multipliers $\lambda$ represents the interaction forces between the subdomains along their common boundary. This approach was introduced by Farhat and Roux (1991) and was implemented in a variety of problems in an extensive study reported by Farhat and Roux (1994). It is usually referred in the literature as the method of Finite Element Tearing and Interconnecting (FETI) because of its resemblance on tearing methods for electric circuits models.

*THE HANDLING OF THE INDEFINITE PROBLEM*

The augmented matrix of Eq.(3.42) is indefinite but non-singular and has a unique solution. Care, however, has to be taken on the handling of floating subdomains which do not have enough prescribed displacements to eliminate the local rigid body modes and are characterized by a singular stiffness matrix $K^{(j)}$. Thus, Eq.(3.42) may be explicitly written on fixed subdomains as

The augmented matrix of Eq.(3.42) is indefinite but non-singular and has a unique solution. Care, however, has to be taken on the handling of floating subdomains which do not have enough prescribed displacements to eliminate the local rigid body modes and are characterized by a singular stiffness matrix $K^{(j)}$. Thus, Eq.(3.42) may be explicitly written on fixed subdomains as

$$u^{(j)} = K^{(j)^{-1}} (f^{(j)} - B^{(j)^t} \lambda) \tag{3.43a}$$

and on floatings subdomains as

$$u^{(j)} = K^{(j)^+} (f^{(j)} - B^{(j)^t} \lambda) + R^{(j)} \gamma^{(j)} \tag{3.43b}$$

with

$$\sum_{j=1}^{s} B^{(j)} u^{(j)} = 0 \tag{3.43c}$$

$K^{(j)^+}$ is a generalized inverse of $K^{(j)}$, $R^{(j)}$ corresponds to the rigid body modes of the floating subdomain $j$, and $\gamma^{(j)}$ specifies a linear combination of these. If $K^{(j)}$ of a floating subdomain is partitioned as

$$K^{(j)} = \begin{bmatrix} K_{11}^{(j)} & K_{12}^{(j)} \\ K_{12}^{(j)^t} & K_{22}^{(j)} \end{bmatrix} \tag{3.44}$$

where $K_{11}^{(j)}$ has full rank, then $K^{(j)^+}$ and $R^{(j)}$ are given by

$$K^{(j)^+} = \begin{bmatrix} K_{11}^{(j)^{-1}} & 0 \\ 0 & 0 \end{bmatrix} \tag{3.45}$$

$$R^{(j)} = \begin{bmatrix} -K_{11}^{(j)^{-1}} K_{12} \\ I \end{bmatrix} \tag{3.46}$$

The additional equations required for determining $\gamma^{(j)}$ are provided by the zero energy condition of the rigid body modes

$$R^{(j)^t} K^{(j)} u^{(j)} = 0 \tag{3.47}$$

The combination of Eqs.(3.43) and (3.47) gives the following interface problem

$$\begin{bmatrix} F_I & -G_I \\ -G_I^t & 0 \end{bmatrix} \begin{bmatrix} \lambda \\ \gamma \end{bmatrix} = \begin{bmatrix} f_\lambda \\ f_\gamma \end{bmatrix} \tag{3.48}$$

where

$$F_I = \sum_{j=1}^{s} B^{(j)} K^{(j)+} B^{(j)^t}$$

$$G_I = \begin{bmatrix} B^{(1)} R^{(1)} \dots B^{(s_f)} R^{(s_f)} \end{bmatrix}$$

$$\gamma = \begin{bmatrix} \gamma^{(1)} \dots \gamma^{(s_f)} \end{bmatrix}^t$$

$$f_\lambda = \sum_{j=1}^{s} B^{(j)} K^{(j)+} f^{(j)}$$

$$f_\gamma = \begin{bmatrix} f^{(1)^t} R^{(1)}, \dots, f^{(s_f)^t} R^{(s_f)} \end{bmatrix}^t$$

where $K^{(j)+} = K^{(j)-1}$ if $j$ is a fixed subdomain, $K^{(j)+}$ is a generalized inverse of $K^{(j)}$ if $j$ is a floating subdomain and $s_f$ is the total number of floating subdomains. The interface operator $F_I$ can also be written as

$$F_I = \sum_{j=1}^{s} B^{(j)} \begin{bmatrix} 0 & 0 \\ 0 & S^{(j)+} \end{bmatrix} B^{(j)^t} \tag{3.49}$$

where $S^{(j)+}$ is the generalized inverse of the local Schur complement matrix of Eq.(3.22), provided that $K^{(j)}$ is partitioned according to

$$K^{(j)} = \begin{bmatrix} K_{ii}^{(j)} & K_{ib}^{(j)} \\ K_{ib}^{(j)^t} & K_{bb}^{(j)} \end{bmatrix}$$

as in the PSI approach. In that case and for mesh partitions without cross points $B^{(j)} = [0\ I]$, and the interface operator becomes

$$F_I = \sum_{j=1}^{s} S^{(j)+} \tag{3.50}$$

Eq.(3.50) justifies the term dual subdomain implementation since this method and the classical (primal) interface subdomain method are dual methods (by Farhat and Roux 1994).

## THE SOLUTION OF THE INDEFINITE PROBLEM

For large-scale problems the most appropriate algorithm for solving Eq.(3.48), is the PCG-method which does not require the submatrix $F_I$ to be assembled explicitly. The interface problem (3.48) can be rewritten as: $min\Pi(\lambda) = \frac{1}{2}\lambda^t F_I \lambda - \lambda^t (G_I \gamma + f_\lambda)$ subject to the constraint $G_I^t \lambda = -f_\gamma$, which can be solved by a Preconditioned Conjugate Projected Gradient (PCPG) algorithm, using the orthogonal projector (Farhat and Roux 1994)

$$P = I - G_I (G_I^t G_I)^{-1} G_I^t \qquad (3.51)$$

This projector removes from the improved solution $\lambda_m$ the component that does not satisfy the constraint. $G_I^t G_I$ is a sparse symmetric positive definite matrix of maximum dimension $3s_f \times 3s_f$ for two-dimensional problems, and $6s_f \times 6s_f$ for three-dimensional ones.

Based on the previous definitions the PCPG algorithm for solving Eq.(3.48) with preconditioning matrix $C_F$ may be stated as follows:

- initialise:

$$\lambda_o = G_I (G_I^t G_I)^{-1} \left[ f^{(1)^t} R^{(1)}, \ldots, f^{(s_f)^t} R^{(s_f)} \right]$$

$$r_o = f_\lambda - F_I \lambda_o$$

- for $m = 0, 1, \ldots$ until convergence

  - Projection matrix-vector operation:

  $$w_m = P r_m$$

  - preconditioning step:

  $$z_m = C_F^{-1} w_m$$

  - Projection matrix-vector operation:

  $$y_m = P z_m$$

  - vector operation:

  $$\beta_{m+1} = \frac{y_m^t w_m}{y_{m-1}^t w_{m-1}}, \qquad (\beta_1 = 0) \qquad (3.52)$$

  $$d_{m+1} = y_m + \beta_{m+1} d_m$$

  - coefficient matrix-vector operation:

  $$\{d^e\}_{m+1} = B d_{m+1} = \left[ d^{(1)} d^{(2)} \ldots d^{(s)} \right]_{m+1}^t$$

$$\{v^{(j)}\}_{m+1} = K^{(j)^+} \{d^{(j)}\}_{m+1} \qquad\qquad j = 1, 2, \ldots, s$$

$$\{v^e\}_{m+1} = \left[ v^{(1)} v^{(2)} \ldots v^{(s)} \right]^t_{m+1}$$

$$v_{m+1} = B^t \{v^e\}_{m+1}$$

– vector operations:

$$\alpha_{m+1} = \frac{y_m^t w_m}{d_{m+1}^t v_{m+1}}$$

$$\lambda_{m+1} = \lambda_m + \alpha_{m+1} d_{m+1}$$

$$r_{m+1} = r_m - \alpha_{m+1} v_{m+1}$$

– convergence check
• end for

The two projection steps are needed to preserve symmetry. Once $\lambda$ is computed, $\gamma$ is obtained from

$$\gamma = \left( G_I^t G_I \right)^{-1} G_I^t \left( F_I \lambda - f_\lambda \right)$$

and $u^{(j)}$ from Eqs.(3.43a) and (3.43b). Each projection matrix-vector operation can be evaluated in three steps as follows:

• compute:

$$\bar{r}_m^{(j)} = R^{(j)^t} B^{(j)^t} r_m \qquad\qquad j = 1, \ldots, s_f$$

• solve:

$$\left[ G_I G_I^t \right] \left[ \tilde{r}^{(1)} \ldots \tilde{r}^{(s_f)} \right]^t_m = \left[ \bar{r}^{(1)} \ldots \bar{r}^{(s_f)} \right]^t_m \qquad (3.53)$$

• compute:

$$w_m = r_m - \sum_{j=1}^{s_f} B^{(j)} R^{(j)} \tilde{r}_m^{(j)}$$

The matrix $G_I^t G_I$ is a sparse symmetric positive definite matrix which can be formed and factored only once and the solution step in Eqs.(3.53) can be carried out via a forward and backward substitution.

## DSI PRECONDITIONERS

As can be derived from Eq.(3.49) a good preconditioner for the interface DSI-PCG method can be constructed by assembling the primal subdomain operators as follows:

$$C_F^{D^{-1}} = \sum_{j=1}^{s} B^{(j)} \begin{bmatrix} 0 & 0 \\ 0 & S^{(j)} \end{bmatrix} B^{(j)^t} \qquad (3.54)$$

where the $S^{(j)}$ is the Schur complement matrix corresponding to subdomain $j$. This preconditioner is called the optimal Dirichlet preconditioner from its analogy to the Dirichlet problem. It was shown numerically (Farhat and Roux 1994) that the interface DSI-PCG method with the Dirichlet preconditioner is an optimal subdomain based PCPG method. The implementation of this preconditioner, however, requires additional storage and computational effort since $K_{ii}^{(j)}$ and $K^{(j)}$, required for $S^{(j)}$ and $F_I$ respectively, have different optimal orderings and thus they have to be stored and factored twice, while at each PCPG iteration an additional pair of forward/backward substitution on the interface d.o.f. must be performed.

A more economical preconditioner is given by

$$C_F^{L^{-1}} = \sum_{j=1}^{s} B^{(j)} \begin{bmatrix} 0 & 0 \\ 0 & K_{bb}^{(j)} \end{bmatrix} B^{(j)t} \qquad (3.55)$$

which requires almost no additional storage and involves only matrix-vector products on the interface level. This lumped preconditioner often outperforms the optimal but expensive Dirichlet preconditioner.

One additional advantage in using the lumped operator is that the convergence of the PCPG algorithm can be monitored on the global level without additional operations. It has been observed (Farhat and Roux 1994) that for most structural problems the norm of the global residuals is typically $10^2$ to $10^3$ larger than the norm of the residuals of Eq.(3.48) which means that the convergence check of the PCPG algorithm should be based on the global convergence check of GSI-PCG algorithm. Since the global residual is not readily available, its evaluation would require additional computation for the determination of $u_m$ and the global residuals at each iteration. This additional computational cost can be avoided with the lumped preconditioner since the global residual norm can be reasonably estimated by the norm of the preconditioned interface residuals

$$\|Ku_m - f\| \simeq \|C_F^{L^{-1}} (F_I\lambda_m - G_I\gamma - f_\lambda)\| \qquad (3.56)$$

The right-hand side vector of Eq.(3.56) is readily available during the PCPG iterations and is given by vector $z_m$. Thus, the convergence check becomes $\|z_m\| \leq \epsilon\|f\|$.

## THE SOLUTION OF THE 'LOCAL' PROBLEM

In the interface DSI-PCG method the solution of the local problem of Eqs.(3.43a) and (3.43b) is required, as in the implicit interface PSI-PCG method, but this time associated with both internal and interface d.o.f. of each subdomain. The solution of the local problem by a direct Cholesky solver is most advantageous for two reasons: (i) A solution of the type of Eqs.(3.43a) or (3.43b) has to be performed at each PCPG iteration within the matrix-vector multiplication $K^{(j)+}\{d^{(j)}\}_m$ for $j = 1, \ldots, s$. This means that once the factorized triangular matrix is computed and stored the subsequent solutions are performed with much less expensive forward/backward substitutions. (ii) The computation of the rigid body modes of the floating subdomains $R^{(j)}$ is performed as a by-product of the factorization procedure (Farhat and Roux 1994). When a zero pivot is encountered, the zero pivot is set to one and the reduced column above is copied into an extra right-hand side, while the coefficients in the skyline corresponding to that pivotal equation are set to zero. This extra right-hand side corresponds to a forward reduction with $K_{12}^{(j)} u_2^{(j)}$ as right-hand side, where $u_2^{(j)}$ corresponds to the redundant d.o.f. of Eq.(3.44). At the end of the factorization process, the non-labeled equations define the full rank matrix $K_{11}^{(j)}$, while the backward substitution is extended into the extra right-hand sides associated with the number of rigid body modes in order to recover $u_1^{(j)} = -K_{11}^{(j)}{}^{-1} K_{12}^{(j)} u_2^{(j)}$ corresponding to subdivision of Eq.(3.44).

For large subdomains, however, a direct solver may need excessive storage. An alternative way of treating the local problem is to use a PCG solver with an incomplete factorization-based preconditioning matrix. The preconditioning matrix is computed by an incomplete factorization of the stiffness matrix: $LDL^t = K - E$, where $E$ is an error matrix depending on a rejection parameter $\psi$ $(0 \le \psi < 1)$ which does not have to be formed (Bitoulas and Papadrakakis 1994). The convergence behaviour of the PCG algorithm is largely determined by the accuracy of computation of the residual vector and by the selection of the preconditioning matrix. A study performed by Papadrakakis and Bitoulas (1993) revealed that the computation of the residual vector from its defining formula $r_m = K u_m - f$ offers no improvement in the accuracy of the computed results. In fact, it was found that contrary to previous recommendations, the calculation of the residuals by the recursive expression of algorithm (3.2) produces a more stable and well-behaving iterative procedure. Based on this observation a mixed precision PCG implementation is proposed in which all computations are performed in single precision, except for double precision computation of the matrix-vector multiplication occurring during the recursive evaluation of the residual vector. This implementation is a robust and reliable solution procedure even for handling large and ill-conditioned problems, while it is also computer storage-effective. It was also demonstrated to be more cost-effective, for the same storage demands, than double precision arithmetic calculations.

The alternative way of solving the local problem for each subdomain by the PCG method would require the computation of the rigid body modes which cannot be obtained as a by-product of the factorization procedure as previously described. Papadrakakis and Bitzarakis (1995) have proposed an analytical handling of the rigid body modes which permits the incorporation of the PCG algorithm with a strong preconditioner for the solution of the local problem in the framework of the interface DSI-PCG method. Under this formulation the floating subdomains are fixed with constraints equal to the number of local singularities which in general are 3 for two-dimensional problems and 6 for three-dimensional ones. The rigid body modes are then computed as follows: For a 3D problem discretized with solid finite elements, the rigid body modes associated with the 3 translations and 3 rotations produce the following displacements at node $i$ of subdomain $j$:

$$R_i^{(j)} = [\, R_1 \quad R_2 \quad R_3 \quad R_4 \quad R_5 \quad R_6 \,]^t \qquad (3.57)$$

with

$$R_1 = \begin{bmatrix} 1 \\ 0 \\ 0 \end{bmatrix}, \; R_2 = \begin{bmatrix} 0 \\ 1 \\ 0 \end{bmatrix}, \; R_3 = \begin{bmatrix} 0 \\ 0 \\ 1 \end{bmatrix}, \; R_4 = \begin{bmatrix} 0 \\ -z_i \\ y_i \end{bmatrix}, \; R_5 = \begin{bmatrix} z_i \\ 0 \\ -x_i \end{bmatrix}, \; R_6 = \begin{bmatrix} -y_i \\ x_i \\ 0 \end{bmatrix}$$

where $x_i, y_i, z_i$ are the coordinates of node $i$. The rigid body mode vector $R^{(j)}$ required for the computation of $G_I$ is formed by

$$R^{(j)} = \left[ R_1^{(j)} R_2^{(j)} \ldots R_n^{(j)} \right]^t \tag{3.58}$$

where $n$ is the total number of nodes of the subdomain. Similar expressions for $R_i^{(j)}$ may be easily defined for plane stress and plate problems.

The 3 translational and 3 rotational rigid body modes of subdomain $j$, discretized by shell elements with 5 d.o.f. per node, produce the following 5 displacements (3 translations on the global coordinate system and 2 rotations on the nodal coordinate system) at node $i$ of subdomain $j$:

$$R_i^{(j)} = [\, R_1 \quad R_2 \quad R_3 \quad R_4 \quad R_5 \quad R_6 \,]^t \tag{3.59}$$

with

$$R_1 = [\, 1 \quad 0 \quad 0 \quad 0 \quad 0 \,]^t$$

$$R_2 = [\, 0 \quad 1 \quad 0 \quad 0 \quad 0 \,]^t$$

$$R_3 = [\, 0 \quad 0 \quad 1 \quad 0 \quad 0 \,]^t$$

$$R_4 = [\, 0 \quad -z_i \quad y_i \quad v_{2i}^x \quad v_{1i}^x \,]^t$$

$$R_5 = [\, z_i \quad 0 \quad -x_i \quad v_{2i}^y \quad v_{1i}^y \,]^t$$

$$R_6 = [\, -y_i \quad x_i \quad 0 \quad v_{2i}^z \quad v_{1i}^z \,]^t$$

where $x_i, y_i, z_i$ are the coordinates of node $i$ and $v_{ki}^x, v_{ki}^y, v_{ki}^z$, $(k = 1, 2)$ are the direction cosines of the nodal coordinate system.

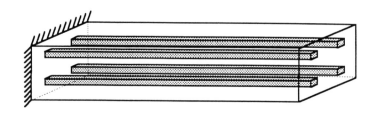

**Figure 3.15** Geometric representation of the solid cantilever beam (modulus of elasticity ratio $\frac{E_1}{E_2} = 100$ )

**Table 3.4**  Example 4 - Test 1: Internal d.o.f. for a typical subdomain and interface d.o.f.

| subdomains | 4 | 8 | 12 |
|---|---|---|---|
| Total d.o.f. | 86,400 | 86,400 | 86,400 |
| Internal d.o.f.* | 22,800 | 12,000 | 8,400 |
| Interface d.o.f. | 3,600 | 8,400 | 13,200 |

\* of the larger subdomain

**Table 3.5**  Example 4 - Test 1: Performance of DSI-PCG algorithms in 8 processors

| $\nu = 0.3$ | no-reorthogonalization | | with-reorthogonalization | |
|---|---|---|---|---|
| method | iterations | time (s) | iterations | time (s) |
| DSI-PCG1 | 29 | 554 | 29 | 641 |
| DSI-PCG2 | 29 | 541 | 29 | 627 |
| DSI-PCG3 | 29 | 530 | 29 | 541 |
| $\nu = 0.4995$ | no-reorthogonalization | | with-reorthogonalization | |
| method | iterations | time (s) | iterations | time (s) |
| DSI-PCG1 | 159 | 988 | 133 | 1,061 |
| DSI-PCG2 | 161 | 964 | 133 | 1,037 |
| DSI-PCG3 | 160 | 926 | 134 | 984 |
| $\nu = 0.4999$ | no-reorthogonalization | | with-reorthogonalization | |
| method | iterations | time (s) | iterations | time (s) |
| DSI-PCG1 | 382 | 1,983 | 272 | 1,776 |
| DSI-PCG2 | 363 | 1,897 | 259 | 1,709 |
| DSI-PCG3 | 366 | 1,810 | 260 | 1,577 |

## NUMERICAL RESULTS

The interface DSI-PCG algorithm is applied to the cantilever beam of Fig.3.15 (Example 4). Test 1 of Example 4 is discretized with $72 \times 15 \times 24$ solid elements resulting in 86,400 d.o.f.. Different modulus of elasticity and Poisson ratio are used for the four shaded bars and for the surrounding material to increase the ill-conditioning of the problem. The characteristic d.o.f. for 4, 8 and 12 subdomains are depicted in Table 3.4. The numerical tests for Example 4 are performed on a SG Power Challenge XL with 14 R8000 processors, with termination criterion $\varepsilon = 10^{-3}$. Three versions of the DSI-PCG algorithm are tested and compared. The DSI-PCG1 is the original method proposed by Farhat and Roux (1994) where the computation of the rigid body modes of the floating subdomains is performed as a by-product of the factorization procedure. The DSI-PCG2, where the computation of the rigid body modes is performed explicitly with Eq.(3.57). The DSI-PCG3, where the rigid body modes are computed explicitly and the local problem is solved via the PCG algorithm with preconditioner the complete factorized stiffness matrix stored in single precision arithmetic. The effect of reorthogonalization discussed in Section 3.5 is also examined.

The performance of the three DSI-PCG versions as well as the effect of the reorthogonalization procedure is shown in Table 3.5 for different degrees of conditioning of the stiffness matrix as affected by the Poisson ratio of the surrounding material. Table 3.6 shows the performance of the DSI-PCG versions in 4, 8 and 12 processors.

**Table 3.6** Example 4 - Test 1: Performance of DSI-PCG algorithms ($\nu$=0.4995, with reorthogonalization)

| method | 4 processors | | | 8 processors | | | 12 processors | | |
|---|---|---|---|---|---|---|---|---|---|
| | iter. | time (s) | storage (Mbyte) | iter. | time (s) | storage (Mbyte) | iter. | time (s) | storage (Mbyte) |
| DSI-PCG1 | 108 | 3,769 | 686 | 133 | 1,061 | 384 | 150 | 475 | 288 |
| DSI-PCG2 | 105 | 3,643 | 686 | 133 | 1,037 | 384 | 150 | 479 | 288 |
| DSI-PCG3 | 109 | 3,638 | 386 | 134 | 984 | 244 | 150 | 491 | 185 |

**Table 3.7** Example 4 - Test 2: Performance of GSI-PCG, PSI-PCG, DSI-PCG algorithms in 8 processors ($\nu$=0.4995, without reorthogonalization)

| method | iterations | time (s) | storage (Mbytes) |
|---|---|---|---|
| GSI-D | 9,088 | 1,413 | 14 |
| GSI-AP | 5,058 | 1,243 | 21 |
| PSI-D | 874 | 882 | 284 |
| PSI-AP | 521 | 761 | 285 |
| DSI-PCG1 | 93 | 557 | 157 |
| DSI-PCG3 | 92 | 511 | 96 |

Finally, Table 3.7 depicts the performance of GSI-PCG, PSI-PCG and DSI-PCG algorithms in Example 4 - Test 2 which is discretized with $80 \times 12 \times 12$ solid elements resulting in 40,560 d.o.f. ($\varepsilon = 10^{-3}$). The PSI-PCG algorithm is implemented with an explicitly formulated Schur complement and the tolerance criterion is applied on the interface d.o.f..The results shown in Table 3.7 correspond to $\nu$=0.4995 and 8 processors.

## 3.5 THE SOLUTION OF LINEAR SYSTEMS WITH MULTIPLE RIGHT-HAND SIDES

One of the main shortcomings of iterative solution methods is encountered when a sequence of right-hand sides has to be processed. In such cases the whole iterative work has to be repeated from the beginning for every right-hand side. Here, direct methods possess a clear advantage over iterative methods since the computationally most intensive part associated with the factorization of the stiffness matrix is not repeated and only one forward and one backward substitution is required for each subsequent right-hand side. The method of minimized iterations proposed by Lanczos (1952) has been used in the past for treating a sequence of right-hand sides. Parlett (1980) and Simon (1984) implemented Lanczos method with partial reorthogonalization by keeping the tridiagonal matrix and the orthonormal basis produced by the algorithm in fast or secondary storage in order to compute good starting values of the solution vectors for the subsequent right-hand sides. A different implementation was proposed by Papadrakakis and Smerou (1990) which handles all approximations to the solution vectors simultaneously without the necessity for storing the tridiagonal matrix and the orthonormal basis. In the latter procedure, when the first solution vector has arrived to the required accuracy good approximations to the remaining solution vectors have simultaneously been obtained. It then takes a few iterations to reach the final accuracy by working separately on each of the remaining vectors. This storage effective implementation can only be applied when all right-hand sides are simultaneously present during the solution phase of the first right-hand side.

A methodology for extending the interface DSI-PCG method to problems with multiple and/or repeated right-hand sides is proposed by Farhat et al. (1994). It is based on the PCG method incorporating a reorthogonalization procedure in connection with the $K$-conjugate property of the search directions $d_m : d_m^t K d_i = 0$ for $m \neq i$. The Krylov subspaces generated by the direction vectors $d_m$ during the PCG solution of all previous right-hand sides are used to obtain initial estimates for the solution vector associated with the subsequent right-hand side. Thus, once the problem $K u^{(i)} = f^{(i)}$ has been solved for the first $i$ right-hand side vectors and the Krylov subspace associated with all direction vectors previously computed is stored in

$$D_k = \left[ d_1^{(1)} \ldots d_{m1}^{(1)} \; \vdots \; d_1^{(2)} \ldots d_{m2}^{(2)} \; \vdots \; \ldots \; \vdots \; d_1^{(i)} \ldots d_{mi}^{(i)} \right] \qquad (3.60)$$

where $m_i$ is the number of iterations for convergence of the PCG algorithm for the $i$-th right-hand side, an initial estimate $u_o$ of the solution vector for the problem $K u^{(i+1)} = f^{(i+1)}$ is given by

$$u_o^{(i+1)} = D_k^t x \qquad (3.61)$$

The vectors $x$ is the trivial solution of the problem

$$D_k^t K D_k \, x = D_k^t f^{(i+1)} \qquad (3.62)$$

since $D_k^t K D_k$ is a diagonal matrix following the $K$-orthogonality property of the search directions. This implementation is equivalent to the two step procedure of the Lanczos method in which $u_o^{(i+1)} = Qx$ and $Tx = Q f^{(i+1)}$, where $Q$ and $T$ are the orthonormal basis and the tridiagonal matrix, respectively (Papadrakakis and Smerou 1990).

Once the augmented Krylov subspace $D_k$ is stored, an explicit reorthogonalization procedure is easily applied at each PCG iteration in order to alleviate the influence of roundoff errors to the convergence properties of the method. This entails the computation of modified search directions $d'_{m+1}$ according to

$$d'_{m+1} = d_{m+1} - \sum_{i=1}^{m} \frac{d_i^t K d_{m+1}}{d_i^t K d_i} d_i \qquad (3.63)$$

which enforces explicitly the orthogonality condition $d'_{m+1} K d_i = 0$, $i = 1, \ldots, m$.

## SOLVING MULTIPLE RIGHT-HAND SIDES WITH THE INTERFACE PSI-PCG AND DSI-PCG METHODS

The implementation of the augmented Krylov subspace for reorthogonalization and the computation of starting values of the solution vectors for the subsequent right-hand sides can be impractical when applied to the GSI problem $K u^{(i)} = f^{(i)}$. For large-scale problems the cost of reorthogonalization usually offsets the benefits of convergence acceleration, while the storage of the augmented Krylov subspace imposes a tremendous burden on the storage resources required by the method. This methodology, however, can be efficiently combined with interface PSI-PCG and DSI-PCG methods where the size of the PCG or PCPG solution methods is determined by the size of the interface problem which can be order(s) of magnitude less than the size of the global problem. Thus, the cost of reorthogonalization is negligible compared to the cost of the solution of the local problems associated with the matrix-vector products of $S$ and $F_I$ with the interface direction vectors of PSI-PCG and DSI-PCG method respectively, while the additional memory requirements for storing the augmented Krylov subspaces are proportional to the reduced size of the interface problem multiplied by the number of iterations required for convergence.

For the interface PSI-PCG method the reorthogonalization equation and the procedure for the initial estimate of the solution vector are analogous to Eqs.(3.63) and (3.61) in which $K$ and $d$ are replaced by $S$ and $d_b$ respectively. For the interface DSI-PCG method the modified search direction of the PCPG algorithm is given (Roux 1995) by

$$d'_{m+1} = d_{m+1} - \sum_{i=1}^{m} \frac{d_i^t F_I d_{m+1}}{d_i^t F_I d_i} d_i \qquad (3.64)$$

and the initial estimate $\lambda_o^{(i+1)}$ of the solution vector of the problem

$$\begin{bmatrix} F_I & -G_I \\ G_I^t & 0 \end{bmatrix} \begin{bmatrix} \lambda^{(i+1)} \\ \gamma^{(i+1)} \end{bmatrix} = \begin{bmatrix} f_\lambda^{(i+1)} \\ f_\gamma^{(i+1)} \end{bmatrix} \qquad (3.65)$$

is given by

$$\lambda_o^{(i+1)} = D_k^t x + x' \qquad (3.66)$$

where $x$ is the trivial solution of the problem

$$D_k^t F_I D_k \, x = D_k^t \left( f_\lambda^{(i+1)} - F_I x' \right) \qquad (3.67)$$

and

$$x' = G_I \left( G_I^t G_I \right)^{-1} f_\gamma^{(i+1)} \qquad (3.68)$$

$D_k$ is the augmented Krylov subspace of the direction vectors of Eq.(3.64), $x'$ is the value of the initial component of $\lambda$ which satisfies the constraint relation of Eq.(3.65) and $f_\lambda^{(i+1)} - F_I x'$ is the value of the initial unprojected residual due to $x'$.

## NUMERICAL RESULTS

The procedure described in this section is applied in Example 4 - Test 1 for 10 repeated load cases. The results obtained for DSI-PCG1 and DSI-PCG3 algorithms on SG Power Challenge XL computer are given in Table 3.8. The load cases were selected to excite different modes of the structure and this explains the rather small improvement in performance observed.

**Table 3.8**  Example 4 - Test 1: Performance of GSI-PCG algorithms for 10 repeated right-hand sides ($\nu=0.4995$)

| right-hand sides | 1 | 2 | 3 | 4 | 5 | 6 | 7 | 8 | 9 | 10 | | |
|---|---|---|---|---|---|---|---|---|---|---|---|---|
| method (4 proccesors) | iterations | | | | | | | | | | time (s) | storage (Mbytes) |
| DSI-PCG1 | 107 | 77 | 76 | 70 | 61 | 64 | 59 | 49 | 53 | 41 | 11,050 | 720 |
| DSI-PCG3 | 107 | 77 | 76 | 70 | 61 | 64 | 59 | 49 | 53 | 41 | 10,101 | 420 |
| method (8 proccesors) | | | | | | | | | | | | |
| DSI-PCG1 | 127 | 92 | 88 | 75 | 75 | 69 | 59 | 41 | 38 | 35 | 3,192 | 469 |
| DSI-PCG3 | 127 | 93 | 88 | 77 | 75 | 69 | 59 | 41 | 38 | 35 | 2,876 | 325 |

## 3.6  SOLVING TOPOLOGY OPTIMIZATION AND SENSITIVITY ANALYSIS PROBLEMS

The objective of this section is to investigate the efficiency of hybrid solution methods, when incorporated into large-scale topology and shape optimization problems and to demonstrate their influence on the overall performance of the optimization procedure.

## TOPOLOGICAL DESIGN OF STRUCTURES

Topology optimization is a tool which assists the designer in the selection of suitable structural topologies by removing or redistributing material of the initial structural domain. The procedure starts with an initial layout of the structure followed by a gradual 'removal' of a small portion of low stressed material which is being used inefficiently. This procedure is a typical case of structural reanalysis, which results in small variations of the stiffness matrix between two optimization steps. The topological or layout optimization tests conducted in this study are based on the fully stressed design technique of Hinton and Sienz (1993) which proceeds by removing the redundant material in an evolutionary process. The algorithm is based on the simple principle that material which has small stress levels is used inefficiently and therefore it can be removed. Thus, by removing small amounts of material at each optimization step the layout of the structure evolves gradually. After the generation of the finite element mesh the evolutionary fully stressed design cycle is activated where a linear elastic finite element stress analysis is carried out. The stress level for each element $\sigma_{evo}$ is computed and compared with the maximum stress level $\sigma_{max}$ of the elements in the structure at the current optimization step. All elements that fulfill the condition $\sigma_{evo} < ratre * \sigma_{max}$ are switched-off, where $ratre$ is the rejection rate. The elements are removed by assigning them a relatively small elastic modulus which is typically $E_{off} = 10^{-5} E_{on}$. The iterative process of element removal and addition is continued until one of several specified convergence criteria are met: (i) All stress levels are larger than a certain percentage value of the maximum stress; (ii) the number of active elements is smaller than a specified percentage of the total number of elements; (iii) when element growth is allowed the evolutionary process is completed when more elements are switched on than those switched-off.

## SHAPE OPTIMIZATION

The shape optimization method used in the present study is based on the previous work for treating two-dimensional problems of Hinton and Sienz (1996). It consists of the following steps: (i) shape generation and control; (ii) mesh generation; (iii) adaptive finite element analysis; (iv) sensitivity analysis and (v) optimization using mathematical programming techniques. Sensitivity analysis is the most important and time-consuming part of this procedure. Although mostly mentioned in the context of structural optimization it has evolved into a research topic of its own. The methods for sensitivity analysis can be divided into discrete and variational methods. The computer implementation for discrete methods is simpler than for variational methods. In the discrete method the derivatives or the sensitivities of the characteristic parameters, i.e. displacements, stresses, volume, etc., which define the objective and constraint functions of the discretized structure, are evaluated using the finite element equations. Two discrete sensitivity analysis methods have been implemented in this study, the global finite difference method and the semi-analytical method.

### The global finite difference (GFD) sensitivity analysis method

In shape optimization the stiffness matrix $K$ and the loading vector $f$ are functions of the design variables $s_k(k = 1, ..., n)$. The primary objective is to compute the derivative of the displacement field with respect to $s_k$, called sensitivities coefficients. The GFD method requires the solution of a linear system $Ku = f$ for the original design variables $s_k$, and for each perturbed design variable $s_k^p = s_k + \Delta s_k$, where $\Delta s_k$ is the magnitude of the perturbation. The design sensitivities $\frac{\vartheta u}{\vartheta s_k}$ and $\frac{\vartheta \sigma}{\vartheta s_k}$ for displacements and stresses are computed using a forward difference scheme

$$\frac{\vartheta u}{\vartheta s_k} \approx \frac{\Delta u}{\Delta s_k} = \frac{u\left(s_k + \Delta s_k\right) - u\left(s_k\right)}{\Delta s_k} \qquad (3.69)$$

$$\frac{\vartheta \sigma}{\vartheta s_k} \approx \frac{\Delta \sigma}{\Delta s_k} = \frac{\sigma \left( s_k + \Delta s_k \right) - \sigma \left( s_k \right)}{\Delta s_k} \tag{3.70}$$

The perturbed displacement vector $u(s_k + \Delta s_k)$ is estimated by

$$K \left( s_k + \Delta s_k \right) u \left( s_k + \Delta s_k \right) = f \left( s_k + \Delta s_k \right) \tag{3.71}$$

and the perturbed stresses are computed from $\sigma(s_k + \Delta s_k) = \mathcal{D}\mathcal{B}(s_k + \Delta s_k)u(s_k + \Delta s_k)$, where $\mathcal{D}$ and $\mathcal{B}$ are the elasticity and deformation matrices respectively. The GFD scheme is usually sensitive to the accuracy of the computed perturbed displacement vectors which is dependent on the magnitude of the perturbation.

## The semi-analytical (SA) sensitivity analysis method

The SA method is based on the chain rule differentiation of $Ku = f$ which is written as

$$K \frac{\vartheta u}{\vartheta s_k} = f_k^* \tag{3.72}$$

where

$$f_k^* = \frac{\vartheta f}{\vartheta s_k} - \frac{\vartheta K}{\vartheta s_k} u \tag{3.73}$$

$f_k^*$ represents a pseudo-load vector. The derivatives $\frac{\vartheta f}{\vartheta s_k}$ and $\frac{\vartheta K}{\vartheta s_k}$ are computed for each design variable by recalculating the new values of $K \left( s_k + \Delta s_k \right)$ and $f \left( s_k + \Delta s_k \right)$ for a small perturbation $\Delta s_k$ of the design variable $s_k$, while the stiffness matrix remains unchanged throughout the whole sensitivity analysis.

In the conventional semi-analytical (CSA) method the derivatives of Eqs.(3.72) and (3.73) are calculated by applying the forward difference approximation

$$\frac{\vartheta K}{\vartheta s_k} \approx \frac{\Delta K}{\Delta s_k} = \frac{K \left( s_k + \Delta s_k \right) - K \left( s_k \right)}{\Delta s_k} \tag{3.74}$$

$$\frac{\vartheta f}{\vartheta s_k} \approx \frac{\Delta f}{\Delta s_k} = \frac{f \left( s_k + \Delta s_k \right) - f \left( s_k \right)}{\Delta s_k} \tag{3.75}$$

Stress gradients, required for stress constraints, are calculated by differentating $\sigma = \mathcal{D}\mathcal{B}u$ as follows

$$\frac{\vartheta \sigma}{\vartheta s_k} = \mathcal{D}\frac{\vartheta \mathcal{B}}{\vartheta s_k}u + \mathcal{D}\mathcal{B}\frac{\vartheta u}{\vartheta s_k} \tag{3.76}$$

where $\frac{\vartheta u}{\vartheta s_k}$ may be computed from Eq.(3.72), while the term $\frac{\vartheta \mathcal{B}}{\vartheta s_k}$ is computed using a forward finite difference scheme.

The CSA method may suffer some drawbacks in particular types of shape optimization problems. This is due to the fact that in the numerical differentiation of the element stiffness matrix with respect to shape design variables, the components of the pseudo load vector associated with the rigid body rotation do not vanish. A similar inaccuracy problem is not occuring if the analytical or the global finite difference method is employed. The solution suggested by Olhoff et al. (1992) alleviates this problem by performing an 'exact' numerical differentiation of the elemental stiffness matrix based on computationally inexpensive first order derivatives. These derivatives can be obtained on the element level as follows

$$\frac{\vartheta K^{(el)}}{\vartheta s_k} = \sum_{j=1}^{n} \frac{\vartheta K^{(el)}}{\vartheta \alpha_j} \frac{\vartheta \alpha_j}{\vartheta s_k} \tag{3.77}$$

where $\alpha_j$ is the nodal coordinate of the element, $n$ is the number of moving nodal coordinates for a given element and $s_k$ is the perturbed design variable. The derivative $\frac{\vartheta \alpha_j}{\vartheta s_k}$ can be computed 'exactly' by means of a forward difference scheme

$$\frac{\vartheta \alpha_j}{\vartheta s_k} \approx \frac{\Delta \alpha_j}{\Delta s_k} = \frac{\alpha_j \left( s_k + \Delta s_k \right) - \alpha_j \left( s_k \right)}{\Delta s_k} \tag{3.78}$$

while the derivative $\frac{\vartheta K^{(el)}}{\vartheta \alpha_j}$ is computed by differentiating the element stiffness matrix expression $K^{(el)} = \int_\Omega \mathcal{B}^t \mathcal{D} \mathcal{B} |\mathcal{J}| d\Omega$ with respect to the nodal coordinate $\alpha_j$. This 'exact' semi-analytical (ESA) method is computationally more expensive than the CSA method. This overhead however, is counterbalanced by the gains in accuracy and the improvement on the overall optimization procedure.

## SOLUTION METHODS FOR TOPOLOGY AND SENSITIVITY ANALYSIS PROBLEMS

The most computational intensive part in the sensitivity analysis problem using either CSA or ESA sensitivity analysis methods is the solution of Eq.(3.72), where the coefficient matrix remains constant and the right-hand side vector is changing. This is the typical case of solving linear systems with multiple right-hand sides discussed in Section 3.5. In topology optimization and sensitivity analysis problem using the GFD method, a typical reanalysis problem of the form

$$\left( K_o + \Delta K_i \right) u_i = f_i \tag{3.79}$$

needs to be solved. In the case of topology optimization $\Delta K_i$ corresponds to the modification due to the switched-off elements at step $i$ with $f_i$ constant. In the case of GFD sensitivity analysis $\Delta K_i$ corresponds to the modification induced by the perturbation of the design variable $s_k^p = s_k + \Delta s_k$, while $f_i$ is given in Eq.(3.71). In both cases $\Delta K_i$ is relatively small compared to $K$. For the solution of this problem two-level PCG methods have been proposed (Papadrakakis and Kotsopulos 1995, Papadrakakis and Tsompanakis 1996) which are based on a combination of the Global Subdomain Implementation and the Dual Subdomain Implementation, namely the GSI-DSI-PCG and GSI-DSI-NCG methods.

### The GSI-DSI-PCG method

The GSI-DSI-PCG method as applied in topology optimization and GFD sensitivity analysis is based on the PCG method with an incomplete Cholesky factorization of the stiffness matrix as preconditioner. Such an approach leads to the incomplete factorization of the stiffness matrix $K_o + \Delta K_i : LDL^t = K_o + \Delta K_i - E$, where

$E$ is an error matrix which does not have to be formed. (Papadrakakis et al. 1995, Papadrakakis and Papadopoulos 1996). $E$ is defined by the computed positions of 'small' elements in $L$ which do not satisfy a specified magnitude criterion and therefore are discarded. For the case of Eq.(3.79), $E$ is taken as $\Delta K_i$. Thus, the preconditioning matrix becomes the complete factorized initial stiffness matrix $K_o$. The solution of the preconditioning step in PCG algorithm now involves the operation $z_{m+1} = K_o^{-1} r_{m+1}$, which can be effortlessly performed, once $K_o$ is factorized, by a forward and backward substitution.

The parallel implementation of this preconditioning step is hindered by the inherent scalability problems associated with direct methods of solution in parallel computing environments. For this reason the following hybrid implementation of the PCG method is proposed. A Global Subdomain Implementation (GSI) of the PCG method is performed as described in Section 3.2, in which the preconditioning step becomes

$$z_{m+1} = K_o^{-1} r_{m+1} \tag{3.80}$$

This solution step, which has to be performed at each GSI-PCG iteration, can be solved in parallel, using the same decomposition of the domain employed by the external GSI, with the interface DSI-PCG method for treating the repeated solutions required in Eq.(3.80) as discussed in Section 3.5 for the solution of linear systems with multiple right-hand sides. The solution of Eq.(3.80) is performed $n_i * n_r$ times, where $n_i$ and $n_r$ correspond to the number of GSI-PCG iterations and the number of reanalysis or GFD sensitivity analysis steps, respectively.

## The GSI-DSI-NCG method

The GSI-DSI-NCG method associated with the solution of Eq.(3.79) is also based on the PCG method. In this case, in order to improve the quality of the preconditioning matrix, the inverse approximation of the preconditioning matrix is computed via a Neumann series expansion (Papadrakakis et al. 1995, Papadrakakis and Papadopoulos 1996). The preconditioning matrix is now defined as the complete stiffness matrix $K_o + \Delta K_i$, but the solution for $z_{m+1}$ of Eq.(3.80) is performed approximately using a truncated Neumann series expansion. Thus, the preconditioned vector $z_{m+1}$ is obtained at each iteration by

$$z_{m+1} = \left(I + K_o^{-1} \Delta K_i\right)^{-1} K_o^{-1} r_{m+1} \tag{3.81}$$

The term in parenthesis can be expressed in a Neumann expansion giving

$$z_{m+1} = \left(I - N + N^2 - N^3 + \ldots\right) K_o^{-1} r_{m+1} \tag{3.82}$$

with $N = K_o^{-1} \Delta K_i$. The preconditioned vector can now be computed by the following series

$$z_{m+1} = z_0' - z_1' + z_2' - z_3' \ldots \tag{3.83}$$

with

$$z_o' = K_o^{-1} r_{m+1} \tag{3.84a}$$

$$z'_j = K_o^{-1}\left(\Delta K_i z'_{j-1}\right) \qquad , j = 1, 2, \ldots \qquad (3.84b)$$

The incorporation of the Neumann series expansion in the preconditioned step of the PCG algorithm can be seen from two different perspectives. From the PCG point of view, an improvement of the quality of the preconditioning matrix is achieved by computing a better approximation to the solution of $x = (K_o + \Delta K)^{-1} f$ during the preconditioning step than the one provided by the preconditioning matrix $K_o$. From the Neumann series expansion point of view, the inaccuracy entailed by the truncated series is alleviated by the conjugate gradient iterative procedure.

The GSI-DSI-NCG method for solving Eqs.(3.79) is implemented as follows: A GSI-PCG method is carried out as described in Section 3.2 in which the preconditioning

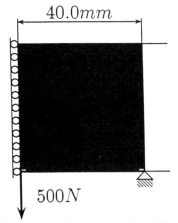

**Figure 3.16** Example 5: Initial layout

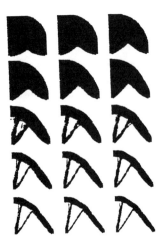

**Figure 3.17** Example 5: Evolution history of topology optimization

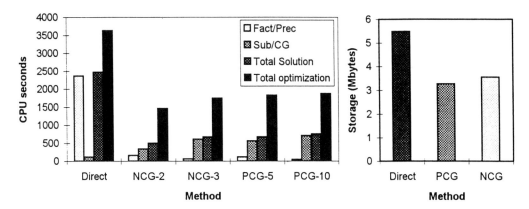

**Figure 3.18** Example 5: Performance of the methods. Computing time in seconds and storage requirements in Mbytes

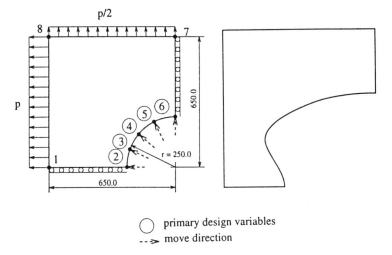

○ primary design variables
--→ move direction

**Figure 3.19** Example 6: (a) initial shape, (b) final shape

step is performed according to Eq.(3.82). Eqs.(3.84) can be solved in a parallel computing environment with the interface DSI-PCG method utilizing the same decomposition of the domain adopted for the external GSI-PCG method. The solution of Eqs.(3.84) is performed $n_j * n_i * n_r$ times, where $n_j, n_i, n_r$ correspond to the number of terms in the Neumann series expansion, the number of GSI-PCG iterations and the number of reanalyses or GFD sensitivity analysis steps, respectively.

## NUMERICAL TESTS

The performance of PCG, NCG, DSI-PCG and GSI-DSI-PCG methods for solving Eq.(3.79) in topology optimization and in shape optimization with GFD and ESA sensitivity analysis is demonstrated for two benchmark test examples (Papadrakakis et al. 1995). PCG corresponds to a global PCG implementation with an incomplete factorized matrix as preconditioner, while NCG is also a global PCG implementation in which the preconditioning step is performed according to Eq.(3.82). The methods are compared with the direct skyline solver. The convergence tolerance varies according to the problem. In topology optimization the convergence tolerance for the hybrid solution methods is taken $10^{-1}$ since only a crudely optimized distribution of the material of the initial structure is needed before the final refinement of the layout of the structure is performed during the shape optimization phase. In shape optimization, a higher level of convergence tolerance ($10^{-3}$) is chosen since the efficiency of the optimizer, as well as the quality of the final solution, is highly dependent on the attained accuracy of the sensitivity analysis phase. For the test examples considered in this section plane stress conditions and isotropic material properties are assumed (elastic modulus $E = 210,000 N/mm^2$ and Poisson's ratio $\nu = 0.3$).

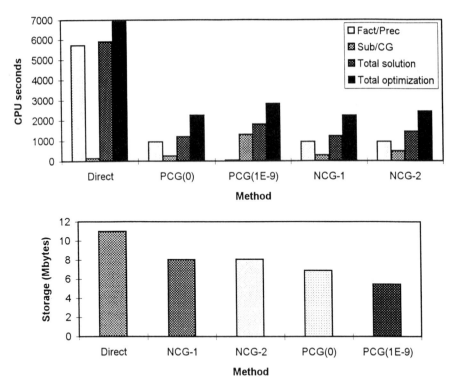

**Figure 3.20** Example 6: Performance of the methods. Computing time and storage requirements

The following abbreviations are used for the topology optimization problem: *Direct* is the conventional skyline direct solver; *PCG-n* is the PCG solver in which the preconditioning matrix is formed with a complete Cholesky factorization ($\psi = 0$)

performed in single precision arithmetic, updated when the number of PCG iterations becomes greater than n; *NCG-n* is the NCG solver in which a refactorization in single precision arithmetic of the stiffness matrix is performed when the number of PCG iterations becomes greater than n.

**Table 3.9**  Example 6 - Test 2: Characteristic d.o.f.

| subdomains | 4 | 8 |
|---|---|---|
| Total d.o.f. | 38,800 | 38,800 |
| Internal d.o.f.* | 9,738 | 5,122 |
| Interface d.o.f. | 998 | 2,290 |

* of the larger subdomain

**Table 3.10**  Example 6 - Test 2: Performance of the methods for one function evaluation of ESA sensitivity analysis (five design variables - one processor - 4 subdomains)

| method (4 subdomains-1 processor) | time (s) | storage (Mbytes) |
|---|---|---|
| Direct skyline | 499 | 95 |
| DSI-PCG1-reorth | 486 | 43 |
| DSI-PCG2-reorth | 466 | 43 |
| DSI-PCG3-reorth | 414 | 26 |

Example 5 is a thick beam fixed at the two bottom corners with a vertical point load applied at the centre of the bottom edge (Fig.3.16). In this study the topology optimization problem is solved using 5,833 plane stress elements with 8,000 d.o.f. A typical evolution history is shown in (Fig.3.17). All analyses performed produced similar topologies. NCG-3 requires the smallest number of optimization steps (Nsteps=89) to converge whereas the direct solver requires the largest number of optimization steps to converge (Nsteps=109). Fig.3.18 depicts the solution time, the total optimization time and the storage requirements as obtained on a SG Indigo R4000 workstation. It can be seen that NCG-2 is 5 times faster than the direct method and reduces the overall optimization time by a factor of 65%. In terms of computing storage PCG requires 40% and 10% less storage than the direct and NCG methods, respectively.

Example 6 is a square plate with a central cut-out. The problem definition is given in Fig.3.19a where due to symmetry only a quarter of the plate is modeled, and the final shape is shown in Fig.3.19b. The objective of this shape optimization problem is to minimize the volume of the structure subject to a limit on the equivalent stress of $\sigma_{max} = 7.0 \ N/mm^2$. The design model consists of 8 key points and 5 design variables (2, 3, 4, 5, 6) which can move along radial lines. The movement directions are indicated by the dashed arrows. The stress constraint is imposed as a global constraint for all the Gauss points and as a key point constraint for the key points 2, 3, 4, 5, 6 and 8. Test 1 of Example 6 is analyzed with an initial mesh of 11,000 d.o.f. The optimization algorithm employed to solve the shape optimization problem is a SQP algorithm from the DoT package (DoT 1994). The GFD method is used to compute the sensitivities with $\Delta s = 10^{-5}$. The abbreviation *NCG-i* corresponds now to the NCG solver with $i$ terms of the Neumann series expansion, while *PCG(0)* uses a complete Cholesky factorization performed in single precision arithmetic as preconditioner. The term *function evaluation* is related to the computations required for one finite element and sensitivity analysis cycle, as well as for the calculation of the values of the objective and constraint functions and their derivatives. The initial volume of the full problem is $149.37mm^3$ and becomes $112mm^3$ at the end of the shape optimization procedure after 7 optimization steps and 12 function evaluations. Fig.3.20 demonstrates the solution time, the total optimization time and the storage requirements as obtained on a SG Indigo R4000 workstation. NCG-1 and PCG(0) are 5 times faster than the direct solver with regard to the solution time and reduce by 65% the overall optimization time, while in terms of computer storage PCG(0) is 40% and 10% better than the direct and the NCG methods, respectively.

**Table 3.11**   Example 6 - Test 2: Performance of the methods for one function evaluation of ESA sensitivity analysis and five design variables

| right-hand sides | 1 | 2 | 3 | 4 | 5 | 6 | | |
|---|---|---|---|---|---|---|---|---|
| method (4 processors) | iterations | | | | | | time (s) | storage (Mbytes) |
| DSI-PCG1 | 67 | 60 | 65 | 51 | 53 | 53 | 420 | 41 |
| DSI-PCG1-reorth | 33 | 16 | 13 | 10 | 9 | 8 | 150 | 43 |
| DSI-PCG3 | 67 | 60 | 65 | 51 | 53 | 53 | 306 | 24 |
| DSI-PCG3-reorth | 33 | 16 | 13 | 10 | 9 | 8 | 120 | 26 |
| method (8 processors) | | | | | | | | |
| DSI-PCG1 | 271 | 266 | 253 | 219 | 199 | 204 | 398 | 20 |
| DSI-PCG1-reorth | 64 | 24 | 18 | 14 | 11 | 11 | 92 | 23 |
| DSI-PCG3 | 269 | 267 | 253 | 220 | 199 | 205 | 290 | 13 |
| DSI-PCG3-reorth | 64 | 24 | 18 | 14 | 11 | 11 | 70 | 16 |

The ESA sensitivity analysis methodology based on repeated solutions of Eq.(3.72) and solved by the DSI-PCG algorithms is applied (Papadrakakis and Tsompanakis 1996) to Example 6 - Test 2. Test 2 is now discretized with a finer mesh of 38,800 d.o.f.. The characteristic d.o.f. for 4 and 8 subdomains are depicted in Table 3.9. The performance of DSI-PCG algorithms, as obtained on a SG Power Challenge XL computer, is depicted in Tables 3.10 and 3.11 during one function evaluation using five design variables. Table 3.10 presents the results in one processor operated with 4 subdomains, while Table 3.11 presents the performance of the algorithms in 4 and 8 processors operated in 4 and 8 subdomains, respectively.

**Table 3.12**  Example 6 - Test 2: Performance of the methods for one function evaluation of GFD sensitivity analysis (five design variables - one processor - 4 subdomains)

| method (4 subdomains) | time (s) | storage (Mbytes) |
|---|---|---|
| Direct skyline | 2,790 | 95 |
| DSI-PCG1-reorth | 2,108 | 43 |
| DSI-PCG3-reorth | 1,782 | 26 |
| GSI-DSI-PCG3-reorth | 795 | 27 |
| GSI-DSI-NCG3-reorth | 779 | 29 |

**Table 3.13**  Example 6 - Test 2: Performance of the methods for one function evaluation of GFD sensitivity analysis and five design variables

| right-hand sides | 1 | 2 | | 3 | | 4 | | 5 | | 6 | | | |
|---|---|---|---|---|---|---|---|---|---|---|---|---|---|
| method (4 processors) | iterations | | | | | | | | | | | time (s) | storage (Mbytes) |
| DSI-PCG1-reorth | 33 | 33 | | 33 | | 33 | | 33 | | 33 | | 667 | 43 |
| DSI-PCG3-reorth | 33 | 33 | | 33 | | 33 | | 33 | | 33 | | 574 | 26 |
| GSI-DSI-PCG3-reorth | 33 | 17 | 13 | 12 | 10 | 10 | 8 | 8 | 7 | 7 | 6 | 256 | 27 |
| GSI-DSI-NCG3-reorth | 33 | 17 | 14 | 12 | 11 | 11 | 9 | 7 | 7 | 6 | 6 | 249 | 29 |
| method (8 processors) | iterations | | | | | | | | | | | | |
| DSI-PCG1-reorth | 64 | 64 | | 64 | | 64 | | 64 | | 64 | | 396 | 23 |
| DSI-PCG3-reorth | 64 | 64 | | 64 | | 64 | | 64 | | 64 | | 332 | 16 |
| GSI-DSI-PCG3-reorth | 64 | 24 | 19 | 17 | 14 | 12 | 11 | 9 | 9 | 8 | 7 | 165 | 17 |
| GSI-DSI-NCG3-reorth | 64 | 24 | 18 | 16 | 14 | 13 | 11 | 10 | 9 | 8 | 7 | 163 | 18 |

Finally Tables 3.12 and 3.13 depict the performance of GSI-DSI-PCG method when applied to the GFD sensitivity analysis for one function evaluation using five design variables. The number of iterations shown in Table 3.13 for the two-level PCG method corresponds to the DSI-PCG iterations for the solution of the preconditioning step, while the iterations performed by the external GSI-PCG method are restricted to two for each of the repeated solutions corresponding to the five design variables.

## 3.7 SOLVING STOCHASTIC FINITE ELEMENT ANALYSIS PROBLEMS WITH MONTE CARLO SIMULATION

Stochastic analysis involves the estimation of the response variability and/or reliability of a stochastic system defined as a structural system that possesses uncertainties in its material and/or geometric properties. The Stochastic Finite Element (SFE) method offers the appropriate platform for treating realistic problems in which the representation of stochastic fields is described as a series of random variables. Among a number of existing SFE approaches, the weighted integral method, developed by Shinozuka and Deodatis (1988) and Deodatis (1991) offers certain advantages since it does not require discretization of the random field and its accuracy is independent of the chosen mesh. Moreover, the coupling of Monte Carlo Simulation (MCS) with SFE approaches has the major advantage that accurate solutions can be obtained for any problem whose deterministic solution is known either numerically or analytically. In fact, MCS is the only method available to solve stochastic problems involving nonlinearities, dynamic loading, stability effects, parametric excitations etc. The disadvantage of the standard MCS is that it is usually extremely computationally demanding. In this section the weighted integral method and MCS are used in conjunction with the innovative solution strategies discussed in Sections 3.5 and 3.6 to produce robust and efficient solutions for stochastic finite element analysis of structural problems.

## THE WEIGHTED INTEGRAL METHOD

The weighted integral method has been applied to formulate the stochastic element stiffness matrix when the modulus of elasticity is assumed to vary randomly along the element area. For beam elements the modulus of elasticity is assumed to vary according to

$$E^{(e)}(x) = E_o \left[ 1 + f^{(e)}(x) \right] \tag{3.85}$$

where $E_o$ is the mean value and $f^e(x)$ is an one-dimensional univariate (1D-1V) zero mean homogeneous stochastic field. The stochastic element stiffness matrix may be expressed as

$$K^{(el)} = K_o^{(el)} + X_o^{(el)} \Delta K_o^{(el)} + X_1^{(el)} \Delta K_1^{(el)} + X_2^{(el)} \Delta K_2^{(el)} \tag{3.86}$$

or

$$K^{(el)} = K_o^{(el)} + \Delta K^{(el)} \tag{3.87}$$

$K_o^{(el)}$ and $\Delta K^{(el)}$ are denoting the stationary and the fluctuating part of the stochastic element stiffness matrix, respectively. $X_o^{(el)}$, $X_1^{(el)}$ and $X_2^{(el)}$ are the so-called weighted integrals which are random variables defined as

$$X_o^{(el)} = \int_0^{L^{(e)}} f^{(e)}(x) \, dx$$

$$X_1^{(el)} = \int_0^{L^{(e)}} x f^{(e)}(x) \, dx \tag{3.88}$$

$$X_2^{(el)} = \int_0^{L^{(e)}} x^2 f^{(e)}(x) \, dx$$

$K_o$ is the mean value of $K^{(el)}$ since the weighted integrals have zero mean and $\Delta K_o$, $\Delta K_1$, $\Delta K_2$ are deterministic matrices (Deodatis 1991).

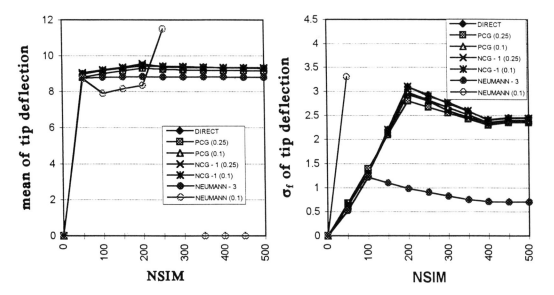

**Figure 3.21** Example 7: Mean value and standard deviation $\sigma_f$ of deflection at top storey

## REPRESENTATION OF THE STOCHASTIC FIELD

Since MCS techniques are used to calculate the response variability of the stochastic structural system, it is necessary to digitally generate sample functions of the 1D-1V stationary Gaussian zero mean homogeneous stochastic field $f(x)$. This is done using the spectral representation method (Shinozuka and Deodatis 1991) taking advantage of the Fast Fourier Transform (FFT) technique in order to reduce the computational effort of the simulation. This is achieved using the formula

$$f^{(j)}(p\Delta x) = Re\left\{\sum_{n=0}^{M-1} B_n e^{inp\frac{2\pi}{M}}\right\} \qquad \begin{array}{l} p = 0, 1, \ldots, M-1 \\ j = 1, 2, \ldots, NSAMP \end{array}$$

$$(3.89)$$

where $Re$ indicates the real part, $M$ defines the number of points at which $f(x)$ process is realized along the element length, $NSAMP$ is the number of samples to be generated. $B_n$ is given by

$$B_n = \sqrt{2}A_n e^{i\phi_n^{(j)}} \qquad n = 0, 1, \ldots, M-1$$

$$(3.90)$$

where $\phi_n^{(j)}$ represents the j-th realizations of the independent random phase angles uniformly distributed in the range $[0 - 2\pi]$, $A_n$ is defined as

$$A_n = (2S_{ff}(n\Delta k)\Delta k)^{1/2}, \qquad n = 0, 1, \ldots, M-1 \qquad (3.91)$$

and

$$\Delta k = \frac{k_u}{N}$$

$k_u$ is the upper cut-off wave number and $N$ is the number of intervals in the discretization of the spectrum. $S_{ff}$ is the two sided power spectral density function defined as

$$S_{ff} = \frac{1}{4}\sigma_f^2 b_f^3 k^2 e^{-b_f k} \tag{3.92}$$

where $\sigma_f$ denotes the standard deviation of the stochastic field, $b_f$ denotes the parameter that influences the shape of the spectrum and hence the scale of the correlation and $k = n\Delta k$.

Using Eq.(3.89) a large number of samples functions (NSIM) is produced for each element of the structure generating a set of stochastic stiffness matrices. The associated structural problem is solved NSIM times, while the response variability can finally be calculated by taking the response statistics of the NSIM simulations.

## SOLUTION METHODS FOR SFE ANALYSIS USING MCS

Eq.(3.87) suggests that the linear problem that has to be solved for each Monte Carlo Simulation may be expressed as

$$(K_o + \Delta K_i)\, u_i = f_i, \qquad i = 1, \ldots, NSIM \tag{3.93}$$

which has the form of the typical reanalysis problem of Eq.(3.79). This problem can be solved in parallel computing environments by both two-level methods discussed in Section 3.6, namely the GSI-DSI-PCG or GSI-DSI-NCG method. It has to be mentioned that the solution of the preconditioning step of Eqs.(3.80) or (3.84) has to be performed several hundred or even thousand times depending on the number of GSI iterations performed and the number of Monte Carlo simulations required.

## NUMERICAL TESTS

The performance of PCG and NCG methods for solving Eq.(3.93) is demonstrated for a twenty storey space frame with 2,400 d.o.f. (Papadrakakis and Papadopoulos 1996). The loads considered here are deterministic and consist of vertical forces equivalent to uniform load of 100 psf ($4.788kN/m^2$) and a basic horizontal pressure of 20 psf ($0.956kN/m^2$). The modulus of elasticity is considered to be a 1D-1V zero mean homogeneous stochastic field with a standard deviation 0.25. The performance of PCG and NCG methods compared to the MCS with a direct skyline solver and the Neumann expansion method without correction is examined in terms of both accuracy and computational efficiency. The tests were performed sequentially on a SG R4000 Indigo workstation.

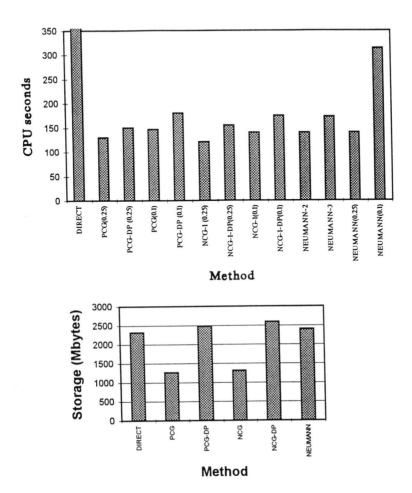

**Figure 3.22** Example 7: (a) total CPU seconds after 100 simulations, (b) storage requirements

Fig.3.21 shows the convergence behaviour and the attained accuracy of the methods for predicting the mean value and the standard deviation of the tip deflection of the frame. The PCG and NCG methods are implemented with mixed precision arithmetic in which all computations are performed in single precision except for the calculation of the residual vector. Numbers in parentheses correspond to the value of the convergence tolerance of the iterative procedure, while the number following the abbreviated name of the Neumann-type method gives the order of expansion of the Neumann series. Fig.3.22a shows the performance of the methods after 100 simulations, while Fig.3.22b depicts the storage requirements for each of the above methods where the abbreviation DP stands for Double Precision arithmetic. Finally, Fig.3.23 gives the number of PCG iterations performed inside a typical Monte Carlo simulation.

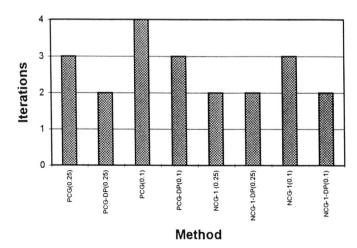

**Figure 3.23** Example 7: CG iterations in a typical Monte Carlo simulation

## ACKNOWLEDGEMENTS

The author acknowledges partial support by the Human Capital and Mobility project of the European Union under grant XCT 93-0390, and partial support by PENED/91ED105, PAVE 92 BE297 of the General Secretariat of Science and Technology of Greece. He also is very grateful to C. Apostolopoulou, S. Bitzarakis, Y. Tsompanakis, V. Papadopoulos, A. Kotsopulos and D. Harbis for their cooperation in implementing the algorithms and obtaining the numerical results presented in this chapter.

## REFERENCES

Bitoulas N. and Papadrakakis M. 1994. An optimized computer implementation of the Incomplete Cholesky Factorization. *Comp. Systems in Engrg.*, 5(3): 265–274.

Concus P., Golub G.H. and Meurant G. 1985. Block Preconditioning for the Conjugate Gradient Method. *SIAM J. Sci. Stat. Comput.*, 6: 220–252.

Deodatis G. 1991. Weighted Integral Method I: Stochastic Stiffness Matrix. *J. Engrg. Mech. ASCE*, 117: 1851–1864.

DoT, 1994. *User manual, VMA Engineering.*

Farhat C. and Roux F.-X. 1991. A Method of Finite Element Tearing and Interconnecting and its Parallel Solution Algorithm. *Internat. J. Numer. Meth. Engrg.*, 32: 1205–1227.

Farhat C., Crivelli L. and Roux F.-X. 1994. Extending Substructure Based Iterative Solvers to Multiple Load and Repeated Analyses. *Comput. Meths. Appl. Mech. Engrg.*, 119: 195–209.

Farhat C. and Roux F.-X. 1994. Implicit Parallel Processing in Structural Mechanics. *Computational Mechanics Advances*, 2: 1–124.

George A. and Liu J.W. 1981. Computer Solution of Large Sparse Positive Definite Systems, *Prentice-Hall Inc., Englewood Cliffs, N.J.*

Han T.Y. and Abel J.F. 1984. Substructure Condensation Using Modified Cholesky Decomposition. *Internat. J. Numer. Meths. Engrg.*, 20: 1959–1964.

Hinton E. and Sienz J. 1993. Fully Stressed Topology Design of Structures Using an Evolutionary Procedure. *Engineering Computations*, 12: 229–244.

Hinton E. and Sienz J. 1996. Adaptive Finite Element Analysis and Shape Optimization, *Saxe-Coburg Publications, Edinburg, U.K.*

Jennings A. and Malik G.M. 1978. The Solution of Sparse Linear Equations by the Conjugate Gradient Method. *Internat. J. Numer. Meth. Engrg.*, 12: 141–158.

Johnson O.G., Michelli C.A. and Paul G. 1983. Polynomial Preconditioners for Conjugate Gradient Calculations. *SIAM J. Numer. Anal.*, 20: 362–376.

Lanczos C. 1952. Solution of Systems of Linear Equations by Minimized Iterations. *J. Res. Nat. Bur. Stands.*, 49: 33–53.

Olhoff N., Rasmussen J. and Lund E. 1992. Method of Exact Numerical Differentiation for Error Estimation in Finite Element Based Semi-Analytical Shape Sensitivity Analyses *Special Report No 10, Institute of Mechanical Engineering, Aalborg University, Aalborg, DK.*

Papadrakakis M. and Bitoulas N. 1993. Accuracy and Effectiveness of PCG Method for Large and Ill-Conditioned Problems. *Comput. Methods Appl. Mech Engrg.*, 109: 219–232.

Papadrakakis M. and Bitzarakis S. 1995. A Dual Substructure Method for Solving Large-Scale Problems on Parallel Computers. *Report 95-1, Institute of Structural Analysis and Seismic Research, NTUA, Athens, Greece.*

Papadrakakis M. and Bitzarakis S. 1996. Domain Decomposition PCG Methods for Serial and Parallel Processing. *Advances in Engineering Software*, 25: 291–307.

Papadrakakis M. and Kotsopulos A. 1995. Domain Decomposition Parallel Techniques for Stochastic Finite Element Analysis. *Report 95-3. Institute of Structural Analysis and Seismic Research, NTUA, Athens, Greece.*

Papadrakakis M. and Papadopoulos V. 1996, Efficient Solution Procedures for the Stochastic Finite Element Analysis Using Monte Carlo Simulation, *Comput. Methods Appl. Mech. Engrg.*, in press.

Papadrakakis M. and Smerou S. 1990. A New Implementation of the Lanczos Method in Linear Problems. *Internat. J. Numer. Meths. Engrg.*, 29: 141–159.

Papadrakakis M. and Tsompanakis Y. 1996. Domain Decomposition Methods for Parallel Solution of Sensitivity Analysis Problems. *Report 96-1, Institute of Structural Analysis and Seismic Research, NTUA, Athens, Greece.*

Papadrakakis M., Tsompanakis Y., Hinton E. and Sienz J. 1996. Advanced Solution Methods in Topology Optimization and Shape Sensitivity Analysis. *Engineering Computations Vol.13*, 5: 57–90.

Parlett B.N. 1980. A New Look at the Lanczos Algorithm for Solving Symmetric Systems of Linear Equations. *Linear Algebra Applics*, 29: 323–346.

Przemieniecki J. S. 1963. Matrix structural analysis of substructures. *AIAA J.*, 1: 138–147

Roux F.-X. 1995. Private Communication.

Saad Y. 1985. Practical Use of Polynomial Preconditionings for the Conjugate Gradient Method. *SIAM J. Sci. St. Comput.* 6: 865–881.

Shinozuka M. and Deodatis G. 1988. Response Variability of Stochastic Finite Element Systems. *J. Engrg. Mech. ASCE* 114: 499–519.

Shinozuka M. and Deodatis G. 1991. Simulation of Stochastic Processes by Spectral Representation Method. *Appl. Mech. Rev.*, 44: 191–203.

Simon M.D. 1984. The Lanczos Algorithm with Partial Reorthogonalization. *Math. Comput.*, 42: 115–142.

# 4

# Parallel Adaptive Multigrid Methods for Elasticity, Plasticity and Eigenvalue Problems

I. D. Parsons[1]

## 4.1 INTRODUCTION

The emergence of multigrid methods represents one of the more interesting recent developments in numerical analysis. Their evolution can be regarded as part of the wider quest to find efficient iterative algorithms capable of solving practical problems. The research community has already used this family of algorithms to solve a wide range of problems in science and engineering. In fact, multigrid methods have generated a cult following over the last three decades, with specialized books, conferences and digital libraries available to anyone wishing to study and utilize their potential.

The algorithm is based on a simple observation: basic iterative methods are quick to reduce the high frequency components of the error in an approximate solution, but slow to reduce the low frequency components. These low frequency components can be represented on a coarse grid, where solution is relatively inexpensive. The combination of fine grid smoothing and coarse grid correction forms the basis of multigrid methodology. In addition, we can use the multigrid method to solve the coarse grid correction equation, since the smooth fine mesh errors have a significant amount of high frequency content on the coarse grid. This recursive approach has the appealing feature that a problem can be solved in $O(n)$ operations, where $n$ is the number of unknowns on the finest grid. This makes it particularly suitable for large-scale discretizations.

An interesting historical account of the algorithm's early development can be found in (Stuben & Trottenberg 1982). Southwell (1940, 1946) was one of the first researchers to attempt improvements to the basic Jacobi and Gauss-Seidel iteration (or relaxation) schemes; special block and group relaxation methods were developed, involving simultaneous relaxation of small groups of unknowns. Fedorenko (1961) and Bachvalov (1966) were the first authors to propose a true multigrid algorithm explicitly

---

[1] Department of Civil Engineering, University of Illinois, Urbana, IL 61801, U.S.A.

*Parallel Solution Methods in Computational Mechanics* edited by M. Papadrakakis

incorporating a coarse grid. However, it was not until Brandt published his landmark paper (1977) that the true efficiency of the method was understood. Since then, the numbers of papers, conferences and books on the general subject have been growing at an increasing rate.

Today, the field has advanced to a point where it is impossible for a single work to adequately review all of the relevant literature. The reader is directed to the bibliography (Douglas & Douglas 1996) and digital library (Douglas 1996) as starting points for further information. Currently, (Douglas & Douglas 1996) contains a list of over 2,500 articles related to multigrid, multilevel and domain decomposition methods. A brief survey of this list demonstrates the broad range of multigrid research, which includes: mathematical analyses of multigrid methods applied to generic equations (e.g., elliptic, parabolic, hyperbolic, Poisson, biharmonic, Navier-Stokes, Helmholtz), various applications (e.g., oil reservoir simulation, transonic flow, image processing), strategies for parallel processing and algebraic multigrid (multigrid without the geometric grids). However, relatively little work has been published concerning applications specifically directed toward solid mechanics. Interestingly, only about 1% of the references in (Douglas & Douglas 1996) explicitly cover the combination of multigrid and solid mechanics.

The author recommends (Brandt 1977), (Briggs 1987), (Hackbusch 1985) and (Stuben & Trottenberg 1982) as essential reading for the researcher who is interested in studying the basics of multigrid methods. The earliest papers to consider the use of multigrid methods to solve solid mechanics problems were (Barrett *et al.* 1985) and (Craig & Zienkiewicz 1985), appearing in the same conference proceedings. The literature in the realm of solid mechanics today ranges from rigorous mathematical studies (e.g., Brenner & Sung 1992, Kocvara & Mandel 1987, Peisker 1991), to articles concerned with the development of methods for solving practical problems (e.g., Fish *et al.* 1995, Hwang & Parsons 1992a, 1992b, 1992c, 1994, Kacou & Parsons 1993, Kocvara 1993, Mahnken 1995, Misra & Parsons 1994, Parsons & Hall 1990a, 1990b, Ronnau & Parsons 1995). However, we should note that the extensive theory and applications in other fields represent a resource of knowledge that can be used to great advantage in the area of computational solid mechanics.

In this chapter, we present some methods the author has used to solve a variety of solid mechanics problems using multigrid methods. We begin with a discussion of the basic multigrid algorithm used to solve linear matrix equations. This algorithm is then embedded within Newton-Raphson iteration to produce a method for solving history-dependent nonlinear problems. Several methods for computing eigensolutions are then presented, followed by a discussion of multigrid methods incorporating adaptive mesh refinement. Finally, some concluding comments are made regarding the status of multigrid methods as practical tools for the structural engineer.

## 4.2 SOLUTION OF LINEAR MATRIX EQUATIONS

One of the fundamental problems in numerical analysis is the solution of the generic linear matrix equation. In this section, we present the basic multigrid method for solving this problem, followed by some details of its implementation on vector and parallel machines. A comparison is also made between the multigrid method and several preconditioned conjugate gradient methods often used to solve large-scale problems.

## THE BASIC MULTIGRID ALGORITHM

As we have already indicated, the basic idea behind any multigrid algorithm is fairly simple: fine mesh relaxation coupled with coarse mesh correction produce a fast solution to the problem at hand. The basic method for solving linear matrix equations is illustrated below by considering two meshes; subscripts $m$ and $m-1$ denote the fine and coarse meshes, respectively. We want to solve the linear matrix equation

$$\mathbf{K}_m \mathbf{x}_m = \mathbf{f}_m \qquad (4.1)$$

on the fine mesh to a prescribed tolerance, where $\mathbf{K}$, $\mathbf{x}$, and $\mathbf{f}$ are the stiffness matrix, displacement vector and load vector, respectively. One cycle of a two mesh method is as follows.

(i) *Fine mesh relaxation.* Apply a small number ($\mu_1$) of relaxation sweeps to the initial fine mesh approximation $\mathbf{x}_m^{(k)}$ to obtain a new approximate solution, $\bar{\mathbf{x}}_m^{(k)}$. Relaxation is performed so that the error associated with $\bar{\mathbf{x}}_m^{(k)}$ will be dominated by low frequency components, and therefore can be approximated on the coarse mesh.

(ii) *Fine-to-coarse mesh restriction.* The fine mesh residual,

$$\mathbf{r}_m^{(k)} = \mathbf{f}_m - \mathbf{K}_m \bar{\mathbf{x}}_m^{(k)}, \qquad (4.2)$$

is restricted to the coarse mesh to obtain the coarse mesh residual

$$\mathbf{r}_{m-1}^{(k)} = \mathcal{I}_m^{m-1} \mathbf{r}_m^{(k)}, \qquad (4.3)$$

where $\mathcal{I}_m^{m-1}$ is the fine-to-coarse mesh restriction operator.

(iii) *Coarse mesh correction.* The coarse mesh correction, $\Delta\mathbf{x}_{m-1}^{(k)}$, is computed by solving the coarse mesh correction equation

$$\mathbf{K}_{m-1} \Delta\mathbf{x}_{m-1}^{(k)} = \mathbf{r}_{m-1}^{(k)}, \qquad (4.4)$$

where $\mathbf{K}_{m-1}$ is the coarse mesh stiffness matrix. The correction $\Delta\mathbf{x}_{m-1}^{(k)}$ should be a good approximation to the smooth error associated with $\bar{\mathbf{x}}_m^{(k)}$.

(iv) *Coarse-to-fine mesh interpolation.* The coarse mesh correction is interpolated to the fine mesh to obtain the fine mesh correction

$$\Delta\mathbf{x}_m^{(k)} = \mathcal{I}_{m-1}^m \Delta\mathbf{x}_{m-1}^{(k)}, \qquad (4.5)$$

where $\mathcal{I}_{m-1}^m$ is the coarse-to-fine mesh interpolation operator.

(v) *Fine mesh relaxation.* The new fine mesh approximate solution, $\hat{\mathbf{x}}_m^{(k)}$, is computed using the fine mesh correction, i.e.,

$$\hat{\mathbf{x}}_m^{(k)} = \bar{\mathbf{x}}_m^{(k)} + \Delta\mathbf{x}_m^{(k)}. \qquad (4.6)$$

A further small number ($\mu_2$) of relaxation sweeps are performed on the fine mesh starting with $\hat{\mathbf{x}}_m^{(k)}$ to reduce any high frequency error introduced by the interpolation. This produces $\mathbf{x}_m^{(k+1)}$, the new approximate solution to eq. (4.1).

Steps (i) to (v) above are repeated until a converged solution is obtained. Usually, we check whether the approximate solution $\mathbf{x}_m^{(k+1)}$ has converged by computing the residual $\mathbf{r}_m^{(k+1)}$, and stop iterating when

$$\frac{\|\mathbf{r}_m^{(k+1)}\|}{\|\mathbf{f}_m\|} \leq \epsilon_{res}, \tag{4.7}$$

where $\|.\|$ represents the Euclidean norm of a vector and $\epsilon_{res}$ is a user specified tolerance (often $10^{-6}$). We will propose a different termination criterion later in this chapter.

A true multigrid method is obtained by noting that the coarse mesh correction can be computed by introducing a still coarser mesh, and then using $\gamma$ cycles of the basic two mesh method described above to obtain an approximate solution to eq. (4.4). Fig. (4.1) summarizes this basic algorithm. The recursive application of the two mesh method is demonstrated in Fig. (4.2). Different multigrid algorithms can be defined, depending on the number of meshes used and the value of $\gamma$ (i.e., the number of multigrid cycles used to approximate the solution of the coarse mesh correction equation), and are named for their graphical representation. For instance, single cycles of three grid methods with $\gamma = 1$ and 2 are called V- and W-cycles, respectively.

## MULTIGRID COMPONENTS

The linear multigrid method consists of four basic components: the relaxation scheme, the restriction operator, the interpolation operator, and the coarse mesh correction equation. If the method is to succeed, then each one of these components must perform its function correctly. As we will see, the failure of the method in difficult situations can often be traced to the failure of one of the components. This can lead to proposed solutions and improved multigrid performance. We now outline each of the basic components in turn.

---

Given  $\mathbf{x}_m^{(0)}, \mu_1, \mu_2, \gamma$

(a)  $\bar{\mathbf{x}}_m^{(k)} = relax_{MG}^{\mu_1}(\mathbf{x}_m^{(k)})$

(b)  $\mathbf{r}_m^{(k)} = \mathbf{f}_m - \mathbf{K}_m \bar{\mathbf{x}}_m^{(k)}$

(c)  $\mathbf{r}_{m-1}^{(k)} = \mathcal{I}_m^{m-1} \mathbf{r}_m^{(k)}$

(d)  $\Delta \mathbf{x}_{m-1}^{(k)} = MG_{m-1}^{\gamma}$

(e)  $\hat{\mathbf{x}}_m^{(k)} = \bar{\mathbf{x}}_m^{(k)} + \mathcal{I}_{m-1}^{m} \Delta \mathbf{x}_{m-1}^{(k)}$

(f)  $\mathbf{x}_m^{(k+1)} = relax_{MG}^{\mu_2}(\hat{\mathbf{x}}_m^{(k)})$

---

**Figure 4.1** A Multigrid Method for Linear Matrix Equations

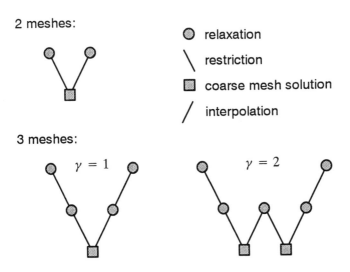

2 meshes:

○ relaxation

\ restriction

■ coarse mesh solution

/ interpolation

3 meshes:

$\gamma = 1$         $\gamma = 2$

**Figure 4.2** Graphical Representations of the Basic Multigrid Method

## The Relaxation Scheme

The function of the relaxation scheme is to rapidly eliminate the high frequency error components from the current approximate fine mesh solution. Many basic iterative

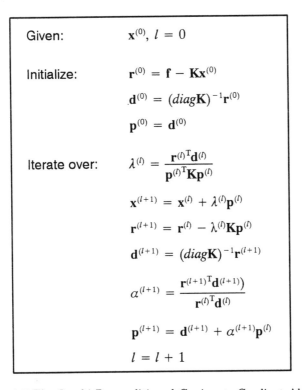

Given:            $\mathbf{x}^{(0)}$, $l = 0$

Initialize:       $\mathbf{r}^{(0)} = \mathbf{f} - \mathbf{K}\mathbf{x}^{(0)}$

                  $\mathbf{d}^{(0)} = (diag\mathbf{K})^{-1}\mathbf{r}^{(0)}$

                  $\mathbf{p}^{(0)} = \mathbf{d}^{(0)}$

Iterate over:     $\lambda^{(l)} = \dfrac{\mathbf{r}^{(l)T}\mathbf{d}^{(l)}}{\mathbf{p}^{(l)T}\mathbf{K}\mathbf{p}^{(l)}}$

                  $\mathbf{x}^{(l+1)} = \mathbf{x}^{(l)} + \lambda^{(l)}\mathbf{p}^{(l)}$

                  $\mathbf{r}^{(l+1)} = \mathbf{r}^{(l)} - \lambda^{(l)}\mathbf{K}\mathbf{p}^{(l)}$

                  $\mathbf{d}^{(l+1)} = (diag\mathbf{K})^{-1}\mathbf{r}^{(l+1)}$

                  $\alpha^{(l+1)} = \dfrac{\mathbf{r}^{(l+1)T}\mathbf{d}^{(l+1)})}{\mathbf{r}^{(l)T}\mathbf{d}^{(l)}}$

                  $\mathbf{p}^{(l+1)} = \mathbf{d}^{(l+1)} + \alpha^{(l+1)}\mathbf{p}^{(l)}$

                  $l = l + 1$

**Figure 4.3** The Jacobi Preconditioned Conjugate Gradient Algorithm

methods perform this function for well-conditioned problems, including the classical Jacobi and Gauss-Seidel iterative methods, and various preconditioned conjugate gradient methods. We have adopted the Jacobi preconditioned conjugate gradient method as our primary smoother, although some of the classical methods will be mentioned in this chapter. Fig. (4.3) contains a summary of this method. Usually, between 5 and 10 relaxation cycles are required to economically smooth fine mesh errors.

### The Coarse-to-Fine Mesh Interpolation Operator

The coarse-to-fine mesh interpolation operator $\mathcal{I}_{m-1}^m$ is associated with the coarsening procedure. The fine meshes used by the author are generated by the refinement (uniform or adaptive) of coarse meshes. For example, this means that a four node quadrilateral element on a coarse mesh can be subdivided into four fine mesh elements, as shown in Fig. (4.4). This procedure enables the interpolation operator to be defined by applying constraints to the fine mesh degrees-of-freedom.

Fig. (4.4) shows sixteen fine mesh quadrilateral elements grouped to form four coarse mesh elements. The constraints that define the fine mesh degrees-of-freedom in terms of the coarse mesh degrees-of-freedom can be written as

$$
\begin{aligned}
x_m^a &= x_{m-1}^a, \\
x_m^e &= \tfrac{1}{2}(x_{m-1}^a + x_{m-1}^b), \\
x_m^f &= \tfrac{1}{4}(x_{m-1}^a + x_{m-1}^b + x_{m-1}^c + x_{m-1}^d),
\end{aligned}
\tag{4.8}
$$

where $x_m^\alpha$ denotes a fine mesh degree-of-freedom at node $\alpha$, and $x_{m-1}^\beta$ denotes the corresponding coarse mesh degree-of-freedom at node $\beta$. This approach can be readily extended to three dimensional elements. Thus, the interpolation operator can be

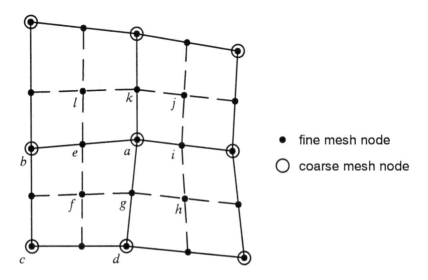

● fine mesh node

○ coarse mesh node

**Figure 4.4** Coarsening Procedure for Four Node Quadrilateral Elements

written in matrix form as

$$\mathcal{I}_{m-1}^m = \mathbf{T}, \tag{4.9}$$

where $\mathbf{T}$ is an interpolation matrix assembled on a coarse mesh element level using expressions such as eq. (4.8).

## The Fine-to-Coarse Mesh Restriction Operator

The restriction operator $\mathcal{I}_m^{m-1}$ transfers residuals (i.e., generalized forces) from the fine mesh to the coarse mesh. If we require that forces on the fine mesh do the same work when they are restricted to the coarse mesh, it follows that the restriction operator can be taken as the transpose of the interpolation operator, i.e.,

$$\mathcal{I}_m^{m-1} = \mathbf{T}^T. \tag{4.10}$$

## The Coarse Mesh Correction Equation

The coarse mesh stiffness matrix must be computed in order to construct and solve the coarse mesh correction equation. This matrix may be defined from the corresponding fine mesh stiffness matrix using the interpolation and restriction operators as

$$\mathbf{K}_{m-1} = \mathbf{T}^T \mathbf{K}_m \mathbf{T}. \tag{4.11}$$

An alternative approach is to assemble the coarse mesh stiffness matrix from individual element stiffness matrices. The construction of $\mathbf{K}_{m-1}$ in this case is independent of $\mathbf{K}_m$, and is assembled from coarse mesh element matrices using the standard form

$$\mathbf{K}_{m-1}^e = \int_{V^e} \mathbf{B}^{e\,T} \mathbf{D}^e \mathbf{B}^e \, dV, \tag{4.12}$$

where the superscript $e$ denotes element quantities, and $\mathbf{B}^e$ and $\mathbf{D}^e$ are the appropriate coarse mesh element strain-displacement and material matrices, respectively.

The coarse mesh correction equation is only solved exactly on the coarsest mesh; two or more cycles of the basic multigrid method are used to approximate the solution to this equation on all of the other coarse meshes. Solution of the coarsest mesh correction equation can be performed using a direct method, or an iterative method such as the relaxation scheme used on the fine meshes.

## 4.3 ALGORITHM BEHAVIOR AND IMPLEMENTATION

Having outlined the basic multigrid method as a technique for solving linear matrix equations, we discuss some important features of the algorithm before considering more complex problems. We briefly examine some problems associated with ill-conditioning, describe an efficient implementation of the algorithm and compare its performance with other iterative methods.

## EFFECTS OF ILL-CONDITIONING

Ill-conditioning has always spelt trouble for iterative methods, causing an increase in the number of cycles required for convergence, or even a complete failure to converge. Multigrid methods are no exception to this rule. Mathematically, ill-conditioning can be measured using $\kappa(\mathbf{A})$, the spectral condition number of a matrix A, i.e.,

$$\kappa(\mathbf{A}) = \|\mathbf{A}\|^* \|\mathbf{A}^{-1}\|^*, \qquad (4.13)$$

where $\|.\|^*$ denotes the norm of a matrix. An ill-conditioned matrix is then one with a high condition number (although high is a subjective term). Physically, ill-conditioning is caused by the presence of stiffnesses of widely varying orders of magnitude. Common causes of ill-conditioning are: small elements, nearly incompressible materials, thin shells and heterogeneous material properties. Here, we will extend our understanding of the term ill-conditioned to include any problem dependent property causing a deterioration in the performance of the multigrid method. It is worth noting that high condition numbers do not necessarily mean trouble for multigrid. For example, successive refinements of a given mesh produce little or no increase in the number of multigrid cycles required, even though the smaller elements on each successive mesh result in higher condition numbers.

A study of some of these factors was presented in (Parsons & Hall 1990a). Conditions of near incompressibility (i.e., values of Poisson's ratio close to $1/2$ for linear elastic materials) were found to slow multigrid convergence. The cause was a deterioration in the smoothing effect of the relaxation; solutions included using different relaxation methods (Jacobi preconditioned conjugate gradient instead of Gauss-Seidel), and using more relaxation cycles on each mesh. Coarse mesh locking was also observed when the coarse mesh stiffness matrix was computed from eq. (4.11), even though reduced integration was used to cure fine mesh locking. This locking is caused by the association of the isochoric modes of the coarse mesh element with non-volume preserving deformation of the constituent fine mesh elements. We therefore recommend explicit computation of the coarse mesh stiffness matrix, particularly when the element level implementation of the method discussed below is considered. Deformations consisting of large amounts of bending deformation are also problematic, because the coarse meshes are generally too stiff. Increasing $\gamma$ or using better elements (e.g., Belytschko & Bindeman 1993) help to cure this problem. Nonuniform meshes tend to require more multigrid cycles than uniform meshes; however, nonuniform meshes should still be used, since greater accuracy can be obtained per unit cost.

We will revisit the deterioration of the smoothing procedure later in this chapter. We will also observe the effect fine discretizations have on the multigrid method. However, one other important source of weakness for multigrid algorithms will not be discussed here: heterogeneous material properties. If the heterogeneity is such that a coarse element consists of fine elements with widely varying material properties, simple interpolation procedures such as those given in eq. (4.8) will not accurately couple the coarse mesh to the fine mesh. Special interpolation methods are then required; the reader is referred to (Behie & Forsyth 1983) and (de Zeeuw 1990) for some examples.

## EFFICIENT IMPLEMENTATION OF THE ALGORITHM

Although it is interesting to study the convergence and behavior of multigrid algorithms, it is equally important to efficiently implement the methods on state-of-the-art computers. At the time of writing, this includes vector and shared memory parallel computers, and scalar RISC workstations. We sidestep the complex issue of implementation on distributed memory computers, noting that the literature contains various proposals that may prove successful on future architectures. We will focus on efficient implementation on vector computers, since experience teaches that similar principles apply to both shared memory machines and RISC workstations.

Examination of the basic components of the method shows that it consists of four basic operations: matrix-vector multiplications, scalar-vector multiplications, vector-vector additions, and vector dot products. The last three operations are found in the relaxation scheme, the residual computation and the addition of the interpolated coarse mesh correction to the fine mesh approximate solution. Contemporary compiler technology is such that optimization of these operations on the machines in question is a non-issue. The matrix-vector multiplications constitute the basis of the restriction and interpolation procedures, and comprise a significant part of the relaxation method. Efficient implementation of this operation is the key to an efficient multigrid code, and must therefore be done with care.

### Element Level Computations

So, our discussion must focus on the calculation of the matrix-vector product $\mathbf{Ay}$, where $\mathbf{A}$ may be the interpolation ($\mathbf{T}$), restriction ($\mathbf{T}^T$), or stiffness ($\mathbf{K}_m$ or $\mathbf{K}_{m-1}$) matrix, and $\mathbf{y}$ may be the fine mesh residual ($\mathbf{r}_m^{(k)}$), coarse mesh correction ($\Delta\mathbf{x}_{m-1}^{(k)}$), or the search direction during conjugate gradient iteration ($\mathbf{p}^{(l)}$). In particular, the most time consuming portion of the multigrid algorithm is the computation of $\mathbf{Kp}^{(l)}$ on the finest mesh. The simplest approach to the computation of $\mathbf{Ay}$ would involve the assembly of $\mathbf{A}$ in either banded or skyline form, followed by the direct computation of $\mathbf{Ay}$. This would produce easily vectorizable code, but would waste a great deal of storage, since many of the terms in the skyline of $\mathbf{A}$ would be zero (especially for large three dimensional problems). A compact storage scheme could be used to eliminate the need to store the zeroes of the assembled $\mathbf{A}$, but this would produce code that could not be vectorized. There remain two other closely related approaches that we will examine in this chapter: element-by-element matrix-vector multiplications, and matrix-free element level computations.

The basic concept behind both of these approaches is to note that $\mathbf{A}$ is an assembly, or summation, of element matrices, which implies that the product $\mathbf{Ay}$ can be written as

$$\mathbf{Ay} = \sum_e \mathbf{A}^e \mathbf{y}^e, \tag{4.14}$$

where the superscript $e$ denotes element versions of the matrix and vector, and summation is implied over all of the elements in the mesh. Three distinct stages are necessary to perform the calculation in this manner: first, gather the components of $\mathbf{y}^e$ from $\mathbf{y}$; second, compute $\mathbf{A}^e \mathbf{y}^e$; third, scatter $\mathbf{A}^e \mathbf{y}^e$ into $\mathbf{Ay}$. Fig. (4.5) demonstrates the approach. Usually, the elements are processed in blocks chosen to maximize code

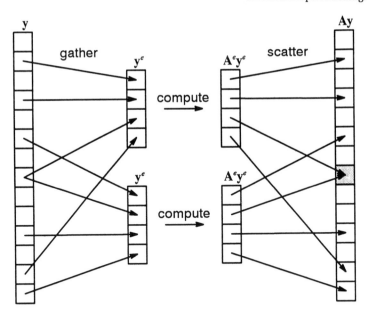

**Figure 4.5** Element Level Computation of Matrix-Vector Products

performance. The gathering and scattering of global and local vectors is accomplished using assembly arrays.

The difference between the two approaches lies in the computation of $\mathbf{A}^e \mathbf{y}^e$. The individual element matrices $\mathbf{A}^e$ are calculated and stored when using the element-by-element method. Actually, only the upper halves of the stiffness matrices are required, and the element forms of the interpolation and restriction matrices are readily available. The product $\mathbf{A}^e \mathbf{y}^e$ is then directly computed. An alternative approach is available when considering products involving stiffness matrices. In this case, matrix-free element level computations are possible, for which the individual element matrices are not explicitly formed and stored. Instead, $\mathbf{K}^e \mathbf{p}^e$ is computed using the structure of the element stiffness matrix given in eq. (4.12) through a three stage process:

(i)    calculate pseudo-strains $\tilde{\epsilon}^e$ at each element Gauss integration point as

$$\tilde{\epsilon}_l^e = \mathbf{B}_l^e \mathbf{p}^e; \tag{4.15}$$

(ii)   calculate pseudo-stresses $\tilde{\sigma}^e$ at the Gauss points as

$$\tilde{\sigma}_l^e = \mathbf{D}_l^e \tilde{\epsilon}_l^e; \tag{4.16}$$

(iii)  calculate the product $\mathbf{K}^e \mathbf{p}^e$ as

$$\mathbf{K}^e \mathbf{p}^e = \sum_l w_l \mathbf{B}_l^{eT} \tilde{\sigma}_l^e, \tag{4.17}$$

where $w_l$ are the Gauss quadrature weights, and summation is implied over all of the Gauss points.

The prefix pseudo is used to emphasize that these stresses and strains are associated with the vector **p**, and are not directly related to the deformation state on the mesh. Great savings in computational time can be realized by using a reduced integration scheme (e.g., one point integration for eight node brick elements), and, if necessary, using a suitable hourglass control scheme, e.g., (Belytschko & Bindeman 1993), which is also a convenient way of avoiding any problems caused by element locking.

It is instructive to compare the computational work involved in these two different strategies by considering the eight node brick element and a linear elastic, homogenous, isotropic material. Explicit computation of the upper half of $\mathbf{K}^e$ requires 6,624 operations if full integration is employed, where one operation represents the combined work of one multiplication and one addition. Computation of $\mathbf{K}^e\mathbf{p}^e$ requires 576 operations. However, computing $\mathbf{K}^e\mathbf{p}^e$ using the matrix-free technique requires 156 operations per Gauss point. So, 1,248 operation are required if full integration is used, but only 312 operations if reduced integration is employed (assuming the stabilization operator requires the same amount of work as a Gauss point computation). Thus, matrix-free computations have great potential for reducing the total computational time. We will see later that the matrix-free method can perform even faster than indicated by a simple operation count, since it preserves locality of data to a greater extent than the element-by-element approach.

## The Grouped Element-by-Element Strategy

Element level computation of matrix-vector products has one important drawback that must be addressed: data dependencies in the scatter phase of the computations that may arise when the algorithm is implemented on vector or shared memory parallel computers. In this phase, the terms in the individual element, or local, vectors are scattered into the appropriate locations of the structure, or global, vector. Fig. (4.5) shows local vectors being scattered into the global vector. A potential data dependency is also shown, whereby two terms from the elements are simultaneously added to the same location in the global vector. This is not possible on many vector machines, and will lead to an error. Therefore, careful consideration must be given to the implementation of the element level computations.

The method used to generate the structured hierarchy of increasing finer meshes required by the multigrid algorithm makes it particularly simple to overcome this problem. The fine meshes are formed by successive refinement of an initial coarse mesh. For example, each quadrilateral element to be refined in the coarse mesh is divided into four elements on the next fine mesh. The data dependency problem is treated by forming groups of elements that are data independent. As a direct result of the mesh refinement, such a group of elements on the coarsest mesh will always produce a group of independent elements on each of the fine meshes. These groups of elements are then processed in vector-concurrent mode by subdividing each group into blocks.

The independent element groups can be formed by requiring that each group contain no elements having the same local equation number assigned to a shared global equation. This avoids any geometric data dependencies caused by the geometry of the mesh under consideration. An example is shown in Fig. (4.6). Three elements $E1$, $E2$ and $E3$ are shown, together with the numbering assigned to the local nodes. In this

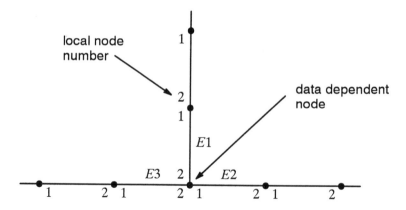

**Figure 4.6** A Finite Element Mesh of Beam Elements with Data Dependencies

case, a data dependency can arise during the scatter phase if elements $E1$ and $E3$ are processed in the same block in vector mode, since the elements have the same local node numbers assigned to a shared node. The solution is to place these two elements in separate groups so that they are considered separately. Thus, this problem can be easily overcome by forming the independent groups on the coarsest mesh. Fig. (4.7) demonstrates how the elements on the coarsest mesh used to model a stiffened plate can be divided into three groups to avoid data dependencies. The reader should note that we need only form data independent groups on the coarsest mesh; refinement of these groups preserves data independence on the fine meshes.

Another source of data dependency arises from the synchronization problems among different processors when parallel processing is invoked. Fig. (4.8) shows four elements ($E1$, $E2$, $E3$ and $E4$) and their local node numbers. If these elements are processed in one group on multiple processors, a data dependency could arise if the code is not designed correctly. For example, simply looping over the elements and scattering local data into the global vector will cause problems if local node 3 for element $E1$ is processed at the same time as local node 1 for element $E3$. The solution we adopt is

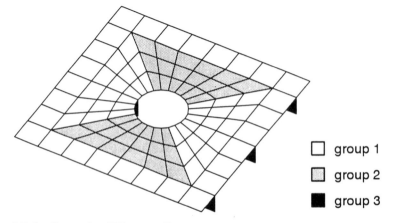

□ group 1
□ group 2
■ group 3

**Figure 4.7** An Example of Element Groups Arranged to Avoid Data Dependencies

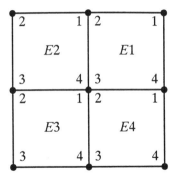

**Figure 4.8** Potential Data Dependency Caused by Processor Synchronization

to scatter information to each of the local nodes in turn, i.e., to scatter data to all the local 1 nodes, then the 2 nodes, and so forth. This enables the elements within a group to be processed in parallel.

## Special Considerations for Interpolation and Restriction

We have focussed on the computation of **Kp**. The interpolation and restriction operations are also expressed as matrix-vector products, and are performed using element-by-element matrix-vector multiplications by dealing with blocks of coarse mesh elements. The contribution to the interpolation matrix **T** from each coarse mesh element is simply a matrix containing the coefficients given in eq. (4.8). Substitution of local quantities into the global vector during the scatter phase is required instead of addition. The implementation of the restriction operator on the element level differs from the interpolation procedure because the residual assigned to a coarse mesh degree-of-freedom will depend on the values of the fine mesh residual at degrees-of-freedom in different coarse mesh elements. For example, consider the plane elements shown in Fig. (4.4). The coarse mesh residual at node $a$ is given by

$$r_{m-1}^a = r_m^a + \tfrac{1}{2}(r_m^e + r_m^g + r_m^i + r_m^k) + \tfrac{1}{4}(r_m^f + r_m^h + r_m^j + r_m^l). \qquad (4.18)$$

This expression can be evaluated for each coarse mesh degree-of-freedom as a summation over all of the coarse mesh elements,

$$\begin{aligned}
r_{m-1}^a =& (r_m^a + \tfrac{1}{2}r_m^e + \tfrac{1}{2}r_m^g + \tfrac{1}{4}r_m^f) + \\
& (r_m^a + \tfrac{1}{2}r_m^g + \tfrac{1}{2}r_m^i + \tfrac{1}{4}r_m^h) + \\
& (r_m^a + \tfrac{1}{2}r_m^i + \tfrac{1}{2}r_m^k + \tfrac{1}{4}r_m^j) + \\
& (r_m^a + \tfrac{1}{2}r_m^k + \tfrac{1}{2}r_m^e + \tfrac{1}{4}r_m^l),
\end{aligned} \qquad (4.19)$$

and zeroing out components of the fine mesh residual as soon as they have been processed for the first time. This element-by-element implementation is simple to vectorize, and produces exact values for the coarse mesh residual in all cases of uniformly refined meshes, and many cases of adaptively refined meshes.

## COMPARISON WITH OTHER ITERATIVE METHODS

We briefly present a comparison of the computational requirements of the multigrid method with other iterative methods by considering the three dimensional Boussinesq problem (which has been used by other workers to benchmark the performance of equation solvers, Hughes *et al.* 1987). An elastic half-space is modeled using a sequence of uniformly refined cubes. Uniform refinement of the coarsest $2 \times 2 \times 2$ element mesh produces a series of meshes that can be used to examine the effect of problem size on the speed of the multigrid method (V-cycles using all of the available coarse meshes and 5 cycles of Jacobi preconditioned conjugate gradient iteration at each smoothing stage), and two preconditioned conjugate gradient algorithms: Jacobi and Crout (Hughes *et al.* 1987) preconditioning. We use the notation $MG$, $JCG$, and $CROUT$ to identify these three algorithms.

We used both element-by-element and matrix-free element level computations for the multigrid and Jacobi preconditioned conjugate gradient methods, denoted by suffixes $(EBE)$ and $(MF)$, respectively. We used reduced integration with the hourglass control described in (Belytschko & Bindeman 1993) for the matrix-free computations. Since Crout preconditioning requires formation and factorization of the element matrices, we computed matrix-vector products explicitly element-by-element when using this preconditioner. All computations were performed on an IBM 780 RISC workstation with 512 MB of RAM.

Figs. (4.9) and (4.10) shows the CPU time and storage requirements of the various algorithms. The matrix-free multigrid method is clearly the fastest, and requires only slightly more storage than the matrix-free Jacobi preconditioned conjugate gradient method. We should note an interesting feature of the matrix-free computations. Although they require 54% of the operations of the element-by-element

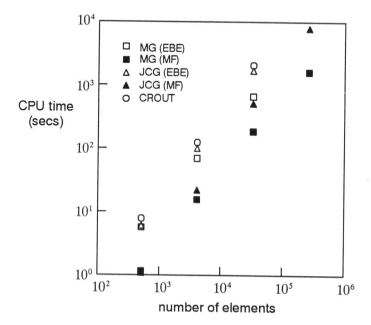

**Figure 4.9** CPU Time Requirements for the Three Dimensional Boussinesq Problem

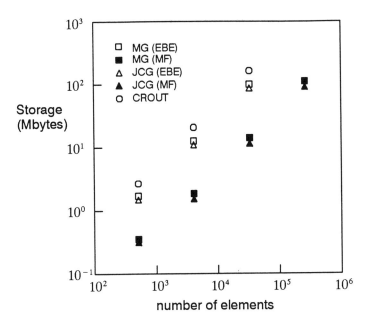

**Figure 4.10** Memory Requirements for the Three Dimensional Boussinesq Problem

multiplications, the CPU time required by the matrix-free multigrid method was never more than 28% of the time required by the element-by-element multigrid method. In other words, the matrix-free technique produced code that was faster by about a factor of 2 than predicted by a simple operation count. This was probably due to the high locality of data inherent with this approach. In addition, the largest problem (262,144 elements, over 800,000 degrees-of-freedom) could not be solved on the IBM when the stiffness matrices were stored, due to the excessive memory requirements.

As well as observing the absolute time and storage required by these different algorithms, we also measured the relationship between time, storage, and problem size. This was done by performing linear regression on the data shown in Figs. (4.9) and (4.10); the results are shown in table 4.1 (where storage $\propto n_e^{\alpha_s}$, time $\propto n_e^{\alpha_t}$, $n_e$ = number of elements). We observe that the storage requirements for all of the iterative methods grow linearly with problem size. However, the CPU

**Table 4.1**   Relationship Between Computational Effort and Problem Size

| Method | $\alpha_s$ | $\alpha_t$ |
|---|---|---|
| MG (EBE) | 0.979 | 1.14 |
| MG (MF) | 0.926 | 1.16 |
| JCG (EBE) | 0.977 | 1.35 |
| JCG (MF) | 0.912 | 1.43 |
| CROUT | 0.987 | 1.34 |

time required by the multigrid methods grows significantly slower than the time required by the conjugate gradient methods. In fact, we see that an almost linear relationship exists between time and problem size for the multigrid algorithms.

## 4.4 TREATMENT OF MATERIAL NONLINEARITIES

We now turn our attention to the solution of problems involving history dependent material nonlinearities. The multigrid method can be used in two basic ways to solve nonlinear problems: either as an equation solver embedded inside Newton iteration, or applied directly to the nonlinear problem (the full approximation storage method, Stuben & Trottenberg 1982). In this chapter, we restrict attention to the combination of the linear multigrid solver with the Newton-Raphson iteration procedure. We should emphasize that this approach involves linearization of the problem on the finest mesh and multigrid solution of the resulting linear matrix equation. The reader is referred to (Fish *et al.* 1995) for a discussion of a full approximation storage multigrid method applied to elasto-plasticity problems, in which no such linearization is required.

*THE NEWTON-MULTIGRID APPROACH*

The algorithm combines the linear multigrid method with an incremental Newton-Raphson solution procedure. Proportional loading is considered; the size of the load steps are controlled by a loading parameter $\lambda$. For each increment of loading, the linear equation

$$\mathbf{K}_T \left( \sigma_i^{(NR)} \right) \Delta \mathbf{U}_i^{(NR+1)} = \lambda_i \mathbf{P}_0 - \mathbf{I}_i^{(NR)} \tag{4.20}$$

is solved for $\Delta \mathbf{U}_i^{(NR+1)}$ using the multigrid method, and a new set of displacements

$$\mathbf{U}_i^{(NR+1)} = \mathbf{U}_i^{(NR)} + \Delta \mathbf{U}_i^{(NR+1)} \tag{4.21}$$

are computed. This procedure is repeated until the internal forces are in equilibrium with the externally applied loads to within some specified tolerance. The subscript $i$ indicates the load step, while the superscript $(NR)$ denotes the Newton-Raphson iteration number. The parameter $\lambda_i$ represents the applied load level and $\mathbf{P}_0$ is the loading pattern. The matrix $\mathbf{K}_T \left( \sigma_i^{(NR)} \right)$ is the updated tangent stiffness matrix (note that we use full Newton iteration), $\Delta \mathbf{U}_i^{(NR+1)}$ is the displacement increment, $\mathbf{U}_i^{(NR)}$ and $\mathbf{U}_i^{(NR+1)}$ are the displacements before and after the current Newton iteration and $\mathbf{I}_i^{(NR)}$ represents the internal forces associated with $\mathbf{U}_i^{(NR)}$. These internal forces are calculated on an element basis as

$$\mathbf{I}_i^{(NR)} = \sum_e \int_{V^e} \mathbf{B}^{eT} \sigma_i^{(NR)\,e} \, dV. \tag{4.22}$$

In order to calculate the internal forces and the tangent stiffness matrix, the stresses must also be computed after each Newton iteration. To demonstrate our method, we restrict attention to a rate independent von Mises material model with the associated flow rule, although other material models can easily be incorporated into this approach. Since we are only considering small deformations, the element tangent stiffness matrix can be written as

$$\mathbf{K}_T^e = \int_{V^e} \mathbf{B}^{eT} \mathbf{D}_T^e \mathbf{B}^e \, dV, \tag{4.23}$$

where a consistent element material tangent matrix, $\mathbf{D}_T^e$, described in (Dodds 1987, Simo & Taylor 1985) is used to preserve the quadratic convergence rate present in the full Newton iteration. The matrix $\mathbf{D}_T^e$ depends on the elastic properties of the material and a number of state variables containing the necessary history of the deformation at each material point. In this case, the state variables depend on the normal to the yield surface in shifted deviatoric stress space, the deviatoric part of the shifted stress, the trial stress based on an assumption of elastic behavior, the instantaneous plastic modulus and the shear modulus. The reader is referred to (Kacou & Parsons 1993) for further details. All that is needed to implement the Newton-multigrid method is an effective technique for determining the coarse mesh stiffness matrices.

## THE COARSE MESH STIFFNESS MATRICES

The coarse mesh stiffness matrix $\mathbf{K}_{m-1}$ depends on the straining history of the material on the finest mesh. In order to compute $\mathbf{K}_{m-1}$, the state variables must be available on the coarse mesh so that $\mathbf{D}_T^e$ can be formed for each coarse mesh element. We directly interpolate the fine mesh variables to the coarse mesh. This means that

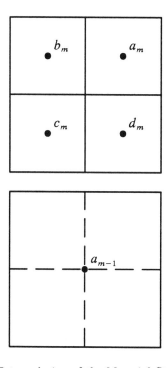

**Figure 4.11** Interpolation of the Material State Variables

the material constitutive law is only considered on the finest mesh employed. This approach can be demonstrated by considering the collection of quadrilateral elements shown in Fig. (4.11). The four fine mesh elements forming a single coarse mesh element are depicted, together with the $1 \times 1$ Gauss integration points on both meshes. The coarse mesh state variables $\phi_{a_{m-1}}$ are obtained by averaging the state variables at the neighboring fine grid integration points located at the fine mesh element centers as shown in Fig. (4.11). Hence,

$$\phi_{a_{m-1}} = \frac{1}{4}(\phi_{a_m} + \phi_{b_m} + \phi_{c_m} + \phi_{d_m}). \tag{4.24}$$

Prior to the determination of the associated state variables, a coarse grid integration point must be specified as being either plastic or elastic. We define a coarse mesh point as being plastic only if all of the neighboring fine mesh points are also plastic. A characteristic history function $h(M)$ is introduced at point $M$ such that

$$h(M) = \begin{cases} 0 & \text{if } M \text{ elastic,} \\ 1 & \text{if } M \text{ plastic.} \end{cases} \tag{4.25}$$

Thus, the state variables at a coarse mesh integration point $M$ only need to be computed if $h(M) = 1$. The history function is also used in the computation of $\mathbf{K}_T \mathbf{p}$ on the element level on each mesh. To avoid troublesome IF statements during the second stage of the matrix-free multiplications (see eq. (4.16)), the material tangent matrix, $\mathbf{D}_T^e$, is rewritten as a function of $h$, so that $h = 0$ produces the elastic $\mathbf{D}_T^e$, and $h = 1$ gives the correct $\mathbf{D}_T^e$ for the local plastic deformation (Kacou & Parsons 1993).

## CONVERGENCE CRITERIA

The accuracy and speed of the Newton-multigrid method depend upon various convergence criteria. Two iteration loops can be identified: one outer loop of Newton-Raphson iterations, and an inner loop over the multigrid cycles. The convergence test applied to the outer loop determines the accuracy of the solution to the nonlinear problem, and examines the ratio of the Euclidean norm of the residual load vector to that of the applied load increment, i.e., Newton-Raphson iteration is terminated when

$$\frac{\|\lambda_i \mathbf{P}_0 - \mathbf{I}_i^{(NR)}\|}{\|\lambda_i \mathbf{P}_0\|} < \epsilon_o, \tag{4.26}$$

where $\epsilon_o$ is the outer loop tolerance specified by the user. The multigrid iterations applied to the solution of the linear matrix eq. (4.20) are terminated when the inner loop convergence criterion is satisfied, i.e., when

$$\frac{\|\lambda_i \mathbf{P}_0 - \mathbf{I}_i^{(NR)} - \mathbf{K}_T \left(\sigma_i^{(NR)}\right) \Delta \mathbf{U}_i^{(k+1)}\|}{\|\lambda_i \mathbf{P}_0 - \mathbf{I}_i^{(NR)}\|} < \epsilon_i, \tag{4.27}$$

where $\Delta \mathbf{U}_i^{(k+1)}$ is the current approximation to $\Delta \mathbf{U}_i^{(NR+1)}$. This latter criterion is equivalent to eq. (4.7) in the linear case.

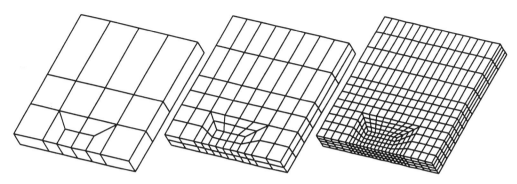

**Figure 4.12** Meshes Used to Model the Three Dimensional Crack Problems

## ALGORITHM PERFORMANCE

We examine the behavior and performance of the Newton-multigrid algorithm by solving a three dimensional, center-cracked panel problem. Fig. (4.12) shows the first three meshes used to model one eighth of the geometry. Continued uniform refinement of these meshes enabled the problem to be solved on a variety of discretizations. We consider isotropic hardening using the standard exponential power law hardening rule with the von Mises material model. The applied load was chosen to be approximately equal to the limit load of the finest mesh shown in Fig. (4.12). All computations were performed on a Convex C240 in vector mode, unless otherwise noted.

Choosing the outer and inner loop tolerances must be done with care. Numerical experiments suggested that the optimum choice for $\epsilon_i$ appears to be between $10^{-1}$ and 1 in many cases. In other words, the total solution time is minimized by using a slack inner tolerance, which has the effect of reducing the number of multigrid cycles used for each linear solve. The reader should note that only the outer tolerance governs the overall accuracy of the nonlinear simulation.

We already have noted that conditions of near incompressibility can adversely effect the performance of the multigrid solver, since the relaxation method takes longer to sufficiently smooth the fine mesh error. This feature is particularly important when elasto-plastic material behavior is modeled, since plastic flow is assumed to be

**Table 4.2**  Effect of Plastic Flow on Multigrid Performance

| CPUsecs | Hardening Exponent | | |
|---|---|---|---|
| $\mu_1, \mu_2$ | 5 | 10 | 20 |
| 1 | 163 | 174 | 340 |
| 5 | 103 | 116 | 150 |
| 10 | 114 | 97 | 159 |
| 20 | 114 | 127 | 152 |

isochoric. With the material model we employed, the hardening exponent controls the amount of plastic flow for a fixed load. Table 4.2 shows the CPU time required to solve the two dimensional version of the three dimensional panel problem using the Newton-multigrid method. Different values of the hardening exponent and different numbers of relaxation cycles were considered. The table clearly shows that the computational effort increases as the plastic flow increases, and that this effect can be somewhat ·reduced by carefully choosing $\mu_1$ and $\mu_2$.

Finally, we examine the parallel performance of the element level implementation of the multigrid method. Fig. (4.13) shows the speed-ups measured on four processors of the Convex when the Newton-multigrid method was used to solve both the two and three dimensional panel problems. The speed-up on $p$ processors, $s_p$, is defined as

$$s_p = \frac{t_1}{t_p},\tag{4.28}$$

where $t_1$ and $t_p$ are the measured computation times on 1 and $p$ processors, respectively. A maximum speed-up of 4.86 was measured for the two dimensional problem. We also observe that the parallel performance increases with the problem size mainly because the higher number of elements offsets the overhead associated with parallelizing the element level computations. Another measure of the parallel performance of a code is the fraction of the code that runs in parallel, $f$, defined from Amdahl's law as

$$f = \left(\frac{s_p - 1}{s_p}\right)\left(\frac{p}{p-1}\right).\tag{4.29}$$

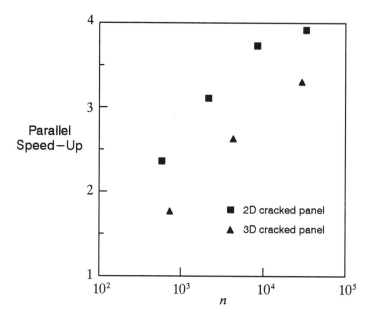

**Figure 4.13** Measured Parallel Speed-Ups for the Cracked Panels

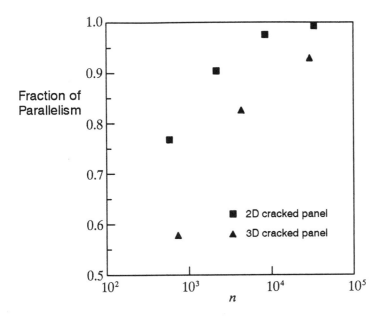

**Figure 4.14** Measured Fraction of Parallelism for the Cracked Panels

Fig. (4.14) shows the fraction of parallelism computed using the speed-ups in Fig. (4.13). As much as 99% of the multigrid solver was executed in parallel on the Convex.

## 4.5 ALGORITHMS FOR EIGENVALUE PROBLEMS

As with nonlinear problems, the basic multigrid idea of fine mesh relaxation followed by coarse mesh correction can be applied to eigenvalue problems in different ways. Two approaches are summarized here: implicit methods using the linear multigrid solver in conjunction with the subspace and Lanczos algorithms, and an explicit multigrid method applying fine mesh relaxation directly to the eigenvalue problem, followed by a rather complicated coarse mesh correction procedure.

Specifically, we are interested in solving the generalized symmetric eigenvalue problem

$$\mathbf{K}\phi = \lambda\mathbf{M}\phi \tag{4.30}$$

encountered in structural dynamics applications. The stiffness and mass matrices of the structure under consideration (both symmetric and positive semi-definite) are $\mathbf{K}$ and $\mathbf{M}$, respectively, and $(\lambda, \phi)$ are the eigensolutions. Usually, only a few ($q$) of the lowest modes of the structure are of interest (i.e., $q \ll n$, where $n$ is the number of degrees-of-freedom in the mesh).

### IMPLICIT MULTIGRID METHODS

Perhaps the most straightforward way in which the multigrid idea can be used to solve eigenvalue problems is to incorporate the multigrid method as a linear matrix

Given $\mathbf{V}_0$

For $j = 1, 2, 3, \ldots$

   (a)  solve $\mathbf{K}\hat{\mathbf{V}}_j = \mathbf{M}\mathbf{V}_{j-1}$ for $\hat{\mathbf{V}}_j$

   (b)  $\mathbf{K}_j^* = \hat{\mathbf{V}}_j^T \mathbf{K}\hat{\mathbf{V}}_j$

   (c)  $\mathbf{M}_j^* = \hat{\mathbf{V}}_j^T \mathbf{M}\hat{\mathbf{V}}_j$

   (d)  solve $\mathbf{K}_j^* \mathbf{\Phi}_j = \Lambda_j \mathbf{M}_j^* \mathbf{\Phi}_j$ for $\mathbf{\Phi}_j, \Lambda_j$

   (e)  $\mathbf{V}_j = \hat{\mathbf{V}}_j \mathbf{\Phi}_j$

   (f)  check for convergence using $\Lambda_j, \mathbf{V}_j$

**Figure 4.15** The Subspace Iteration Method

equation solver in standard eigensolution procedures. We call this family of algorithms implicit methods, and consider both the subspace and Lanczos methods.

### The Subspace Iteration Method

Fig. (4.15) lists the steps in the subspace iteration method (Bathe 1982). A block of $p$ vectors, $\mathbf{V}_j$, are updated until the required $q$ eigenvectors have converged. Often, the block size $p$ is chosen to be greater than $q$ to improve the convergence of the method. Each cycle of the algorithm requires the solution of the $p$ matrix equations

$$\mathbf{K}\hat{\mathbf{V}}_j = \mathbf{M}\mathbf{V}_{j-1} \qquad (4.31)$$

for $\hat{\mathbf{V}}_j$, which will be achieved using the basic multigrid method described previously.

### The Block Lanczos Method

Lanczos algorithms have recently been developed to solve the eigenproblems encountered in structural dynamics. Fig. (4.16) outlines a block Lanczos method applied to eq. (4.30), which is an extension of the single vector method described by Nour-Omid (1987). Given a block of $p$ starting vectors forming the columns of an $n \times p$ matrix $\mathbf{R}$, a sequence of subspaces $\left(\mathbf{R}, \mathbf{K}^{-1}\mathbf{M}\mathbf{R}, (\mathbf{K}^{-1}\mathbf{M})^2\mathbf{R}, \ldots, (\mathbf{K}^{-1}\mathbf{M})^j\mathbf{R}\right)$ is generated. This sequence of $j + 1$ subspaces is then used to produce Ritz vectors to estimate the required eigensolutions using the Rayleigh-Ritz method. To do this, a block of Lanczos vectors, $\tilde{\mathbf{Q}}_j$, must be generated that is associated with each new subspace $(\mathbf{K}^{-1}\mathbf{M})^j\mathbf{R}$ so that the $n \times pj$ matrix

$$\mathbf{Q}_j = \left[\tilde{\mathbf{Q}}_1 | \tilde{\mathbf{Q}}_2 | \ldots | \tilde{\mathbf{Q}}_j\right] \qquad (4.32)$$

is an $\mathbf{M}$-orthonormal basis for the generated sequence of the $j$ subspaces. As a result of the orthogonality properties of the Lanczos vectors, the Rayleigh-Ritz method requires

Given  $\mathbf{R}_0$,  $\tilde{\mathbf{Q}}_0 = 0$

For  $j = 1, 2, 3, \ldots$

(a)  compute  $\mathbf{b}_j$  from  $\mathbf{b}_j \mathbf{b}_j^T = \mathbf{R}_{j-1}^T \mathbf{M} \mathbf{R}_{j-1}$

(b)  $\tilde{\mathbf{Q}}_j = \mathbf{R}_{j-1} \mathbf{b}_j^{-T}$

(c)  solve  $\mathbf{K} \bar{\mathbf{R}}_j = \mathbf{M} \tilde{\mathbf{Q}}_j$  for  $\bar{\mathbf{R}}_j$

(d)  $\hat{\mathbf{R}}_j = \bar{\mathbf{R}}_j - \tilde{\mathbf{Q}}_{j-1} \mathbf{b}_j$

(e)  $\mathbf{a}_j = \tilde{\mathbf{Q}}_j^T \mathbf{M} \hat{\mathbf{R}}_j$

(f)  check for convergence

(g)  $\mathbf{R}_j = \hat{\mathbf{R}}_j - \tilde{\mathbf{Q}}_j \mathbf{a}_j$

**Figure 4.16** The Block Lanczos Algorithm

the solution of the reduced eigenvalue problems

$$\mathbf{T}_j \mathbf{s}_i^{(j)} = \theta_i^{(j)} \mathbf{s}_i^{(j)}, i = 1, \ldots, pj. \tag{4.33}$$

The new approximations to the required eigensolutions are then $(\frac{1}{\theta_i^{(j)}}, \mathbf{Q}_j \mathbf{s}_i^{(j)})$. The block tridiagonal matrix $\mathbf{T}_j$ is defined as

$$\mathbf{T}_j = \begin{pmatrix} \mathbf{a}_1 & \mathbf{b}_2 & & & \\ \mathbf{b}_2^T & \mathbf{a}_2 & \mathbf{b}_3 & & \\ & \ddots & \ddots & \ddots & \\ & & \mathbf{b}_{j-1}^T & \mathbf{a}_{j-1} & \mathbf{b}_j \\ & & & \mathbf{b}_j^T & \mathbf{a}_j \end{pmatrix}, \tag{4.34}$$

where the matrices $\mathbf{a}_k$, $\mathbf{b}_k$, $k = 1, \ldots, j$ are defined in Fig. (4.16).

The algorithm is repeated until converged eigensolutions have been obtained using the columns of $\mathbf{Q}_j$ as trial vectors in the Rayleigh-Ritz procedure. We use a full reorthogonalization procedure, whereby the columns of $\mathbf{R}_j$ are reorthogonalized with respect to the columns of $\mathbf{Q}_j$ after each Lanczos cycle. The multigrid method is used to solve the linear matrix equation (c) in Fig. (4.16), i.e., to solve

$$\mathbf{K} \bar{\mathbf{R}}_j = \mathbf{M} \tilde{\mathbf{Q}}_j \tag{4.35}$$

for $\bar{\mathbf{R}}_j$. In contrast to the subspace iteration method, both the number of Lanczos vectors and the dimension of the reduced system keep growing from step to step. In other words, the storage required by the Lanczos vectors significantly increases as the number of Lanczos steps increases. Therefore, we employ the $s$-step block Lanczos method (Golub & van Loan 1983), in which the block Lanczos algorithm is restarted after every $s$ steps of iteration (i.e., after some eigenvectors have converged), using the vectors that have not yet converged as the initial vectors.

## AN EXPLICIT MULTIGRID METHOD

An alternative to the implicit algorithms is to directly apply fine mesh relaxation and coarse mesh correction to eq. (4.30). This explicit multigrid algorithm is described in detail in (Hwang & Parsons 1992a, 1992b), and is based on an algorithm presented in (Hackbusch 1985). In summary, $\nu_1$ cycles of a suitable relaxation method are applied to an approximate eigensolution $(\lambda_m^{(k)}, \mathbf{v}_m^{(k)})$. This produces a smooth error on the fine mesh, which is also orthogonal to $\mathbf{M}_m \phi_m$. A coarse mesh correction, $\Delta \mathbf{v}_m^{(k)}$, approximating the smooth fine mesh error can be obtained by solving the coarse mesh correction equation

$$(\mathbf{K}_{m-1} - \lambda_{m-1}\mathbf{M}_{m-1})\,\Delta\mathbf{v}_{m-1}^{(k)} = \mathbf{r}_{m-1}^{(k)} - \left({\mathbf{r}_{m-1}^{(k)}}^T \phi_{m-1}\right)\mathbf{M}_{m-1}\phi_{m-1}. \qquad (4.36)$$

Note that the operator in this equation is singular. Eq. (4.36) can be solved to give a unique coarse mesh correction $\Delta\mathbf{v}_{m-1}^{(k)}$ because the right-hand side is orthogonal to $\phi_{m-1}$ and $\Delta\mathbf{v}_{m-1}^{(k)}$ is orthogonal to $\mathbf{M}_{m-1}\phi_{m-1}$. Eq. (4.36) requires the coarse mesh residual (which is restricted from the fine mesh as before), and the coarse mesh eigensolution $(\lambda_{m-1}, \phi_{m-1})$. Once the coarse mesh correction has been computed, it is interpolated to the fine mesh to give a new approximation to the fine mesh eigenvector. This vector can be used with the Rayleigh quotient to give a new fine mesh eigenvalue. The two mesh method is repeated until converged fine mesh eigensolutions have been computed. The algorithm can be extended to compute the first $q$ eigensolutions of eq. (4.30) (Hwang & Parsons 1992a, 1992b).

Fig. (4.17) lists the steps involved in one cycle of this explicit multigrid method to solve for the first eigensolutions $(\lambda_m, \phi_m)$ on a mesh $m$. The fine mesh

---

Given $\nu_1, \mu_1, \mu_2, \gamma$

(a) $\bar{\mathbf{v}}_m^{(k)} = relax_{EMG}^{\nu_1}(\mathbf{v}_m^{(k)}, \lambda_m^{(k)})$

(b) $\bar{\lambda}_m^{(k)} = \dfrac{{\bar{\mathbf{v}}_m^{(k)}}^T \mathbf{K}_m \bar{\mathbf{v}}_m^{(k)}}{{\bar{\mathbf{v}}_m^{(k)}}^T \mathbf{M}_m \bar{\mathbf{v}}_m^{(k)}}$

(c) $\mathbf{r}_m^{(k)} = -(\mathbf{K}_m - \bar{\lambda}_m^{(k)}\mathbf{M}_m)\bar{\mathbf{v}}_m^{(k)}$

(d) $\mathbf{r}_{m-1}^{(k)} = \mathcal{I}_m^{m-1}\mathbf{r}_m^{(k)}$

(e) $\Delta\mathbf{v}_{m-1}^{(k)} = SMG_{m-1}^\gamma$

(f) $\mathbf{v}_m^{(k+1)} = \bar{\mathbf{v}}_m^{(k)} + \mathcal{I}_{m-1}^m \Delta\mathbf{v}_{m-1}^{(k)}$

(g) $\lambda_m^{(k+1)} = \dfrac{{\mathbf{v}_m^{(k+1)}}^T \mathbf{K}_m \mathbf{v}_m^{(k+1)}}{{\mathbf{v}_m^{(k+1)}}^T \mathbf{M}_m \mathbf{v}_m^{(k+1)}}$

(h)  check for convergence using $(\lambda_m^{(k+1)}, \mathbf{v}_m^{(k+1)})$

---

**Figure 4.17** The Explicit Multigrid Method for Eigenvalue Problems

Given $\mathbf{x}_m^{(0)}, \mu_1, \mu_2, \gamma$

(a) $\overline{\mathbf{w}}_m^{(j)} = relax_{SMG}^{\mu_1}(\mathbf{w}_m^{(j)})$

(b) $\overline{\mathbf{r}}_m^{(j)} = \hat{\pi}_m(\mathbf{r}_m^{(k)}) - (\mathbf{K}_m - \lambda_m \mathbf{M}_m)\overline{\mathbf{w}}_m^{(j)}$

(c) $\mathbf{r}_{m-1}^{(j)} = \mathcal{I}_m^{m-1}\overline{\mathbf{r}}_m^{(j)}$

(d) $\Delta\mathbf{w}_{m-1}^{(j)} = SMG_{m-1}^{\gamma}$

(e) $\hat{\mathbf{w}}_m^{(j)} = \overline{\mathbf{w}}_m^{(j)} + \mathcal{I}_{m-1}^{m}\Delta\mathbf{w}_{m-1}^{(j)}$

(f) $\tilde{\mathbf{w}}_m^{(j)} = relax_{SMG}^{\mu_1}(\hat{\mathbf{w}}_m^{(j)})$

(g) $\mathbf{w}_m^{(j+1)} = \pi_m(\tilde{\mathbf{w}}_m^{(j)})$

**Figure 4.18** A Multigrid Method for Singular Matrix Equations

relaxation method is represented as $relax_{EMG}^{\nu_1}$; the method used in this study is the preconditioned conjugate gradient method described by Papadrakakis & Yakoumidakis (1987). The singular coarse mesh correction equation can be solved in several ways. One alternative is to use a suitable iterative method on the coarse mesh. Another approach is to introduce a still coarser mesh, and to use $\gamma$ cycles of a multigrid method developed specifically to solve singular matrix equations. This method is denoted as $SMG_{m-1}^{\gamma}$ in Fig. (4.17), and is outlined in Fig. (4.18).

The algorithm shown in Fig. (4.18) is basically the same as the regular multigrid method for linear equations in Fig. (4.1). However, projection operators have been added in equations (b) and (g) to enforce the necessary orthogonality conditions. The projection operators $\pi_m(.)$ and $\hat{\pi}_m(.)$ are defined as

$$\pi_m(\mathbf{x}_m) = \mathbf{x}_m - \left(\mathbf{x}_m^T \mathbf{M}_m \phi_m\right)\phi_m, \qquad (4.37)$$

producing a vector orthogonal to $\mathbf{M}_m\phi_m$, and

$$\hat{\pi}_m(\mathbf{x}_m) = \mathbf{x}_m - \left(\mathbf{x}_m^T \phi_m\right)\mathbf{M}_m\phi_m, \qquad (4.38)$$

producing a vector orthogonal to $\phi_m$. The singular multigrid method recursively solves the coarse mesh correction equation on mesh $m - 1$. Note that, because of the construction of the coarse mesh correction eq. (4.36), it is necessary to determine the eigensolutions on all of the coarse meshes involved in the solution of the eigenvalue problem on the finest mesh.

## ALGORITHM PERFORMANCE

We now attempt to compare the performance and behavior of these multigrid methods for solving eigenvalue problems. We first examine the effect of a commonly encountered source of ill-conditioning, then compare the computational resources required by these methods, and finally examine the parallel performance of the schemes.

## Effects of Ill-Conditioning

We have already noted that thin shells produce ill-conditioned stiffness matrices. Here, we examine the deterioration in the performance of the multigrid eigensolvers caused by this type of ill-conditioning. As with conditions of near incompressibility, the performance of the relaxation component of the algorithm is adversely effected, and can be restored by increasing the number of relaxation cycles used at each level. To demonstrate this, a simple curved shell problem was solved using all three of the methods. The value of $R/t$ (i.e., the ratio of shell radius to shell thickness) was varied from 50 to 200; higher values of $R/t$ increase the ill-conditioning of the problem. Two mesh algorithms employing $16 \times 16$ and $32 \times 32$ meshes were used to compute the first four eigensolutions of the curved shell. The computations were performed in vector mode on the Convex C240. Table 4.3 shows the total solution times required by the implicit subspace multigrid algorithm for different values of $\mu_1$, $\mu_2$ and $R/t$. Clearly, as $R/t$ is increased, more time is required to solve the problem for fixed values of $\mu_1$ and $\mu_2$. The optimum values for these relaxation parameters increases as $R/t$ increases, i.e., as the problem becomes more ill-conditioned. This is a result of the decrease in the smoothing effect of the relaxation methods caused by the coupling of the different stiffnesses. However, the decrease in the smoothing effect of the relaxation methods is not disastrous, and reasonable performance can still be obtained if the relaxation parameters are adjusted accordingly. Similar observations hold for the implicit Lanczos and explicit multigrid algorithms.

## Parallel Performance

We now examine the parallel performance of the multigrid eigensolvers by computing the first eigensolution of a clamped, simply-supported square plate discretized with a sequence of uniform meshes. The problem was solved on two machines: the four processor Convex C240, and an eight processor Cray Y-MP8/832. Figs. (4.19) and (4.20) show the speed-ups and fraction of parallelism measured on the two machines when the explicit eigensolver was used. Once again, the element level implementation of the multigrid method parallelized extremely well. Interestingly, the calculated fraction of parallelism is consistent between the two computers.

**Table 4.3**   Effect of Thin Shells on Multigrid Performance

| CPUsecs | $R/t$ | | |
|---|---|---|---|
| $\mu_1, \mu_2$ | 50 | 100 | 200 |
| 5 | 288 | 1,517 | 7,505 |
| 10 | 298 | 1,148 | 4,414 |
| 20 | 344 | 1,024 | 2,655 |
| 40 | 497 | 1,370 | 2,412 |
| 80 | 591 | 1,602 | 2,674 |

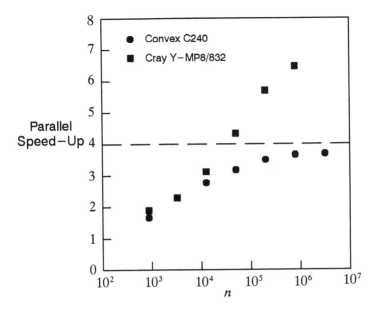

**Figure 4.19** Measured Parallel Speed-Ups for the Square Plate

Tables 4.4 and 4.5 contain additional measures of the performance of the multigrid eigensolver. Here, the stiffened plate originally shown in Fig. (4.7) was used. Fig. (4.21) shows the first three meshes in the hierarchy of uniformly refined meshes generated to model this problem. The tables show the storage and CPU times required to

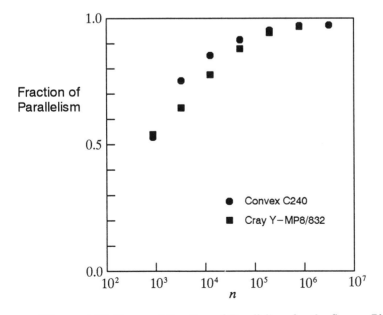

**Figure 4.20** Measured Fraction of Parallelism for the Square Plate

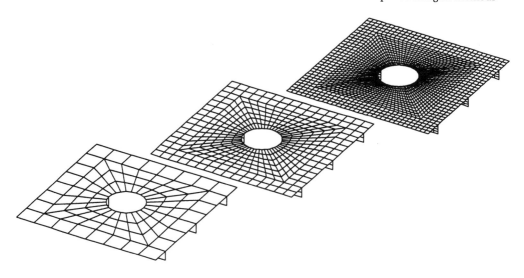

**Figure 4.21** Hierarchy of Meshes for the Stiffened Plate

compute the first eigensolution of the plate on the Convex and Cray. All three multigrid eigensolvers were used on the Convex; only the explicit solver was used on the Cray. The observed computational rate is shown for the Cray (the peak performance on a single processor of the Cray is 330 Mflops), together with the speed-ups measured on both machines. We were sometimes unable to solve all of the problems using the multigrid code on 1 processor (due to the excessive time requirements). We therefore estimate the speed-up by calculating the machine speed-up, $\bar{s}_p$, obtained from the CPU time required by all of the processors, $\bar{t}_p$. The machine speed-up is defined as

$$\bar{s}_p = \frac{\bar{t}_p}{t_p}. \tag{4.39}$$

As a result of overhead associated with multitasking and degraded vector performance, $\bar{t}_p$ is always greater than $t_1$. This means that the machine speed-up $\bar{s}_p$ is greater than the actual speed-up $s_p$. Clearly, the data demonstrate that we were able to solve some extremely large problems using the multigrid eigensolvers, and obtain excellent performance on the shared-memory parallel computers.

**Table 4.4**   Performance of the Explicit Multigrid Method on the Cray Y-MP8/832

| Problem | $n$ | Storage (Mbytes) | $t_1$ (secs) | $t_8$ (secs) | Mflops$_1$ | Mflops$_8$ | $s_8$ |
|---|---|---|---|---|---|---|---|
| Stiffened Plate | 768,000 | 135 | 274 | 50.8 | 152 | 817 | 5.40 |

**Table 4.5** Performance of the Multigrid Methods on the Convex C240

| Problem | Method | $n$ | Storage (Mbytes) | $t_1$ (secs) | $t_4$ (secs) | $s_4$ | $\bar{t}_4$ (secs) | $\bar{s}_4$ |
|---|---|---|---|---|---|---|---|---|
| Stiffened Plate | subspace | 3,059,712 | 322 | – | 6,253 | – | 24,481 | 3.92 |
| Stiffened Plate | Lanczos | 3,059,712 | 462 | – | 34,500 | – | 135,439 | 3.93 |
| Stiffened Plate | explicit | 3,059,712 | 368 | 17,492 | 4,759 | 3.68 | 18,661 | 3.92 |

## Comparison of the Multigrid Methods

Table 4.5 suggests that the explicit multigrid eigensolver is the fastest of the three methods described in this chapter. Fig. (4.22) confirms this observation by presenting the CPU times required to extract the first eigensolution of the stiffened plate on four processors of the Convex using the three methods. Over the range of problem sizes considered, the explicit method is always the fastest. To verify we were obtaining the theoretical optimum performance of the multigrid method, linear regression was applied to the times measured for problems having more than 100,000 degrees-of-freedom. The solution times required by the implicit subspace, implicit Lanczos and explicit methods were found to be proportional to $n^{0.88}$, $n^{0.97}$ and $n^{1.07}$, respectively. Once again, we are able to observe the optimum linear proportionality between time and problem size for the multigrid algorithms.

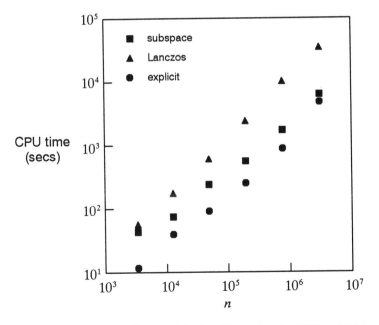

**Figure 4.22** Convex Solution Times for the Stiffened Plate

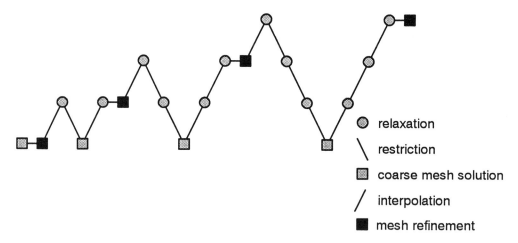

**Figure 4.23** Graphical Representation of the Adaptive Multigrid Method

## 4.6 ADAPTIVE MULTIGRID METHODS

So far, we have used hierarchies of uniformly refined meshes to demonstrate the implementation and performance of various multigrid methods. However, it has widely been recognized that multigrid methods offer the greatest potential when combined with adaptive mesh refinement schemes to create the necessary meshes (Brandt 1977). Here, we describe some of our experiences with multigrid methods used to solve linear problems using $h$−adaptivity; more details are given in (Misra & Parsons 1994).

### H-ADAPTIVE MULTIGRID METHODS

Fig. (4.23) shows a schematic representation of the $h$−adaptive multigrid algorithm used in this study. This diagram is essentially an extension of the fixed mesh methods shown in Fig. (4.2). V-cycles are used to solve the problem on a given mesh. Once the solution has been calculated, an error estimator is used to produce a new mesh. This procedure is continued until the problem has been solved on a mesh that yields a solution within the user specified tolerance.

### Coarse Mesh and Grid Selection

Fig. (4.24) shows an example of the meshes generated and used by the adaptive multigrid method for a hypothetical two dimensional problem. The top row of the diagram represents the sequence of finite element meshes generated by the mesh refinement scheme. Each mesh may contain elements produced at a different level of mesh refinement, i.e., the elements in a particular mesh may be of different generation levels. For example, the third mesh consists of zero, first and second generation elements. Note that the generation level does not necessarily correspond to the refinement level (e.g., second generation elements are produced during the third refinement). As we will see later, it is sometimes more effective to build a sequence

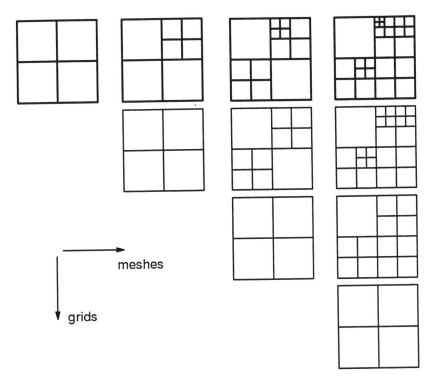

**Figure 4.24** Sequence of Meshes and Grids Used in the Adaptive Multigrid Algorithm

of grids for each actual finite element mesh by regular unrefinement of the original mesh. We now use the term grid to mean a finite element mesh produced by unrefinement of an actual mesh, rather than a mesh produced by refinement of an existing mesh. These grids are shown below the four different finite element meshes.

## Full and Selective Relaxation

During a multigrid cycle, relaxation is used on each level to smooth the error associated with the current mesh equation. We have used two different relaxation strategies: full and selective relaxation. Full relaxation involves modifying all of the degrees-of-freedom in the finite element mesh. On the other hand, only degrees-of-freedom associated with elements not present on the coarser meshes are changed when selective relaxation is employed. Relaxation cycles over the entire mesh may be a waste of effort in cases where the mesh refinement is localized, particularly since the solution will be smooth everywhere except in the refined regions. This relaxation scheme is illustrated in Fig. (4.25), showing newly created elements as shaded regions over which relaxation is performed.

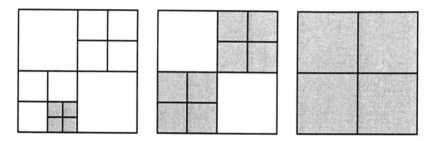

Figure 4.25 Subdomains for Selective Relaxation on Different Meshes

## Convergence Criteria

The adaptive multigrid algorithm requires two different convergence criteria: one to terminate mesh refinement, and a second to terminate solution of the finest mesh equation using the multigrid method. Mesh refinement proceeds until the error estimator on mesh $m$, $\eta_m$, is less than a user specified tolerance. Iterative solution of the finest mesh matrix equation is stopped using two different methods.

The method commonly used to terminate an iterative method was described previously in this chapter. As shown in eq. (4.7), the residual on the finest mesh is computed and compared with the applied loads. Iteration is terminated when the ratio of residual to load is less than a specified tolerance. Using this approach, solution is often possible to within machine accuracy. An alternative approach used here is to compute the error estimator associated with the current approximate solution on the current mesh,

$$\eta_m^{(k)} = \eta_m(\mathbf{x}_m^{(k)}), \tag{4.40}$$

and to stop iterating when $\eta_m^{(k)}$ is minimized. This will yield a solution to within the discretization error on the current mesh.

## PARALLEL IMPLEMENTATION OF THE ALGORITHM

The adaptive multigrid algorithm can be efficiently implemented using the same element level philosophy described earlier. However, one important point must be given careful consideration: the application of constraints. Fig. (4.26) shows a simple mesh produced by $h$−refinement. The nodes (and degrees-of-freedom) can be divided into two groups: nodes constrained to ensure interelement displacement continuity, and unconstrained nodes. The constraints must be enforced when computing matrix-vector products on the element level.

All of the degrees-of-freedom in the mesh, $\mathbf{x}_{all}$, can be partitioned into constrained ($\mathbf{x}_{con}$) and unconstrained, or active, ($\mathbf{x}_{act}$) degrees-of-freedom so that

$$\mathbf{x}_{all} = \left( \frac{\mathbf{x}_{act}}{\mathbf{x}_{con}} \right) = \mathcal{T}\mathbf{x}_{act}, \tag{4.41}$$

where the matrix $\mathcal{T}$ represents the necessary constraints. Then the stiffness matrix for the active degrees-of-freedom, $\mathbf{K}_{act}$, can be written as

$$\mathbf{K}_{act} = \mathcal{T}^T \mathbf{K}_{all} \mathcal{T} = \mathcal{T}^T \sum_e \mathbf{K}_{all}^e \mathcal{T}, \tag{4.42}$$

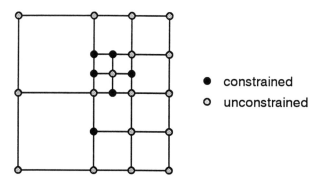

**Figure 4.26** Constrained and Unconstrained Nodes

where $\mathbf{K}_{all}$ is the stiffness matrix for the entire mesh. The matrix-vector product $\mathbf{K}_{act}\mathbf{x}_{act}$ which must be computed during relaxation can then be written as

$$\mathbf{K}_{act}\mathbf{x}_{act} = \mathcal{T}^T \sum_e \mathbf{K}^e_{all} \mathcal{T}\mathbf{x}_{act}. \tag{4.43}$$

This product can be evaluated in parallel by working from right to left in eq. (4.43) using the following three steps:

(i)   compute the vector $\mathbf{p}$

$$\mathbf{p} = \mathcal{T}\mathbf{x}_{act} \tag{4.44}$$

    by looping over all of the constrained degrees-of-freedom;

(ii)  compute the vector $\mathbf{q}$

$$\mathbf{q} = \sum_e \mathbf{K}^e_{all}\mathbf{p}^e \tag{4.45}$$

    by looping over all of the elements in the mesh (either using element-by-element or matrix-free methods);

(iii) compute the vector $\mathbf{K}_{act}\mathbf{x}_{act}$

$$\mathbf{K}_{act}\mathbf{x}_{act} = \mathcal{T}^T\mathbf{q} \tag{4.46}$$

    by once again looping over all of the constrained degrees-of-freedom.

## ALGORITHM PERFORMANCE

We used the standard L-shape problem employed by many researchers to examine the performance of the parallel adaptive multigrid algorithm. Fig. (4.27) shows some of the meshes we employed: the coarsest mesh, and the meshes after five and ten refinements. The Zienkiewicz-Zhu error estimator (Zienkiewicz & Zhu 1987) terminated mesh refinement and identified elements for refinement. All of the results presented below were obtained using a four processor Convex C240. Unless noted otherwise, we adopted a full relaxation scheme, and used the meshes created during the refinement process, rather than the grids described above.

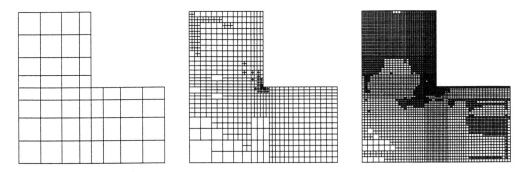

Figure 4.27 Sample Meshes Used for the L-Shaped Problem

Figs. (4.28) and (4.29) show the convergence behavior of the multigrid algorithm in terms of the residual norm and the error estimator. Fig. (4.28) shows that the residual is reduced to six orders of magnitude less than the force vector after about four multigrid cycles on each mesh considered. Fig. (4.29), on the other hand, demonstrates that the error estimator associated with the approximate solution reaches a minimum value after just a single multigrid cycle; i.e., the multigrid algorithm produces a solution to within discretization error in just one cycle. This important property of the multigrid method can be used to minimize the total solution time, since all approximate solutions within the discretization error associated with a given mesh are equally valid.

Fig. (4.30) shows plots of the CPU times required to solve the L-shaped problem on the increasingly finer meshes. The times required to solve the problem using the

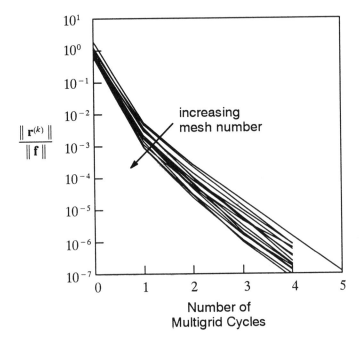

Figure 4.28 Convergence in the Residual for Different Meshes

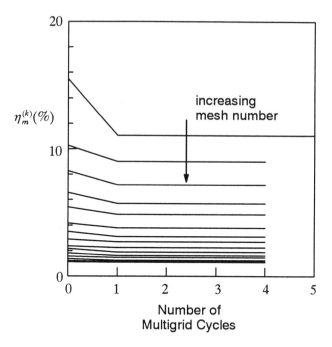

**Figure 4.29** Convergence in the Error Estimator for Different Meshes

residual norm with $\epsilon_{res}$ set at $10^{-6}$ (denoted as $MG$), and the times required when only one multigrid cycle was employed on each mesh (denoted as $MG_1$) are both shown. Clearly, a great deal of time can be saved using the single cycle strategy. Solution on the finest mesh ($n=$ 141,158) required 809 and 204 CPUsecs using the $MG$ and $MG_1$ methods, respectively. A least squares fit was applied to the measured solution times for the problems with more than 10,000 degrees-of-freedom to estimate the relationship between problem size and solution time. The solution times for the single and multiple cycle schemes were found to be both proportional to $n^{0.88}$. This sub-linear dependence on the problem size is due to the parallel performance of the algorithm (the larger problems gain more from the parallelism than the smaller ones).

Table 4.6 contains data comparing the performance of different solvers used to obtain a solution to the L-shaped problem with an estimated error of less than 2%. The problem size $n$, the CPU time and the measured speed-ups on four processors are shown for four different methods. Three multigrid methods were employed: full relaxation using the coarse meshes ($MG_1$), full relaxation using the coarse grids ($MG_1 - F$) and selective relaxation using the coarse grids ($MG_1 - S$). The subscript 1 indicates that only one multigrid cycle was performed on each mesh. In addition, the Jacobi preconditioned conjugate gradient method ($JCG$) was used to solve the matrix equation on each mesh to a tolerance $\epsilon_{res}$ of $10^{-6}$. The speed of the multigrid solvers is apparent in comparison to the conjugate gradient method. The $MG_1$ and $MG_1 - S$ methods appear to be the fastest; the additional time spent building the coarse grids in the $MG_1 - S$ method seems to be roughly balanced by the extra relaxation time associated with the $MG_1$ method.

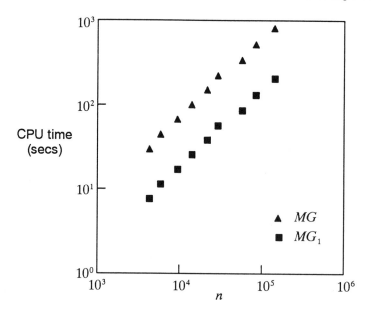

**Figure 4.30** CPU Times for Different Refinement Levels

## 4.7 CONCLUDING REMARKS

In this chapter, we have attempted to demonstrate the implementation of some multigrid methods for solving a variety of structural mechanics problems. Often, a linear relationship is observed between problem size and computational effort (CPU time and memory), making this approach ideal for large-scale simulations. Element level implementation of the crucial matrix-vector products appears to be optimal for a variety of computer architectures currently in use. Unfortunately, the need to use a sequence of meshes, rather than a single fine mesh, makes the multigrid approach inconvenient. However, the combination of adaptive mesh refinement and multigrid solution would seem to lessen the burden of coarse mesh production, since the meshes would be automatically generated by the refinement scheme. As practicing structural engineers continue to make increased demands for accurate simulation of complex structural behavior, multigrid methods will undoubtedly become more widely used, perhaps even prompting commercial developers to invest the effort in rewriting trusted code in order to benefit from the performance of these methods.

**Table 4.6**   Comparison of Different Solvers

|           | $MG_1$  | $MG_1 - F$ | $MG_1 - S$ | $JCG$  |
|-----------|---------|------------|------------|--------|
| $n$       | 29,394  | 29,394     | 29,394     | 30,470 |
| CPU secs  | 56      | 132        | 45         | 2,839  |
| Speed–Up  | 3.47    | 3.70       | 3.42       | 3.82   |

## ACKNOWLEDGEMENTS

The author is grateful for the financial support provided by the National Science Foundation through grant numbers EET-8808190 and ECS-9157304; the support of program manager Dr. George Lea is particularly appreciated. Dr. Bahram Nour-Omid provided helpful advice regarding the block Lanczos algorithm. Computer time was provided by the Pittsburgh Supercomputing Center, the Computing Services Office at the University of Illinois and the Center for Supercomputing Research and Development, also at the University of Illinois. This chapter would not have been possible without the efforts of Professor John F. Hall, Dr. Tsanhuei Hwang, Dr. Sylvain Kacou, Dr. Himanshu Misra, Andrew Ronnau, Ali Namazifard and Hector Estrada.

## REFERENCES

N. S. Bachvalov, 1966. On the convergence of a relaxation method with natural constraints on the elliptic operator. *USSR Computational Mathematics and Mathematical Physics* 6(5), 101-135.

K. E. Barrett, D. M. Butterfield, S. E. Ellis, C. J. Judd and J. H. Tabor, 1985. Multigrid analysis of linear elastic stress problems. *Multigrid methods for integral and differential equations*, D. J. Paddon and H. Holstein, eds., The Institute of Mathematics and its Applications Conference Series 3, Clarendon Press, 263-282.

K. J. Bathe, 1982. *Finite element procedures in engineering analysis.* Prentice-Hall.

A. Behie and P. A. Forsyth, 1983. Multi-grid solution of the pressure equation in reservoir simulation. *Society of Petroleum Engineers Journal* 23, 623-632.

T. Belytschko and L. P. Bindeman, 1993. Assumed strain stabilization of the eight node hexahedral element. *Computer Methods in Applied Mechanics and Engineering* 105, 225-260.

A. Brandt, 1977. Multi-level adaptive solutions to boundary-value problems. *Mathematics of Computation* 31, 333-390.

S. C. Brenner and L. Y. Sung, 1992. Linear finite element methods for planar linear elasticity. *Mathematics of Computation* 59, 321-338.

W. L. Briggs, 1987. *A multigrid tutorial.* SIAM.

A. W. Craig and O. C. Zienkiewicz, 1985. A multigrid algorithm using a hierarchical finite element basis. *Multigrid methods for integral and differential equations*, D. J. Paddon and H. Holstein, eds., The Institute of Mathematics and its Applications Conference Series 3, Clarendon Press, 301-312.

R. H. Dodds, 1987. Numerical techniques for plasticity computations in finite element analysis. *Computers and Structures* 26, 767-779.

C. C. Douglas, 1996. *MGNet.* http://na.cs.yale.edu/mgnet/www/mgnet.html.

C. C. Douglas and M. B. Douglas, 1996. *MGNet bibliography.* mgnet/bib/mgnet.bib on anonymous ftp server casper.cs.yale.edu. Yale University, Department of Computer Science, New Haven, CT.

R. P. Fedorenko, 1961. A relaxation method for solving elliptic difference equations. *USSR Computational Mathematics and Mathematical Physics* 1(4), 1092-1096.

J. Fish, M. Pandheeradi and V. Belsky, 1995. An efficient multilevel solution scheme for large scale non-linear systems. *International Journal for Numerical Methods in Engineering* 38, 1597-1610.

G. H. Golub and C. F. van Loan, 1983. *Matrix computations.* The Johns Hopkins University

Press

W. Hackbusch, 1985. *Multi-grid methods and applications.* Springer.

T. J. R. Hughes, R. M. Ferencz and J. O. Hallquist, 1987. Large-scale vectorized implicit calculations in solid mechanics on a Cray X-MP/48 utilizing EBE preconditioned conjugate gradients. *Computer Methods in Applied Mechanics and Engineering* 61, 215-248.

T. Hwang and I. D. Parsons, 1992a. A multigrid method for the generalized symmetric eigenvalue problem: part I - algorithm and implementation. *International Journal for Numerical Methods in Engineering* 35, 1663-1676.

T. Hwang and I. D. Parsons, 1992b. A multigrid method for the generalized symmetric eigenvalue problem: part II - performance evaluation. *International Journal for Numerical Methods in Engineering* 35, 1677-1696.

T. Hwang and I. D. Parsons, 1992c. Multigrid solution procedures for structural dynamics eigenvalue problems. *Computational Mechanics* 10, 247-262.

T. Hwang and I. D. Parsons, 1994. Parallel implementation and performance of multigrid algorithms for solving eigenvalue problems. *Computers and Structures* 50, 325-336.

S. Kacou and I. D. Parsons, 1993. A parallel multigrid method for history dependent elastoplasticity computations. *Computer Methods in Applied Mechanics and Engineering* 108, 1-21.

M. Kocvara and J. Mandel, 1987. A multigrid method for three-dimensional elasticity and algebraic convergence estimates. *Applied Mathematics and Computation* 23, 121-135.

M. Kocvara, 1993. Adaptive multigrid technique for three-dimensional elasticity. *International Journal for Numerical Methods in Engineering* 36, 1703-1716.

R. Mahnken, 1995. Newton-multigrid algorithm for elasto-plastic/viscoplastic problems. *Computational Mechanics* 15, 408-425.

H. Misra and I. D. Parsons, 1994. Adaptive finite element multigrid methods on parallel computers. *Advances in parallel and vector processing for structural mechanics,* B. H. V. Topping and M. Papadrakakis, eds., 75-82.

B. Nour-Omid, 1987. The Lanczos algorithm for solution of large generalized eigenproblems. Chapter 6 of *The finite element method,* T. J. R. Hughes, Prentice-Hall.

M. Papadrakakis and M. Yakoumidakis, 1987. A partial preconditioned conjugate gradient method for large eigenproblems. *Computer Methods in Applied Mechanics and Engineering* 62, 195-207.

I. D. Parsons and J. F. Hall, 1990a. The multigrid method in solid mechanics: part I - algorithm description and behaviour. *International Journal for Numerical Methods in Engineering* 29, 719-738.

I. D. Parsons and J. F. Hall, 1990b. The multigrid method in solid mechanics: part II - practical applications. *International Journal for Numerical Methods in Engineering* 29, 739-754.

P. Peisker, 1991. A multigrid method for Reissner-Mindlin plates. *Numerische Mathematik* 59, 511-528.

A. Ronnau and I. D. Parsons, 1995. Suppressing singularities when computing critical points using multigrid methods. *Computational mechanics '95,* S. N. Atluri, G. Yagawa, and T. A. Cruse, eds., Proceedings of the International Conference on Computational Engineering Science, Springer, 1517-1522.

J. C. Simo and R. L. Taylor, 1985. Consistent tangent operators for rate-independent elastoplasticity. *Computer Methods in Applied Mechanics and Engineering* 48, 1101-1118.

R. V. Southwell, 1940. *Relaxation methods in engineering science.* Oxford University Press

R. V. Southwell, 1946. *Relaxation methods in theoretical physics.* Clarendon Press.

K. Stuben and U. Trottenberg, 1982. Multigrid methods: fundamental algorithms, model problem analysis and applications. *Multigrid methods,* W. Hackbusch and U. Trottenberg, eds., Lecture Notes in Mathematics 960, Springer, 1-176.

P. M. de Zeeuw, 1990. Matrix-dependent prolongations and restrictions in a blackbox multigrid solver. *Journal of Computational and Applied Mathematics* 33, 1-27.

O. C. Zienkiewicz and J. C. Zhu, 1987. A simple error estimator and adaptive procedure for practical engineering analysis. *International Journal for Numerical Methods in Engineering* 24, 337-357.

# 5

# Accuracy, Stability and Parallelization of Multi-Time Step Intergration Schemes in Explicit Structural Dynamic Analysis

Sanjeev Gupta[1] and Martin Ramirez[2]

## ABSTRACT

In this chapter, an accuracy or error analysis of multi-time step explicit integration schemes using nodal partitioning is presented for structural dynamic systems. The results of numerical experimentation are provided, which suggest that multi-time step integration should not be used indiscriminately. In general, it is observed that a subcycling solution can offset the accuracy otherwise achieved by the introduction of small elements into the mesh. In many cases, the accuracy of a solution for a mixed element mesh, integrated with multi-time step schemes may even be *less* than the accuracy of a solution for a mesh with large elements and integrated with the large time step only. Thus, while selected finer spatial discretization is used to further reduce the errors in the solution, selected coarser temporal discretization should be used very carefully so as to not induce those errors back into the solution. A methodology is also discussed for implementing multi-time step algorithm on parallel computers.

## 5.1 INTRODUCTION

In explicit finite element analysis, because of the stability criterion, the time step is always limited by the critical time step - governed by the stiffest element i.e., the element having the maximum frequency (Cook, Malkus, and Plesha (1989)). Typical large-scale finite element meshes are composed of variable size elements. Usually, due to the configuration of the structure,

---

[1] Sr. Project Engineer, Engineering Analysis Services, Inc., Bingham Center, 30800 Telegrah Road, Suite 3700, Bingham Farms, MI 48025, USA. He received his Ph.D. from The Johns Hopkins University, Baltimore, MD 21218, USA.

[2] Visiting Scholar, Engineering System Research Center, 3112 Etcheverry Hall, University of California, Berkeley, CA 94720, USA.; also Visiting Professor, The Johns Hopkins University, Baltimore, MD 21218, USA.

*Parallel Solution Methods in Computational Mechanics* edited by M. Papadrakakis
© 1997 John Wiley & Sons Ltd

geometric or material heterogeneity, or localization of high activity areas, one or more regions of the mesh require small or stiff elements for better accuracy. In both explicit and implicit analyses, areas of the mesh where stress concentration occurs may require small elements and small time steps. However the requirements of one region limit the time step for the rest of the mesh where the elements are relatively flexible; this increases the computational time of the whole mesh significantly. Subcycling of time steps for hyperbolic as well as parabolic systems has been proposed in recent years to reduce some of the computational demand (Belytschko, Yen and Mullen (1979)).

In an *explicit-explicit* subcycling procedure, the mesh is divided into two or more submeshes and different time steps are used in each submesh. For each submesh, the time step is determined based upon its own critical time step. The computational advantage results mainly by eliminating many computations in a submesh with a large time step. Such procedures are also known as multi-time step integration methods or variable explicit algorithms. For both first and second order systems, subcycling algorithms can be classified in two general classes as defined by the approach used to partition the finite element mesh. In a nodal partition, various submeshes are made by grouping nodes of the mesh such that different groups can be integrated by different time steps. Updating a particular node's temperature or displacement is dependent on the associated nodal critical time step. In an element partition, the mesh is divided into groups of elements. Element contributions to the system of equations during a particular time step are determined using the critical time step of that element.

## LITERATURE

In general, a subcycling procedure can have implicit, explicit or both algorithms working on differents parts of a finite element mesh. Each part has its own time step which is usually based on a stability criterion if an explicit algorithm is used and on an accuracy criterion if an implicit algorithm is used. Belytschko and Mullen (1976) introduced an *implicit-explicit* mixed time integration method for hyperbolic systems. In this method, they used a nodal partition and both implicit and explicit methods were employed for time integration. Later Belytschko and Mullen (1978) conducted the stability analysis of this method. A stability analysis was also conducted by Park (1980) (also see Liu and Belytschko (1982) and Liu, Belytschko, and Zhang (1984)). Descriptions and analyses of such methods for parabolic systems can be found in Belytschko (1981), Belytschko, Liu, and Smolinski (1983), and Smolinski, Belytschko, and Neal (1988). Convergence of such methods is discussed by Smolinski (1990) and stability by Smolinski (1991).

An *implicit-explicit* method for hyperbolic systems based on element partitioning was discussed by Hughes and Liu (1978b). The stability theory behind the method was discussed by Hughes and Liu (1978a). Element partitioning methods for parabolic systems are also discussed by Belytschko (1981), Liu (1983), Liu and Zhang (1983), and Belytschko, Smolinski, and Liu (1984). Stability of these techniques is discussed by Liu and Lin (1982), Belytschko, Smolinski, and Liu (1985), and Smolinski, Belytschko, and Liu (1987), and convergence is discussed by Hughes, Belytschko, and Liu (1987), and Smolinski (1990). The difference between nodal and element partitioning lies in how the equations are discretized to implement subcycling. The purpose of both nodal as well as element partitioning methods is the same; they are used to reduce the overall number of computations and inturn the computation time.

In an *explicit-explicit* mixed time integration method, the same explicit method is used on all parts of the mesh but different time steps are used. One such method based on nodal partitioning was introduced by Belytschko, Yen and Mullen (1979). Further developments are given by Smolinski, Belytschko and Neal (1988), Smolinski (1992a) and Hulbert (1992). One major drawback of various methods proposed above is that the ratio between the time steps of different partitions is always an integer. This leads to restrictions on the division of the mesh into more than just a few groups, thereby reducing the efficiency of the mixed time integration methods. Mizukami (1986) relaxed this restriction for first order ordinary differential equations which result from semidiscretization of the diffusion equation. His *explicit-explicit* method was based on nodal partitioning and allowed greater flexibility while assigning time steps to various groups. His stability proof was limited to a particular class of elements which was later extended to arbitrary elements by Donea and Laval (1988). Multi-time step methods for structural dynamic systems with non-integer time step ratios are discussed by Neal and Belytschko (1989) and Belytschko and Lu (1993b).

As far as feasibility, convergence, stability and accuracy of subcycling methods are concerned, many researchers have discussed these issues in many of the above mentioned references. Smolinski (1992b) and Belytschko and Lu (1993a,b) have also conducted convergence and stability analyses of subcycling schemes. However, in many cases only preliminary results were given because of the complexity of these schemes. It may be noted that subcycling schemes eliminate many floating point operations by using different time steps. The corresponding tradeoff is the need of interpolation at nodes where boundaries of various submeshes meet. The effect of interpolation on the accuracy of the solution is not fully understood. This issue has received little attention. An accuracy assessment of the Newmark method was made in a reference by Wang, Murti and Valliappan (1992). However it was limited to the conventional transient analysis of wave propagation problems.

In this chapter, an effort is made to better understand the interaction between subcycling methods and corresponding accuracy and stability. The accuracy of the subcycling solution of a mixed element mesh is compared with 3 other possible cases :

1. A mesh with uniform, large elements integrated with large time steps.
2. A mesh with uniform, small elements integrated with small time steps.
3. A mesh with mixed elements integrated with only small time steps.

It can be expected that accuracy would be a minimum for Case 1 and maximum for Case 2. For Case 3, accuracy can be expected to lie somewhere between that of Cases 1 and 2. A multi-time stepping solution shall be acceptable only if its accuracy is reasonably close or better than that of Case 3. As discussed by Gupta and Ramirez (1994,1995), it will be shown in this chapter, this is not always so and thus one must be careful before employing the multi-time step schemes.

It will be shown in this chapter that the solution not only can exhibit less accuracy but also for a long duration simulation, the total error in the solution increases without bounds at an increasing rate. The rate of increase depends upon the type of method used for interpolation. In this chapter, an in-depth look is taken into the behavior of multi-time step methods. Various numerical examples are used to exhibit the problem and the type of divergence induced into the system. Use of numerical damping is proposed to stabilize the multi-time step solution (Gupta (1995)).

The layout of the chapter is as follows. In Section 5.2 we discuss briefly the steps involved in a subcycling procedure. Error terms are defined in Section 5.3. With the help of a few classical example problems, we compare the accuracy of multi-time step algorithm with other cases in Section 5.4. Using one such problem, apparent instabilities induced into the system are shown

Section 5.5. A further in-depth look is taken at the divergence of the solution in Section 5.6. This is achieved by conducting an analysis in the frequency domain. In Section 5.7, $\alpha$ and $\gamma$ damping procedures are discussed. Their applicability and effectiveness in the explicit central difference method is also discussed in this section. The solutions via standard explicit integration and multi-time step methods, when used in conjunction with proper amount of damping are compared in Section 5.8. Parallel implementation of subcycling algorithm is discussed in Section 5.9. Finally, in Section 5.10 conclusions are drawn, and future research directions are discussed.

## 5.2 MULTI-TIME STEP ALGORITHM

For the central difference method, the steps involved in standard time integration can be summarized as follows:

For all nodes:

$$a_n = (f_n^{ext} - f_n^{int})/M \tag{5.1}$$
$$v_{n+1/2} = v_{n-1/2} + \Delta t a_n \tag{5.2}$$
$$d_{n+1} = d_n + \Delta t v_{n+1/2} \tag{5.3}$$

where $d_n$, $v_n$, and $a_n$ are displacement, velocity, and acceleration for a node respectively for the $n_{th}$ time step. To employ multi-time step schemes in the central difference method, all the nodes of the mesh can be categorized into 3 groups. Group A contains all nodes that are integrated with time step $\Delta t$. Group B is formed with those nodes that use time step $m\Delta t$ for integration. For the sake of simplicity in this study we restrict our attention to integer values of $m$, the time step ratio. However, conclusions drawn can be easily extended to non-integer values of time step ratios. All other nodes that act as interface nodes between group A and B comprise group C. These nodes are integrated with time step $m\Delta t$ and intermediate values of displacement are interpolated for the benefit of those nodes in group A that are next to interface nodes. The two types of methods proposed by researchers for interpolation are the constant velocity method (Smolinski, Belytschko and Neal (1988)) and the constant acceleration method (Belytschko and Lu (1993b)). The order of displacement field between $d_n$ and $d_{n+1}$ dictates the type of method. In the constant velocity method, this displacement field is assumed linear while in constant acceleration method, it is assumed quadratic. A summary of steps leading to multi-time step integration is as follows:

Define

$$i = mod(n, m) \tag{5.4}$$

For nodes in group A :

$$a_n = (f_n^{ext} - f_n^{int})/M \tag{5.5}$$
$$v_{n+1/2} = v_{n-1/2} + \Delta t a_n \tag{5.6}$$
$$d_{n+1} = d_n + \Delta t v_{n+1/2} \tag{5.7}$$

For nodes in group B :

Only when i = 1

$$a_n = (f_n^{ext} - f_n^{int})/M \tag{5.8}$$

$$v_{n+m/2} = v_{n-m/2} + m\Delta t a_n \tag{5.9}$$

$$d_{n+m} = d_n + m\Delta t v_{n+m/2} \tag{5.10}$$

For nodes in group $C$ :
Only when i = 1

$$a^{intf} = a_n = (f_n^{ext} - f_n^{int})/M \tag{5.11}$$

$$v^{intf} = v_{n+m/2} = v_{n-m/2} + m\Delta t a_n \tag{5.12}$$

$$d^{intfold} = d_n \tag{5.13}$$

Always

$$d_{n+1} = d^{intfold} + i\Delta t v^{intf} - [i(m-i)\Delta t^2 a^{intf}/2] \tag{5.14}$$

In equation 5.14, the term in [ ] is used only if interpolation is by the constant acceleration method. In this chapter we further restrict ourselves to a maximum of 3 groups (i.e., only one value of time step ratio $m$ in one mesh), since the results can be easily generalized. Unless otherwise stated, interpolation for interface nodes has been done by the constant velocity method for various examples.

## 5.3 DEFINITION OF ERROR TERMS

We define, instantaneous error in displacement at a node at time step $n$ as

$$e_n = d_n^{fem} - d_n^{exact} \tag{5.15}$$

where $d_n^{fem}$ is the solution obtained through finite element analysis and $d_n^{exact}$ is the exact analytical solution at time step $n$ i.e. at time $n\Delta t$. We further define the second order norm of cumulative error over time (Ramirez and Belytschko (1995)) as

$$E_n = \sqrt{\sum_{i=1}^{n} e_i^2 \Delta t} \tag{5.16}$$

dividing cumulative error by the second order norm of cumulative displacement, we get the normalized cumulative error norm as

$$E_{norm} = \frac{E_n}{\sqrt{\sum_{i=1}^{n} d_i^2 \Delta t}} \tag{5.17}$$

To differentiate between errors due to spatial discretization, temporal discretization, and both together, the *modal* solution (analytical integration of discrete equations of motion corresponding to a spatial discretization) was obtained through the use of a symbolic manipulator for a few example problems. Similar error expressions can be obtained by appropriately replacing terms pertaining to finite element solution or exact solution in the above expression by the modal solution. However, normalization is always done with respect to the exact solution. In general

modal - exact   $\Rightarrow$   errors due to spatial discretization, $E_{\Delta x}$

fem - modal   $\Rightarrow$   errors due to temporal discretization, $E_{\Delta t}$

fem - exact   $\Rightarrow$   errors due to both spatial and temporal
                              discretizations, $E_{\Delta x + \Delta t}$

It may be noted that the explicit finite element solution also contains small errors due to roundoff of floating point numbers, which will also be part of $E_{\Delta t}$ and $E_{\Delta x + \Delta t}$.

## 5.4  NUMERICAL EXPERIMENTATION

*DEFINITION OF EXAMPLE PROBLEM*

Although a theoretical accuracy analysis of multi-time step methods would be ideal, this work concentrates on the development of an experimental base to facilitate future theoretical research. We thus present a study based upon various numerical simulations.

Figure 5.1: Example Problem — Bar subjected to a Heaviside load.

We consider the classical problem depicted in Fig. 5.1. In order to minimize the dependence on external variables, we employ the same structure for most of the chapter. This allows for an objective base of comparison. Various parameters such as time step ratio, size of elements, spatial and temporal discretizations, etc., and a few other conditions such as type of loading, interpolation method, etc., are changed appropriately for different simulations.

The structure consists of a bar of length 400 in., rigidly fixed at the right end. For all simulations where 2 different size elements are used, the first half of the bar, i.e. from x = 0 to x = 200 in., consists of uniformly distributed small elements while the large elements belong to the second half of the bar, i.e. from x = 200 to x = 400. For those cases where subcycling is used, a node that has no small element connected to itself is designated as an interface node. To have maximum efficiency in the solution, the interface node must be as near to the small elements group as possible. For our example problem, this puts the interface node next to the node at the center of the bar, in the positive X direction.

In most of the simulations, a Heaviside loading of constant force of 100 lb is applied at the left end. In some simulations, where initial velocity is given to all nodes, this force is not applied and the structure is considered to be free of any external force. Modulus of Elasticity of the bar is $3 \times 10^7$ psi and density of the material is .003 lbf sec$^2$/in.$^4$. For all simulations, results are plotted at x = 0 since this is the point which is often a focus of interest in calculating response. Both the cumulative error as well as the normalized cumulative error have been shown for better comparison for each simulation.

*ACCURACY DEPENDENCE ON TIME STEP RATIO*

**Example 1: Bar with Heaviside Loading,** *m* = 2

First we consider the problem defined above for which various combinations of spatial and temporal discretization are defined in Table 5.1. Various cases of spatial discretization are shown in Fig. 5.2. The number of elements varies from 20 to 40. In Case 1, there are 20 uniform elements in the bar and in Case 2 there are 40 uniform elements. In general, accuracy of the finer mesh can be expected to be higher than that of the coarser mesh. Case 3 is a combination of the above 2 cases with 30 elements. The first half of the bar consists of 20 uniform elements that are of the same size as the elements in Case 2. The second half has 10 elements which are of the size of elements in Case 1. The accuracy for this case can be expected to be somewhere between those of Case 1 and 2. Finally Case 4 has exactly same spatial discretization as Case 3, but subcycling is used in this case for time integration. The purpose of this simulation is to investigate the behavior of errors induced in a multi-time step finite element algorithm, when compared with the standard finite element algorithm as well as with finer and coarser meshes. This example should be able to provide some information about where the subcycled solution ranks.

Table 5.1: Various combinations of discretization and integration schemes for example problem shown in Fig. 5.1 and results plotted in Figs. 5.4 and 5.5.

| Case | No. of | Length | Range (in.) | | Nodes in Group | | | | $\Delta t$ |
|------|--------|--------|------|-----|------|-------|-----|-----|--------|
| No. | Elements | (in.) | From | To | A | B | C | $m$ | (msec) |
| 1 | 20 | 20 | 0 | 400 | 1-21 | – | – | – | .190 |
| 2 | 40 | 10 | 0 | 400 | 1-41 | – | – | – | .095 |
| 3 | 20 | 10 | 0 | 200 | 1-31 | – | – | – | .095 |
| | 10 | 20 | 200 | 400 | | | | | |
| 4 | 20 | 10 | 0 | 200 | 1-21 | 23-31 | 22 | 2 | .095 |
| | 10 | 20 | 200 | 400 | | | | | |

    Exact solution of Displacement time history at x = 0 is given in Fig. 5.3. Displacement time histories for various finite element meshes are similar in nature but are not shown. For most of the examples, error in displacement time history has been compared instead. For Cases 1-4, the error in displacement at x = 0 is shown in Figs. 5.4 and 5.5. The error in Case 2 is almost half the error in Case 1. As expected, error in Case 3 lies between that of Case 1 and 2. However in Case 4, where multi-time stepping is used, the error behavior is quite peculiar. Initially, it follows very closely the error in the 40 element mesh i.e. Case 2, but as the wave reflects back and forth at the ends of the bar, the rate of error increases. After time 0.0635 sec, the error becomes *more* than that in the 20 element mesh, integrated with the large time step.

    Note that the primary objective of having finer spatial discretization (i.e., to refine the mesh from 20 elements to 30 elements) is to have higher accuracy in the solution. However subsequent coarser time integration schemes, i.e. multi-time stepping if not used cautiously, can cause the error to increase at a much higher rate, thereby decreasing the accuracy even *below* the accuracy of a coarser mesh being integrated with large time steps for longer simulation times.

## Example 2: Bar with Heaviside Load, $m = 4$

In the above example, a time step ratio of 2 was used for the subcycling solution. The error in the subcycling solution was found to increase more rapidly as compared to both coarser as well as finer meshes integrated with the non-subcycling procedure. A similar or maybe more intensified behavior can be expected for higher time step ratios because amount of interpolation increases with the time step ratio. In order to generalize the results for other time step ratios, this simulation was done for the cases defined in Table 5.2 where the time step ratio is 4. Here the

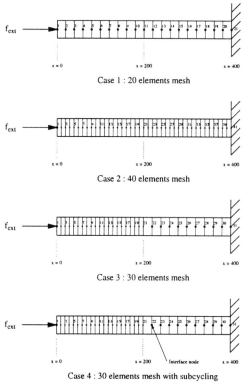

Case 1 : 20 elements mesh

Case 2 : 40 elements mesh

Case 3 : 30 elements mesh

Case 4 : 30 elements mesh with subcycling

Figure 5.2: Spatial discretization for cases defined in Table 5.1 for Example 1.

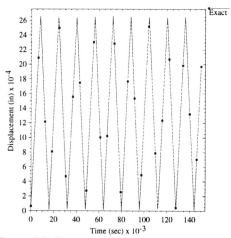

Figure 5.3: Exact solution of displacement time history @x=0.

structure shown in Fig. 5.1 is discretized into 20 to 80 elements. In Case 1, there are 20 uniform elements, hence it is essentially the same case as Case 1 in the previous example. Case 2 has 80 equal elements of length 5 in. Cases 3 and 4 follow the same type of distribution of elements as in the previous example. There are 40 small elements in the left half of the bar and 10 large elements in the right half. Since the length of the small elements is one quarter of that of the large elements, a time step ratio of 4 is used in Case 4 for maximum efficiency.

The results for the tip of the bar are shown in Figs. 5.6 and 5.7. Again, similar behavior can be observed for higher time step ratios. Like Example 1, for the first few time steps, the errors for Cases 3 and 4 are almost as low as the error for the mesh with 80 uniform elements. Error in the 80 element mesh is almost one quarter of the error in the 20 element mesh. For the mixed element mesh i.e. Case 3, accuracy lies between that of the 20 and 80 element meshes. Accuracy of the multi-time time step solution (i.e., Case 4) is almost as high as of the 80 element mesh but only initially. Like the previous example, slowly as the wave reflects at the ends, error increases in the solution. However, it follows the course of the standard explicit solution (i.e., Case 3) for a longer time when compared with similar cases in Example 1. Then error increases rapidly in the subcycled solution, and the accuracy soon becomes even *less* than the coarser mesh discretized with *only* 20 elements.

The crossover between Case 1 and 4 occurs at about 0.08 sec, which is more than the crossover time in Example 1 for the similar cases. This could be attributed to the fact that the minimum time step is one half of the time step in the previous example. Hence, in this simulation, the multi-time step solution is more accurate for a longer time but on the other

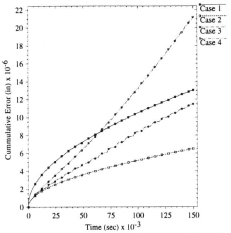

Figure 5.4: Error comparison for cases defined in Table 5.1.

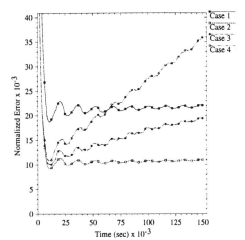

Figure 5.5: Normalized Error comparison for cases defined in Table 5.1.

Table 5.2: Various combinations of discretization and integration schemes for example problem shown in Fig. 5.1 and results plotted in Figs. 5.6 and 5.7.

| Case No. | No. of Elements | Length (in.) | Range (in.) From | Range (in.) To | Nodes in Group A | Nodes in Group B | Nodes in Group C | $m$ | $\Delta t$ (msec) |
|---|---|---|---|---|---|---|---|---|---|
| 1 | 20 | 20 | 0 | 400 | 1-21 | – | – | – | .1900 |
| 2 | 80 | 5 | 0 | 400 | 1-81 | – | – | – | .0475 |
| 3 | 40 | 5 | 0 | 200 | 1-51 | – | – | – | .0475 |
|   | 10 | 20 | 200 | 400 |  |  |  |  |  |
| 4 | 40 | 5 | 0 | 200 | 1-41 | 43-51 | 42 | 4 | .0475 |
|   | 10 | 20 | 200 | 400 |  |  |  |  |  |

hand, the rate of error increase is also higher even when compared with the corresponding case in Example 1, where the mesh is coarser and time step is higher but time step ratio is lower. Overall, we can deduce that large time ratios have a more severe effect on the accuracy of the solution.

## EFFECTS OF MESH REFINEMENT

### Example 3: Fine Mesh of Bar with Heaviside Load, $m = 2$

Based on the previous two simulations, it can be generalized that accuracy of the solution integrated with multi-time step methods is only acceptable in the initial or early part of the simulation. As time progresses, more and more errors are introduced into the system as a result of subcycling. Since this study is based on results observed from numerical simulations, we run this simulation to see if the element size is a factor in this behavior. Our purpose is to find if the additional errors due to subcycling, are increased, decreased, or remain same when element size is changed. Intuitively, we do not expect any change in the behavior.

The same example problem shown in Fig. 5.1 is discretized into a very fine mesh, the properties of which are defined in Table 5.3. Now in Case 1, there are 1000 uniform elements of length 0.4 in. Case 2 is comprised of 2000 uniform elements of length 0.2 in. The spatial

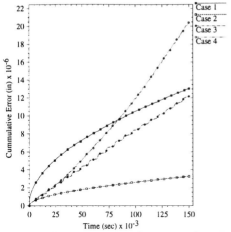

Figure 5.6: Error comparison for cases defined in Table 5.2.

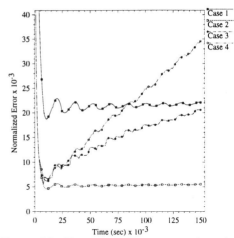

Figure 5.7: Normalized Error comparison for cases defined in Table 5.2.

discretization in Cases 3 and 4 is similar to that in the previous simulations. There are 1000 small elements in the first half and 500 large elements in the second half of the bar. Node 1002 acts as the interface node and the time step ratio is 2.

Table 5.3: Various combinations of discretization and integration schemes for example problem shown in Fig. 5.1 and results plotted in Figs. 5.8 and 5.9.

| Case No. | No. of Elements | Length (in.) | Range (in.) From | Range (in.) To | Nodes in Group A | Nodes in Group B | Nodes in Group C | $m$ | $\Delta t$ ($\mu$sec) |
|---|---|---|---|---|---|---|---|---|---|
| 1 | 1000 | 0.4 | 0 | 400 | 1-1001 | – | – | – | 3.8 |
| 2 | 2000 | 0.2 | 0 | 400 | 1-2001 | – | – | – | 1.9 |
| 3 | 1000 | 0.2 | 0 | 200 | 1-1501 | – | – | – | 1.9 |
|   | 500 | 0.4 | 200 | 400 | | | | | |
| 4 | 1000 | 0.2 | 0 | 200 | 1-1001 | 1003-1501 | 1002 | 2 | 1.9 |
|   | 500 | 0.4 | 200 | 400 | | | | | |

The results for this mesh are plotted in Figs. 5.8 and 5.9 for x = 0. Comparing these figures with Figs. 5.4 and 5.5, it becomes quite clear that in both the simulations, errors follow almost the same pattern. The only difference is that in the finer mesh, errors are reduced for all cases by approximately a factor of 50 when compared with the coarser mesh in Example 1. The ratio of element sizes between the 2 simulations is also 50. The crossover between Case 1 and Case 4 also occurs at about 0.06 sec in both simulations, even though for the finer mesh, the solution has progressed over 30000 time steps while this number is only about 600 for the coarser mesh.

Overall this simulation suggests that accuracy of the subcycling solution does not depend on the number of time steps but on the real time duration of the simulation. Hence, additional errors in the multi-time step solution propagate with the traveling wave and accumulate at a point when the wave passes that point. A basic inference, drawn from Examples 1, 2, and 3, is that the subcycled solution in the long term is invariably *less* accurate relative to the standard solution irrespective of the number of elements, element size, or time step ratio.

Total error in a finite element solution essentially arises due to *spatial* discretization and *temporal* discretization through time integration. Since multi-time step methods only affect the

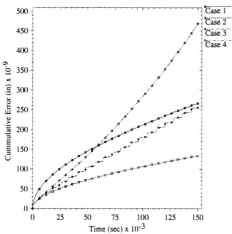

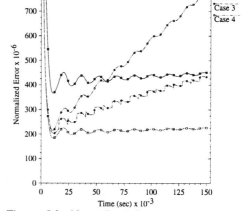

Figure 5.8: Error comparison for cases defined in Table 5.3.

Figure 5.9: Normalized Error comparison for cases defined in Table 5.3.

error due to temporal discretization, in the next 2 simulations we look at these two kinds of errors separately to gain further insight into the behavior of these methods. These simulations should also be able to help us in identifying the reasons for the continuous increase of normalized error in Case 3 of all previous simulations.

## SEPARATION OF ERRORS

### Examples 4 and 5: Modal Solution, Spatial and Temporal Errors

So far we have considered the effect of multi-time step methods on total errors only, since they are of primary interest in a solution. It can be safely assumed that error due to spatial discretization is not affected by subcycling. We assume this because same mesh is used in standard and subcycling solution. At this point, we also assume that temporal discretization error is not influenced from different spatial discretizations. It will be shown later numerically that this is the case. A further insight into the behavior of multi-time step schemes can be gained by separating the errors due to spatial and temporal discretization. This might also provide better understanding as to the means to increase the accuracy of these schemes.

We revisit the problem of Example 1, where Case 1 has 20 elements of length 20 in. ($\Delta t = .190$ msec), Case 2 has 40 elements of length 10 in. ($\Delta t = .095$ msec), and Case 3 has 20 elements of length 10 in. and 10 elements of length 20 in. ($\Delta t = .095$ msec). Total error for these cases is plotted in Figs. 5.4 and 5.5. The effect of subcycling on total errors has already been discussed above. Here we calculate the *modal* solution for Cases 1-3 through the use of a symbolic manipulator (Char et al. (1992)). In general, finding modal solution requires finding the eigenvalues which involves solving an $n^{th}$ degree polynomial equation. Since for $n \geq 5$ explicit algebraic solutions cannot in general be found, the solution is often numerical (Wolfram (1988)). Since Case 4, where multi-time stepping is used, has exactly the same spatial discretization as Case 3, calculating its modal solution is redundant. Once we have all the exact, finite element, and modal solutions, spatial as well as temporal discretization errors can be calculated using the method discussed earlier in Section 5.3.

The spatial discretization errors at x = 0 are shown in Figs. 5.10 and 5.11. The temporal discretization errors are shown in Figs. 5.12 and 5.13. A few observations made from these results are as follows:

- Like total errors, both spatial as well as temporal errors in the 40 element mesh are about half of the respective errors in the 20 element mesh.
- Both spatial and temporal errors are *more* than the total errors for 20 as well as 40 element meshes.
- Like total error, spatial error in the 30 element mesh is between that of 20 and 40 element meshes.
- Unlike total error, spatial error in the 30 element mesh does not increase towards intersecting with the 20 element mesh error.
- Temporal error in Case 4 is initially as low as that in the 40 element mesh but then increases at a higher rate.
- Compared to total error, for temporal error, crossover between Case 1 and Case 4 occurs at a *later* time.

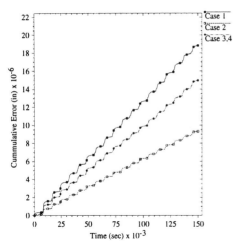
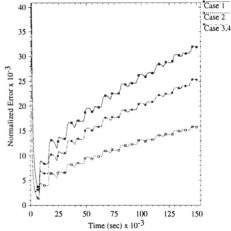

Figure 5.10: $\Delta x$ Error comparison for cases defined in Table 5.1.

Figure 5.11: $\Delta x$ Normalized error comparison for cases defined in Table 5.1.

In Figs. 5.12 and 5.13, temporal error in the 30 element mesh (Case 3) is almost as low as that in the 40 element mesh (Case 2). In both cases, the same time step is used, i.e. $\Delta t = .095$ msec. We believe this to be a conclusive proof of our earlier assumption that the spatial discretization has little if any influence on the $\Delta t$ error.

An interesting point may be noted by comparing Figs. 5.5, 5.11, and 5.13. For both Cases 1 and 2, the spatial (Fig. 5.11) as well as the temporal (Fig. 5.13) normalized errors increase with time whereas the total (Fig. 5.5) normalized error approaches a constant value. This could be attributed to the fact that in an explicit central difference method, these two kinds of errors are compensatory in nature (Krieg and Key (1973)). In Case 1, where element size $l = 20$ in. and the time step $\Delta t = .190$ msec, the rate of increase in $\Delta x$ error nullifies the rate of increase in the $\Delta t$ error, and the total error approaches a constant value. When going from Case 1 to Case 2, due to mesh refinement ($l = 10$ in.), a reduction can be observed in the rate of increase in $\Delta x$ error. Simultaneously, due to stability criterion, the time step is also reduced to $\Delta t = .095$ msec for the finer mesh in Case 2. This causes a corresponding reduction in the rate of increase of

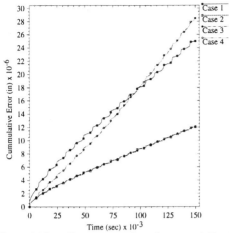
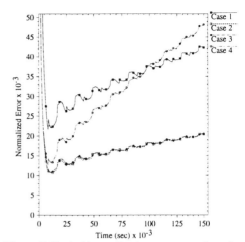

Figure 5.12: $\Delta t$ Error comparison for cases defined in Table 5.1.

Figure 5.13: $\Delta t$ Normalized error comparison for cases defined in Table 5.1.

$\Delta t$ error. Both reductions are such that the rates of increase of $\Delta x$ and $\Delta t$ errors still cancel each other, and the total error apparently approaches a constant value.

If for some reason, the rate of increase of $\Delta x$ and $\Delta t$ errors do not cancel each other completely, then total error will not be able to approach a constant but will show a steady rate of increase. This would happen if one of them is changed while the other one is either kept constant or not changed as much. Even worse results can be expected if one is increased while the other one is decreased. This happens to be the case in Case 3, where total normalized error does not approach a constant value but keeps on increasing (Fig. 5.5).

In Case 3, where there are 30 elements, the $\Delta x$ error is between those of Cases 1 and 2. This is as expected because half of the mesh is coarse (similar to Case 1) and half of the mesh is fine (similar to Case 2). However, since the time step in Case 3 is the same as in Case 2, $\Delta t$ error is almost the same in both cases. Considering the error in Case 1 as a baseline, in Case 3 the rate of $\Delta t$ error is reduced as much as in Case 2, while the rate of spatial error is not reduced as much. In Case 1, the rate of increase of both $\Delta x$ and $\Delta t$ errors is high, resulting in the convergence (loosely speaking) of the total normalized error. In Case 2, the rate of increase of both $\Delta x$ and $\Delta t$ errors is low, again resulting in the convergence of the total normalized error. In Case 3, the rate of increase of the $\Delta x$ error is medium while that of the $\Delta t$ error is low, resulting in an increase of total normalized error.

It can be inferenced from the above experiment, that for a finite element mesh total error cannot be reduced only by lowering the time step much below the critical time step. This only reduces the temporal error and causes the total normalized error to follow an upward path rather than converging to a constant value. If a simulation is run for a long time, the total error for a mesh integrated with a time step much smaller than the critical time step, catches up with the total error in the same mesh integrated with a time step close to the critical time step. As far as Case 3 is concerned, it may be possible to get the total normalized error to approach a constant value, if we increase the rate of increase of $\Delta t$ error appropriately. Convergence of the total normalized error in both the coarse as well as the fine meshes suggests that it may be possible to achieve such accuracy if the coarse part of the mesh is integrated with large time steps and the fine part with small time steps. In fact multi-time step methods could increase the $\Delta t$ error thereby helping to reduce the rate of increase of the total error.

It may be noted from Figs. 5.12 and 5.13, that in Case 4 where the multi-time step method is used for time integration, the rate of $\Delta t$ error does increase when compared with Case 3. However, this increase happens to be much more than we desire. In Case 4, the rate of increase of $\Delta t$ error becomes even more than that in Case 1. Again considering error in Case 1 as a baseline, in Case 4 the rate of $\Delta t$ error is increased while the rate of the $\Delta x$ error is decreased. This causes the total error in Case 4 to increase even more rapidly than in Case 3 (Figs. 5.4 and 5.5). Consequently, for total error crossover occurs earlier compared to the $\Delta t$ error *alone*. To increase the accuracy in a solution, both spatial as well as temporal errors must decrease simultaneously and appropriately. Although multi-time step schemes only affect the temporal discretization error, total error increases even more in the solution because now spatial discretization error has been reduced.

As discussed above, multi-time step schemes can be very useful if their temporal discretization error can be kept between those of Case 1 and Case 3 such that after compensatory effects of spatial discretization error, total normalized error approaches a constant value. Unmodified multi-time step schemes do not achieve such accuracy in the long term.

## INITIAL VALUE (UNFORCED) PROBLEM

### Example 6: Bar with Initial Velocity, $m = 2$

So far we have conducted the accuracy investigation of multi-time step methods by analyzing numerical simulations in which a Heaviside load is applied on the structure. In all cases, the initial conditions were quiescent, i.e., displacement and velocity were zero at time $t = 0$. To further generalize the results for non zero initial conditions, we revisit the problem in Example 1 but apply an initial velocity on the structure. No external force is applied on any node of the mesh. The details of the discretization are given in Table 5.1 and were discussed in Example 1. The initial velocity is considered to be

$$v(x, 0) = 10 \cos \frac{\pi x}{2L} \tag{5.18}$$

Results for Cases 1-4 are shown in Figs. 5.14 and 5.15. Once again, the 40 element mesh produces more accurate results than the 20 element mesh. In Case 3, initially the error is as low as the error in the 40 element mesh but for a very short time. Then it increases at a much higher rate when compared with the error in the similar case in Example 1 where Heaviside loading is applied. A crossover occurs between Case 3 and Case 1 at a very early stage of the solution.

As far as the effect of subcycling is concerned on the solution, there seems to be little change as compared to the previous examples. Initially, error in Case 4 is comparable to the error in the finer mesh. However this behavior is short-lived and the error increases at a rate much higher than even that in Case 3. Shortly thereafter, the accuracy in Case 4 becomes *less* than the accuracy in Case 1. At 0.015 sec, the 30 element mesh multi-time step solution normalized error is about 4 times that of the standard explicit finite element solution.

## EFFECTS OF INTERPOLATION SCHEMES

### Example 7: Bar with Initial Velocity, $m = 10$ (Constant Acceleration Method)

In all simulations conducted thus far, we have discussed the effects of multi-time step schemes using a constant velocity method for interpolation at the interface between nodal groups. Overall

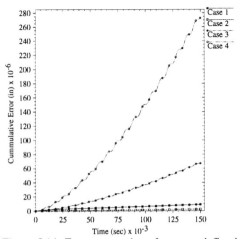

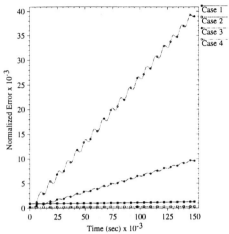

Figure 5.14: Error comparison for cases defined in Table 5.1 (initial velocity problem).

Figure 5.15: Normalized error comparison for cases defined in Table 5.1 (initial velocity problem).

it is established that for boundary-initial value problems, accuracy of multi-time step methods can be less than the accuracy of standard explicit methods. Earlier, researchers have used other interpolation schemes to improve accuracy in a subcycling solution by making better estimates for displacement at the interface node. One such scheme is the constant acceleration method which is discussed in Belytschko and Lu (1993b). With the help of an example problem, it was shown that this method yields much better results than the constant velocity method.

Keeping the earlier discussion in perspective, consider the nature of the multi-time step solution. It tends to produce quite accurate results in the initial part of the simulation, but then it begins to exhibit increasingly higher error. Hence, it is possible that the above conclusion that the constant acceleration interpolation method is more accurate than the constant velocity method, was obtained based on the initial part of the simulation.

Here, we rerun the above mentioned example (i.e., Example B1, Section 7.2) of Belytschko and Lu (1993b) but for a longer period of time. For the sake of completeness, we define the problem again. Essentially it is the same structure as shown in Fig. 5.1, but has different dimensions and material properties. Now the bar has a length of 100 in. The modulus of elasticity is $3 \times 10^7$ psi and density of the material is .00073 lbf sec$^2$/in.$^4$. The bar has 10 elements of length 5 in. from x = 0 to 50 in., 10 elements of length 0.5 in. from x = 50 to 55 in., and 9 elements of length 5 in. from x = 55 to 100 in. The integration is done with $\Delta t = 2$ μsec, and the time step ratio for subcycling is 10. Nodes 1 - 9 and 23 - 30 are integrated with a time step of 10$\Delta t$, nodes 11 - 21 are integrated with a time step of $\Delta t$, and nodes 10 and 22 are interface modes. For the case where no subcycling is used, all the nodes 1 - 30 are integrated with a time step of $\Delta t$. An initial velocity is applied as per Eq. 5.18, and no external load is applied.

Instantaneous error in the velocity at x = 50 in. (i.e., node 11) is plotted in Fig. 5.16 for 3 different cases. In the first case, interpolation is done by the constant acceleration method. In the second case, it is by the constant velocity method. The third case is the standard case where no subcycling is used. The results for time t = 0 to 0.01 sec, are in complete agreement with the results obtained earlier in Belytschko and Lu (1993b). In that time span, the constant acceleration method yields more accurate results than the constant velocity method. In fact, it is even more accurate than the standard explicit finite element solution where no subcycling is used.

The error in the constant velocity solution is many times more than the error in the standard explicit method. However, in both cases, the errors exhibit a sinusoidal nature and a linearly increasing envelope at least in the time span considered here (i.e., from t = 0 to 0.017 sec). The error in the constant acceleration solution initially exhibits the same nature but after about 0.007 sec, oscillations start appearing which increase in magnitude exponentially. At the end of the simulation, i.e. at t = 0.017 sec, the error behavior is highly oscillatory, and the rate of increase in its magnitude is extremely high.

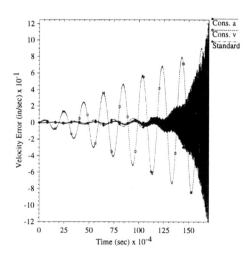

Figure 5.16: Instantaneous velocity error comparison for cases defined in Example 7.

Hence, even though the constant acceleration interpolation solution is initially more accurate, its use is not recommended without some modifications as it appears to become highly unstable in the later stages of the solution. The peculiar behavior of this method prompted us to perform the same analysis for this method as for the constant velocity method, done earlier in Example 1. This is the topic of the next example.

## Example 8: Bar with Heaviside Load, $m = 2$ (Constant Acceleration Method)

In the previous example, we compared the instantaneous error in velocity for a bar integrated with the multi-time step method where the constant acceleration method was used for interpolation of displacements at interface nodes. To compare the accuracy of this method with other standard cases (i.e., finer mesh, coarser mesh, and standard explicit soution), we rerun the example problem shown in Fig. 5.1 for cases defined in Example 1. A Heaviside force of 100 lb is applied at the left end. However, this time the constant acceleration method is used for interpolation. We expect the same behavior in this problem as observed in the previous example. However, this comparison should shed some light on whether extra initial accuracy is worth the price we pay later.

The cumulative errors and normalized cumulative errors for Cases 1-4 are shown in Figs. 5.17 and 5.18 respectively. Compared to the constant velocity method (i.e., Case 4a), this time error in Case 4b follows very closely the error in Case 3 for a much longer time. In fact until about 0.06 sec, it is slightly less than the error in the standard explicit solution. Then as expected, it increases rapidly and crosses the error in Case 3 first and then the error in the coarser mesh integrated with large time step (i.e., Case 1). For the constant acceleration method, the crossover with Case 1 occurs at a later time when compared with the constant velocity method. However, the latter method has a much lower rate of error growth at the point of the crossover. Error in constant acceleration method keeps increasing along with its rate of increase. At the end of the simulation, the error in Case 4b is as high as that in Case 4a but has a very high rate of increase.

Overall this example suggests that the constant acceleration method can be acceptable only if the duration of the simulation is small. Otherwise it can induce highly undesirable errors

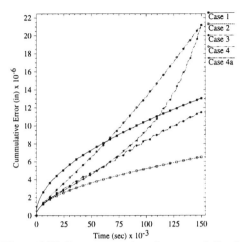

Figure 5.17: Error comparison for cases defined in Table 5.1 (constant acceleration interpolation).

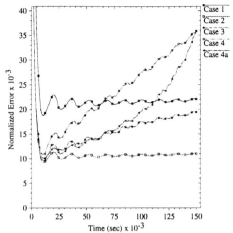

Figure 5.18: Normalized error comparison for cases defined in Table 5.1 (constant acceleration interpolation).

into the solution. Although the constant velocity method seems quite stable in all the previous simulations, the exponential increase in the error due to the constant acceleration method made us wonder why the same couldn't happen with the constant velocity method. This is the topic of our discussion for the next and final example.

## LONG DURATION SIMULATION

### Example 9: Bar with Heaviside load, $m = 2$

In all the previous examples, the constant velocity method has shown a linear growth for cumulative errors as well as for normalized cumulative errors. On the other hand, errors in the constant acceleration method have shown an exponential growth in the later part of the simulation. Consider Example 7, where the simulation duration of 0.01 sec was too short for this phenomenon to occur. By increasing the duration of the simulation by 70%, we were able to capture this effect. This leads us to think that the same could be true with the constant velocity method. Maybe the duration of 0.15 sec is too small to see any instabilities that may occur later on. To answer these questions, we rerun the previous example for a duration of 1 sec.

Results pertaining to these 2 types of interpolation methods are shown in Figs. 5.19 and 5.20. Results for other cases are not shown here as comparisons with those cases have already been discussed. Surprisingly, the constant velocity method also exhibits the same behavior as the constant acceleration method but at a later time and at a slower speed. Initially, the error increases linearly but then shows exponential growth. At the end, the errors approach infinity and the solution diverges for both multi-time step schemes, even for m = 2. In short, it can be deduced that the constant acceleration method is more accurate initially but diverges sooner than the constant velocity method which is less accurate initially.

This simulation suggests that irrespective of the type of interpolation method, the solution starts exhibiting instabilities at some point in time. If the desirable duration of simulation is such that simulation ends before the errors become sufficiently large, then multi-time step methods can be used. However this poses a serious restriction on the user as the exponential growth occurs at different times for different problems. Without taking any extra measure to control the

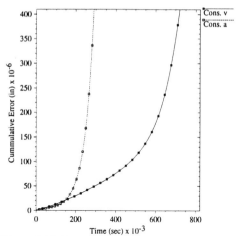

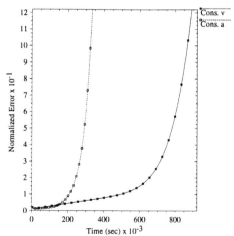

Figure 5.19: Error comparison in a long simulation for Case 4 defined in Table 5.1.          Figure 5.20: Normalized error comparison in a long simulation for Case 4 defined in Table 5.1.

errors induced into the solution due to subcycling, multi-time step methods do not seem to be practical for all problems.

## 5.5 INSTABILITIES IN MULTI-TIME STEP SCHEMES

In the previous section, it is shown that the error in multi-time step methods increases without bounds when the simulation is run for a long time. Before we attempt to find a remedy to this problem, a question we must ask is, what is the source of the solution divergence? Unless we know the reasons for this type of behavior, a direction cannot be set towards finding a solution. To analyze the problem, it becomes necessary to take a deeper look at the time history. We revisit the example problem depicted in Fig. 5.1. We look at the displacement time history first and if the need arises, velocity and acceleration time histories can also be used. Temporal and spatial discretizations of the bar are defined in Table 5.4.

Table 5.4: Spatial and temporal discretizations for example problem shown in Fig. 5.1.

| Case | No. of | Length | Range (in.) | | Nodes in Group | | | | $\Delta t$ |
| No. | Elements | (in.) | From | To | A | B | C | $m$ | (msec) |
|---|---|---|---|---|---|---|---|---|---|
| 1 | 20 | 10 | 0 | 200 | 1-31 | – | – | – | .095 |
| | 10 | 20 | 200 | 400 | | | | | |
| 2 | 20 | 10 | 0 | 200 | 1-21 | 23-31 | 22 | 2 | .095 |
| | 10 | 20 | 200 | 400 | | | | | |

For Cases 1 and 2, the behavior of displacement at x = 0 is shown in Figs. 5.21 and 5.22. It can be seen that the solution for the case where no subcycling is used (i.e., the standard finite element solution), remains accurate. On the other hand, in Case 2 where subcycling is used, solution remains accurate only for about 4 msec and then it starts to diverge. Magnified views of first, middle, and last segment of Fig. 5.22 are shown in Figs. 5.23, 5.24, and 5.25, respectively. Initially the displacement solution is free of any noise (Fig. 5.23). After some time, small disturbances start appearing in the solution (Fig. 5.24). These disturbances are about the

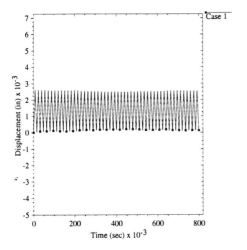

Figure 5.21: Displacement time history @x=0 for standard solution.

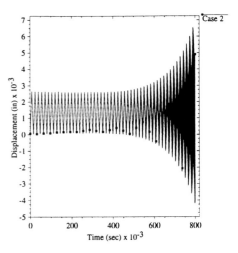

Figure 5.22: Displacement time history @x=0 for subcycled solution.

standard solution, thereby on average the solution follows the same pattern as the no subcycling case. However, the last segment in Fig. 5.25 shows that these disturbances grow in magnitude very rapidly and dominate the complete solution; finally the disturbances become so large that the solution becomes divergent.

To confirm that this behavior is not restricted solely to the Heaviside loading case, the same problem was run again with the loading function as

$$f_{ext}(t) = 100 \, sin50t \qquad (5.19)$$

For Cases 1 and 2, results for the tip of the bar for sinusoidal loading are plotted in Figs. 5.26 and 5.27. Again for Case 1, displacement time history remains accurate for the duration of the simulation. However,

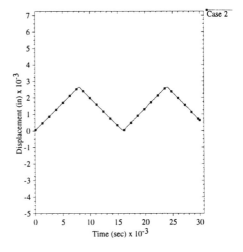

Figure 5.23: Zoom of initial part of displacement time history @x=0 for subcycled solution.

in Case 2, multi-time step solution exhibits the same behavior for sinusoidal loading as for the Heaviside loading.

## 5.6 FREQUENCY DOMAIN ANALYSIS

It is observed in the previous section that errors in the time history are fluctuating about the standard solution. Also with time, errors grow in magnitude exponentially and completely dominate the solution. Another way to look at this behavior is by conducting an analysis in the frequency domain. Such analysis delineates the frequencies that are participating in the solution. In this section, we take the Fast Fourier Transform (FFT) of displacement time history which

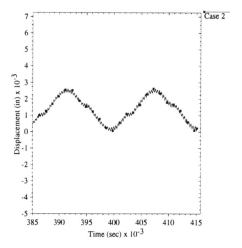
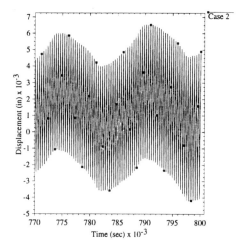

Figure 5.24: Zoom of middle part of displacement time history @x=0 for subcycled solution.

Figure 5.25: Zoom of last part of displacement time history @x=0 for subcycled solution.

reveals the frequencies that are being excited during the course of the simulation. The purpose is to find if there are any spurious frequencies that are being excited or produced by the multi-time step scheme. The FFT is conducted through the use of the SDRC-IDEAS package.

Consider the example problem shown in Fig. 5.1 with sinusoidal loading at the tip of the bar as given by Eq. 5.19. The spatial and temporal discretizations of the bar are given in Table 5.4. For the standard case with no subcycling, an FFT of the displacement time history for the first 10000 time steps is shown in Fig. 5.28. Only low frequencies in the range of 0-500 Hz are being excited in the solution. The solution is free of any high frequency modes. For the multi-time step case, an FFT of displacement history is shown in Figs. 5.29, 5.30, and 5.31. Figure 5.29 shows the spectrum for first 4000 time steps. It can be seen that high frequencies in the range of 1500-3500 Hz start appearing, however their participation is relatively low as compared to the first and second modes. The spectrum for first 5000 time steps is shown in Fig. 5.30: a substantial increase occurs in the participation of high frequencies in the solution. A corresponding increase in errors can be seen in the time history during that time period in Fig. 5.27. Finally Fig. 5.31 shows the spectrum for the first 6000 steps. It can be seen that participation of high frequencies becomes almost as high as the participation of low frequencies and by this time the solution is completely taken over by the errors introduced by such frequencies.

## 5.7 DISSIPATIVE ALGORITHM, $\alpha$ FAMILY OF METHODS

In the previous section it is shown that if no measures are taken to curb the participation of high frequencies, the solution slowly becomes highly erroneous and then diverges exponentially. An analogy can be drawn between the behavior of multi-time step methods and many implicit time integration methods. Because of large time steps used in implicit methods, excitation of high frequencies has been a problem for researchers over the past few decades (Hilber, Hughes and Taylor (1977)). Researchers have proposed the use of numerical damping to dissipate the inherently inaccurate high frequencies.

The central difference method is a representative of explicit scheme while the trapezoidal method is a representative implicit scheme. Both of these methods do not possess any kind of

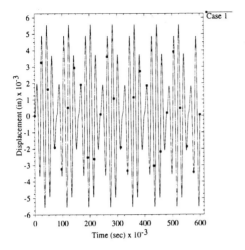

Figure 5.26: Displacement time history @x=0 for standard solution (sinusoidal loading).

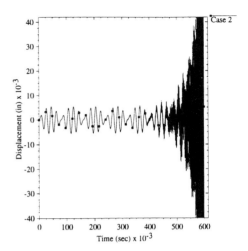

Figure 5.27: Displacement time history @x=0 for subcycled solution (sinusoidal loading).

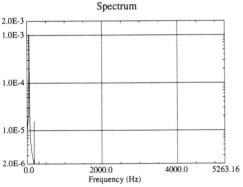

Figure 5.28: Spectrum for standard solution for first 10000 time steps.

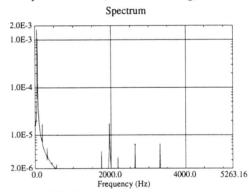

Figure 5.29: Spectrum for multi-time step solution for first 4000 time steps.

numerical damping. Since the time step in explicit methods is restricted by the critical time step, explicit solutions are usually free of high frequencies, and so they do not require damping. Hence, development of time integration schemes possessing numerical damping has mainly focussed on implicit methods. Some of such methods are Houbolt's method, Wilson θ-method and the Newmark family of methods with $\gamma > 1/2$ and $\beta \geq (\gamma + 1/2)^2/4$. The requirements that a good dissipative algorithm should possess (Hilber, Hughes and Taylor (1977)) are given as follows:

1. We should be able to control the amount of numerical damping by a parameter other than the time step.
2. A chosen value of this parameter should not significantly affect the lower modes.
3. For implicit algorithms there is another requirement of unconditional stability.

Since our primary concern is dissipation in the central difference method, stability is always conditional. However, a dissipative explicit method should not decrease the stability time limit considerably when compared with the stability limit of the central difference method.

Spectrum

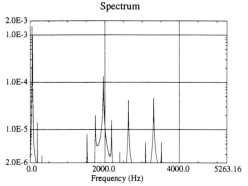

Spectrum

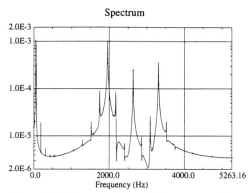

Figure 5.30: Spectrum for multi-time step solution for first 5000 time steps.

Figure 5.31: Spectrum for multi-time step solution for first 6000 time steps.

In Hilber, Hughes and Taylor (1977), dissipation in the above mentioned implicit methods was compared. It was shown that all of these methods possess some drawbacks. Another family of methods called $\alpha$ methods was proposed where positive $\gamma$ damping was used in conjunction with negative $\alpha$ damping (also see Hilber and Hughes (1978)). They showed that this dissipation scheme works better than positive $\gamma$ damping or positive $\alpha$ damping alone. In this section we look at the possibility of using $\alpha$ methods in explicit analysis.

## FORMULATION

### Implicit Method

The governing equations of motion for the linear finite element model are

$$\mathbf{M}\ddot{\mathbf{u}} + \mathbf{K}\mathbf{u} = \mathbf{f} \tag{5.20}$$

with initial conditions

$$\mathbf{u}_0 = \mathbf{u}(0) = \mathbf{d}, \ \dot{\mathbf{u}}_0 = \dot{\mathbf{u}}(0) = \mathbf{v} \tag{5.21}$$

The $\alpha$ family of methods is defined as follows:

$$\mathbf{M}\mathbf{a}_{n+1} + (1 + \alpha)\mathbf{K}\mathbf{d}_{n+1} - \alpha\mathbf{K}\mathbf{d}_n = \mathbf{f}_{n+1} \tag{5.22}$$

$$\mathbf{d}_{n+1} = \mathbf{d}_n + \Delta t\mathbf{v}_n + \Delta t^2[(\frac{1}{2} - \beta)\mathbf{a}_n + \beta\mathbf{a}_{n+1}] \tag{5.23}$$

$$\mathbf{v}_{n+1} = \mathbf{v}_n + \Delta t[(1 - \gamma)\mathbf{a}_n + \gamma\mathbf{a}_{n+1}] \tag{5.24}$$

where $\alpha$, $\beta$ and $\gamma$ are the free parameters and $\mathbf{a}_n$, $\mathbf{v}_n$ and $\mathbf{d}_n$ are approximations for $\ddot{\mathbf{u}}(t_n)$, $\dot{\mathbf{u}}(t_n)$ and $\mathbf{u}(t_n)$, respectively in which $t_n = n\Delta t$. The $\alpha$ family reduces to the Newmark family if $\alpha = 0$. In addition, $\beta = 0$ and $\gamma = 1/2$, yield the central difference method, and $\beta = 1/4$ and $\gamma = 1/2$, yield the trapezoidal method. For a method to be explicit, $\beta$ must always be equal to 0.

The characteristics of the $\alpha$ methods can be analyzed by considering the single degree of freedom model equation

$$M\ddot{u} + Ku = 0 \tag{5.25}$$

In an analogous notation, the $\alpha$ family of methods becomes

$$Ma_{n+1} + (1+\alpha)Kd_{n+1} - \alpha Kd_n \;=\; 0 \tag{5.26}$$

$$d_{n+1} \;=\; d_n + \Delta t v_n + \Delta t^2[(\tfrac{1}{2} - \beta)a_n + \beta a_{n+1}] \tag{5.27}$$

$$v_{n+1} \;=\; v_n + \Delta t[(1-\gamma)a_n + \gamma a_{n+1}] \tag{5.28}$$

Equations 5.26-5.28 can be written in recursive form as

$$\mathbf{X}_{n+1} = \mathbf{A}\mathbf{X}_n \tag{5.29}$$

where

$$\mathbf{X}_n = (d_n, \Delta t v_n, \Delta t^2 a_n)^{\mathrm{T}} \tag{5.30}$$

Here $\mathbf{A}$ is the amplification matrix whose eigenvalues dictate the stability and dissipation in an algorithm. For $\alpha$ methods, the amplification matrix is

$$\mathbf{A} = \frac{1}{D}\begin{bmatrix} 1 + \alpha\beta\Omega^2 & 1 & \tfrac{1}{2} - \beta \\ -\gamma\Omega^2 & 1 - (1+\alpha)(\gamma-\beta)\Omega^2 & 1 - \gamma - (1+\alpha)(\tfrac{1}{2}\gamma - \beta)\Omega^2 \\ -\Omega^2 & -(1+\alpha)\Omega^2 & -(1+\alpha)(\tfrac{1}{2} - \beta)\Omega^2 \end{bmatrix} \tag{5.31}$$

in which

$$D \;=\; 1 + (1+\alpha)\beta\Omega^2 \tag{5.32}$$

$$\Omega \;=\; \omega\Delta t \tag{5.33}$$

$$\omega \;=\; (K/M)^{1/2} \tag{5.34}$$

For calculating the eigenvalues of $\mathbf{A}$, the characteristic equation is

$$|\mathbf{A} - \lambda\mathbf{I}| = \lambda^3 - 2A_1\lambda^2 + A_2\lambda - A_3 = 0 \tag{5.35}$$

where $A_1$, $A_2$, and $A_3$ are invariants of $\mathbf{A}$, $\lambda$ is an eigenvalue of $\mathbf{A}$, and $\mathbf{I}$ is an identity matrix. Invariants of $\mathbf{A}$ can be computed as

$$\left.\begin{aligned} A_1 &= \tfrac{1}{2}\text{trace } \mathbf{A} &= 1 - \Omega^2[(1+\alpha)(\gamma + \tfrac{1}{2}) - \alpha\beta]/2D \\ A_2 &= \Sigma\text{principal minors of } \mathbf{A} &= 1 - \Omega^2[\gamma - \tfrac{1}{2} + 2\alpha(\gamma - \beta)]/D \\ A_3 &= \text{determinant } \mathbf{A} &= \alpha\Omega^2(\beta - \gamma + \tfrac{1}{2})/D \end{aligned}\right\} \tag{5.36}$$

## Explicit Method

Since we are primarily interested in explicit integration methods, the above set of equations can be simplified by substituting $\beta$ equal to 0, giving

$$D \;=\; 1 \tag{5.37}$$

$$\mathbf{A} \;=\; \begin{bmatrix} 1 & 1 & \tfrac{1}{2} \\ -\gamma\Omega^2 & 1 - \gamma(1+\alpha)\Omega^2 & 1 - \gamma - \tfrac{1}{2}\gamma(1+\alpha)\Omega^2 \\ -\Omega^2 & -(1+\alpha)\Omega^2 & -\tfrac{1}{2}(1+\alpha)\Omega^2 \end{bmatrix} \tag{5.38}$$

Invariants of $\mathbf{A}$ are also simplified as

$$\left.\begin{array}{rcl} A_1 &=& 1 - \Omega^2[(1+\alpha)(\gamma + \frac{1}{2})]/2 \\ A_2 &=& 1 - \Omega^2[\gamma - \frac{1}{2} + 2\alpha\gamma] \\ A_3 &=& \alpha\Omega^2(\frac{1}{2} - \gamma) \end{array}\right\} \tag{5.39}$$

To better understand the dependence of $\mathbf{A}$ on $\alpha$ and $\gamma$, we substitute

$$\gamma = \frac{1}{2} + \delta \tag{5.40}$$

into equations (5.38) and (5.39), giving

$$\mathbf{A} = \begin{bmatrix} 1 & 1 & \frac{1}{2} \\ -(\frac{1}{2}+\delta)\Omega^2 & 1 - (\frac{1}{2}+\delta)(1+\alpha)\Omega^2 & \frac{1}{2} - \delta - \frac{1}{2}(\frac{1}{2}+\delta)(1+\alpha)\Omega^2 \\ -\Omega^2 & -(1+\alpha)\Omega^2 & -\frac{1}{2}(1+\alpha)\Omega^2 \end{bmatrix} \tag{5.41}$$

and

$$\left.\begin{array}{rcl} A_1 &=& 1 - \Omega^2[(1+\alpha)(1+\delta)]/2 \\ A_2 &=& 1 - \Omega^2[\alpha + \delta + 2\alpha\delta] \\ A_3 &=& -\alpha\delta\Omega^2 \end{array}\right\} \tag{5.42}$$

It can be seen from the above equations that $\alpha$ and $\delta$ are interchangeable (due to symmetry). Since the eigenvalues of $\mathbf{A}$ depend only on the invariants of $\mathbf{A}$, $\alpha = a$ and $\delta = b$ would produce the same eigenvalues as $\alpha = b$ and $\delta = a$. Furthermore, stability and dissipation are directly dependent on the eigenvalues of amplification matrix, they also have the same dependency on $\alpha$ as $\delta$. It may be noted however that $\alpha$ and $\delta$ are not interchangeable in $\mathbf{A}$ itself. The effect of this property is not fully known.

## STABILITY AND DISSIPATION ANALYSIS

### Case 1: $\alpha = 0$

### Stability

Let us first look at the stability of $\alpha$ family of methods when $\alpha$ is 0. In such case, the invariants of $\mathbf{A}$ become

$$\left.\begin{array}{rcl} A_1 &=& 1 - (1+\delta)\Omega^2/2 \\ A_2 &=& 1 - \delta\Omega^2 \\ A_3 &=& 0 \end{array}\right\} \tag{5.43}$$

The characteristic equation becomes

$$\lambda(\lambda^2 - 2A_1\lambda + A_2) = 0 \tag{5.44}$$

The non-trivial eigenvalues are

$$\lambda_{1,2} = A_1 \pm \sqrt{A_1^2 - A_2} \tag{5.45}$$

After substituting above equations and simplifying, we get

$$\lambda_{1,2} = 1 - (1+\delta)\frac{\Omega^2}{2} \pm \Omega\sqrt{\left[(1+\delta)\frac{\Omega}{2}\right]^2 - 1} \qquad (5.46)$$

For stability, the eigenvalues must be complex, we get

$$\Omega < \frac{2}{1+\delta} \qquad (5.47)$$

Furthermore, the spectral radius must also be bounded by 1. The spectral radius of A is

$$\rho = |\lambda_1| = |\lambda_2| = \sqrt{A_2} = \sqrt{1 - \delta\Omega^2} \qquad (5.48)$$

Hence for stability we should also have $\delta \geq 0$.

Dissipation

The spectral radius of A can be expanded using Taylor series as

$$\rho = 1 - \frac{1}{2}\delta\Omega^2 - \frac{1}{8}\delta^2\Omega^4 + O(\delta^3\Omega^6) \qquad (5.49)$$

Hence, if $\delta\Omega^2 \ll 1$, $\rho$ varies linearly with $\delta$ and quadratically with $\Omega$. In terms of frequencies, it can be said that higher modes are more severely affected than lower modes. For larger values of $\delta\Omega^2$, the next one or more higher order terms become significant and dissipation becomes progressively higher for higher modes. In any case, from Eqs. 5.47 and 5.48

$$\Omega_{max} = \frac{2}{1+\delta} \qquad (5.50)$$

$$\rho_{min} = \frac{1-\delta}{1+\delta} = \Omega_{max} - 1 \qquad (5.51)$$

Exact values of $\rho$ as a function of $\Omega$ are plotted in Fig. 5.32 for various values of $\delta$.

**Case 2: $\delta = 0$**

It may be noted that the analysis conducted in Case 1 is also true for Case 2 if we replace $\delta$ with $\alpha$.

**Case 3: $\alpha \neq 0, \delta \neq 0$**

Dissipation

When neither $\alpha$ nor $\delta$ is equal to 0, the invariants of A are as given in Eq. 5.42. The major difference as compared to previous cases is that here $A_3$ is not 0. Hence, Eq. 5.35 has three roots out of which two are principal roots and one a spurious root.

As obtaining the eigenvalues in this case is quite involved, a symbolic manipulator was used; the spectral radius of the amplification matrix is plotted in Fig. 5.33 for $\delta + \alpha = 0.1$, $\delta = -0.1$ to 0.2, and $\Omega = 0$ to 1.5. We restrict the maximum value of $\Omega$ to be 1.5 rather than 2.0

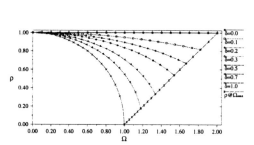

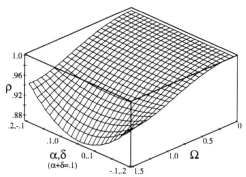

Figure 5.32: Dissipation in explicit $\alpha$ methods when $\alpha = 0$.

Figure 5.33: Dissipation in explicit $\alpha$ methods when $\alpha + \delta = 0.1$.

because positive $\alpha$ and $\delta$ reduce the stability limit. As is clear from this figure that while $\delta + \alpha$ is kept constant, the maximum dissipation is attained in higher modes when $\alpha = \delta$.

In an implicit scheme where $\gamma = 1/2 - \alpha$ (i.e., $\delta = -\alpha$) and $\beta = (1/2 - \alpha)^2/4$, $\alpha = 0$ to $-.33$ produces low dissipation in the lower modes (Hilber, Hughes, and Taylor (1977)). It may be noted this is achieved by having negative $\alpha$ in conjunction with positive $\delta$ (i.e., $\gamma > 1/2$). In an explicit case $\beta$ is always zero, hence the above range of $\alpha$ is not applicable. Here, negative $\alpha$ produces low dissipation in the higher modes which is not desirable.

For $\delta + \alpha = 0.2$, $\delta = -0.1$ to $0.3$, and $\Omega = 0$ to $1.5$, spectral radius is plotted in Fig. 5.34. Again maximum dissipation (or minimum $\rho$) is achieved when $\delta = \alpha = 0.1$. Negative values of $\alpha$ tend to decrease the amount of dissipation in higher modes. Furthermore, negative $\alpha$ (or $\delta$) causes the spectral radius curve to be flattened for higher values of $\Omega$. Such behavior is more prominent in Fig. 5.35 where $\rho$ vs. $\Omega$ is plotted for different values of $\delta$.

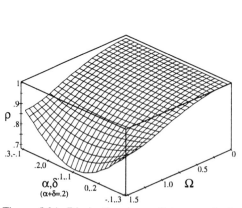

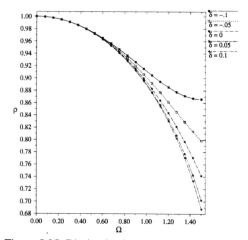

Figure 5.34: Dissipation in explicit $\alpha$ methods when $\alpha + \delta = 0.2$.

Figure 5.35: Dissipation in explicit $\alpha$ methods for various values of $\delta$ when $\alpha + \delta = 0.2$.

Stability

Since the eigenvalues of **A** are quite complex in this case, it is not possible to obtain a measure for stability in discrete form. We resort to numerical experimentation to examine how the eigenvalues of the amplification matrix behave. Since the maximum dissipation is attained when $\alpha = \delta$, we would restrict ourselves to this case only. For $\delta = 0.1$, the real and imaginary parts of $\lambda_1$, $\lambda_2$ and $\lambda_3$ are plotted in Fig. 5.36. Overall, it can be seen that eigenvalues have three ranges. In the first range, $\lambda_2$ and $\lambda_3$ are complex conjugates. In the second range, imaginary parts of both $\lambda_2$ and $\lambda_3$ vanish, however the real part of $\lambda_2$ increases with $\Omega$ and of $\lambda_3$ decreases with $\Omega$. In the third range, the real part of $\lambda_2$ suddenly drops to the value of the real part of $\lambda_3$. Henceforth, both $\lambda_2$ and $\lambda_3$ are complex conjugates again. The value of $\lambda_1$ remains quite small in the first range. It shows a increasing trend in the second range and finally in the third range it becomes quite large, increasing at a very high rate.

It may be noted that when $\alpha$ or $\delta$ is zero, only the first two ranges exist. The first range extends from $\Omega = 0$ to $2/(1+\delta)$, and the second range is from $2/(1 + \delta)$ to $\infty$; $\lambda_1$ is also zero in such a case. In the present case, the second range starts shrinking from $\infty$ as $\delta$ increases. It is found that in the second range, the real part of $\lambda_2$ has more growth while in the third

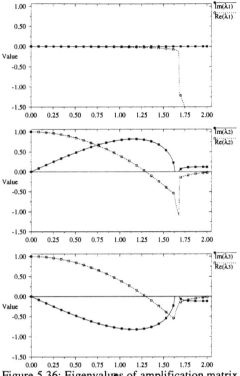

Figure 5.36: Eigenvalues of amplification matrix when $\alpha = \delta = 0.1$.

range $\lambda_1$ has more growth. For smaller values of $\delta$, the second range is relatively large but has large $\lambda_2$, hence becomes unstable. For larger values of $\delta$, the second range becomes very small, and the third range starts just after the first range ends. Because of the presence of large $\lambda_1$, the third range is also unstable. Hence, the practical range for stable computation is the first range. Although this stability range is not available in discrete form, from numerical experimentation, for positive $\alpha$ and $\delta$, it can be approximated as

$$\Omega < \frac{2}{(1 + \alpha)(1 + \delta)} \tag{5.52}$$

## CHOICE OF DISSIPATION METHOD

It is shown above that positive values of $\delta$ and $\alpha$ introduce damping into the explicit system. A combination of $\alpha = \delta$ is found to produce maximum dissipation in higher modes. However, when both $\delta$ and $\alpha$ are non zero, the amplification matrix has 3 eigenvalues , one of which is spurious ($\lambda_1$). The stability limit is also not defined clearly in this case. Although $\alpha = 0$ and $\delta = 2a$ (or vice versa) has slightly less dissipation in higher modes when compared with $\alpha = \delta = a$, it has a few advantages viz.

- the stability limit is clearly defined
- the spurious eigenvalue is not present

- implementation requires minimal changes to the code.

Hence, we choose the dissipation method as positive $\delta > 0$ and $\alpha = 0$.

## *IMPLEMENTATION*

Dissipation in explicit multi-time step methods is implemented as follows
Define

$$i = mod(n, m) \tag{5.53}$$

For nodes in group $A$ :

$$a_n = (f_n^{ext} - f_n^{int})/M \tag{5.54}$$

$$v_{n+1/2-\delta} = v_{n-1/2-\delta} + \Delta t a_n \tag{5.55}$$

$$d_{n+1} = d_n + \Delta t v_{n+1/2-\delta} + \delta \Delta t^2 a_n \tag{5.56}$$

For nodes in group $B$ :
Only when i = 1

$$a_n = (f_n^{ext} - f_n^{int})/M \tag{5.57}$$

$$v_{n+m/2-m\delta} = v_{n-m/2-m\delta} + m\Delta t a_n \tag{5.58}$$

$$d_{n+m} = d_n + \Delta t v_{n+m/2-m\delta} + \delta(m\Delta t)^2 a_n \tag{5.59}$$

For nodes in group $C$ :
Only when i = 1

$$a^{intf} = a_n = (f_n^{ext} - f_n^{int})/M \tag{5.60}$$

$$v^{intf} = v_{n+m/2-m\delta} = v_{n-m/2-m\delta} + m\Delta t a_n \tag{5.61}$$

$$d^{intfold} = d_n \tag{5.62}$$

Always

$$d_{n+1} = d^{intfold} + i\Delta t v^{intf} [-i(m-i)\Delta t^2 a^{intf}/2] + \delta m i \Delta t^2 a^{intf} \tag{5.63}$$

In equation 5.63, the term in [ ] is used only if interpolation is by the constant acceleration method.

## 5.8 NUMERICAL EXAMPLES

In this section we apply the dissipative algorithm discussed above to multi-time step schemes. Various example problems, that were shown to diverge earlier are run again after introducing $\delta$ damping into the system. The example problem shown in Fig. 5.1 and described in Table 5.4 is run with $\delta = .004$ for both the standard and multi-time step solution. The displacement time history for the standard solution is shown in Fig. 5.37. Effects of dissipation is evident in this figure as amplitude decay occurs in the time history. The displacement time history for multi-time solution is shown in Fig. 5.38. It can be seen that the solution remains accurate and free of any noise or high frequencies. Amplitude decay is slightly higher than the standard solution. Other than this, multi-time step solution is quite comparable with the standard solution. It may be noted here that the value of $\delta = 0.004$ is achieved by numerical experimentation and is found to be sufficient to dissipate the higher modes. A value of $\delta = .003$ was found to be not sufficient to stabilize the solution. To validate the method for long term simulations, the standard and multi-time step solutions are also run for 100,000 time steps with $\delta = 0.004$. The plot for about final 5000 time steps is shown in Figs. 5.39 and 5.40 for standard and multi-time step method respectively. It can be seen that the solution remains within bounds.

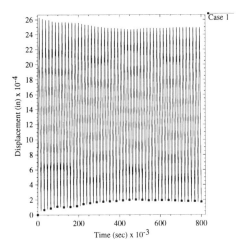

Figure 5.37: Displacement time history for damped standard solution with Heaviside loading, $\delta = .004$.

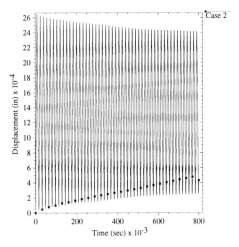

Figure 5.38: Displacement time history for damped multi-time step solution with Heaviside loading, $\delta = .004$.

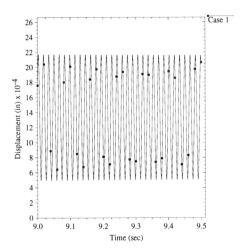

Figure 5.39: Displacement time history at the end of 100,000 time steps for damped standard solution with Heaviside loading, $\delta = .004$.

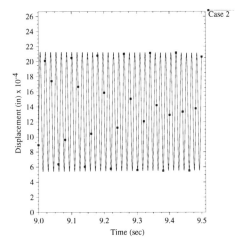

Figure 5.40: Displacement time history at the end of 100,000 time steps for damped multi-time step solution with Heaviside loading, $\delta = .004$.

Next we run the above problem with a sinusoidal loading applied at the tip of the bar. The loading function is as given in Eq. 5.19. The displacement time histories for the standard and multi-time step solutions without damping are plotted in Figs. 5.26 and 5.27, respectively. In this case, a value of $\delta = .006$ is found to be sufficient to damp out higher frequencies. Time history curves for displacement are plotted in Figs. 5.41 and 5.42 for damped standard and multi-time step solutions, respectively. Comparing Fig. 5.42 with Fig. 5.27, it can be seen that the multi-time step solution with the introduction of small damping remains within bounds, while without damping, it diverges. Again the standard and multi-time step solutions are quite comparable. This problem is also run for 100,000 time steps to validate the method for long term simulation. The plot for the final stages of the displacement time history for standard and multi-time step solution is shown in Figs. 5.43 and 5.44 respectively. It can be seen that the solution remains within bounds at the end of simulation also. The results of standard and multi-time step solutions are also quite comparable.

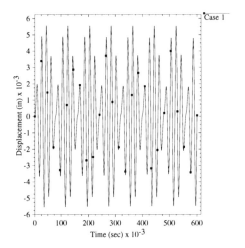

Figure 5.41: Displacement time history for damped standard solution with sinusoidal loading, $\delta = .006$.

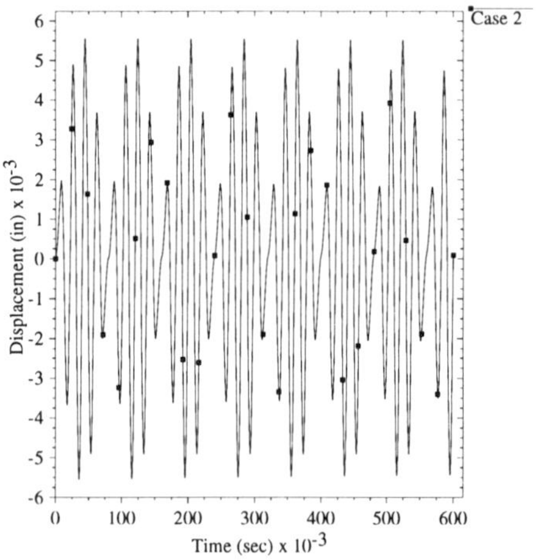

Figure 5.42: Displacement time history for damped multi-time solution with sinusoidal loading, $\delta = .006$.

## 5.9 PARALLEL IMPLEMENTATION

Another avenue where performance of finite element simulations can be enhanced is by exploiting the architectures of parallel computers. The performance enhancements of the currently dominant sequential Von Neumann computer architectures are approaching an asymptote, hence placing a fundamental physical limit on their potential computational power. Such a limit is the speed of light ($3 \times 10^8$ m/sec in a vacuum) which results in a signal transmission speed for silicon of at most $3 \times 10^7$ m/sec (DeCegama (1989)). Hence, a 3 cm diameter chip can propagate a signal in $10^{-9}$ seconds. The maximum power of a computer built with such chips is at most $10^9$ floating-point operations per second (1 GFLOPS). A similar limit can be estimated for other materials such as gallium arsenide, which has a lower signal propagation time. Since the speed of light is the governing limit, sequential processors will reach the maximum speed physically possible in the near future. It is clear that computer architectures other than the classical serial Von Neumann model must be exploited to have significant improvement in computational

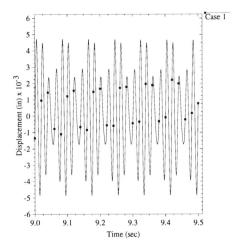

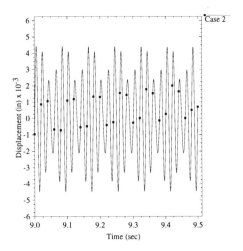

Figure 5.43: Displacement time history at the end of 100,000 time steps for damped standard solution with sinusoidal loading, $\delta = .006$.

Figure 5.44: Displacement time history at the end of 100,000 time steps for damped multi-time solution with sinusoidal loading, $\delta = .006$.

speeds. Parallel computer architectures provide a first step improvement over the confines of a uniprocessor model.

Another reason for considering parallel machines is the enormous cost associated with uniprocessor supercomputers. Grosch's law (DeCegama (1989)) states that the computing power of a single processor increases in proportion to the square of the cost. However, this law no longer applies to computer systems built with many inexpensive processors. The relationship between extra cost and increase in the speed of a processor is governed by the law of diminishing returns. As we move towards the underlying physical limit, an increasingly larger amount of money needs to be spent for increasingly smaller gains in computational processor power. Processors having a comparatively low speed are available at a very low cost. A parallel machine comprised of many such processors not only has more speed than a uniprocessor supercomputer but also has less cost (Dongarra and Duff (1992)). This makes parallel computers more accessible as well as economical.

Fox (1991) discusses the achievements and prospects of parallel computing (also see Messina (1991)). For an introduction to parallel programming, see Brawer (1989). High performance Fortran programming is explained by Levesque and Williamson (1989) and portable tools for parallel Fortran programming are discussed by Kumar and Philips (1991). The latest release, Fortran 90, which is suitable for parallel computers, is explained by Metcalf and Reid (1990). The most significant new features are the ability to handle arrays using a concise but powerful notation, and the ability to define and manipulate user-defined data types. A bibliography on finite elements and supercomputing is given by White and Abel (1988).

The latest in parallel computer architectures are the MIMD (*Multiple Instruction, Multiple Data Stream*) architectures. This classification is based on Flynn (1972). In such machines, each processor has its own independent instruction and data streams. In general, such a computer contains several processing elements operating in an asynchronous manner on different data streams coming from a shared or local memory. What differentiates MIMD from independent processors is that in an MIMD architecture, processors can communicate with each other.

Based on how memory is accessed by each processor, MIMD machines can be classified into two categories.

1. **Tightly-Coupled Parallel Processors.** All processors share the same memory (Shared Memory) and have equal access to it at all times. Various processors are synchronized by using shared variables for interprocessor communication.

2. **Loosely-Coupled Parallel Processors.** All processors have their own local memory and no other processor has any access to it. The synchronization between the processors is achieved by passing messages via the interconnection network.

Distributed machines with loosely coupled processors have the facility of transmitting messages among its processors. For a distributed network of workstations, the P4 package as described by Boyle et. al. (1987) and Butler and Lusk (1992) is used for interprocessor communication. In the Intel Touchstone parallel computers, communication is done by calling various subroutines and functions (Ragsdale (1991)). The code is portable as the two message-passing libraries are identical in function and differ only in semantics. A processor sends a message by calling a *send* subroutine which reads the data from the buffer and transmits it to the receiving processor. A processor receives a message by calling a *receive* subroutine which takes the incoming message and stores the data in the buffer. These message passing calls can be synchronous and asynchronous. On Intel parallel computers, synchronous subroutines for message passing include **csend()** and **crecv()**. Execution of a process resumes only after message buffer has been completely sent or received by the processor issuing the call. Corresponding asynchronous subroutines include **isend()** and **irecv()**. In this case, process resumes execution as soon as command is issued. A message ID is issued to later check if the command has been completed. The number of message IDs is limited, and an error occurs when no message IDs are available for a requested asynchronous call. Therefore, message IDs need to be released as soon as possible by calling **msgwait()**, **msgdone()**, or **msgcancel()**. Details of these subroutines are given in the Touchstone Delta Fortran System Calls Reference Manual and Using the Intel iPSC/860 (Fortran Version).

The biggest challenge lies in the substitution of the key algorithms in an application program with redesigned algorithms which exploit the new architectures and use better or more appropriate numerical techniques. A great deal of complexity can be avoided in parallelizing the multi-time step algorithm if the advances made in the area of regular parallel finite element analysis can be exploited. Hence, our aim is to decouple the implementation of parallel processing and subcycling. Subcycling affects part of the finite element code where kinematic quantities, i.e., displacements, velocities or accelerations, are calculated. Hence if parallel processing is implemented without touching the segments of the code where kinematic quantities are calculated, the goal of decoupling the problem would be achieved. One such methodology is discussed by Plaskacz, Ramirez and Gupta (1992,1994). In these papers, parallel processing in a finite element code is achieved by exchanging internal forces only. Performance exceeding 1 gigaflop was attained. For a problem consisting 16384 elements, a speedup of 333.15 was obtained by running the problem on 512 processors of the Intel Delta. This translates to 99.89% of the code running completely in parallel and only 0.11% of the code running in serial.

An essential element in conducting finite element analyses on parallel computers is domain decomposition. For most of the problems, domain decomposition is used as a first step towards conquering the problem. Methodologies proposed by various researchers such as Farhat (1988), Malone (1988), Al-Nasra and Nguyen (1991), Simon (1991), Hsieh and Abel (1993) and Gupta and Ramirez (1995b) can be used for domain decomposition. Subdomains for parallel processing can overlap subdomains for subcycling, because the two do not affect each other. It may be noted that for maximum efficiency, some enhancements may be required to the domain decomposition algorithms. A subcycling problem in general can have various groups of nodes and elements that are integrated at different time steps. To attain a good degree of load balancing, domain decomposition program should be able to identify each of such groups and decompose them into desired number of subdomains while minimizing the length of interface boundaries. The worst scenario is where a subdomain has only the elements that are being integrated at large time step.

An algorithm for parallelizing the code using an asynchronous *send* and a synchronous *receive* is given in Table 5.5. Further details on the consistency of the algorithm can be found in Plaskacz, Ramirez and Gupta (1994). Similar performance figures as obtained in this referance for standard finite element analysis are expected. Exchange of internal forces at each time step does include some redundant exchange for nodes that are integrated by large time step. Such redundancy can be removed by assembling the *send* buffer for only those nodes that require an update at a particular time step. However, part of the time gain achieved in transmission of the message would be offset by the extra time spent in preparing the message buffer or in decoding the message buffer by the receiving processor. Overall gain or loss would depend on speed of the CPU, transmission speed, geometry of the mesh, and multi-time step ratio.

Table 5.5: Message passing algorithm with **isend()** and **crecv()**.

---

1. Time step i=1.
2. Calculate internal forces at nodes.
3. Exchange forces (send); Loop over all the neighboring subdomains.
    Use buffer *afint* and *bfint* for odd and even time steps respectively.
    a. Copy identity of current subdomain in next element of buffer.
    b. Store this location in *flag*.
    c. Copy forces pertaining to the interface nodes into the next
        $6 \times N$ elements of buffer [*may skip nodes that are not being
        updated at this time step*].
    d. If time step is not 1 or 2 release the id, assigned by the **isend()**
        call for $(i-2)^{th}$ time step, by calling **msgdone()**.
    e. Call **isend()** to send buffer from location *flag* to the neighboring
        subdomain and store the id assigned by **isend()** in *aid* for odd
        time steps or in *bid* for even time steps.
    End sending forces.
4. Exchange forces (receive); Loop over all the neighboring subdomains.
    a. Call **crecv()** to receive a message and store it in buffer *fintex*.
    b. Get the identity of the sending processor.
    c. Identify the interface nodes.
    d. Update the internal forces [*skip nodes that are not being
        updated at this time step, if and only if they are skipped in 3c*].
    e. Use *fintex* again for the next neighboring subdomain.
    End receiving forces.
5. i ← i+1; Go to 2.

---

During time integration, at each time step, subdomain computations proceed independently in parallel in accordance with the serial finite element algorithm. First, the element internal forces are computed, followed by nodal force-vector assembly, and for parallel computations, an exchange of element internal forces for nodes at subdomain interfaces. Finally, the equations of motion are integrated. The need for synchronization of processes arises from the fact that not all the steps of time integration can be run in parallel independently. At some point during each time step, messages must be exchanged to update various arrays. The exchange of internal forces pertaining to interface nodes is required because contributions from elements in neighboring subdomains are not added in the calculation of nodal internal force. No kinematic quantities, i.e., displacements, velocities or accelerations, are exchanged. The equations of motion for the interface nodes are integrated redundantly. Each subdomain is in itself a complete finite element mesh. Thus, in addition to requiring minimal communication, the exchange algorithm has the advantage of allowing analysts to use the same post-processing software for each subdomain as they would normally do for a job executing on a single processor machine.

## 5.10  CONCLUSION

A study is presented to investigate the accuracy of the multi-time step schemes for second order equations. In general it is found that subcycling can induce errors in the solution that may not be acceptable for long simulations. Accuracy of a finer mesh integrated with the multi-time step scheme is found to be lower than the accuracy of the coarser mesh integrated with a large time step. Even for $m = 2$ the multi-time step solution exhibits is very inaccurate, and the accuracy is expected to decrease for higher and non-integer values of the time step ratio. Both constant velocity and constant acceleration interpolation methods are studied. The constant acceleration method is more accurate in the initial part of the solution, while the constant velocity method appears to remain stable for a longer period of time. For all cases the subcycling solution for long simulations diverges rapidly.

An inference drawn from this study is that multi-time step solutions, in their present form, cannot be used for *all* problems. Users must consider the behavior of their particular problem before employing them. Although this study is for structural dynamic systems, it raises a need for the accuracy analysis of other subcycling schemes such as for heat conduction systems. Further research is also needed to find better interpolation schemes for the interface nodes.

Without preventive measures, multi-time step schemes in their present form tend to diverge if the simulation is run for a long time. An effort is undertaken to pin-point the reason for divergence exhibited by subcycling procedures and to find procedures to control the growth of the errors. Frequency domain analysis reveals that subcycling gives rise to higher modes. Because of these higher modes small fluctuations in the time history are introduced. Furthermore, it is found that these higher modes become dominant with time and completely take over the solution.

The central difference method is an explicit method and does not possess any numerical damping. To introduce numerical damping into the central difference method, the $\alpha$ family of methods is investigated; the parameter $\alpha$ as well as the analogous parameter $\delta$ ($\delta = \gamma - 1/2$) are found to have the same amount of dissipation in an explicit method which can be considered to be a damped central difference method. Stability and dissipation criterions are investigated for explicit $\alpha$ methods. It is shown that the combination $\alpha = \delta$ produces maximum disspation in higher frequencies. However, the presence of both of these parameter gives rise to a spurious eigenvalue of the amplification matrix. Stability limits are also not easily computable in this case. Since, $\delta = 2a$ and $\alpha = 0$ is only slightly less dissipative than when $\alpha = \delta = a$, the former method is chosen for developing a damped multi-time step scheme. It is shown that the introduction of a small amount of numerical damping into the system is sufficient to damp out the spurious higher modes. The damped multi-time step solution is found to be quite comparable to the damped standard solution. Small values of $\delta$ also have little effect on the stability limits. However, a value of $\delta$ that is sufficient to dissipate higher modes in a particular problem, could only be found by numerical experimentation. Future research is required to find an analytical relationship between the minimum $\delta$ required and the parameters of the spatial and temporal discretization.

Finally, parallel implementation of the subcycling algorithm is discussed. It is shown that parallelization and subcycling affect different parts of the algorithm and the two can be decoupled from each other. Hence, methodologies of standard parallel finite element analysis can be applied to subcycled parallel finite element analysis. However, further work is required to enhance domain decomposition algorithms such that they address the subcycling needs more effectively and efficiently.

# REFERENCES

T. Belytschko and R. Mullen, "Mesh Partitions of Explicit-Implicit time Integrators," *Formulations and Computational Algorithms in Finite Element Analysis*, eds. K. J. Bathe, J. T. Oden and W. Wunderlich, pp. 673-690, MIT Press, Cambridge, MA, 1976.

T. Belytschko and R. Mullen, "Mesh Partitions of Explicit-Implicit time Integrators," *Formulations and Computational Algorithms in Finite Element Analysis*, eds. K. J. Bathe, J. T. Oden and W. Wunderlich, pp. 673-690, MIT Press, Cambridge, MA, 1976.

T. Belytschko and R. Mullen, "Stability of Explicit-Implicit Mesh Partitions in Time Integration," *International Journal for Numerical Methods in Engineering*, 12, pp. 1575-1586, 1978.

T. Belytschko, H.-J. Yen, and R. Mullen, "Mixed Methods for Time Integration," *Computer Methods in Applied Mechanics and Engineering*, 17/18, pp. 259-275, 1979.

T. Belytschko, "Partitioned and Adaptive Algorithms for Explicit Time Integration," *Nonlinear Finite Element Analysis in Structural Mechanics*, eds. W. Wunderlich, W. Stein, and K. J. Bathe, pp. 572-584, Springer-Verlag, Berlin, 1981.

T. Belytschko, W. K. Liu, and P. Smolinski, "Computational Methods for Analysis of Transient Response," *Recent Advances in Engineering Mechanics and Their Impact on Civil Engineering Practice, Proceedings of the 4th Engineering Mechanics Division Speciality Conference*, eds. W. F. Chen and A. D. M. Lewis, pp. 1-17, New York, ASCE, 1983.

T. Belytschko, P. Smolinski, and W. K. Liu, "Multi-Stepping Implicit-Explicit Procedures in Transient Analysis," *Innovative Methods for Nonlinear Problems*, eds. W. K. Liu et al., pp. 135-153, Pineridge Press, Swansea, U.K., 1984.

T. Belytschko, P. Smolinski, and W. K. Liu, "Stability of Multi-Time Step Partitioned Integraters for First-Order Finite-Element Systems," *Computer Methods in Applied Mechanics and Engineering*, 49, pp. 281-297, 1985.

T. Belytschko and Y. Y. Lu, "Convergence and Stability Analyses of Multi-time Step Algorithm for Parabolic Systems," *Computer Methods in Applied Mechanics and Engineering*, 102, pp. 179-198, 1993a.

T. Belytschko and Y. Y. Lu, "Explicit Multi-Time Step Integration for First and Second Order Finite Element Semidiscretizations," *Computer Methods in Applied Mechanics and Engineering*, 108, pp. 353-383, 1993b.

J. Boyle, R. Butler, T. Disz, B. Glickfeld, E. Lusk, R. Overbeek, J. Patterson, and R. Stevens, *Portable Programs for Parallel Processors*, Holt, Rinehart and Winston, New York, 1987.

S. Brawer, *Introduction to Parallel Programming*, Academic Press, San Diego, 1989.

R. Butler and E. Lusk, *User's Guide to the p4 Programming System*, Technical Report ANL-92/17, Mathematics and Computer Science Division, October 1992.

Bruce W. Char, Keith O. Geddes, Gaston H. Gonnet, Benton L. Leong, Michael B. Monagan, and Stephen M. Watt, *First Leaves: A Tutorial Introduction to Maple V*, Springer-Verlag, New York, 1992.

R. D. Cook, D. S Malkus, and M. E. Plesha, *Concepts and Applications of Finite Element Analysis*, 3rd ed., Wiley, New York, 1987.

A. L. DeCegama, *The Technology of Parallel Processing: Parallel Processing Architectures and VLSI Hardware*, Vol. 1, Prentice Hall, 1989.

J. Dongarra and I. S. Duff, "Advance Architecture Computers," *Supercomputing in Engineering Analysis*, ed. H. Adeli, Marcel Dekker, 1992.

J. Donea and H. Laval, "Nodal Partioning of Explicit Finite Element Methods for Unsteady Diffusion Problems," *Computer Methods in Applied Mechanics and Engineering*, 68, pp. 189-204, 1988.

C. Farhat, "A Simple and Efficient Automatic FEM Domain Decomposer," *Computers & Structures*, 28, pp. 579-602, 1988.

M. J. Flynn, "Some Computer Organizations and Their Effectiveness," *IEEE Transactions on Computers*, C-21, pp. 948-960, 1972.

Geoffrey C. Fox, "Achievements and Prospects for Parallel Computing," *Concurrency: Practice and Experience*, 3(6), pp. 725-739, 1991.

Sanjeev Gupta and Martin R. Ramirez, "An Accuracy Analysis of Multi-Time Step Integration Schemes in Explicit Structural Dynamic Analysis," *Proceedings of the IACM Third World Congress on Computational Mechanics*, Chiba, Japan, August 1994.

Sanjeev Gupta, *Frontiers in High Performance Computing in Finite Element Analysis*, Ph.D. Dissertation, The Johns Hopkins University, Baltimore, MD 21218, June 1995.

Sanjeev Gupta and Martin R. Ramirez, "Comparison of Standard and Subcycling Integration Schemes for Second Order Equations," *Proceedings of the Third U.S. National Congress on Computational Mechanics*, Dallas, Texas, June 1995a.

Sanjeev Gupta and Martin R. Ramirez, "A Mapping Algorithm for Domain Decomposition in Massively Parallel Finite Element Analysis," *Computing Systems in Engineering: An International Journal*, 6(2), pp. 111-150, 1995b.

Hans M. Hilber, Thomas J. R. Hughes and Robert L. Taylor, "Improved Numerical Dissipation for Time Integration Algorithms in Structural Dynamics," *Earthquake Engineering and Structural Dynamics*, 5, pp. 283-292, 1977.

Hans M. Hilber and Thomas J. R. Hughes, "Collocation, Dissipation and 'Overshoot' for the Integration Schemes in Structural Dynamics," *Earthquake Engineering and Structural Dynamics*, 6, pp. 99-117, 1978.

Shang-Hsien Hsieh and John F. Abel, "Use of Networked Workstations for Parallel Nonlinear Structural Dynamic Simulations of Rotating Bladed-Disk Assemblies," *Computing Systems in Engineering*, 4(4-6), pp. 521-530, 1993.

T. J. R. Hughes and W. K. Liu, "Implicit-Explicit Finite Elements in Transient Analysis: Stability Theory," *Journal of Applied Mechanics*, 45, pp. 371-374, 1978a.

T. J. R. Hughes and W. K. Liu, "Implicit-Explicit Finite Elements in Transient Analysis: Implementation and Numerical Examples," *Journal of Applied Mechanics*, 45, pp. 375-378, 1978b.

T. J. R. Hughes, T. Belytschko, and W. K. Liu, "Convergence of an Element-Partitioned Subcycling Algorithm for the Semidiscrete Heat Equation," *Numerical Methods for Partial Differential Equations*, 3, pp. 131-137, 1987.

Gregory M. Hulbert, "Second-Order-Accurate Explicit Subcycling Algorithm for Unsteady Heat Conduction," *Numerical Heat Transfer, Part B*, 22, pp. 199-209, 1992.

Raymond D. Krieg and Samuel W. Key, "Transient Shell Response by Numerical Time Integration," *International Journal for Numerical Methods in Engineering*, 7, pp. 273-286, 1973.

Swarn P. Kumar and Ivor R. Philips, "Portable Tools for Fortran Parallel Programming," *Concurrency: Practice and Experience*, 3(6), pp. 559-572, 1991.

John M. Levesque and Joel W. Williamson, *A Guidebook to Fortran on Supercomputers*, Academic Press, Inc., 1989.

W. K. Liu and T. Belytschko, "Mixed-time Implicit Explicit finite Elements for Transient Analysis," *Computers & Structures*, 15, pp. 445-450, 1982.

W. K. Liu and J. Lin, "Stability of Mixed time Integration Schemes for Transient Thermal Analysis," *Numerical Heat Transfer*, 5, pp. 211-222, 1982.

W. K. Liu, "Development of Mixed Time Partition Procedures for Thermal Analysis of Structures," *International Journal for Numerical Methods in Engineering*, 19, pp. 125-140, 1983.

W. K. Liu and Y. F. Zhang, "Improvement of Mixed Time Implicit-Explicit Algorithms for Thermal Analysis of Structures," *Computer Methods in Applied Mechanics and Engineering*, 37, pp. 207-223, 1983.

W. K. Liu, T. Belytschko, and Y. F. Zhang, "Implementation and Accuracy of Mixed-Time Implicit-Explicit Methods for Structural Dynamics," *Computers & Structures*, 19, pp. 521-530, 1984.

J. G. Malone, "Automated Mesh Decomposition and Concurrent Finite Element Analysis for Hypercube Multiprocessor Computers," *Computer Methods in Applied Mechanics and Engineering*, 70, pp. 27-58, 1988.

Paul C. Messina, "Parallel Computing in the 1980s — One Person's View," *Concurrency: Practice and Experience*, 3(6), pp. 501-524, 1991.

Michael Metcalf and John Reid, *Fortran 90 Explained*, Oxford University Press, 1990.

A. Mizukami, "Variable Explicit Finite Element Methods for Unsteady Heat Conduction Equations," *Computer Methods in Applied Mechanics and Engineering*, 59, pp. 101-109, 1986.

M. Al-Nasra and D. T. Nguyen, "An Algorithm For Domain Decomposition In Finite Element Analysis," *Computers & Structures*, 39, pp. 277-289, 1991.

Mark O. Neal and Ted Belytschko, "Explicit-Explicit Subcycling with Non-Integer Time Step Ratios for Structural Dynamic Systems," *Computers & Structures*, 31, No. 6, pp. 871-880, 1989.

K. C. Park, "Partitioned Transient Analysis Procedure for coupled field Problems: Stability Analysis," *Journal of Applied Mechanics*, 47, pp. 370-376, 1980.

Edward J. Plaskacz, Martin R. Ramirez, and Sanjeev Gupta, "On Distributed Processing Applications in Finite Element Analysis," *Proceedings of the Engineering Mechanics Division*, ASCE, May 1992.

Edward J. Plaskacz, Martin R. Ramirez, and Sanjeev Gupta, "Nonlinear Explicit Finite Element Analysis on the Intel Delta," *Computing Systems in Engineering: An International Journal*, 5(1), pp. 1-17, 1994.

S. Ragsdale, ed. *Parallel Programming*, Intel McGraw-Hill, 1991.

Martin R. Ramirez and Ted Belytschko, "Analysis of Error in Forced Systems," *submitted for publication*, 1995.

H. D. Simon, "Partitioning of Unstructured Problems for Parallel Processing," *Computing Systems in Engineering: An International Journal*, 2(2), pp. 135-148, 1991.

P. Smolinski, T. Belytschko, and W. K. Liu, "Stability of Multi-Time Step Partitioned Transient Analysis for First-Order Systems of Equations," *Computer Methods in Applied Mechanics and Engineering*, 65, pp. 115-125, 1987.

P. Smolinski, T. Belytschko, and M. Neal, "Multi-time-step Integration Using Nodal Partitioning," *International Journal for Numerical Methods in Engineering*, 26, pp. 349-359, 1988.

P. Smolinski, "Convergence of Multi-Time Step Integration Methods," *Computers & Structures*, 35, pp. 719-724, 1990.

P. Smolinski, "Stability of Variable Explicit Time Integration for Unsteady Diffusion Problems," *Computer Methods in Applied Mechanics and Engineering*, 93, pp. 247-252, 1991.

P. Smolinski, "An Explicit Multi-Time Step Integration Method for Second Order Equations," *Computer Methods in Applied Mechanics and Engineering*, 94, pp. 25-34, 1992a.

P. Smolinski, "Stability Analysis of a Multi-Time Step Explicit Integration Method," *Computer Methods in Applied Mechanics and Engineering*, 95, pp. 291-300, 1992b.

Yue-Cong Wang, Viriyawan Murti, and Somasundaram Valliappan, "Assessment of the Accuracy of the Newmark Method in Transient Analysis of Wave Propagation Problems," *Earthquake Engineering and Structural Dynamics*, 21, pp. 987-1004, 1992.

Donald W. White and John F. Abel, "Bibliography on Finite Elements and Supercomputing," *Communications in Applied Numerical Methods*, 4, pp. 279-294, 1988.

Stephen Wolfram, *Mathematica: A System for Doing Mathematics by Computers*, 2nd ed., Addison-Wesley Publishing Company, New York, 1992.

*A Touchstone DELTA System Description*, Intel Supercomputer Systems Division, Intel Corporation, 1991.

*Using the Intel Touchstone DELTA System*, ANL/MCS-TM-152, Mathematics and Computer Science Division, June 1991.

# 6

# Krylov Subspace Methods on Parallel Computers

Yousef Saad[1]

Krylov subspace methods have enjoyed a growing popularity in computational sciences and engineering in the last decade. They are likely to make further inroads in various applications mainly because of the advances made in sophisticated modeling techniques, particularly three-dimensional ones. Another important reason for the progress of iterative methods, is the inherent difficulty in implementing full-scale direct methods on parallel computers. It is likely that direct solvers will progressively be used as local solvers in conjunction with iterative accelerators. In this overview we discuss some of the issues related to the implementations of parallel iterative solvers, focussing on implementations on distributed parallel environments.

**Key words:** Large linear systems; Krylov subspace methods; Iterative methods; Preconditioned Conjugate Gradient; Multicoloring; Incomplete LU preconditioning.

## 6.1 INTRODUCTION

The last two decades have been characterized by two important developments in scientific computing. First, efficient methods, both iterative and direct, have been developed for solving large sparse linear systems. Second, simultaneously to progress made in algorithms, hardware components and architectures made substantial gains through parallel and vector processing. One to two orders of magnitude in speed resulted from architecture alone. However, progress in the implementations on parallel computer platforms is relatively slow in coming. The main stumbling blocks are the lack of fixed and standard parallel computing paradigms and the difficulties in developing highly parallel and easily portable techniques.

In this paper we discuss some trends in the solution of general sparse matrix problems in parallel environments. There are essentially two general approaches that have been taken so far. The first is to extract parallelism from a standard iterative solver.

---

[1]University of Minnesota, department of computer science, Minneapolis, Minnesota 55455. Research supported by the National Science Foundation under grant number NSF/CCR-9214116, by ARPA under grant number NIST 60NANB2D1272, and by the Minnesota Supercomputer Institute.

In principle, this involves only implementing the standard operations such as triangular solves, inner products, and other vector operations, in parallel. The second class of
methods consists of algorithms with enhanced parallelism. Specifically, either an intrinsically parallel preconditioner is employed or a strategy, such as reordering, is used
to enhance parallelism of the preconditioner. These techniques are developed mostly
with the data parallel model in mind but can of course be exploited on any parallel platform. They include multicoloring techniques and polynomial preconditioners
among others. Finally, a class of methods geared towards MIMD environments, especially message passing architectures, has emerged from domain decomposition ideas. A
sparse matrix is distributed among processors in a certain fashion and then adequate
strategies for preconditioning the matrix are derived by exploiting the distributed
data structure. These methods constitute a fairly general and convenient paradigm
for solving Partial Differential Equations (PDEs) on parallel computers. The idea is
to partition a physical domain into several pieces (sub-domains) and then use a certain strategy to recover the global solution by a succession of solutions of independent
subproblems associated with the subdomains. Each processor handles one or several
subdomains in the partition. The partial solutions can be combined, in various ways,
to deliver an approximation to the global system. The rationale of this approach is
that each processor can do a large part of the work independently, and the work to
construct the global solution from the partial solutions is not too expensive relative
to that of obtaining these partial solutions.

   This paper is a brief overview of these three different approaches. We will consider
the problem of implementing Preconditioned Conjugate Gradient type methods for
solving linear systems. The framework is that of large sparse linear systems that are
not necessarily well structured. Section 2 is a summary of Krylov subspace methods.
Section 3 considers their standard parallel implementations. Section 4 describes the
fine-grain or data-parallel preconditioners. Section 5 discusses methods for distributed
sparse matrices.

## 6.2 KRYLOV SUBSPACE METHODS

Consider the linear system,

$$Ax = b, \qquad (6.1)$$

in which $A$ is an $n \times n$ nonsingular and real matrix. Given an initial guess $x_0$ to this
linear system a general *projection method* seeks an approximate solution $x_m$ from an
affine subspace $x_0 + K_m$ of dimension $m$ by imposing the Petrov-Galerkin condition

$$b - Ax_m \perp L_m \qquad (6.2)$$

where $L_m$ is another subspace of dimension $m$. A Krylov subspace method is a method
for which the subspace $K_m$ is the Krylov subspace

$$K_m(A, r_0) = span\{r_0, Ar_0, A^2 r_0, \ldots, A^{m-1} r_0\}, \qquad (6.3)$$

in which $r_0 = b - Ax_0$. When there is no ambiguity we will denote $K_m(A, r_0)$ by $K_m$.
The different versions of Krylov subspace methods arise from different choices of the

subspaces $K_m$ and $L_m$ and from the ways in which the system is preconditioned. Note that from the approximation theory point of view, a member of the Krylov subspace is of the form $p(A)r_0$ and so these methods attempt to approximate the solution vector $x_0 + A^{-1}r_0$ by a vector of the form $x_0 + p(A)r_0$, where $p$ is a polynomial of degree $m - 1$. This parallel with approximation theory is useful in many instances.

There are essentially three classes of Krylov subspace methods. In the first class a standard Galerkin approach is used in which $K_m$ and $L_m$ are the same and equal to $K_m(A, r_0)$. The conjugate gradient method is a particular instance of this method when the matrix is symmetric positive definite. Another method in this class is the Full Orthogonalization Method (FOM) (Saad Y. 1981) which is closely related to Arnoldi's method for solving eigenvalue problems (Arnoldi W. E. 1951). ORTHORES (Jea K. C. and Young D. M. 1980) developed by Jea and Young is mathematically equivalent to FOM as is another method derived by Axelsson (Axelsson O. 1980). By definition, the final residual vector is orthogonal to all of the Krylov subspace. Thus, these methods are sometimes termed orthogonal residual methods. They have the important property of minimizing the $A$-norm of the error over $x_0 + K_m$ when the matrix $A$ is symmetric positive. This property does not extend to nonsymmetric problems.

In the second class of methods, $L_m$ is taken to be equal to $AK_m$ where $K_m = K_m(A, r_0)$. It can be shown, see e.g., (Saad Y. 1995) that in this case the approximate solution $x_m$ minimizes the residual norm $\|b - Ax\|_2$ over all candidate vectors in $x_0 + K_m$. Methods in this category are often termed Minimal Error methods. Quite a few methods have been developed in this class (Axelsson O. 1987),(Jea K. C. and Young D. M. 1980),(Elman H. C. 1982), (Saad Y. and Schultz M. H. 1986).

Finally, among oblique projection methods, are the methods based on bi-conjugacy and the Lanczos biorthogonalization method. In these methods, $L_m = K_m(A^T, r_0)$ and $K_m = K_m(A, r_0)$. In the nonsymmetric case, the biconjugate gradient method (BCG) due to Lanczos (Lanczos C. 1952) and Fletcher (Fletcher R. 1975) is a good representative of this class. There are various mathematically equivalent formulations of the biconjugate gradient method some of which are more numerically viable than others. The QMR algorithm (Freund R. W. and Nachtigal N. M. 1991) ignores the non-orthogonality of the Lanczos basis and attempts to minimize the 2-norm of the expression of the residual in this basis.

An efficient variation on the BCG algorithm, called CGS (Conjugate gradient squared) and proposed by Sonneveld (Sonneveld P. 1989), (Polak S. J., Heijer C. Den , Schilders W. H. A. and Markowich P.. 1987), avoids the use of the transpose of $A$. This gave rise to a whole class of efficient schemes named transpose-free techniques, two notable examples of which are the BiConjugate Gradient Stabilized (BICGSTAB) algorithm of van der Vorst (van der Vorst H. A. 1992), and the Transpose-Free QMR (TFQMR) algorithm of Freund (Freund Roland W. 1993).

The Krylov subspace methods mentioned above are often referred to as *accelerators*. More important than the accelerator for the success of a Krylov subspace method is the preconditioning technique used. Preconditioning consists of replacing the original linear system (6.1) by, for example, the equivalent system

$$M^{-1}Ax = M^{-1}b, \tag{6.4}$$

where $M$ is the preconditioning matrix. In one of the simplest cases, $M$ is the iteration matrix in the SSOR iteration, namely,

$$M = (D - E)D^{-1}(D - F).$$

Here $D$ is the main diagonal of $A$, $D - E$ is the lower triangular part of $A$, and $D - F$ its upper triangular part. Thus, $M$ is of the form $M = LU$ where $L$ and $U$ have the same structure as the lower and upper triangular parts of $A$ respectively. A slightly more advanced technique, known as ILU(0), obtains the preconditioning matrix $M$ in the same form. However, in this case the factors L and U are obtained from an incomplete LU factorization resulting from performing the standard LU factorization of $A$ and dropping all fill-in elements that are generated during the process. A fill-in element is a nonzero element which is introduced in a position which initially contains a zero element. More elaborate factorizations allow some limited fill-in. For example, in the structured case fill-in is allowed along specific diagonals to get more accurate incomplete factorizations, denoted by ILU(k).

For the standard ILU preconditioners, solving a linear system with the matrix $M$, requires performing a forward and a backward triangular system solution at every step of the Krylov subspace method. This may constitute the main bottleneck in iterative methods if not carefully implemented on supercomputers and will be discussed shortly.

In the rest of the paper we will illustrate the concepts with the preconditioned conjugate gradient method, a version of which is described below. Here $M$ represents the preconditioning matrix.

ALGORITHM **6.1** *Preconditioned Conjugate Gradient*

    1.  Compute $r_0 := b - Ax_0$, $z_0 = M^{-1}r_0$, and $p_0 := z_0$
    2.  For $j = 0, 1, \ldots$, until convergence Do:
    3.      $\alpha_j := (r_j, z_j)/(Ap_j, p_j)$
    4.      $x_{j+1} := x_j + \alpha_j p_j$
    5.      $r_{j+1} := r_j - \alpha_j Ap_j$
    6.      $z_{j+1} := M^{-1}r_{j+1}$
    7.      $\beta_j := (r_{j+1}, z_{j+1})/(r_j, z_j)$
    8.      $p_{j+1} := z_{j+1} + \beta_j p_j$
    9.  EndDo

This algorithm is obtained by writing the conjugate gradient method applied to the linear system $M^{-1}Ax = M^{-1}b$ in which the standard Euclidean inner product is replaced by the inner product $(x, y)_M = (Mx, y)$. The matrix $M^{-1}A$ is self-adjoint with respect to this inner product. The preconditioned residual vector $M^{-1}(b - Ax_i)$ obtained at step $i$ is M-orthogonal to all the previous preconditioned residual vectors.

So far there has been just a few general principles used to develop parallel preconditioners. The first idea is to exploit 'wavefronts' or 'level-scheduling' (Anderson E. C. and Saad Y. 1989),(Saad Y. 1989),(Saltz J. H. 1987), (Wing O. and Huang J. W. 1980),(van der Vorst H. A. 1986), (van der Vorst H. A. 1987),(Greenbaum A., Li C. and Chao H. Z. 1989), (Saad Y. and Schultz M. H. 1986),(Ashcraft C. C. and Grimes R. G. 1988) for solving the triangular systems that arise in many preconditioners. These are essentially parallel implementations of the forward and backward triangular solves and can always be applied to any ILU-type preconditioner, without changing the preconditioner itself. Other techniques involve modifying a given preconditioner to increase parallelism or using special methods specifically geared toward parallel computers.

## 6.3 PARALLEL IMPLEMENTATIONS

When implementing the PCG algorithm in a parallel platform, it is important to start by identifying the main operations that constitute the algorithm. These are

1. Setting up of the preconditioner;

2. Matrix vector multiplications;

3. Vector updates;

4. Dot products;

5. Preconditioning operations.

There is inherent parallelism in each of these operations. However, the preconditioner set-up and triangular solves in the preconditioning operation require special attention. In this paper we only discuss the preconditioning operations for standard preconditioning operations such as ILU. The traditional way of implementing the algorithm in a shared memory environment is to unravel parallelism in the forward and backward triangular solves for these standard preconditioners.

The main loop in solving a (unit) triangular system $Lx = b$ is the following:

$$x_i := b_i - \sum_{j=1}^{i-1} L_{ij} x_j.$$

Note that when $L$ is sparse, most of the $L_{ij}$'s are zero. It may appear at first that this loop is inherently sequential, namely that all $x_j$'s for $j = 1, \ldots, i-1$ must be determined before $x_i$ can be computed. In fact, because $L$ is sparse only a few of the previous $x_j$'s are actually needed. As a result there are usually many sets of equation that can be solved in parallel. Consider for example a five-point matrix, using natural ordering as shown in Figure 6.1. The associated $L$-matrix is shown on the left side of Figure 6.2.

The dependence of $x_i$ upon the other $x_j$'s is shown by the stencil of the matrix $L$: $x_i$ can be determined once $x_{i-1}$ and $x_{i-m}$ are known, where $m$ is the number of mesh-points on the $x$-direction. In the figure, the unknown $x_1$ can be determined immediately. Then $x_2$ and $x_5$ are computable at the same time, and then the group of unknowns $x_3, x_6, x_9$. As can be seen these groups are along 'wavefronts'. The last three wavefronts are $\{4, 7, 10\}$, $\{8, 11\}$ and $\{12\}$. Unknowns of the same front can be solved in parallel. If we permute the $L$-matrix using this wavefront ordering we would get the matrix shown at the right side of Figure 6.2. Therefore, wavefront ordering (also known as level-scheduling) is simply a technique for finding an ordering of the triangular matrix which keeps the matrix triangular and such that its diagonal blocks are diagonal matrices.

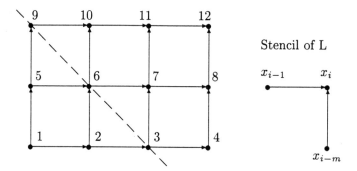

Figure 6.1: A 5-point mesh and the corresponding stencil.

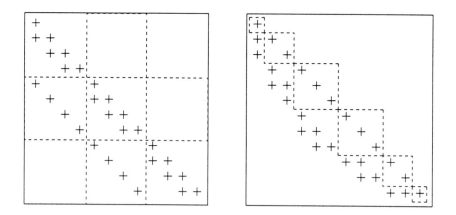

Figure 6.2: The $L$ matrix before and after wavefront ordering.

An important improvement to the performance can be achieved by exploiting efficient matrix-vector product operations. Indeed, each group of unknowns belonging to the same wavefront is determined by an operation of the form

$$x_G := b_G - E_G x$$

where $x_G$ and $b_G$ are the vector of unknowns and right-hand side components belonging to the group $G$ and $E_G$ is the submatrix of $L$ consisting of all nonzero $L_{ij}$ for $j \in G$ and $i \neq j$. This can be easily understood by considering the reordered matrix on the right-side of Figure 6.2.

The idea of wavefronts can easily be extended to general sparse matrices. Here, an inexpensive preprocessing is required to obtain the levels; for details see (Anderson E. C. and Saad Y. 1989), (Saad Y. 1995). Performance can be more than adequate on vector processors, by exploiting suitable storage formats and efficient matrix-vector multiplications. For typical problems a gain of a factor of 3 to 5 is easily achievable when compared with a standard implementation. However, this approach has two drawbacks. First, it is typical that there is a substantial number of small level-sets. In the above example, level-sets start with a size of one then increase by one for each next level. These small levels can hamper performance considerably for massively parallel machines (a consequence of Amdahl's law (Kumar V., Grama A., Gupta A. and Kapyris G. 1994)). The second difficulty is that the maximum degree of parallelism can be rather small. For rectangular 2-dimensional meshes it is of the order of the smaller of the number of mesh points in the $x$- and $y$-directions. These two difficulties can be remedied by using multi-coloring which is examined next.

## 6.4 DATA-PARALLEL PRECONDITIONERS

In early implementations of PCG on supercomputers, researchers abandoned the standard ILU-type preconditioners in favor of techniques which relied more on matrix-by-vector products. For example, the inverse of a sparse triangular matrix can be approximated using a simple polynomial expansion. Solutions of bidiagonal systems which are required in the forward-backward solutions for structured matrices are approximated similarly (van der Vorst H. A. 1982). At one extreme, the inverse of $A$ can also be approximated by a polynomial in $A$. Multiplying this polynomial of $A$ by an arbitrary vector can be done using a sequence of matrix-vector products.

In "data-parallel preconditioners," a maximum degree of parallelism is sought. These techniques are very general and can used in data parallel (SIMD) environment but also on MIMD platforms since data parallelism is the most general form of parallelism. We refer to these preconditioners as "data-parallel" only to emphasize the fact that they are characterized by a high degree of fine-grain parallelism.

Among the simplest techniques used in this context are polynomial preconditioners; see, e.g., (Saad Y. 1985),(Johnson O. G., Micchelli C. A. and Paul G. 1983), (Greenbaum A. and Rodrigue G. H. 1977), (Dubois P. F., Greenbaum A. and Rodrigue G. H. 1979). Another strategy which has been exploited is that of multicoloring and independent set orderings. These are discussed next.

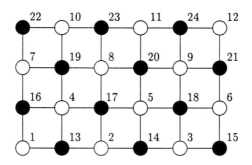

Figure 6.3: Red-black coloring of a 6 × 4 grid.

## MULTI-COLORING

The problem addressed by multicoloring is to determine a coloring of the nodes of the adjacency graph of a matrix such that any two adjacent nodes have different colors. This general idea has been used in many different ways in sparse matrix computations. In particular it is one of the main ingredients used to increase parallelism in relaxation methods or preconditioning techniques. We start by discussing red-black ordering.

For a 2-dimensional finite difference grid (5-point operator), multicoloring can be achieved with two colors, typically referred to as "red" and "black", as is illustrated in Figure 6.3 for a 6 × 4 mesh. If the unknowns are labeled by listing the red unknowns first together, followed by the black ones, then the system that results from this reordering will have the following structure,

$$\begin{pmatrix} D_1 & F \\ E & D_2 \end{pmatrix} \begin{pmatrix} x_1 \\ x_2 \end{pmatrix} = \begin{pmatrix} b_1 \\ b_2 \end{pmatrix} \tag{6.5}$$

in which $D_1$ and $D_2$ are diagonal matrices. This is because the red nodes are not coupled with other red nodes, and, similarly, the black nodes are not coupled with other black nodes.

One way to exploit the red-black ordering is to use the standard SSOR or ILU(0) preconditioners for solving the system (6.5). The resulting preconditioning operations are highly parallel. Thus, the linear system arising from the forward solve in an SSOR preconditioning is of the form

$$\begin{pmatrix} D_1 & O \\ E & D_2 \end{pmatrix} \begin{pmatrix} x_1 \\ x_2 \end{pmatrix} = \begin{pmatrix} b_1 \\ b_2 \end{pmatrix}.$$

This system can be solved by first solving for $x_1$, ($x_1 = D_1^{-1} b_1$), then updating the right-hand side ($\hat{b}_2 := b_2 - E x_1$), and then solving for $x_2$ ($x_2 = D_2^{-1} \hat{b}_2$). This consists of two diagonal scalings and a sparse matrix-by-vector product which means that the degree of parallelism, is at least $n/2$ if an atomic task is considered to be any arithmetic operation. The situation is identical with the ILU(0) preconditioning. However, since the matrix has been reordered before ILU(0) is applied to it, the resulting LU factors are not related in any simple way to those associated with the original matrix. In fact, a simple look at the structure of the ILU factors reveals that many more elements are

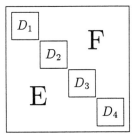

Figure 6.4: A four-color ordering of a general sparse matrix.

dropped with the red-black ordering than with the natural ordering. The result is that the number of iterations to achieve convergence can be much higher with red-black ordering than with the natural ordering.

Another option with the red-black ordering is to solve the reduced system which involves only the black unknowns. Eliminating the red unknowns from (6.5) results in the reduced system

$$(D_2 - ED_1^{-1}F)x_2 = b_2 - ED_1^{-1}b_1.$$

This new system is again a sparse linear system with about half as many unknowns. For some problems the reduced system can often be solved efficiently with only diagonal preconditioning. Forming the reduced system itself is a highly parallel and inexpensive process. It is not even necessary to form the reduced system. This strategy is standard when $D_1$ is not diagonal, such as in domain decomposition methods. It can also be beneficial in other situations. For example, the product $Ax$ can be performed using nearest-neighbor communication, and this can be more efficient than the standard approach of multiplying the vector by the Schur complement matrix $D_2 - ED_1^{-1}F$. It can also save storage, a more critical issue in some cases.

We now discuss the case of a general sparse matrix $A$. There are a number of inexpensive techniques to color an arbitrary graph, see e.g. (Saad Y. 1995). The simplest heuristic used is to traverse the graph in any fashion and for each node visited assign the smallest color number allowable. By allowable, we mean a color number that is not already assigned to any of the vertices that are adjacent to the node being visited. Once the multicolor ordering is applied to the matrix, we obtain a block diagonal matrix whose diagonal blocks are diagonal matrices. This structure generalizes that 2-block matrix obtained for the red-black ordering. The number of blocks is the number of colors. For example, for four colors, a matrix would result with the structure shown in Figure 6.4 where the $D_i$'s are diagonal and $E$, $F$ are general sparse.

Just as for the red-black ordering, ILU(0), SOR, or SSOR preconditioning can be used on this reordered system. The parallelism of SOR/SSOR is now of order $n/p$ where $p$ is the number of colors. Note that the principle of multicoloring is to transform the triangular solve into a sequence of matrix-vector products of reasonably large size, and diagonal scalings. A loss in efficiency may occur since the number of iterations may increase.

A Gauss-Seidel sweep will essentially consist of $p$ scalings and $p - 1$ matrix-by-vector products, where $p$ is the number of colors. For example, it may be convenient to store the matrix $A$ in the Ellpack-Itpack format in which case, the block structure of the permuted matrix can be defined by a pointer array *iptr* which points to the first row in each block. One single step of the multicolor SOR iteration can then be efficiently implemented by accessing these $p$ submatrices in succession.

One of the most promising approaches in the context of parallel preconditioners is to combine multiple-step SOR or SSOR preconditioning with multi-coloring (Saad Y. 1994),(DeLong M. A. and Ortega J. M. 1994), (Adams L. M. and Ortega J. 1982),(Ortega J. 1991), (Poole E. L and Ortega J. M. 1987),(Ortega J. M. 1988). This combines the main advantages of multicoloring, namely ample degree of parallelism, with the robustness afforded by multiple step techniques. A multiple-step relaxation consists simply of performing $s$ steps of the relaxation scheme instead of one. In fact, the number of steps can vary arbitrarily if a "flexible variant" of an accelerator is used. These variants allow a right-preconditioner to be essentially arbitrary; see e.g., (Saad Y. 1993), (Axelsson O. and Vassilevski P. S. 1991),(van der Vorst H. A. and Vuik C. 1994). In (Saad Y. 1994) it was shown by means of experiments that on vector computers, one-step relaxation was often far from being the optimal number of steps to use when preconditioning a linear system. This was confirmed by Delong and Ortega (DeLong M. A. and Ortega J. M. 1994) who show, in addition, that $\omega = 1$ was also far from being optimal. However, in the context of preconditioning, the usual optimal formulas for obtaining the best $\omega$ are no longer valid. The combined gains from using a good value of $\omega$ and multiple steps can be rather substantial.

*ILUM*

A general principle exploited in sparse Gaussian elimination is that one can dynamically find sets of unknowns which are independent, i.e., unknowns that are not coupled by the matrix. A set of such unknowns is called an independent set. Therefore, independent set orderings can be viewed as permutations to put the original matrix in the form

$$\begin{pmatrix} D & E \\ F & C \end{pmatrix} \tag{6.6}$$

in which $D$ is diagonal, but $C$ can be arbitrary. The unknowns associated with the diagonal block $D$ constitute the independent set. This is a less restrictive form of multicoloring discussed earlier. Inexpensive algorithms for finding independent set orderings are easy to develop and are similar to multicoloring algorithms (Saad Y. 1995).

When the rows associated with an independent set are eliminated, we obtain a smaller linear system which is again sparse. Then we can find an independent set for this reduced system and repeat the process of reduction. The process can be repeated recursively a few times. As the level of the reduction increases, the reduced systems gradually lose their sparsity. A direct solution method would continue the reduction until the reduced system is small enough or dense enough to switch to a dense Gaussian elimination to solve it. If a dropping strategy is added to this process, a incomplete factorization can be obtained.

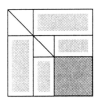 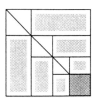

Figure 6.5: Illustration of the processed matrices obtained from three steps of independent set ordering and reductions.

Let $A_j$ be the matrix obtained at the $j$-th step of the reduction, $j = 0, \ldots, nlev$ with $A_0 = A$. An independent set ordering is applied to $A_j$ and the matrix is permuted accordingly to obtain

$$P_j A_j P_j^T = \begin{pmatrix} D_j & F_j \\ E_j & C_j \end{pmatrix} \tag{6.7}$$

where $D_j$ is a diagonal matrix. The unknowns of the independent set are eliminated to get the next reduced matrix,

$$A_{j+1} = C_j - E_j D_j^{-1} F_j. \tag{6.8}$$

Implicitly this results in a block LU factorization

$$P_j A_j P_j^T = \begin{pmatrix} D_j & F_j \\ E_j & C_j \end{pmatrix} = \begin{pmatrix} I & O \\ E_j D_j^{-1} & I \end{pmatrix} \times \begin{pmatrix} D_j & F_j \\ O & A_{j+1} \end{pmatrix}$$

with $A_{j+1}$ defined above. In order to solve a system with the matrix $A_j$, both a forward and a backward substitution need to be performed with the block matrices on the right-hand side of the above system.

These successive reduction steps will give rise to matrices that become denser and denser due to the fill-ins introduced by the elimination process. Some fill-ins can be neglected by using a simple dropping strategy as the reduced systems are formed. For example, any fill-in element introduced is dropped, whenever its size is less than a given tolerance times the 2-norm of the original row. This "approximate" version of the successive reduction steps can be used to provide an approximate solution $M^{-1}v$ to $A^{-1}v$ for any given $v$ which can be used to precondition the original linear system. Conceptually, the modification leading to an "incomplete" factorization replaces (6.8) by

$$A_{j+1} = (C_j - E_j D_j^{-1} F_j) - R_j \tag{6.9}$$

in which $R_j$ is the matrix of the elements that are dropped in this reduction step. Globally, the algorithm can be viewed as a form of incomplete block LU factorization with permutations. An independent set ordering for the new matrix $A_{j+1}$ is then found and reduced again in the same manner. This block factorization can be used recursively until a small enough system is obtained which can be solved with a standard direct or iterative method.

An illustration of three reduction steps of the process is shown in Figure 6.5. We refer to this incomplete factorization as ILUM (ILU with Multi-elimination). The preprocessing phase consists of a succession of *nlev* applications of the following three steps: (1) finding the independent set ordering, (2) permuting the matrix, and (3) reducing it. The implementation of the ILUM preconditioner corresponding to this strategy is rather complicated and involves several parameters.

## APPROXIMATE INVERSE PRECONDITIONERS

It was observed above that the implicit goal of multicoloring and ILUM is to obtain LU factors which have diagonal blocks that are large diagonal matrices. With such a structure, a triangular solve becomes a sequence of matrix multiplications and diagonal scalings. An alternative is to attempt to obtain a preconditioning matrix $M$ which approximates $A^{-1}$ directly. For example we may seek a sparse matrix $M$ to minimize the Frobenius norm of $I - AM$. Techniques of this type have received much attention in recent years (Benson M. W. and Frederickson P. O. 1982), (Grote M. and Simon H. D. 1992), (Cosgrove J. D. F., Díaz J. C. and Griewank A. 1992), (Huckle T. and Grote M. 1994), (Kolotilina L. Yu. and Yeremin A. Yu. 1993), (Kolotilina L. Yu. and Yeremin A. Yu. 1992), (Chow E.and Saad Y. May 1994),(Chow E. and Saad Y. Jan. 1995). One of the reasons for this interest is parallelism. Another is the common failure of ILU-type preconditioners for indefinite problems.

One approach consists of finding a matrix $M$ such that $AM$ is close to the identity. Then the preconditioned system is of the form,

$$AMy = b, \quad x = My .$$

The columns $m_j$ of $M$ can be obtained by approximately solving the linear system $Am_j = e_j$ by an iterative process. It is important to make the following observation. If the initial guess is sparse then the iterates produced by algorithms such as GMRES or MR applied to the system $Am = e_j$ will stay sparse initially. They will become denser as the iteration number progresses. This was exploited either explicitly or implicitly in several papers; see, e.g., (Chow E.and Saad Y. May 1994).

Rough approximations can be obtained inexpensively by performing a very small number of steps starting with the columns of the current $M$. The initial approximation $M$ is often taken to be the identity matrix. Once a first approximation of $M$ is obtained we can improve it by incorporating it in an outer loop which takes as initial guess the columns of the most current $M$. A few steps of sparse-sparse mode iterative technique are used. A potential improvement to this technique is to use the most recent approximate inverse to precondition the system solved when approximating a column. We refer to this as *self-preconditioning*.

There are two opportunities for exploiting parallelism in implementations of approximate inverse techniques. First, the preprocessing to obtain the matrix $M$ can itself be implemented in parallel. This is because the linear systems $Am_j = e_j$ associated with each column are all linearly independent. More important is the fact that the preconditioning operation is now purely a matrix-vector product.

One of the most promising uses of approximate inverse techniques is to combine them with other, e.g. block, factorizations. A number of techniques based on this viewpoint were developed in (Kolotilina L. Yu. and Yeremin A. Yu. 1986) and, more recently, in (Chow E. and Saad Y. Jan. 1995). The main idea in (Chow E. and Saad

Y. Jan. 1995) is to exploit blockings of the original linear system in the block form,

$$\begin{pmatrix} B & F \\ E & C \end{pmatrix} \begin{pmatrix} x \\ y \end{pmatrix} = \begin{pmatrix} f \\ g \end{pmatrix}. \tag{6.10}$$

This blocking may originate from a domain decomposition partitioning for example or may be the original structure of $A$ as in the Navier-Stokes equations for example. The above blocking can be used in several different ways to define preconditioners for $A$. We can, for example, approximate the block LU factorization

$$M = LU$$

in which

$$L = \begin{pmatrix} B & 0 \\ E & S \end{pmatrix} \quad \text{and} \quad U = \begin{pmatrix} I & B^{-1}F \\ 0 & I \end{pmatrix}$$

where $S$ is the Schur complement

$$S = C - EB^{-1}F.$$

Since $F$ is a sparse block, the general approach of approximate inverse techniques can be effectively used to construct a sparse approximation $Y$ to $B^{-1}F$. Once $Y$ is obtained, then $S$ is replaced by $M_S = C - EY$ in the $L$ factor and $B^{-1}F$ is replaced by $Y$ in the $U$ factor. This yields an approximate block-factorization, It has been observed that this technique is remarkably robust, possibly due to the decoupling of the block structure.

## 6.5 DISTRIBUTED SPARSE MATRICES

The first task when solving a linear system on a distributed memory computer is to distribute the data among processors. In contrast with shared memory environments, this must be done explicitly by the user in distributed memory computers. It is natural to map pairs of equations-unknowns to the same processor. This mapping can be determined automatically by a graph partitioner or it can be assigned ad hoc from knowledge of the problem. Once the partitioning is done, we need to distribute the actual data among processors and preprocess it to prepare for operations that are needed in the course of the iteration.

### THE LOCAL DATA STRUCTURE

Assume that the graph has been partitioned in a certain way. For illustration, the matrix under consideration can be viewed as originating from the discretization of a Partial Differential Equation on a certain domain, as illustrated in Figure 6.6. For simplicity we assume that each subgraph (or subdomain, in the PDE literature) is assigned to a different processor.

A local data structure must be set up in each processor to allow the basic operations such as (global) matrix-by-vector products and preconditioning operations to be performed efficiently in the iteration process. The only assumption we make regarding the mapping is that if row number $i$ is mapped into processor $p$, then so is the

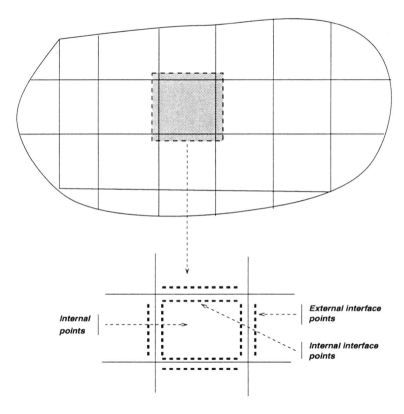

Figure 6.6: Decomposition of physical domain or adjacency graph and the local data structure.

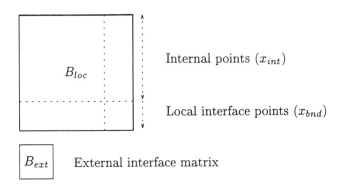

Figure 6.7: The local matrices and data structure associated with each subdomain.

unknown $i$, i.e., the matrix is distributed row-wise across the processors according to the distribution of the variables.

To perform a matrix-by-vector product with a distributed sparse matrix, the matrix consisting of rows that are local to a given processor must be multiplied by some global vector $v$. Some components of this vector will be local, and some components must be brought from external processors. These external variables correspond to interface points belonging to adjacent subdomains. When performing a matrix-by-vector product, neighboring processors must exchange values of their adjacent interface nodes. A preprocessing phase is needed in which the following information must be extracted from the data:

1. List of processors with which communication will take place. These are called "neighboring processors" although they may not be physically nearest neighbors.

2. List of the local nodes that are coupled with external nodes, for each of the neighboring processors. These are the *local interface nodes*.

3. *Local representation of the distributed matrix in each processor.*

Let $A_{loc}$ be the local part of the matrix, i.e., the (rectangular) matrix consisting of all the rows that are mapped to *myproc*. Let $B_{loc}$ the diagonal block of $A$ located in $A_{loc}$, i.e., the submatrix of $A_{loc}$ whose nonzero elements $a_{ij}$ are such that $j$ is a local variable. Similarly, call $B_{ext}$ the "off-diagonal" block, i.e., the submatrix of $A_{loc}$ whose nonzero elements $a_{ij}$ are such that $j$ is not a local variable. To perform a matrix-by-vector product, we start by multiplying the diagonal block $B_{loc}$ by the local variables. Then, we multiply the external variables by the sparse matrix $B_{ext}$. Since the external interface points are not coupled with local internal points, only the rows $n_{int} + 1$ to $n_{nloc}$ in the matrix $B_{ext}$ have nonzero elements. Thus, the matrix-by-vector product can be separated into two such operations, one involving only the local variables and the other involving external variables. These two matrices must be built and a local numbering of the local variables must be defined in order to perform the two matrix-by-vector products efficiently each time.

## DISTRIBUTED SPARSE MATRIX-VECTOR PRODUCTS

To perform a global matrix-by-vector product, with the distributed data structure described above, each processor performs the following operations. First, it multiplies the local variables by the matrix $B_{loc}$. Second, it obtains the external variables from the neighboring processors in a certain order. Third, it multiplies these variables by the matrix $B_{ext}$ and adds the resulting vector to the result obtained from the first multiplication by $B_{loc}$. Note that the first and second steps can be done in parallel. With this decomposition, the global matrix-by-vector product can be implemented as indicated in Algorithm 6.2 below. In what follows, $x_{loc}$ is a vector of variables that are local to a given processor. The components corresponding to the local interface points (ordered to be the last components in $x_{loc}$ for convenience) are called $x_{bnd}$. The external interface points, listed in a certain order, constitute a vector which is called $x_{ext}$. The matrix $B_{loc}$ is a sparse $nloc \times nloc$ matrix which represents the restriction of $A$ to the local variables $x_{loc}$. The matrix $B_{ext}$ operates on the external variables $x_{ext}$ to give the correction which must be added to the vector $B_{loc}x_{loc}$ in order to obtain the desired result $(Ax)_{loc}$.

ALGORITHM **6.2** *Distributed Sparse Matrix Product Kernel*

   1. *Scatter $x_{bnd}$ to neighbors*
   2. *Do Local Matvec: $y = B_{loc}x_{loc}$*
   3. *Gather $x_{ext}$ from neighbors*
   4. *Do External Matvec: $y = y + B_{ext}x_{ext}$*

An important observation is that the matrix-by-vector products in lines 2 and 4 can use any convenient data structure that will improve efficiency by exploiting knowledge on the local architecture. Also recall that steps 1 and 2 are independent and can overlap. An example of the implementation of this operation as it is in PSPARSLIB (Saad Y. and Malevsky A. 1995), is illustrated next:

```
        call MSG_bdx_send(nloc,x,y,nproc,proc,ix,ipr,ptrn,ierr)
c       do local matrix-vector product for local points
        call amux(nloc,x,y,aloc,jaloc,ialoc)
c       receive the boundary information
        call MSG_bdx_receive(nloc,x,y,nproc,proc,ix,ipr,ptrn,ierr)
c       do local matrix-vector product for external points
        nrow = nloc - nbnd + 1
        call amux1(nrow,x,y(nbnd),aloc,jaloc,ialoc(nloc+1))
        return
```

In the above code segment, *MSG_bdx_send and MSG_bdx_receive are the routines which perform the communication tasks. They exchange interface values between nearest neighbor processors. The first call to amux1 performs the operation $y := y + B_{loc}x_{loc}$, where $y$ has been initialized to zero prior to the call. The second call to amux1 performs $y := y + B_{ext}x_{ext}$.*

## PRECONDITIONERS FOR DISTRIBUTED SPARSE MATRICES

We now turn our attention to preconditioning techniques for distributed sparse matrices. Many of the ideas used in preconditioning distributed sparse matrices are borrowed from Domain Decomposition literature; see for example (Farhat C. and Roux J. X. 1994),(LeTallec P. 1994). Among the simplest of these are the use of block preconditionings based on the domains. These are termed Schwarz alternating procedures in the PDE literature. Let $R_i$ be a restriction operator which projects a global vector onto its restriction in subdomain $i$. Algebraically, this is represented by an $n_i \times n$ matrix, consisting of zeros and ones, where $n_i$ is the size of subdomain $i$. The transpose of $R_i$ represents a prolongation operator from subdomain $i$ to the whole space. Let $A_i$ be the local submatrix associated with subdomain $i$. In the Block Jacobi (additive Schwarz procedure), the global solution is updated as,

$$x := x + \sum_{i=1}^{s} R_i^T A_i^{-1} R_i r$$

where $r$ is the residual vector $b - Ax$. Each of the solves is done independently. In each domain a direct or iterative solver must be used. In addition, the subdomain partitions may be allowed to overlap. This technique works reasonably well for a small number of overlapping subdomains as was shown in experiments using a purely algebraic form in (Radicati di Brozolo G. and Robert Y. 1989).

This can be extended to block Gauss-Seidel or Symmetric Gauss-Seidel techniques in which, likewise, a block is associated with a domain. In the PDE framework this is referred to as a multiplicative Schwarz procedure. Here, the algorithm can be described by a simple loop:

> For $i = 1, \ldots, s$ Do:
> $\quad \delta_i := R_i^T A_i^{-1} R_i r$
> $\quad x := x + \delta_i$
> $\quad r := r - A\delta_i$
> EndDo

The only difference with the block Jacobi iteration is that the solution and residual are updated immediately for each local correction $\delta_i$. In the block Jacobi case, these local corrections are all computed using the same residual $r$ and then they are added at the same time to the current solution $x$.

As it was described above, the Gauss-Seidal procedure is sequential. However, multicoloring can exploited to extract parallelism. The subdomains can be colored in such a way that vertices are not coupled with vertices of other subdomains which have the same color. One such coloring is illustrated in Figure 6.8 where the numbers refer to a color. With this coloring, the Gauss-Seidel loop can proceed differently. Since all the vertices of the same color are not connected to vertices of domains having the same color, then the Gauss-Seidel sweep can proceed by color. Thus, in Figure 6.8 all 4 domains of color number 1 can update their variables first at the same time. Then the two domains with color number 2 can do their update, and so on. For this example this color-based sweep will require 4 steps instead of $s = 12$ steps for the domain-based sweep. The color-based loop is as follows.

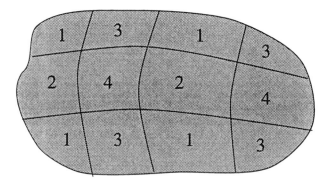

Figure 6.8: Multicoloring of the subdomains for a multicolor Gauss-Seidel sweep.

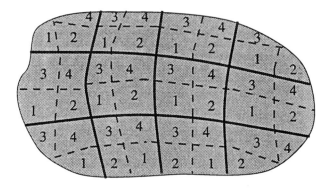

Figure 6.9: Two-level multicoloring of the subdomains for a multicolor Gauss-Seidel sweep.

For $color = 1, \ldots, num_{colors}$ Do:
   If Color(mynode)=$color$ Do:
      $\delta := A_i^{-1} R_i r$
      $x := x + \delta_i$
      $r := r - A\delta_i$
   EndIf
EndDo

    In the above code fragment, Color(mynode) represents the color of the node to which the subdomain is assigned. One important observation is that when one color is active all processors that do not hold a subdomain of this color will be idle. This may lead to substantial loss of efficiency. A remedy is to further partition each subdomain in each processor and then find a global coloring of the new, finer, partitioning. The goal here is to have all colors of this new partitioning represented in each processor. An example of a second level partitioning of the domain in Figure 6.8 is shown in Figure 6.9.

    Consider now an implementation of ILU(0) for distributed sparse matrices. It is sufficient to consider a local view of the distributed sparse matrix, as is illustrated in

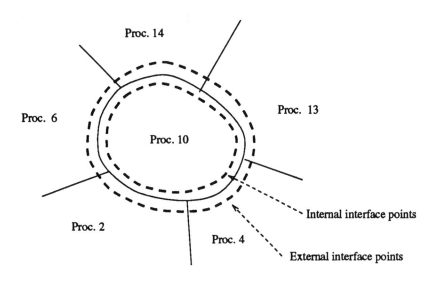

Figure 6.10: A distributed view of ILU(0)

Figure 6.10 in an example. The problem is partitioned into $p$ subdomains or subgraphs using some graph partitioning technique. This results in a mapping of the matrix into processors in which we assume that the $i$-th equation (row) and the $i$-th unknown are mapped to the same processor. We distinguish between *interior* points and *interface* points. The interior points are those nodes that are not coupled with nodes belonging to other processors. Interface nodes are those local nodes that are coupled with at least one node which belongs to another processor.

Thus, processor number 10 in the figure holds a certain number of rows that are local rows. Consider the rows associated with the interior nodes. The unknowns associated with these nodes do not depend on the other variables. As a result they can be eliminated independently and in parallel in the ILU(0) process. The rows associated with the nodes on the interface of the subdomain will require more attention. Recall that an ILU(0) factorization is entirely determined by the order in which we process the rows.

Once the interior nodes have been eliminated we can then eliminate the interface rows *in a certain order*. There are two natural choices. The first would be to impose a global order based on the labels of the processors. Thus in our illustrative Figure we will process the rows belonging to processors 2, 4 and 6 before those in processor 10. The interface rows in Processor 10 must in turn be processed before those of Processors 13 and 14. The local order, i.e., the order in which we process the interface rows in the same processor (e.g. Processor 10) may not be as important. This global order based on PE-number, defines a natural priority graph and parallelism can easily be exploited in a data-driven implementation.

There are alternatives to using the subdomain labels for a global ordering. Observe that we can also define a proper order in which to perform the elimination by *replacing the PE-numbers by any labels provided any two neighboring processors have a different label.* The most natural way of doing this is to perform a multicoloring of the subdomains, and use the colors in exactly the same way we used the labels before to define an order of the tasks. The colors can be defined to be positive integers as was illustrated in Figure 6.8. We can then write the algorithms in this general form, i.e., with a label associated with each processor. Thus, the simplest valid labels are the PE numbers, which leads to the PE-label based order. In the following we will define $Lab_j$ as the label of Processor number $j$. The main loop in this block ILU(0) loop is similar to the one in block Gauss-Seidel.

ALGORITHM **6.3 Distributed ILU(0) factorization in wavefront order**

   *1. In each processor $P_i$, $i = 1, \ldots, p$ Do:*
   *2.   Perform the ILU(0) factorization for interior local rows;*
   *3.   Receive the factored rows from the adjacent processors*
          *j with $Lab_j < Lab_i$;*
   *4.   Perform the ILU(0) factorization for the interface rows with*
          *pivots received from the external processors in step 3;*
   *5.   Perform the ILU(0) factorization for the boundary nodes, with*
          *pivots from the interior rows completed in step 2;*
   *6.   Send the completed interface rows to adjacent processors*
          *j with $Lab_j > Lab_i$;*
   *7. EndDo*

Step 2 of the above algorithm can be performed in parallel as it does not depend on data from other processors.

## 6.6 CONCLUSION

There is no shortage of strategies available for implementing preconditioned Krylov subspace methods on parallel computers. We distinguished three classes of techniques and each of these can be useful in a different situation. Furthermore, a few tools, e.g., graph coloring, find uses across a wide variety of techniques. The bad news is that these methods are often rather complex to implement. In addition, good efficiency is not guaranteed and the developer must pay attention to performance issues related to the implementation on each specific architecture. For example, if a poor mapping of the data is chosen, efficiency may suffer because of communication overhead or load imbalance. These difficulties suggest that software libraries, which include highly tuned solution algorithms as well as performance analysis tools may become quite important for porting iterative solution techniques on parallel computers.

# REFERENCES

Adams L. M. and Ortega J. 1982. A multi-color SOR Method for Parallel Computers. In *Proceedings of the 1982 International Conference on Pararallel Processing*, pages 53–56.

Anderson E. C. and Saad Y. 1989. Solving sparse triangular systems on parallel computers. *International Journal of High Speed Computing*, 1:73–96.

Arnoldi W. E. 1951. The principle of minimized iteration in the solution of the matrix eigenvalue problem. *Quart. Appl. Math.*, 9:17–29.

Ashcraft C. C. and Grimes R. G. 1988. On vectorizing incomplete factorization and SSOR preconditioners. *SIAM J. on Sci. and Stat. Comput.*, 9:122–151.

Axelsson O. 1980. Conjugate gradient type-methods for unsymmetric and inconsistent systems of linear equations. *Lin. Alg. and its Appl.*, 29:1–16.

Axelsson O. 1987. A generalized conjugate gradient, least squares method. *Numer. Math.*, 51:209–227.

Axelsson O. and Vassilevski P. S. 1991. A block generalized conjugate gradient solver with inner iiterations and variable step preconditioning. *SIAM J. on Mat.rix Anal. and Appl.*, 12.

Benson M. W. and Frederickson P. O. 1982. Iterative solution of large sparse linear systems arising in certain multidimensional approximation problems. *Utilitas Math.*, 22:127–140.

Chow E. and Saad Y. Jan. 1995. Approximate inverse techniques for block-partitioned matrices. Technical Report UMSI – 95 – 13, University of Minnesota, Supercomputer Institute, Minneapolis, MN 55415. Submitted.

Chow E.and Saad Y. May 1994. Approximate inverse preconditioners for general sparse matrices. Technical Report UMSI 94-101, University of Minnesota Supercomputer Institute, Minneapolis, MN 55415. Submitted.

Cosgrove J. D. F., Díaz J. C. and Griewank A. 1992. Approximate inverse preconditioning for sparse linear systems. *Intl. J. Comp. Math.*, 44:91–110.

DeLong M. A. and Ortega J. M. 1994. SOR as a preconditioner. Technical Report CS-94-43, Department of Computer Science, University of Virginia, Charlottesville, VA.

Dubois P. F., Greenbaum A. and Rodrigue G. H. 1979. Approximating the inverse of a matrix for use on iterative algorithms on vectors processors. *Computing*, 22:257–268.

Elman H. C. 1982. *Iterative Methods for Large Sparse Nonsymmetric Systems of Linear Equations*. PhD thesis, Yale University, Computer Science Dept., New Haven, CT.

Farhat C. and Roux J. X. 1994. Implicit parallel processing in structural mechanics. *Computational Mechanics Advances*, 2(1):1–124.

Fletcher R. 1975. Conjugate gradient methods for indefinite systems. In G. A. Watson, editor, *Proceedings of the Dundee Biennal Conference on Numerical Analysis 1974*, pages 73–89, New York. University of Dundee,Scotland, Springer Verlag.

Freund R. W. and Nachtigal N. M. 1991. QMR: a quasi-minimal residual method for non-Hermitian linear systems. *Numer. Math.*, 60:315–339.

Freund Roland W. 1993. A Transpose-Free Quasi-Minimal Residual algorithm for non-Hermitian linear systems. *SIAM J. on Sci. Comput.*, 14(2):470–482.

Greenbaum A. and Rodrigue G. H. 1977. The incomplete choleski conjugate gradient for the star (5-point) operator. Technical Report UCID 17574, Lawrence Livermore National Lab., Livermore, California.

Greenbaum A., Li C. and Chao H. Z. 1989. Parallelizing preconditioned conjugate gradient algorithms. *Computer Physics Communications*, 53:295–309.

Grote M. and Simon H. D. 1992. Parallel preconditioning and approximate inverses on the connection machine. In R. F. Sincovec, D. E. Keyes, L. R. Petzold, and D. A. Reed, editors, *Parallel Processing for Scientific Computing – vol. 2*, pages 519–523. SIAM.

Huckle T. and Grote M. 1994. A new approach to parallel preconditioning with sparse approximate inverses. Technical Report SCCM-94-03, Stanford University, Scientific Computing and Computational Mathematics Program, Stanford, California.

Jea K. C. and Young D. M. 1980. Generalized conjugate gradient acceleration of nonsymmetrizable iterative methods. *Lin. Alg. and its Appl.*, 34:159–194.

Johnson O. G., Micchelli C. A. and Paul G. 1983. Polynomial preconditionings for conjugate gradient calculations. *SIAM J. on Numer. Anal.*, 20:362–376.

Kolotilina L. Yu. and Yeremin A. Yu. 1986. On a family of two-level preconditionings of the incomplete block factorization type. *Soviet Journal of Numerical Analysis and Mathematical Modeling*, 1:293–320.

Kolotilina L. Yu. and Yeremin A. Yu. 1992. Factorized sparse approximate inverse preconditionings II. solution of 3d fe systems on massively parallel computers. Technical Report EM-RR 3/92, Elegant Mathematics, Inc., Bothell, Washington.

Kolotilina L. Yu. and Yeremin A. Yu. 1993. Factorized sparse approximate inverse preconditionings I. theory. *SIAM J. on Matrix Anal. and Appl.*, 14:45–58.

Kumar V., Grama A., Gupta A. and Kapyris G. 1994. *Parallel Computing*. Benjamin Cummings, Redwood City, CA.

Lanczos C. 1952. Solution of systems of linear equations by minimized iterations. *J. of Res. of the Nat. Bur. of Stand.*, 49:33–53.

LeTallec P. 1994. Domain decomposition methods in computational mechanics. *Computational Mechanics Advances*, 1(2):121–220.

Ortega J. 1991. Orderings for conjugate gradient preconditionings. *SIAM J. on Sci. and Stat. Comput.*, 12:565–582.

Ortega J. M. 1988. *Introduction to Parallel and Vector Solution of Linear Systems*. Plenum Press, New York.

Polak S. J., Heijer C. Den , Schilders W. H. A. and Markowich P.. 1987. Semiconductor device modelling from the numerical point of view. *Internat. J. Numer. Meth. Eng.*, 24:763–838.

Poole E. L and Ortega J. M. 1987. Multicolor ICCG methods for vector computers. *SIAM J. on Numer. Anal.*, 24:1394–1418.

Radicati di Brozolo G. and Robert Y. 1989. Parallel conjugate gradient-like algorithms for solving sparse non-symmetric systems on a vector multiprocessor. *Parallel Computing*, 11:223–239.

Saad Y. 1981. Krylov subspace methods for solving large unsymmetric linear systems. *Mathematics of Computation*, 37:105–126.

Saad Y. 1985. Practical use of polynomial preconditionings for the conjugate gradient method. *SIAM J. on Sci. and Stat. Comput.*, 6:865–881.

Saad Y. 1989. Krylov subspace methods on supercomputers. *SIAM J. on Sci. and Stat. Comput.*, 10:1200–1232.

Saad Y. 1993. A flexible inner-outer preconditioned GMRES algorithm. *SIAM J. on Sci. and Stat. Comput.*, 14:461–469.

Saad Y. 1994. Highly parallel preconditioners for general sparse matrices. In G. Golub, M. Luskin, and A. Greenbaum, editors, *Recent Advances in Iterative Methods, IMA Volumes in Mathematics and Its Applications*, volume 60, pages 165–199, New York. Springer Verlag.

Saad Y. 1995. *Iterative Methods for Sparse Linear Systems*. PWS publishing, New York.

Saad Y. and Malevsky A. 1995. PSPARSLIB: A portable library of distributed memory sparse iterative solvers. In V. E. Malyshkin et al., editor, *Proceedings of Parallel Computing Technologies (PaCT-95), 3-rd international conference, St. Petersburg, Sept. 1995*.

Saad Y. and Schultz M. H. 1986. GMRES: a generalized minimal residual algorithm for solving nonsymmetric linear systems. *SIAM J. on Sci. and Stat. Comput.*, 7:856–869.

Saad Y. and Schultz M. H. 1986. Parallel implementations of preconditioned conjugate gradient methods. In W. E. Fitzgibbon, editor, *Mathematical and Computational Methods in Seismic Exploration and Reservoir Modeling*, Philadelphia, PA. SIAM.

Saltz J. H. 1987. Automated problem scheduling and reduction of synchronization delay effects. Technical Report 87-22, ICASE, Hampton, VA.

Sonneveld P. 1989. CGS, a fast Lanczos-type solver for nonsymmetric linear systems. *SIAM J. on Sci. and Stat. Comput.*, 10(1):36–52.

van der Vorst H. A. 1982. A vectorizable version of some ICCG methods. *SIAM J. on Sci. and Stat. Comput.*, 3:350–356.

van der Vorst H. A. 1986. The performance of FORTRAN implementations for preconditioned conjugate gradient methods on vector computers. *Parallel Computing*, 3:49–58.

van der Vorst H. A. 1987. Large tridiagonal and block tridiagonal linear systems on vector and parallel computers. *Parallel Computing*, 5:303–311.

van der Vorst H. A. 1992. Bi-CGSTAB: A fast and smoothly converging variant of Bi-CG for the solution of non-symmetric linear systems. *SIAM J. on Sci. and Stat. Comput.*, 12:631–644.

van der Vorst H. A. and Vuik C. 1994. GMRESR: a family of nested GMRES methods. *Numerical Linear Algebra with Applications*, 1:369–386.

Wing O. and Huang J. W. 1980. A computation model of parallel solution of linear equations. *IEEE Transactions on Computers*, C-29:632–638.

# 7

# Parallel and Distributed Solution of Coupled Nonlinear Dynamic Aeroelastic Response Problems

C. Farhat[1]

## SUMMARY

Aeroelasticity studies the mutual interaction between aerodynamic and elastic forces for an aerospace vehicle. The accurate prediction of aeroelastic phenomena such as divergence and flutter is essential in the design of high performance and safe aircrafts. This prediction requires solving *simultaneously* the coupled fluid and structural equations of motion. Therefore, numerical aeroelastic simulations are in general resource intensive. They belong to the family of Grand Challenge engineering problems, and as such, can benefit from the parallel processing technology. This chapter highlights some important aspects of nonlinear computational aeroelasticity. These include a three-field arbitrary Lagrangian-Eulerian (ALE) finite element/volume formulation for coupled transient aeroelastic problems, a rigorous derivation of geometric conservation laws (GCLs) for flow problems with moving boundaries and *unstructured deformable* meshes, the design of a family of staggered procedures for the efficient solution of the coupled fluid/structure partial differential equations, and fast parallel domain decomposition solvers. The solution of the governing three-field equations with mixed implicit/implicit and explicit/implicit staggered procedures are analyzed with particular reference to accuracy, stability, subcycling, distributed computing, I/O transfers, and parallel processing. A general and flexible framework for implementing the partitioned analysis of coupled transient aeroelastic problems with non-matching fluid/structure interfaces on heterogeneous and/or parallel computational platforms is also described. This framework and the staggered solution procedures are demonstrated with examples ranging from the numerical investigation on an iPSC-860 massively parallel processor of the instability of flat panels with infinite aspect ratio in supersonic airstreams, to the solution on the Paragon XP/S, Cray T3D and IBM SP2 parallel systems of three-dimensional wing response problems in the transonic regime.

## 7.1 INTRODUCTION

Aeroelasticity is the study of the effect of aerodynamic forces on elastic bodies. Because these effects have a great impact on performance and safety issues, aeroelasticity has

---

Department of Aerospace Engineering Sciences and Center for Aerospace Structures, University of Colorado at Boulder, Boulder, CO 80309-0429, USA

*Parallel Solution Methods in Computational Mechanics* edited by M. Papadrakakis
© 1997 John Wiley & Sons Ltd

rapidly become one of the most important considerations in aircraft design. The basic mechanism of a fluid/structure interaction phenomenon can be simply explained as follows. The aerodynamic forces acting on an aircraft depend critically on the attitude of its lifting body with respect to the flow, which in turn depends on the flexibility of the aircraft. Therefore, the elastic deformations of a structure play an important role in determining its external loading. Since the magnitude of the aerodynamic forces cannot be known until the elastic deformations are first determined, it follows that the external load cannot be evaluated until the coupled aeroelastic problem is solved.

In general, aeroelastic problems are divided into: (a) *stability*, and (b) *response* problems. Each of these two classes can be further classified into *steady-state* or *static* problems in which the inertia forces may be neglected, and *unsteady*, or *dynamic*, or *transient* problems which are characterized by the interplay of all of the aerodynamic, elastic, and inertia forces. Throughout this chapter, we focus exclusively on dynamic aeroelasticity problems.

If one notes that the external aerodynamic forces acting on an aircraft structure increase rapidly with the flight speed, while the internal elastic and inertial forces remain essentially unchanged, one can easily imagine that there may exist a critical flight speed at which the structure becomes unstable. Such instability may cause excessive structural deformations and may lead to the destruction of some components of the aircraft. Panel or wing *flutter*, which is a sustained oscillation of panels or wings caused by the high-speed passage of air along the panel or around the wing, is an example of such instability problems. *Buffeting*, which is the unsteady loading of a structure by velocity fluctuations in the oncoming flow, is another important example. Because of the potentially disastrous character of these phenomena, aircraft flutter and buffeting speeds must be well outside the flight envelope. In many cases, this requirement is the determining factor in the design of wings and tail surfaces.

An aeroelastic response problem can associate with a stability problem. For example, if a control surface of an aircraft is displaced, or a turbulence in the flow is encountered, the response to be found may be the motion, the deformation, or the stress state induced in the elastic body of the aircraft. When the response of the structure to such an input is finite, the structure is stable and flutter will not occur. When the structure flutters, its response to a finite disturbance is unbounded. However, an aeroelastic response problem can also associate with a performance rather than a stability problem. For example, it is well-known that for transonic flows, small variations in incidence may lead to considerable changes in the pressure distribution, shock position, and shock strength. It is also well-known that there are some margins within the Mach number and incidence that can be varied around the design condition of a supercritical airfoil without a serious deterioration of the favorably low-drag property of the shock-free flow condition (Tijdeman and Seebass 1980). Determining whether an oscillating airfoil is within or outside these margins requires determining its aeroelastic response.

Past literature on aeroelasticity is mostly devoted to linear models where the motion of a gas or a fluid past a structure, the deformation and vibration of that structure, and more importantly the interaction phenomenon itself are described with linear mathematical concepts (Bisplinghoff and Ashley 1962 and Fung 1969). These linear models assume a linearized flow theory and a harmonic motion of the structure with small displacement amplitudes. Usually, a linear aerodynamic operator is computed using the doublet-lattice method (Albano and Rodden 1969) in the subsonic regime, and the potential gradient method (Jones and Appa 1977), or the harmonic gradient method (Chen and Liu 1983), or the piston theory (Bisplinghoff and Ashley 1962) in the supersonic regime. In all cases, the linear theory assumes an inviscid, irrotational, and isentropic flow. In the transonic regime, the mixed subsonic-supersonic flow patterns and shock waves are such that there are no reliable theoretical means for predicting the unsteady aerodynamic forces. In that case, the linear aeroelasticity theory simply breaks down. This is most unfortunate because of the current renewed interest in transonic flight for both military and civil aircraft.

Besides transonic flights, there are many other important cases where the linear aeroelastic theory cannot be used for predicting the dynamic response or stability of an aircraft. These include, to name only a few, problems where the structure undergoes large displacements and/or rotations — as an example, we note that the maximum upward deflection of the wing of the B52 bomber is 22 feet (Bisplinghoff and Ashley 1962) — parachute dynamics, bluff body oscillators, airfoil oscillations in separated flow, buffeting, and high-G and high angle of attack maneuvers such as those performed by the X-31 aircraft. Some of these and related problems are discussed in Dowell and Llgamov (1988) where emphasis is placed on the fundamental understanding of the nonlinear theory of interaction, others are still unresolved. The pressing need for solving and understanding all of these problems is the main motivation for designing a reliable nonlinear transient aeroelastic numerical simulation capability.

## FORMULATION OF COUPLED NONLINEAR AEROELASTIC PROBLEMS

Here, the structure is not restricted to a harmonic motion with small displacement amplitudes. In principle, there is also no reason to confine its constitutive modeling to that of an elastic material. However, while aircraft structures can undergo large displacements and rotations, they seldom experience large strains. Therefore, the nonlinear modeling of the structural behavior can be limited to the proper accounting of nonlinear geometric effects without a serious loss of generality.

More importantly, the aerodynamic forces acting on the structure are not predicted here by the use of a linear aerodynamic operator because of the important limitations associated with such an approach and discussed in the previous section. Rather, these unsteady forces are determined from the solution of the compressible Euler equations when viscous effects are neglected, and the solution of the compressible Navier-Stokes equations otherwise. Furthermore, no restriction is imposed on the nature of the fluid/structure coupling, at least in principle. This coupling is numerically modeled by suitable fluid/structure interface boundary conditions. Clearly, this means that the methodology described here for simulating nonlinear transient aeroelastic problems is based on the simultaneous solution of the governing nonlinear fluid and structure equations, and as such, is computationally intensive and can benefit from parallel processing.

One difficulty in handling numerically the fluid/structure coupling stems from the fact that the structural equations are usually formulated with material (Lagrangian) co-ordinates, while the fluid equations are typically written using spatial (Eulerian) co-ordinates. Therefore, a straightforward approach to the solution of the coupled fluid/structure dynamic equations requires moving at each time-step at least the portions of the fluid grid that are close to the moving structure. This can be appropriate for small displacements of the structure but may lead to severe grid distorsions when the structure undergoes large motion. Several different approaches have emerged as an alternative to partial regridding in transient aeroelastic computations, among which we note the arbitrary Lagrangian/Eulerian (ALE) formulation (Donea 1982, Hughes et al. 1978, and Belytschko and Kennedy 1978), the co-rotational approach (Kandil and Chuang 1988 and Farhat and Lin 1990), dynamic meshes (Batina 1989) which are closely related to ALE concept, interpolation based methods (Guruswamy 1988), and space-time formulations (Masud 1993). All of these approaches treat a computational aeroelastic problem as a coupled two-field problem.

However, a moving mesh (Fig. 7.1) can also be viewed as a pseudo-structural system with its own dynamics (Lesoinne and Farhat 1993), and therefore, the coupled transient aeroelastic problem can be formulated as a *three-* rather than two-field problem: the fluid, the structure, and the dynamic mesh (Fig. 7.2). The semi-discrete equations governing this three-way coupled problem can be written as follows:

$$
\begin{aligned}
\frac{\partial}{\partial t}(\mathbf{V}(x,t)\ \mathbf{W}(x,t)) + \mathbf{F}^c(\mathbf{W}(x,t),x,\dot{x}) \\
= \mathbf{R}(\mathbf{W}(x,t)) \\
\mathbf{M}\frac{d^2\mathbf{q}}{dt^2} + \mathbf{f}^{int}(\mathbf{q}) = \mathbf{f}^{ext}(\mathbf{W}(x,t),x) \\
\widetilde{\mathbf{M}}\frac{d^2\mathbf{x}}{dt^2} + \widetilde{\mathbf{D}}\frac{d\mathbf{x}}{dt} + \widetilde{\mathbf{K}}\mathbf{x} = \widetilde{\mathbf{K}}_c\ \mathbf{q}
\end{aligned}
\tag{7.1}
$$

where $x$ is the *displacement or position, depending on the context of the sentence* of a moving fluid grid point, $\mathbf{W}$ is the fluid state vector, $\mathbf{V}$ results from the finite element/volume discretization of the fluid equations, $\mathbf{F}^c$ is the vector of convective ALE fluxes that depend on the fluid grid velocity, $\mathbf{R}$ is the vector of diffusive fluxes, $\mathbf{q}$ is the structural displacement vector, $\mathbf{f}^{int}$ denotes the vector of internal structural forces that is equal to $\mathbf{Kq}$ in the linear case where $\mathbf{K}$ denotes the finite element stiffness matrix, $\mathbf{f}^{ext}$ the vector of external forces acting on the structure, $\mathbf{M}$ is the finite element mass matrix of the structure, $\widetilde{\mathbf{M}}$, $\widetilde{\mathbf{D}}$, and $\widetilde{\mathbf{K}}$ are fictitious mass, damping, and stiffness matrices associated with the fluid moving grid (Fig. 7.3) and constructed to avoid any parasitic interaction between the fluid and its grid, or the structure and the moving fluid grid (Lesoinne and Farhat 1993), and $\widetilde{\mathbf{K}}_c$ is a transfer matrix that describes the action of the motion of the structural side of the fluid/structure interface on the fluid dynamic mesh (Lesoinne 1994). For example, $\widetilde{\mathbf{M}} = \widetilde{\mathbf{D}} = 0$, and $\widetilde{\mathbf{K}} = \widetilde{\mathbf{K}}^R$ where $\widetilde{\mathbf{K}}^R$ is a rotation matrix corresponds to a rigid mesh motion of the fluid grid around an oscillating airfoil, and $\widetilde{\mathbf{M}} = \widetilde{\mathbf{D}} = 0$ includes as particular cases the spring-based mesh motion scheme introduced in Batina (1989) and the continuum based updating strategy advocated by several investigators.

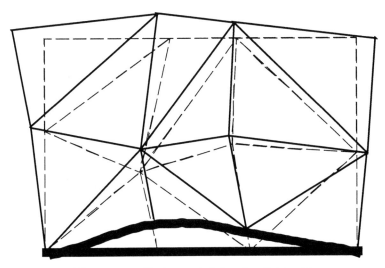

Figure 7.1 Moving and deforming fluid grid

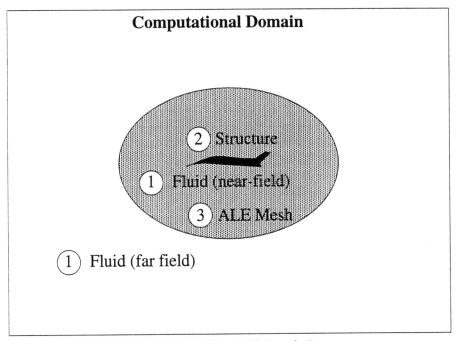

Figure 7.2 Three-field formulation

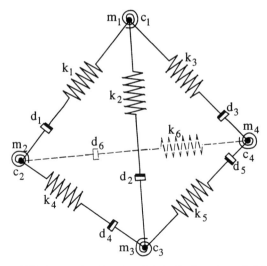

Figure 7.3 A pseudo-structural tetrahedron in a fluid mesh

The first of Eqs. (7.1) is derived in details in Section 7.2. The second of Eqs. (7.1) is the standard nonlinear structural dynamics equation of equilibrium. The notation $\mathbf{f}^{ext}(\mathbf{W}(t), \mathbf{x})$ is used to remind the reader that the external forces acting on the structure include, among others, the aerodynamic forces that are computed from the knowledge of the fluid state vector $\mathbf{W}$ and the motion and deformation of the surface of the structure, which in turn controls the motion $\mathbf{x}(t)$ of the fluid grid. Hence, Eqs. (7.1) are fully coupled.

## OBJECTIVES AND OUTLINE OF THIS PAPER

Each of the three components of the three-way coupled problem described by Eqs. (7.1) has different mathematical and numerical properties, and distinct software implementation requirements. Consequently, the numerical solution of Eqs. (7.1) via a fully coupled monolithic scheme is computationally challenging and software-wise unmanageable.

Alternatively, Eqs. (7.1) can be solved via a *partitioned* analysis or a *staggered* procedure (Park and Felippa 1983, Belytschko et al. 1985, Farhat et al. 1991, and Piperno et al. 1995). This approach offers several appealing features including the ability to use well established discretization and solution methods within each discipline, simplification of software development efforts, and preservation of software modularity.

Traditionally, nonlinear transient aeroelastic problems have been solved via the simplest possible partitioned analysis whose cycle can be described as follows: a) advance the structural system under a given pressure load, b) update the fluid mesh accordingly, and c) advance the fluid system and compute a new pressure load (Batina 1989, Guruswamy 1988, Borland and Rizzetta 1982, Shankar and Ide 1988, Rausch et al. 1989, and Blair et al. 1995). Occasionally, some investigators have advocated the introduction of a few predictor/corrector iterations within each cycle of this three-step staggered integrator in order to improve accuracy (Strganac and Mook 1990), especially when the fluid equations are nonlinear and treated implicitly (Pramono and Weeratunga 1994). However, more efficient staggered solution procedures can and should be devised.

At the heart of nonlinear transient aeroelastic simulations is the computation of unsteady flow problems with moving boundary conditions and dynamic unstructured meshes. In this chapter, we do not discuss the state-of-the-art of unsteady flow solvers, especially that their status seems to be far from satisfactory (Venkatakrishnan 1995). For this specific topic, we refer the reader to references (Venkatakrishnan 1995 and Edwards and Malone 1992). However, we focus in Section 7.2 on the important issues of geometric conservation laws (GCLs) which, in the presence of dynamic meshes, impose important constraints on the algorithms employed for time-integrating the semi-discrete equations governing the fluid and dynamic mesh motions. In particular, we address the problem of satisfying both displacement and velocity continuity constraints between the structure and fluid mesh motions at the fluid/structure interface, and the impact of this problem on the accuracy and stability of the time-integrator selected for predicting the aeroelastic structural response. In Section 7.3, we present a broad family of staggered solution procedures where the fluid flow can be integrated using either an implicit or an explicit scheme, and the structural response is advanced using an implicit one. We address important issues pertaining to numerical stability, subcycling, accuracy vs. speed trade-offs, implementation on heterogeneous computing platforms, and inter-field as well as intra-field parallel processing. Next, we describe in Section 7.4 our particular two- and three-dimensional unsteady flow solvers. In Section 7.5, we overview the solution of the structural dynamics equations. Because our goal is to handle linear as well as nonlinear structural dynamics problems, we opt for a direct time integration method rather than the restrictive modal superposition approach. We employ a substructure based nonlinear time integration implicit algorithm that features second-order accuracy and unconditional stability. For scalability purposes, we also adopt as a linearized solver the substructure based preconditioned conjugate gradient FETI method (Farhat, Mandel and Roux 1994, Farhat, Chen and Mandel 1995) equipped with the projection scheme presented in Farhat, Crivelli and Roux (1994a) for solving iteratively and efficiently systems with repeated right hand sides. In general, the fluid and structure meshes have two independent representations of the physical fluid/structure interface, and do not necessarily match at that interface. We discuss this and other related issues in Section 7.6 where we also describe "Matcher" (Maman and Farhat 1995), a program for generating in parallel the data structures needed for handling arbitrary and non-conforming fluid/structure interfaces in aeroelastic computations. In Section 7.7, we turn to the solution of the equations governing the dynamic motion of the fluid grid. In Section 7.8, we mention a unified and portable approach for parallel fluid/structure computations that is based on the mesh partitioning paradigm. In Section 7.9, we illustrate our framework for computational dynamic aeroelasticity with examples ranging from the numerical investigation on an iPSC-860 massively parallel processor of the instability of flat panels with infinite aspect ratio in supersonic airstreams, to the solution on the Paragon XP/S, Cray T3D and IBM SP2 parallel systems of three-dimensional wing response problems in the transonic regime. Finally, we conclude this chapter in Section 7.10.

*REMARK* : Some of the content of this chapter is based on recent publications by the author and his co-workers. These publications are indicated at the beginning of each section and wherever is appropriate.

## 7.2 GEOMETRIC CONSERVATION LAWS (Lesoinne 1994 and Lesoinne and Farhat 1995)

As stated earlier, the matrices $\widetilde{\mathbf{K}}^c$ and $\widetilde{\mathbf{K}}$ that appear in the third of Eqs. (7.1) are designed to enforce continuity between the grid motion and the structural displacement and/or velocity at the moving fluid/structure boundary $\Gamma_{F/S}(t)$

$$
\begin{aligned}
x(t) &= q(t) \quad \text{on } \Gamma_{F/S}(t) \\
\dot{x}(t) &= \dot{q}(t) \quad \text{on } \Gamma_{F/S}(t)
\end{aligned}
\tag{7.2}
$$

The first of Eqs. (7.1) involves both the position and velocity of the underlying fluid dynamic mesh. These entities are usually obtained from the solution of the second and third of Eqs. (7.1), and optionally from the use of a predictor. When selecting a method for integrating the fluid equations, it is desirable to choose one that preserves the trivial solution of a uniform flow field ( in the absence of other boundary conditions, a uniform flow field is a solution of the Navier-Stokes equations). In this section, we show that this property is verified only when the numerical scheme chosen for solving the fluid equations and the algorithm constructed for updating the mesh position and velocity satisfy a certain condition. We refer to this condition as the Geometric Conservation Law (GCL) because: (a) it can be identified as integrating exactly the area or volume swept by the boundary of a cell in a finite volume formulation, and (b) its principle is similar to the GCL condition that was first pointed out in Thomas and Lombard (1979) for structured grids and finite difference schemes. In the present work, we derive the conditions imposed by the GCL in terms of an appropriate choice of integration points in time, and a consistent scheme for updating the grid point velocities. This is in contrast with previous works (NKonga and Guillard 1994, and Zhang et al. 1993) where the GCL was addressed in terms of *averaged* normal or velocity coefficients for moving finite volume cells. The approach exposed herein for deriving and satisfying a GCL is deemed more general than those previously discussed in the literature. For example, it recovers the results of the normal averaging algorithm recently proposed in (NKonga and Guillard 1994) for finite volume discretizations, and applies as well to finite element methods that are not covered by this normal averaging procedure.

Throughout this section, we consider flow computations using unstructured moving meshes. We focus on the Euler equations, because in our formulation the viscous terms are not explicitly affected by the mesh motion. We derive several GCL conditions for these problems, and discuss their various algorithmic implications. We consider the case where the finite volume method is chosen for the spatial approximation of the flow equations, and the ALE formulation is used for handling dynamic meshes. We investigate the consequences of the GCL condition on the temporal integration of the structural equations of motion. Most importantly, we address the problem of satisfying both displacement and velocity continuity equations between the structure and fluid mesh at the fluid/structure interface, without violating the GCL. Finally, we highlight the importance of the GCL with an illustration of its effect on the computation of the transient aeroelastic response of a flat panel in transonic flow.

*THE FINITE VOLUME METHOD WITH AN ALE FORMULATION*

Let $\Omega(t) \subset \mathcal{R}^n$ $(n = 2, 3)$ be the flow domain of interest and $\Gamma(t)$ be its moving and deforming boundary. We introduce a mapping function between $\Omega(t)$ where time is denoted by $t$ and the grid point coordinates by $x$, and a reference configuration $\Omega(0)$ where time is denoted by $\theta$ and the grid point coordinates by $\xi$ as follows

$$x = x(\xi, \theta); \quad t = \theta \tag{7.3}$$

The conservative form of the equations describing Euler flows can be written in arbitrary Lagrangian-Eulerian (ALE) form as

$$\frac{\partial(JW)}{\partial t}\big|_\xi + J\nabla_x.\mathcal{F}^c(W, \dot{x}) = 0 \tag{7.4}$$
$$\mathcal{F}^c(W, \dot{x}) = \mathcal{F}(W) - \dot{x}W$$

where $J = det(dx/d\xi)$ is the jacobian of the frame transformation $\xi \to x$, $W$ denotes the fluid conservative variables, $\mathcal{F}^c$ denotes the convective ALE fluxes, and $\dot{x} = \frac{\partial x}{\partial \theta}\big|_\xi$ is the ALE grid velocity that may be different from the fluid velocity and from zero.

The finite volume method for unstructured meshes relies on the discretization of the computational domain into control volumes or cells $C_i$ constructed around the vertices $S_i$, with boundaries denoted by $\partial C_i$, and normals to these boundaries denoted by $\nu_i$. Eq. (7.4) can then be integrated over the control cells. In an ALE formulation, these cells move and deform in time. First, integration is performed over a reference cell in the $\xi$ space as follows

$$\int_{C_i(0)} \frac{\partial(JW)}{\partial t}\big|_\xi \, d\Omega_\xi + \int_{C_i(0)} J\nabla_x.\mathcal{F}^c(W, \dot{x}) \, d\Omega_\xi = 0 \tag{7.5}$$

In the above equation, the partial time derivative is evaluated at constant $\xi$; hence, it can be moved outside of the integral sign to obtain

$$\frac{d}{dt} \int_{C_i(0)} W \, J \, d\Omega_\xi + \int_{C_i(0)} \nabla_x.\mathcal{F}^c(W, \dot{x}) \, J \, d\Omega_\xi = 0 \tag{7.6}$$

Switching back to the time-varying cells, Eq. (7.6) above can be rewritten as

$$\frac{d}{dt} \int_{C_i(t)} W \, d\Omega_x + \int_{C_i(t)} \nabla_x.\mathcal{F}^c(W, \dot{x}) \, d\Omega_x = 0 \tag{7.7}$$

Finally, integrating by parts the last term yields the governing integral equation

$$\frac{d}{dt} \int_{C_i(t)} W \, d\Omega_x + \int_{\partial C_i(t)} \mathcal{F}^c(W, \dot{x}).\nu_i \, d\sigma = 0$$

In a finite volume method, the flux through the cell boundary $\partial C_i(t)$ is usually evaluated via a flux splitting approximation (Steger and Warming 1981) as follows

$$F_i^c(\mathbf{W}, \mathbf{x}, \dot{\mathbf{x}}) = \sum_j \int_{\partial C_{i,j}(x)} (\mathcal{F}_+^c(W_i, \dot{x}) + \mathcal{F}_-^c(W_j, \dot{x})).\nu_i \, d\sigma \tag{7.8}$$

where $\partial C_{i,j}$ is the intersection between the boundaries of cells $C_i$ and $C_j$, $W_i$ denotes the average value of $W$ over the cell $C_i$, $\mathbf{W}$ is the vector formed by the collection of $W_i$, and $\mathbf{x}$ is the vector of the time-dependent grid point positions. The numerical flux functions $\mathcal{F}_+^c$ and $\mathcal{F}_-^c$ are designed to make the resulting system stable. An example of such functions can be found in Anderson et al. (1987). For consistency, these numerical fluxes must verify

$$\mathcal{F}_+^c(W, \dot{x}) + \mathcal{F}_-^c(W, \dot{x}) \;=\; \mathcal{F}^c(W, \dot{x}) \tag{7.9}$$

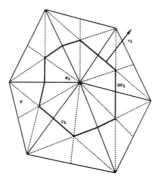

**Figure 7.4** Control volume unstructured two-dimensional mesh

Thus, the resulting discrete equation is

$$\boxed{\frac{d}{dt}(V_i W_i) + F_i^c(\mathbf{W}, \mathbf{x}, \dot{\mathbf{x}}) \;=\; 0} \tag{7.10}$$

where

$$V_i = \int_{C_i(t)} d\Omega_x \tag{7.11}$$

is the area for two-dimensional flow problems, and the volume for three-dimensional flow problems, of cell $C_i$. Collecting all Eqs. (7.11) into a single system yields

$$\boxed{\frac{d}{dt}(\mathbf{V}\mathbf{W}) + \mathbf{F}^c(\mathbf{W}, \mathbf{x}, \dot{\mathbf{x}}) = 0} \tag{7.12}$$

where $\mathbf{V}$ is the diagonal matrix of the cell areas, $\mathbf{W}$ is the vector containing all state variables $W_i$, and $\mathbf{F}^c$ is the collection of the fluxes $F_i^c$. This also completes the derivation of the first of Eqs. (7.1).

**The Geometric Conservation Law**

Let $\Delta t$ and $t^n = n\Delta t$ denote respectively the chosen time-step and the $n$-th time-station. Integrating Eq. (7.11) between $t^n$ and $t^{n+1}$ leads to

$$V_i(\mathbf{x}^{n+1})W_i^{n+1} - V_i(\mathbf{x}^n)W_i^n + \int_{t^n}^{t^{n+1}} F_i^c(\mathbf{W}, \mathbf{x}, \dot{\mathbf{x}})dt = 0 \tag{7.13}$$

The most important issue in the solution of the first of Eqs. (7.1) via an ALE method is the proper evaluation of $\int_{t^n}^{t^{n+1}} F_i^c(\mathbf{W}, \mathbf{x}, \dot{\mathbf{x}})dt$ in Eq. (7.14). In particular, it is crucial to establish where the fluxes must be integrated: on the mesh configuration at $t = t^n$ ($\mathbf{x}^n$), on that at $t = t^{n+1}$ ($\mathbf{x}^{n+1}$), or in between these two configurations. The same questions arise as to the choice of the mesh velocity vector $\dot{\mathbf{x}}$.

Clearly, a proposed numerical algorithm for computing the quantity $\int_{t^n}^{t^{n+1}} F_i^c(\mathbf{w}, \mathbf{x}, \dot{\mathbf{x}})dt$ involving general and arbitrary time dependent fluid state vectors and mesh configurations cannot be acceptable unless it conserves the state of a uniform flow. Let $W^*$ denote a given uniform state of the flow. Substituting $W_k^n = W_k^{n+1} = W^*$ in Eq. (7.14) gives

$$(V_i^{n+1} - V_i^n)W^* + \int_{t^n}^{t^{n+1}} F_i^c(\mathbf{W}^*, \mathbf{x}, \dot{\mathbf{x}}) \, dt = 0 \qquad (7.14)$$

where $\mathbf{W}^*$ is the vector of the state variables when $W_k = W^*$ for all $k$. From Eq. (7.30), it follows that

$$F_i^c(\mathbf{W}^*, \mathbf{x}, \dot{\mathbf{x}}) = \sum_j \int_{\partial C_{i,j}(x)} (\mathcal{F}_+^c(W^*, \dot{x}) + \mathcal{F}_-^c(W^*, \dot{x})).\nu_i \, d\sigma$$

$$= \int_{\partial C_i(\mathbf{X})} (\mathcal{F}(W^*) - \dot{x}W^*).\nu_i \, d\sigma \qquad (7.15)$$

Given that the integral on a closed boundary of the flux of a constant function is identically zero

$$\int_{\partial C_i(\mathbf{X})} \mathcal{F}(W^*).\nu_i \, d\sigma = 0 \qquad (7.16)$$

it follows that

$$F_i^c(\mathbf{W}^*, \mathbf{x}, \dot{\mathbf{x}}) = -\int_{\partial C_i(\mathbf{X})} \dot{x}W^*.\nu_i \, d\sigma \qquad (7.17)$$

Hence, substituting Eq. (7.18) into Eq. (7.15) yields

$$(V_i(\mathbf{x}^{n+1}) - V_i(\mathbf{x}^n))W^* - \left(\int_{t^n}^{t^{n+1}} \int_{\partial C_i(\mathbf{X})} \dot{x}.\nu_i \, d\sigma \, dt\right)W^* = 0 \qquad (7.18)$$

which can be rewritten as

$$\boxed{(V_i(\mathbf{x}^{n+1}) - V_i(\mathbf{x}^n)) = \int_{t^n}^{t^{n+1}} \int_{\partial C_i(\mathbf{X})} \dot{x}.\nu_i \, d\sigma \, dt} \qquad (7.19)$$

Eq. (7.20) above defines a geometric conservation law (GCL) that must be verified by any proposed ALE mesh updating scheme. This law states that the change in area (volume) of each control volume between $t^n$ and $t^{n+1}$ must be equal to the area (volume) swept by the cell boundary during $\Delta t = t^{n+1} - t^n$. Therefore, the updating of $\mathbf{x}$ and $\dot{\mathbf{x}}$ cannot be based on mesh distorsion issues alone when using ALE solution schemes.

The assumption that the numerical method performs exactly the integration of Eq. (7.17) is referred to in Zhang et al. (1993) as the Surface Conservation Law (SCL). Satisfying of this condition is necessary for flow computations on static meshes and is not specific to dynamic ones. Therefore, we do not discuss this condition in this section any further and refer the reader to Zhang et al. (1993) for additional details.

### Implications of the GCL

From the analysis presented in the previous section, it follows that an appropriate scheme for evaluating $\int_{t^n}^{t^{n+1}} F_i^c(\mathbf{W}^*, \mathbf{x}, \dot{\mathbf{x}})dt$ in Eq. (7.15) is a scheme that respects the GCL (7.20). Note that once a mesh updating scheme is given, the left hand side of Eq. (7.20) is always exactly computed. Hence, a proper method for evaluating $\int_{t^n}^{t^{n+1}} F_i^c(\mathbf{W}^*, \mathbf{x}, \dot{\mathbf{x}})dt$ is a method that obeys the GCL and therefore computes exactly the right hand side of Eq. (7.20)— that is, $\int_{t^n}^{t^{n+1}} \int_{\partial C_i(\mathbf{x})} \dot{\mathbf{x}}.\nu_i \, d\sigma \, dt$.

### The Two-Dimensional Case

Given that in two dimensions $\partial C_i$ is the union of segments, it suffices to consider the integration of $\dot{x}.n$ along a segment $[ab]$ with a normal $n$

$$I_{[ab]} = \int_{t^n}^{t^{n+1}} \int_{[ab]} \dot{x}.n \, dsdt \tag{7.20}$$

Let $x_a$ and $x_b$ denote the instantaneous positions of two connected vertices $a$ and $b$ (Fig. 7.5). The position of any point on the edge $[ab]$ during the time-interval $[t^n, t^{n+1}]$ can be parametrized as follows

$$\begin{aligned} x(t) &= \alpha x_a(t) + (1 - \alpha)x_b(t) \\ \dot{x}(t) &= \alpha \dot{x}_a(t) + (1 - \alpha)\dot{x}_b(t) \\ \alpha &\in [0, 1] \quad t \in [t^n, t^{n+1}] \end{aligned} \tag{7.21}$$

where

$$\begin{aligned} x_a(t) &= \delta(t)x_a^{n+1} + (1 - \delta(t))x_a^n \\ x_b(t) &= \delta(t)x_b^{n+1} + (1 - \delta(t))x_b^n \end{aligned} \tag{7.22}$$

and $\delta(t)$ is a real function that satisfies

$$\delta(t^n) = 0; \quad \delta(t^{n+1}) = 1 \tag{7.23}$$

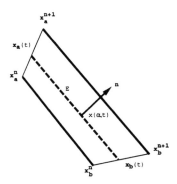

**Figure 7.5** Parametrization of an edge in a two-dimensional space

Substituting Eqs. (7.22,7.23) into Eq. (7.21) yields

$$
\begin{aligned}
I_{[ab]} &= \int_{t^n}^{t^{n+1}} \int_0^1 (\alpha \dot{x}_a + (1-\alpha)\dot{x}_b).n\, l\, d\alpha dt \\
&= \int_{t^n}^{t^{n+1}} \frac{1}{2}(\dot{x}_a + \dot{x}_b).n\, l\, dt \\
&= \int_{t^n}^{t^{n+1}} \frac{1}{2}(\dot{x}_a + \dot{x}_b)H(x_a - x_b)\, dt \\
&= \int_{t^n}^{t^{n+1}} \frac{1}{2}(\dot{x}_a + \dot{x}_b)H(\delta(t)(x_a^{n+1} - x_b^{n+1}) + (1 - \delta(t))(x_a^n - x_b^n))\, dt
\end{aligned}
\tag{7.24}
$$

where $l$ is the length of edge $[ab]$, and $H = \begin{pmatrix} 0 & -1 \\ 1 & 0 \end{pmatrix}$. The mesh velocities $\dot{x}_a$ and $\dot{x}_b$ can be obtained from the differentiation of Eq. (7.2).

$$
\dot{x}_a = \dot{\delta}(t)(x_a^{n+1} - x_a^n) \quad \dot{x}_b = \dot{\delta}(t)(x_b^{n+1} - x_b^n)
\tag{7.25}
$$

and $I_{[ab]}$ can be finally written as

$$
\begin{aligned}
I_{[ab]} &= \frac{1}{2} \int_{t^n}^{t^{n+1}} \dot{\delta}((x_a^{n+1} - x_a^n) + (x_b^{n+1} - x_b^n))H(\delta(x_a^{n+1} - x_b^{n+1}) + (1 - \delta)(x_a^n - x_b^n))\, dt \\
&= \frac{1}{2} \int_0^1 ((x_a^{n+1} - x_a^n) + (x_b^{n+1} - x_b^n))H(\delta(x_a^{n+1} - x_b^{n+1}) + (1 - \delta)(x_a^n - x_b^n))\, d\delta
\end{aligned}
\tag{7.26}
$$

Clearly, the integrand of $I_{[ab]}$ is linear in $\delta$. Therefore, $I_{[ab]}$ can be exactly computed using the *midpoint rule*, provided that Eq. (7.26) holds — that is

$$
\dot{x} = \dot{\delta}(t)(x^{n+1} - x^n) = \frac{\Delta\delta}{\Delta t}(x^{n+1} - x^n)
\tag{7.27}
$$

which in view of Eq. (7.24) can also be written as

$$\dot{x} = \frac{x^{n+1} - x^n}{\Delta t} \tag{7.28}$$

In summary, the GCL derived herein shows that for two-dimensional problems, the integrand of $\int_{t^n}^{t^{n+1}} F_i^c(\mathbf{W}, \mathbf{x}, \dot{\mathbf{x}})\, dt$ in Eq. (7.14) must be evaluated at the midpoint configuration, and that this integral must be computed as follows

$$
\begin{aligned}
\int_{t^n}^{t^{n+1}} F_i^c(\mathbf{W}, \mathbf{x}, \dot{\mathbf{x}})\, dt &= \Delta t F_i^c(\mathbf{W}^k, \mathbf{x}^{n+\frac{1}{2}}, \dot{\mathbf{x}}^{n+\frac{1}{2}}) \\[2mm]
\mathbf{W}^{n+\frac{1}{2}} &= \frac{\mathbf{W}^n + \mathbf{W}^{n+1}}{2} \\[2mm]
\mathbf{x}^{n+\frac{1}{2}} &= \frac{\mathbf{x}^n + \mathbf{x}^{n+1}}{2} \\[2mm]
\dot{\mathbf{x}}^{n+\frac{1}{2}} &= \frac{\mathbf{x}^{n+1} - \mathbf{x}^n}{\Delta t}
\end{aligned}
\tag{7.29}
$$

where the superscript $k$ depends on the time discretization of the flow equation.

## The Three-Dimensional Case

In a three-dimensional space, the boundary of each cell is polygonal and can be decomposed into a set of non overlapping triangular facets. Similarly to the two-dimensional case, let $I_{[abc]}$ denote the flux crossing the facet $[abc]$

$$I_{[abc]} = \int_{t^n}^{t^{n+1}} \int_{[abc]} \dot{x}.n\, d\sigma dt \tag{7.30}$$

Let $x_a$, $x_b$ and $x_c$ denote the instantaneous positions of three connected vertices $a$, $b$ and $c$. The position of any point on the facet can be parametrized as follows (see Fig. 7.6)

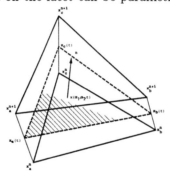

Figure 7.6 Parametrization of a facet in a three-dimensional space

$$x = \alpha_1 x_a(t) + \alpha_2 x_b(t) + (1 - \alpha_1 - \alpha_2) x_c(t)$$
$$\dot{x} = \alpha_1 \dot{x}_a(t) + \alpha_2 \dot{x}_b(t) + (1 - \alpha_1 - \alpha_2) \dot{x}_c(t) \tag{7.31}$$
$$\alpha_1 \in [0, 1]; \quad \alpha_2 \in [0, 1 - \alpha_1]; \quad t \in [t^n, t^{n+1}]$$

where

$$x_a(t) = \delta(t)x_a^{n+1} + (1 - \delta(t))x_a^n$$
$$x_b(t) = \delta(t)x_b^{n+1} + (1 - \delta(t))x_b^n \tag{7.32}$$
$$x_b(t) = \delta(t)x_b^{n+1} + (1 - \delta(t))x_b^n$$

and $\delta(t)$ is given in (7.24). Substituting the above parametrization in (7.31) we obtain

$$
\begin{aligned}
I_{[abc]} &= \int_{t^n}^{t^{n+1}} \int_0^1 \int_0^{1-\alpha_1} (\alpha_1 \dot{x}_a + \alpha_2 \dot{x}_b + (1 - \alpha_1 - \alpha_2)\dot{x}_c).n \, |x_{ac} \wedge x_{bc}| \, d\alpha_2 \, d\alpha_1 \, dt \\
&= \int_{t^n}^{t^{n+1}} \frac{1}{6}(\dot{x}_a + \dot{x}_b + \dot{x}_c).(x_{ac} \wedge x_{bc}) \, dt \\
&= \int_{t^n}^{t^{n+1}} \frac{1}{6}\dot{\delta}(\Delta x_a + \Delta x_b + \Delta x_c).(x_{ac} \wedge x_{bc}) \, dt \\
&= \int_0^1 \frac{1}{6}(\Delta x_a + \Delta x_b + \Delta x_c).(x_{ac} \wedge x_{bc}) \, d\delta
\end{aligned}
\tag{7.33}
$$

with

$$x_{ac} = x_a - x_c; \quad x_{bc} = x_b - x_c$$
$$\Delta x_a = x_a^{n+1} - x_a^n; \quad \Delta x_b = x_b^{n+1} - x_b^n; \quad \Delta x_c = x_c^{n+1} - x_c^n \tag{7.34}$$

Noting that

$$x_{ac} \wedge x_{bc} = (\delta x_{ac}^{n+1} + (1 - \delta)x_{ac}^n) \wedge (\delta x_{bc}^{n+1} + (1 - \delta)x_{bc}^n) \tag{7.35}$$

is a quadratic function of $\delta$, the integrand of $I_{[abc]}$ is clearly quadratic in $\delta$ and therefore can be exactly computed using a *2-point integration rule*, provided that Eq. (7.50) is used to compute $\dot{x}$.

Hence, the proper method for evaluating $\int_{t^n}^{t^{n+1}} F_i^c(\mathbf{W}, \mathbf{x}, \dot{\mathbf{x}}) \, dt$ that respects the GCL (7.20) in the three-dimensional case is

$$
\begin{aligned}
\int_{t^n}^{t^{n+1}} F_i^c(\mathbf{W}, \mathbf{x}, \dot{\mathbf{x}}) \, dt &= \frac{\Delta t}{2}(F_i^c(\mathbf{W}^{k1}, \mathbf{x}^{m1}, \dot{\mathbf{x}}^{n+\frac{1}{2}}) + F_i^c(\mathbf{W}^{k2}, \mathbf{x}^{m2}, \dot{\mathbf{x}}^{n+\frac{1}{2}})) \\
m1 &= n + \frac{1}{2} - \frac{1}{2\sqrt{3}} \\
m2 &= n + \frac{1}{2} + \frac{1}{2\sqrt{3}} \\
\mathbf{W}^{n+\eta} &= \eta \mathbf{W}^{n+1} + (1 - \eta)\mathbf{W}^n \\
\mathbf{x}^{n+\eta} &= \eta \mathbf{x}^{n+1} + (1 - \eta)\mathbf{x}^n \\
\dot{\mathbf{x}}^{n+\frac{1}{2}} &= \frac{\mathbf{x}^{n+1} - \mathbf{x}^n}{\Delta t}
\end{aligned}
\tag{7.36}
$$

where the superscripts $k1$ and $k2$ depend on the time discretization of the flow equation.

## Recovery of the Averaged-Normals Method

In NKonga and Guillard (1994), the convected flux accross the facet $I_{[abc]}$ is computed using

$$
I_{[abc]} = \frac{1}{3}(\Delta x_a + \Delta x_b + \Delta x_c).\eta
$$
$$
\eta = \frac{1}{6}(x_{ac}^n \wedge x_{bc}^n + x_{ac}^{n+1} \wedge x_{bc}^{n+1} + \frac{1}{2}(x_{ac}^n \wedge x_{bc}^{n+1} + x_{ac}^{n+1} \wedge x_{bc}^n))
$$

(7.37)

while the evaluation of Eq. (7.31) using the two-point rule gives

$$
I_{[abc]} = \frac{1}{3}(\Delta x_a + \Delta x_b + \Delta x_c).\eta^*
$$
$$
\eta^* = \frac{1}{4}(x_{ac}^{m1} \wedge x_{bc}^{m1} + x_{ac}^{m2} \wedge x_{ac}^{m2})
$$

(7.38)

Expanding $\eta^*$, we obtain

$$
\eta^* = \frac{1}{4}(((\frac{1}{2} + \frac{1}{2\sqrt{3}})x_{ac}^n + (\frac{1}{2} - \frac{1}{2\sqrt{3}})x_{ac}^{n+1}) \wedge ((\frac{1}{2} + \frac{1}{2\sqrt{3}})x_{bc}^n + (\frac{1}{2} - \frac{1}{2\sqrt{3}})x_{bc}^{n+1})
$$
$$
+ ((\frac{1}{2} - \frac{1}{2\sqrt{3}})x_{ac}^n + (\frac{1}{2} + \frac{1}{2\sqrt{3}})x_{ac}^{n+1}) \wedge ((\frac{1}{2} - \frac{1}{2\sqrt{3}})x_{bc}^n + (\frac{1}{2} + \frac{1}{2\sqrt{3}})x_{bc}^{n+1}))
$$
$$
= \frac{1}{2}(\frac{1}{3}x_{ac}^n \wedge x_{bc}^n + \frac{1}{3}x_{ac}^{n+1} \wedge x_{bc}^{n+1} + \frac{1}{6}(x_{ac}^n \wedge x_{bc}^{n+1} + x_{ac}^{n+1} \wedge x_{bc}^n))
$$

(7.39)

which shows that the proposed GCL (7.20) recovers the same results as the averaged-normals method proposed in [36] for the finite volume discretization of flow equations with moving meshes.

## IMPACT OF THE GCL ON THE TEMPORAL SOLUTION OF AEROELASTIC PROBLEMS

The most remarquable implication of the GCL condition is the constraint it imposes on the mesh velocity computation, independently of the integration formula for the flow equations

$$
\dot{x}^{n+\frac{1}{2}} = \frac{x^{n+1} - x^n}{\Delta t}
$$

(7.40)

This formula is intuitive and has been "naturally" used by several investigators independently from any geometric conservation law (see, for example, Batina (1989)). However, when sophisticated time-integrators are used for the structure and/or the mesh equations, neither the computed mesh velocities $\dot{x}^{n+\frac{1}{2}}$ nor the computed structural velocites on the fluid/structure interface are guaranteed to obey $\dot{x}^{n+\frac{1}{2}} = \frac{x^{n+1} - x^n}{\Delta t}$. In that case, satisfying the GCL requires

- using the mesh velocity $\widetilde{\dot{\mathbf{x}}^{n+\frac{1}{2}}}$ computed by the time-integrator, only for evaluating $\mathbf{x}^{n+1}$.

- using the mesh velocity $\dot{\mathbf{x}}^{n+\frac{1}{2}} = \frac{\mathbf{x}^{n+1} - \mathbf{x}^n}{\Delta t}$ in the evaluation of the fluid fluxes.

This means that it is not always possible to respect the continuity of both the displacement and velocity fields on the fluid structure boundary as prescribed by Eqs. (7.2) without violating the GCL. For example, if the displacement continuity condition $x(t) = q(t)$ is enforced at the fluid/structure interface $\Gamma_{F/S}$, — and that is usually the case — respecting the GCL implies computing a mesh velocity field on $\Gamma_{F/S}$ that is equal to

$$\dot{x}^{n+\frac{1}{2}} = \frac{x^{n+1} - x^n}{\Delta t} = \frac{q^{n+1} - q^n}{\Delta t} \quad \text{on} \quad \Gamma_{F/S} \tag{7.41}$$

In that case, satisfying also the velocity continuity condition $\dot{x}(t) = \dot{q}(t)$ on $\Gamma_{F/S}$ requires that

$$\dot{q}^{n+\frac{1}{2}} = \frac{q^{n+1} - q^n}{\Delta t} \quad \text{on} \quad \Gamma_{F/S} \tag{7.42}$$

which is not enforced by all structural time-integrators. Therefore, it is not always possible to satisfy the continuity between both the displacement and the velocity of the structure, and those of the fluid mesh at the fluid/structure interface, without violating the GCL.

Unfortunately, a discontinuity between the velocity of the structure and that of the fluid mesh at the fluid/structure interface can perturb the energy exchange between the fluid and the structure. However, it can be shown that when the implicit midpoint rule is used for advancing the structure and the displacement condition $x(t) = q(t)$ is enforced on $\Gamma_{F/S}$ using a staggered algorithm, both continuity equations (7.23 ) can be enforced without violating the GCL. The proof goes as follows.

Given some initial conditions $q^0$ and $\dot{q}^0$, suppose that the mesh motion is initialized such that the following holds on the fluid/structure interface

$$x^{-\frac{1}{2}} = q^0 - \frac{\Delta t}{2} \dot{q}^0 \quad \text{on} \quad \Gamma_{F/S} \tag{7.43}$$

Also suppose that at each time-station $t^n$, the continuity of the velocity field is enforced on the fluid/structure boundary

$$\dot{x}^n = \dot{q}^n \quad \text{on} \quad \Gamma_{F/S} \tag{7.44}$$

If the midpoint rule is used for time-integrating the structural equations of motion, and the dynamic fluid mesh is updated consistently with the GCL, it can be proved by induction that

$$x^{n-\frac{1}{2}} = q^n - \frac{\Delta t}{2} \dot{q}^n \quad \text{on} \quad \Gamma_{F/S} \tag{7.45}$$

Indeed, the above relation holds at $n = 0$. Assuming it holds at $n$, it follows that

$$
\begin{aligned}
x^{n+\frac{1}{2}} &= x^{n-\frac{1}{2}} + \Delta t \dot{x}^n \\
&= q^n - \frac{\Delta t}{2} \dot{q}^n + \Delta t \dot{q}^n \\
&= q^n + \frac{\Delta t}{2} \dot{q}^n
\end{aligned}
\tag{7.46}
$$

Since the midpoint rule algorithm applied to the structural equations implies

$$q^{n+1} - q^n = \frac{\Delta t}{2}(\dot{q}^n + \dot{q}^{n+1})$$ (7.47)

it follows that

$$x^{n+\frac{1}{2}} = q^{n+1} - \frac{\Delta t}{2}\dot{q}^{n+1}$$ (7.48)

which completes the proof by induction of Eq. (7.46).

Now, a staggered algorithm for solving the coupled Eqs. (7.1) can be described as follows

1)  using the mesh displacement $\mathbf{x}^{n-\frac{1}{2}}$, and the mesh velocity $\dot{\mathbf{x}}^n$ that matches the structural velocity $\dot{\mathbf{q}}^n$ on $\Gamma_{F/S}$, update the mesh as follows

$$\mathbf{x}^{n+\frac{1}{2}} = \mathbf{x}^{n-\frac{1}{2}} + \Delta t \dot{\mathbf{x}}^n$$ (7.49)

2)  using $\mathbf{x}^{n-\frac{1}{2}}$, $\mathbf{x}^{n+\frac{1}{2}}$ and $\dot{\mathbf{x}}^n$, update the fluid state vector $\mathbf{W}^{n+\frac{1}{2}}$ in a manner that satisfies the GCL

3)  using the pressure computed from $\mathbf{W}^{n+\frac{1}{2}}$, compute $\mathbf{q}^{n+1}$ and $\dot{\mathbf{q}}^{n+1}$ using the midpoint rule

Defining $\mathbf{x}^n$ as

$$\mathbf{x}^n = \frac{\mathbf{x}^{n-\frac{1}{2}} + \mathbf{x}^{n+\frac{1}{2}}}{2}$$ (7.50)

and substituting Eq. (7.50) into Eq. (7.51) leads to

$$\mathbf{x}^n = \mathbf{x}^{n-\frac{1}{2}} + \frac{\Delta t}{2}\,\dot{\mathbf{x}}^n$$ (7.51)

which in view of Eqs. (7.46,7.45) yields

$$x^n = q^n \quad \text{on } \Gamma_{F/S}$$ (7.52)

and demonstrates that, when the midpoint rule is used for time-integrating the structure and a proper staggered procedure is used for solving the coupled fluid/structure problem, the continuity of both the displacement and velocity fields ca be enforced on $\Gamma_{F/S}$ without violating the GCL.

## NUMERICAL EXAMPLE

In order to highlight the impact of the GCL on coupled aeroelastic computations, we consider here the simulation of the two-dimensional transient aeroelastic response of a flexible panel in a transonic regime. The panel is represented by its cross section that is assumed to have a unit length and a uniform thickness and Young modulus, and to be clamped at both ends. This rectangular cross section is discretized into plane strain 4-node elements with perfect aspect ratios. The two-dimensional flow domain around the panel is discretized into triangles, and the Euler equations are used for this computation. The free stream Mach number is set to $M_\infty = 0.8$, and a slip condition is imposed at the fluid/structure boundary. Further details on the specifics of this simulation are deferred to Section 7.9.

Initially, a steady-state flow is computed around the panel at $M_\infty = 0.8$. Next, this flow is perturbed via an initial displacement of the panel that is proportional to its second fundamental mode, and the subsequent panel motion and flow evolution are computed using one of the staggered explicit/implicit fluid/structure procedures described in the following section. Two computed histories of the lift using the same time-step are reported in Fig. 7.7 for the case where the GCL is violated by updating the mesh velocity field at the fluid/structure interface via a higher-order scheme than that given in Eq. (7.41), and in Fig. 7.8 for the case where the GCL is respected. Clearly, this example demonstrates the impact of the GCL on aeroelastic computations as it shows that violating this law leads to undesirable spurious oscillations in the lift prediction.

## 7.3 A FAMILY OF STAGGERED SOLUTION PROCEDURES (Piperno et al. 1995, Farhat, Lesoinne and Maman 1995, and Farhat, Lesoinne, Chen and Lantéri 1995)

In the nonlinear theory, predicting whether an aircraft will flutter or not for a given set of flight conditions is determined by computing the solution of Eqs. (7.1), and establishing numerically whether this solution grows continuously in time or not. In other words, a linear aeroelastic dynamic stability problem can be solved without computing explicitly the response of the structure, but a nonlinear aeroelastic dynamic stability problem is typically solved by simulating a set of corresponding nonlinear response problems. Hence, transient nonlinear aeroelastic investigations are in general computationally intensive. For example, establishing the transonic flutter boundary of an aircraft for a given set of aeroelastic parameters requires about 30 aeroelastic response analyses, which clearly demonstrates the need for a fast capability for solving Eqs. (7.1). Such a capability requires not only powerful supercomputers, but also powerful computational methodologies and algorithms.

One approach for solving the three-way coupled aeroelastic problem described in Eqs. (7.1) is known as the "monolithic augmentation" approach where, as specific problems arise, a large-scale single computer program — for example, a finite element structural analysis code — is expanded to house more interaction effects — for example, fluid/structure interaction. Such an approach poses several difficulties, most of which are related to the fact that each of the three components of the three-way coupled aeroelastic problem described in Eqs. (7.1) has different mathematical and numerical properties, and distinct software implementation requirements. Some of these difficulties are summarized in Park and Felippa (1983). In our opinion, the monolithic augmentation approach is unattractive because once it is implemented, it cannot easily accommodate neither new or improved problem formulations, nor future advances within any of the computational fluid and/or structural dynamics disciplines.

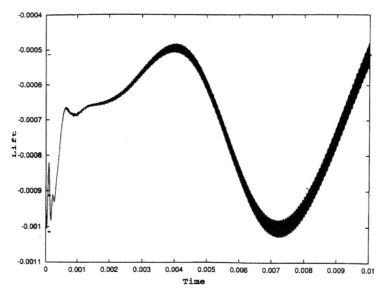

**Figure 7.7** Lift history when the GCL is violated

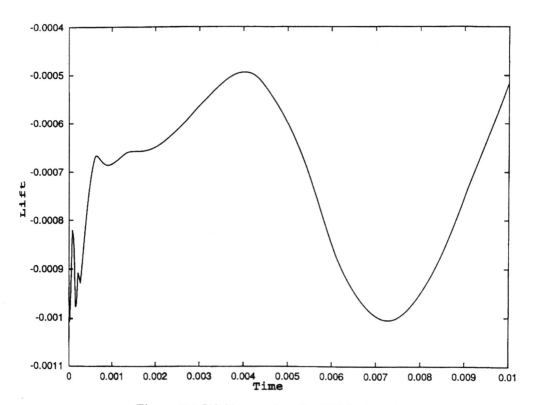

**Figure 7.8** Lift history when the GCL is obeyed

Alternatively, the solution of Eqs. (7.1) can be obtained through a staggered procedure in which separate fluid and structural analysis programs — often called *field analyzers* (Park and Felippa 1983) — execute and exchange data. Such an approach is also known as partitioned analysis. It offers several appealing features, including the ability to use well established discretization and solution methods within each discipline, simplification of software development efforts, reuse of existing and validated code, accommodation of future single discipline improvements, and preservation of software modularity. Traditionally, nonlinear transient aeroelastic problems have been solved via the simplest possible staggered procedure where the separate fluid and structural analysis programs execute in a strictly sequential fashion, and exchange strictly *interface-state data* such as pressures and velocities at *each single time-step* (see, for example, Farhat and Lin (1990), Guruswamy (1988), Borland and Rizzetta (1982), Shanker and Ide (1988), Rausch et al. (1989), and Blair et al. (1995)). The objective of this section is to overview a broader family of more powerful staggered solution procedures that address some important issues related to numerical stability, subcycling, accuracy vs. speed trade-offs, implementation on heterogeneous computing platforms, and inter-field as well as intra-field parallel processing.

## PRELIMINARIES

Of course, the global performance of a partitioned analysis for solving the time-dependent Eqs. (7.1) depends on the local performances of the fluid and structural field analyzers. But more importantly, the global performance also depends on the stability and accuracy properties of the staggered solution procedure itself. For a given prescribed accuracy, the more stable a staggered algorithm is, the larger is the allowable *coupled time-integration step*, and therefore the faster is the total solution time. Hence, our primal goal is to construct partitioned analysis procedures for Eqs. (7.1) with superior stability properties.

Because the aeroelastic response of a structure is often dominated by low frequency dynamics, we consider only implicit schemes for time-integrating the structural displacement field. However, we consider both explicit and implicit time-integrators for advancing the fluid field, as both approaches are popular in computational fluid dynamics. On the other hand, we also note that time-accurate implicit and unstructured flow solvers seem to be less available than their explicit counterparts. In the sequel, we refer to a partitioned analysis procedure as an explicit/implicit one if an explicit time-accurate flow solver is employed, and as an implicit/implicit one if an implicit flow solver is used. In the implicit/implicit case, our goal is to devise an unconditionally stable staggered algorithm, or at least a partitioned procedure that allows a relatively large time step. In the explicit/implicit case, our objective is to design a staggered solution algorithm whose stability limit is not worse than that of the underlying explicit flow solver. These are not trivial tasks because coupling effects can restrict the stability limits of the independent field time-integrators.

Next, we make the following observations

- linear and nonlinear transient fluid /structure interaction problems have one particularity: they possess a wide variety of self-excited vibrations and instabilities. We have already mentioned the flutter problem. Another example of a dynamic instability is that of the vibrations due to Von Kármán vortices (Schlichting

1960). Therefore, when it comes to analyzing the numerical stability of a proposed staggered algorithm for time-integrating fluid/structure interaction problems, it is essential to consider the case where the coupled system is physically stable — that is, when Eqs. (7.1) have a solution that does not grow indefinitely in time.

- when the structure undergoes small displacements, the fluid mesh can be frozen and "transpiration" fluxes can be introduced at the fluid side of the fluid/structure boundary to account for the motion of the structure. In that case, the nonlinear transient aeroelastic problem simplifies from a three- to a two-field coupled problem.

- most fluid/structure instability problems can be analyzed by investigating the response of the coupled system to a perturbation around a steady state. If the response is an amplification of the initial perturbation, it is an indication that the system is unstable. If it is a dissipation of the initial perturbation, it means that the system is stable. This suggests that aeroelastic stability or instability problems can be investigated by linearizing the flow around an equilibrium position $\mathbf{W}_0$, and analyzing the response of the fluid/structure system to a perturbation.

Based on the above observations, Piperno et al. (1995) have constructed a simplified but relevant aeroelastic "test" problem where the coupled fluid/structure system is always physically stable. They have also presented a mathematical framework for analyzing the accuracy and stability properties of staggered procedures applied to the solution of their test problem. Subsequently, this test problem was also shown to be a good model problem for the complex nonlinear aeroelastic systems that we are interested in solving (Piperno et al. 1995, Lesoinne and Farhat 1993, and Lesoinne 1994). In the test problem, the structure is assumed to remain in the linear regime, and the flow is linearized around an equilibrium position of the fluid state vector denoted here by $W_0$. The semi-discrete equations governing this coupled aeroelastic model problem are given by (see Piperno et al. (1995) for details)

$$
\begin{pmatrix} \delta \dot{\mathbf{W}} \\ \mathbf{Q} \end{pmatrix} = \begin{pmatrix} \mathbf{A}^* & \mathbf{B} \\ \mathbf{C} & \mathbf{D}^* \end{pmatrix} \begin{pmatrix} \delta \mathbf{W} \\ \mathbf{Q} \end{pmatrix}
$$
$$
\begin{pmatrix} \delta \mathbf{W} \\ \mathbf{Q} \end{pmatrix}_{(t=0)} = \begin{pmatrix} \delta \mathbf{W} \\ \mathbf{Q} \end{pmatrix}_0
$$

(7.53)

where $\delta \mathbf{W}$ is the perturbed fluid state vector, $\mathbf{Q} = \begin{pmatrix} \mathbf{q} \\ \dot{\mathbf{q}} \end{pmatrix}$ is the structure state vector, $\mathbf{A}^*$ results from the spatial discretization of the flow equations, $\mathbf{B}$ is the matrix induced by the transpiration fluxes at the fluid/structure boundary $\Gamma_{F/S}$, $\mathbf{C}$ is the matrix that transforms the fluid pressure on $\Gamma_{F/S}$ into prescribed structural forces, and $\mathbf{D}^* = \begin{pmatrix} \mathbf{0} & \mathbf{I} \\ -\mathbf{M}^{-1}\mathbf{K} & -\mathbf{M}^{-1}\mathbf{D} \end{pmatrix}$, where as before, $\mathbf{M}$, $\mathbf{D}$, and $\mathbf{K}$ are the structural mass, damping, and stiffness matrices.

In Lesoinne (1994), the aeroelastic model problem described in Eqs. (7.54) has been extended to include the mesh motion of the fluid grid, and therefore to truly represent the three-way coupled aeroelastic problem governed by Eqs. (7.1). More importantly, Lesoinne (1994) discusses a methodology for considering a staggered solution procedure that was designed for solving the three-field equivalent of the model problem (7.54), and extending it to the case of nonlinear transient aeroelastic problems such as those governed by Eqs. (7.1).

In this section, we overview a family of partitioned analysis procedures for solving the nonlinear transient coupled Eqs. (7.1). These algorithms are based on the mathematical results established in Peperno et al. (1995) and Lesoinne (1994), and have recently been described in Farhat, Lesoinne and Maman (1995) and Farhat, Lesoinne, Chen and Lantéri (1995). Rather than discussing mathematical proofs and details that can be found in Peperno et al. (1995) and Lesoinne (1994), we emphasize important computational and implementational issues pertaining to accuracy, stability, distributed computing, I/O transfers, subcycling, and parallel processing.

## EXPLICIT/IMPLICIT PARITIONED PROCEDURES

### ALG0: the basic explicit/implicit staggered solution procedure

In order not to obscure the following discussion by the complex notation needed for three-dimensional viscous flows, we focus here, without any loss of generality, on the case of two-dimensional Euler flows discretized by the finite volume method. For three-dimensional inviscid flows, Eqs. (7.37) should be used instead of Eqs. (7.30).

From the results established in Section 7.2, it follows that the semi-discrete equations governing the three-way coupled aeroelastic problem can be written in that case as

$$
\mathbf{V}(\mathbf{x}^{n+1})\mathbf{W}^{n+1} - \mathbf{V}(\mathbf{x}^{n})\mathbf{W}^{n} + \Delta t \mathbf{F}^{c}(\mathbf{W}^{k}, \mathbf{x}^{n+\frac{1}{2}}, \dot{\mathbf{x}}^{n+\frac{1}{2}}) = 0
$$

$$
\mathbf{x}^{n+\frac{1}{2}} = \frac{\mathbf{x}^{n} + \mathbf{x}^{n+1}}{2}
$$

$$
\dot{\mathbf{x}}^{n+\frac{1}{2}} = \frac{\mathbf{x}^{n+1} - \mathbf{x}^{n}}{\Delta t}
$$

$$
\mathbf{M}\ddot{\mathbf{q}}^{n+1} + \mathbf{f}^{int}(\mathbf{q}^{n+1}) = \mathbf{f}^{ext}(\mathbf{x}^{n+1}, \mathbf{W}^{n+1})
$$

$$
\widetilde{\mathbf{M}}\ddot{\mathbf{x}}^{n+1} + \widetilde{\mathbf{D}}\dot{\mathbf{x}}^{n+1} + \widetilde{\mathbf{K}}\mathbf{x}^{n+1} = \widetilde{\mathbf{K}}_{c}\,\mathbf{q}^{n+1}
$$

(7.54)

where the superscript $k$ depends on the time discretization of the fluid flow equations.

In many aeroelastic investigations such as wing flutter problems, first a steady flow is computed around a structure in equilibrium. Next, the structure is perturbed via an initial displacement and/or velocity and the aeroelastic response of the coupled fluid/structure system is analyzed. This suggests that a natural sequencing for the staggered time-integration of Eqs. (7.55) is

1.  perturb the structure via some initial conditions.

2.  update the fluid grid to conform to the new structural boundary.

3.  advance the flow with the new boundary conditions.

4.  advance the structure with the new pressure load.

5.  repeat from step 2 until the objective of the simulation is reached.

An important feature of partitioned solution procedures is that they allow using existing single discipline software modules. In our work, we have been particularly interested in re-using the massively parallel explicit flow solver described in Farhat, Lantéri and Fezoui (1992), Farhat, Fezoui and Lantéri (1993), Lantéri and Farhat (1993), and Farhat and Lantéri (1994) for two-dimensional problems, and a variant for three-dimensional applications. Therefore, we consider here the case where the semidiscrete fluid equations are integrated with a 3-step variant of the explicit Runge-Kutta algorithm. Of course, other explicit time-integrators can be equally employed. On the other hand, the aeroelastic response of a structure is often dominated by low frequency dynamics. Hence, the structural equations are most efficiently solved by an implicit time-integration scheme. For example, we select to time-integrate the structural motion with the implicit midpoint rule because it allows enforcing both continuity Eqs. (7.2) while still respecting the GCL (see Section 7.2). Consequently, we propose the explicit/implicit solution algorithm (7.55) for solving the three-field coupled problem. In the sequel, we refer to this explicit/implicit staggered solution procedure as ALG0. It is graphically depicted in Fig. 7.9. Extensive numerical simulations using this algorithm have shown that its stability limit is governed by the critical time-step of the explicit fluid solver, and therefore is not worse than that of the underlying fluid explicit time-integrator.

The 3-step Runge-Kutta algorithm is third-order accurate for linear problems and second-order accurate for nonlinear ones. The midpoint rule is second-order accurate. A simple Taylor expansion shows that the partitioned analysis procedure ALG0 is first-order accurate when applied to the linearized Eqs. (7.54). When applied to Eqs. (7.55), its accuracy depends on the solution scheme selected for solving the mesh equations. As long as the time-integrator applied to the last of Eqs. (7.44) is consistent, ALG0 is guaranteed to be at least first-order accurate.

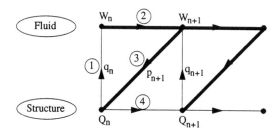

**Figure 7.9** ALG0: the basic staggered algorithm

Given a steady flow and initial structural conditions

1. Update the dynamic fluid grid

$Solve$ $\widetilde{\mathbf{M}}\ddot{\mathbf{x}}^{n+1} + \widetilde{\mathbf{D}}\dot{\mathbf{x}}^{n+1} + \widetilde{\mathbf{K}}\mathbf{x}^{n+1} = \widetilde{\mathbf{K}}_c\,\mathbf{q}^n$

$$\mathbf{x}^{n+\frac{1}{2}} = \frac{\mathbf{x}^n + \mathbf{x}^{n+1}}{2}$$

$$\dot{\mathbf{x}}^{n+\frac{1}{2}} = \frac{\mathbf{x}^{n+1} - \mathbf{x}^n}{\Delta t}$$

2. Advance the fluid system using RK3

$$W_i^{n+1^{(0)}} = W_i^n$$

$$W_i^{n+1^{(k)}} = \frac{V_i(\mathbf{x}^n)}{V_i(\mathbf{x}^{n+1})} W_i^{n+1^{(0)}}$$

$$+ \frac{1}{V_i(\mathbf{x}^{n+1})} \frac{1}{4-k} \Delta t \times F_i^c(W^{(k-1)}, \mathbf{x}^{n+\frac{1}{2}}, \dot{\mathbf{x}}^{n+\frac{1}{2}})$$

$$k = 1,\ 2,\ 3$$

$$W_i^{n+1} = W_i^{n+1^{(3)}}$$

3. Advance the structure using the midpoint rule

$$\mathbf{M}\ddot{\mathbf{q}}^{n+1} + \mathbf{f}^{int}(\mathbf{q}^{n+1}) = \mathbf{f}^{ext}(\mathbf{x}^{n+1}, \mathbf{W}^{n+1})$$

$$\mathbf{q}^{n+1} = \mathbf{q}^n + \frac{\Delta t}{2}(\dot{\mathbf{q}}^n + \dot{\mathbf{q}}^{n+1})$$

$$\dot{\mathbf{q}}^{n+1} = \dot{\mathbf{q}}^n + \frac{\Delta t}{2}(\ddot{\mathbf{q}}^n + \ddot{\mathbf{q}}^{n+1})$$

$$(7.55)$$

### ALG1: subcycling

The fluid and structure fields have often different time scales. For problems in aeroelasticity, the fluid flow usually requires a smaller temporal resolution than the structural vibration. Therefore, if ALG0 is used to solve Eqs. (7.55), the coupling time-step $\Delta t_c$ will be typically dictated by the stability time-step of the fluid system $\Delta t_F$ and not the time-step $\Delta t_S > \Delta t_F$ that meets the accuracy requirements of the structural field.

Using the same time-step $\Delta t$ in both fluid and structure computational kernels presents only minor implementational advantages. On the other hand, subcycling the fluid computations with a factor $n_{S/F} = \Delta t_S/\Delta t_F$ can offer substantial computational advantages, including

- savings in the overall simulation CPU time, because in that case the structural field will be advanced fewer times.

- savings in I/O transfers and/or communication costs when computing on a heterogeneous platform, because in that case the fluid and structure kernels will exchange information fewer times.

However, the computational advantages highlighted above are effective only if subcycling does not restrict the stability region of the staggered algorithm to values of the coupling time-step $\Delta t_c$ that are small enough to offset these advantages. In Piperno et al. (1995), it is shown that for the linearized problem (7.54), the straightforward conventional subcycling procedure — that is, the scheme where at the end of each $n_{S/F}$ fluid subcycles only the interface pressure computed during the last fluid subcycle is transmitted to the structure — lowers the stability limit of ALG0 to a value that is less than the critical time-step of the fluid explicit time-integrator. On the other hand, it is also shown in Piperno et al. (1995) that when solving Eqs. (7.54), the stability limit of ALG0 can be preserved if

- the deformation of the fluid mesh between $t^n$ and $t^{n+1}$ is evenly distributed among the $n_{S/F}$ subcycles.

- at the end of each $n_{S/F}$ fluid subcycles, the average of the interface pressure field $\overline{p}_{\Gamma_{F/S}}$ computed during the subcycles between $t^n$ and $t^{n+1}$ is transmitted to the structure rather than the last computed pressure.

Hence, we propose Eq. (7.56), an explicit/implicit fluid-subcycled partitioned procedure for solving Eqs. (7.55). In the sequel, we refer to this explicit/implicit fluid-subcycled staggered solution procedure presented here as ALG1. It is graphically depicted in Fig. 7.10. Extensive numerical experiments have shown that for medium values of $n_{S/F}$, the stability limit of ALG1 is governed by the critical time-step of the explicit flow solver. However, experience has also shown that there exists a maximum subcycling factor beyond which ALG1 becomes numerically unstable.

From the theory developed in Piperno et al. (1995) for the linearized Eqs. (7.54), it follows that ALG1 is first-order accurate, and that as one would have expected, subcycling amplifies the fluid errors by the factor $n_{S/F}$.

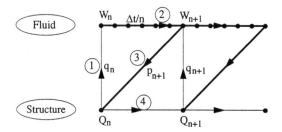

**Figure 7.10** ALG1: subcycling

1. Build $\{q^{n^{(s)}}\}_{s=1}^{s=n_{S/F}}$ / $q^{n^{(n_{S/F})}} = q^n$

$\bar{p}_{\Gamma_{F/S}} = 0$

For $s = 1, ..., n_{S/F}$ {

  Update in stages the dynamic fluid grid

  $Solve$ $\widetilde{\mathbf{M}}\ddot{\mathbf{x}}^{n+1^{(s)}} + \widetilde{\mathbf{D}}\dot{\mathbf{x}}^{n+1^{(s)}} + \widetilde{\mathbf{K}}\mathbf{x}^{n+1^{(s)}} = \widetilde{\mathbf{K}}_c\, \mathbf{q}^{n^{(s)}}$

$$\mathbf{x}^{n+\frac{1}{2}^{(s)}} = \frac{\mathbf{x}^{n^{(s)}} + \mathbf{x}^{n+1^{(s)}}}{2}$$

$$\dot{\mathbf{x}}^{n+\frac{1}{2}^{(s)}} = \frac{\mathbf{x}^{n+1^{(s)}} - \mathbf{x}^{n^{(s)}}}{\Delta t}$$

2. Advance the fluid system using RK3

$$W_i^{n+1^{(0)^{(s)}}} = W_i^{n^{(s)}}$$

$$W_i^{n+1^{(k)^{(s)}}} = \frac{V_i(\mathbf{x}^{n^{(s)}})}{V_i(\mathbf{x}^{n+1^{(s)}})} W_i^{n+1^{(0)^{(s)}}}$$

$$+ \frac{\Delta t^{(s)} F_i^c(W^{(k-1)^{(s)}}, \mathbf{x}^{n+\frac{1}{2}^{(s)}}, \dot{\mathbf{x}}^{n+\frac{1}{2}^{(s)}})}{(4-k)V_i(\mathbf{x}^{n+1^{(s)}})}$$

$$k = 1, 2, 3$$

$$W_i^{n+1^{(s)}} = W_i^{n+1^{(3)^{(s)}}}$$

$$\bar{p}_{\Gamma_{F/S}} = \bar{p}_{\Gamma_{F/S}} + \Delta t^{(s)} p_{\Gamma_{F/S}}^{n+1^{(s)}} \}$$

$\bar{p}_{\Gamma_{F/S}}^{n+1} = \bar{p}_{\Gamma_{F/S}} / \sum \Delta t^{(s)}$

3. Advance the structure using the midpoint rule

$$\mathbf{M}\ddot{\mathbf{q}}^{n+1} + \mathbf{f}^{int}(\mathbf{q}^{n+1}) = \mathbf{f}^{ext}(\bar{p}_{\Gamma_{F/S}}^{n+1})$$

$$\mathbf{q}^{n+1} = \mathbf{q}^n + n_{S/F}\frac{\Delta t}{2}(\dot{\mathbf{q}}^n + \dot{\mathbf{q}}^{n+1})$$

$$\dot{\mathbf{q}}^{n+1} = \dot{\mathbf{q}}^n + n_{S/F}\frac{\Delta t}{2}(\ddot{\mathbf{q}}^n + \ddot{\mathbf{q}}^{n+1})$$

(7.56)

## ALG2-ALG3: inter-field parallelism

ALG0 and ALG1 are inherently sequential. In both partitioned analysis procedures, the fluid system must be updated before the structural system can be advanced. Of course, ALG0 and ALG1 allow intra-field parallelism (parallel computations within each discipline), but they inhibit inter-field parallelism. Advancing the fluid and structural systems simultaneously is appealing because it can reduce the total simulation time.

A simple variant ALG2 of ALG1 — or ALG0 if subcycling is not desired — that allows inter-field parallel processing is given in Eq. (7.57). Clearly, the fluid and structure kernels can run in parallel during the time-interval $[t_n, t_{n+n_{S/F}}]$. Inter-field communication or I/O transfer is needed only at the beginning of each time-interval.

The basic steps of ALG2 are graphically depicted in Fig. 7.11. The theory developed in Piperno et al. (1995) shows that for the linearized Eqs. (7.54), ALG2 is first-order accurate, and parallelism in ALG2 is achieved at the expense of amplified errors in the fluid and structure responses.

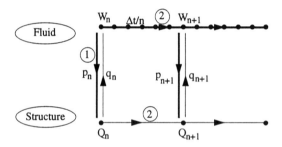

**Figure 7.11** ALG2: subcycling and inter-field parallelism

In order to improve the accuracy of the basic parallel time-integrator ALG2, we propose to exchange correction type information between the fluid and structure kernels at half-step in the following specific manner (ALG3, Eq. (7.58)).

1. Build $\{q^{n^{(s)}}\}_{s=1}^{s=n_{S/F}} \ / \ q^{n^{(n_{S/F})}} = q^n$

$\bar{p}_{\Gamma_{F/S}} = 0$

For $s = 1, \ldots, n_{S/F}$ {

Update in stages the dynamic fluid grid

$$Solve \ \widetilde{\mathbf{M}}\ddot{\mathbf{x}}^{n+1^{(s)}} + \widetilde{\mathbf{D}}\dot{\mathbf{x}}^{n+1^{(s)}} + idetilde\mathbf{K}\mathbf{x}^{n+1^{(s)}} = \widetilde{\mathbf{K}}_c \ \mathbf{q}^{n^{(s)}}$$

$$\mathbf{x}^{n+\frac{1}{2}^{(s)}} = \frac{\mathbf{x}^{n^{(s)}} + \mathbf{x}^{n+1^{(s)}}}{2}$$

$$\dot{\mathbf{x}}^{n+\frac{1}{2}^{(s)}} = \frac{\mathbf{x}^{n+1^{(s)}} - \mathbf{x}^{n^{(s)}}}{\Delta t}$$

2. Advance the fluid system using RK3

$$W_i^{n+1^{(0)(s)}} = W_i^{n^{(s)}}$$

$$W_i^{n+1^{(k)(s)}} = \frac{V_i(\mathbf{x}^{n^{(s)}})}{V_i(\mathbf{x}^{n+1^{(s)}})} W_i^{n+1^{(0)(s)}}$$

$$+ \frac{\Delta t^{(s)} F_i^c(W^{(k-1)^{(s)}}, \mathbf{x}^{n+\frac{1}{2}^{(s)}}, \dot{\mathbf{x}}^{n+\frac{1}{2}^{(s)}})}{(4-k)V_i(\mathbf{x}^{n+1^{(s)}})}$$

$$k = 1, 2, 3$$

$$W_i^{n+1^{(s)}} = W_i^{n+1^{(3)(s)}}$$

$$\bar{p}_{\Gamma_{F/S}} = \bar{p}_{\Gamma_{F/S}} + \Delta t^{(s)} p_{\Gamma_{F/S}}^{n+1^{(s)}} \}$$

$\bar{p}_{\Gamma_{F/S}}^{n+1} = \bar{p}_{\Gamma_{F/S}} / \sum \Delta t^{(s)}$

3. Advance the structure using the midpoint rule

$$\mathbf{M}\ddot{\mathbf{q}}^{n+1} + \mathbf{f}^{int}(\mathbf{q}^{n+1}) = \mathbf{f}^{ext}(\bar{p}_{\Gamma_{F/S}}^n)$$

$$\mathbf{q}^{n+1} = \mathbf{q}^n + n_{S/F}\frac{\Delta t}{2}(\dot{\mathbf{q}}^n + \dot{\mathbf{q}}^{n+1})$$

$$\dot{\mathbf{q}}^{n+1} = \dot{\mathbf{q}}^n + n_{S/F}\frac{\Delta t}{2}(\ddot{\mathbf{q}}^n + \ddot{\mathbf{q}}^{n+1})$$

$$(7.57)$$

$$\bar{p}_{\Gamma_{F/S}} = 0$$

$$For \ s = 1, \ ..., \ \frac{n_{S/F}}{2} - 1$$

$$\{$$

$$Solve \quad \widetilde{\mathbf{M}}\ddot{\mathbf{x}}^{n+1^{(s)}} + \widetilde{\mathbf{D}}\dot{\mathbf{x}}^{n+1^{(s)}} + \widetilde{\mathbf{K}}\mathbf{x}^{n+1^{(s)}} = \widetilde{\mathbf{K}}_c \ \mathbf{q}^{n^{(s)}}$$

$$\mathbf{x}^{n+\frac{1}{2}^{(s)}} = \frac{\mathbf{x}^{n^{(s)}} + \mathbf{x}^{n+1^{(s)}}}{2}; \quad \dot{\mathbf{x}}^{n+\frac{1}{2}^{(s)}} = \frac{\mathbf{x}^{n+1^{(s)}} - \mathbf{x}^{n^{(s)}}}{\Delta t}$$

$$W_i^{n+1^{(0)^{(s)}}} = W_i^{n^{(s)}}$$

$$W_i^{n+1^{(k)^{(s)}}} = \frac{V_i(\mathbf{x}^{n^{(s)}})}{V_i(\mathbf{x}^{n+1^{(s)}})}W_i^{n+1^{(0)^{(s)}}} + \frac{\Delta t^{(s)}F_i^c(W^{(k-1)^{(s)}}, \ \mathbf{x}^{n+\frac{1}{2}^{(s)}}, \ \dot{\mathbf{x}}^{n+\frac{1}{2}^{(s)}})}{(4-k)V_i(\mathbf{x}^{n+1^{(s)}})} \qquad k = 1, 2, \ :$$

$$W_i^{n+1^{(s)}} = W_i^{n+1^{(3)^{(s)}}}; \quad \bar{p}_{\Gamma_{F/S}} = \bar{p}_{\Gamma_{F/S}} + \Delta t^{(s)}p_{\Gamma_{F/S}}^{n+1^{(s)}}$$

$$\}$$

$$\bar{p}_{\Gamma_{F/S}}^{n+\frac{1}{2}} = \bar{p}_{\Gamma_{F/S}}/\sum \Delta t^{(s)}$$

$$\mathbf{M}\ddot{\bar{q}}^{n+1} + \mathbf{f}^{int}(\bar{q}^{n+1}) = \mathbf{f}^{ext}(\bar{p}_{\Gamma_{F/S}}^{n+\frac{1}{2}})$$

$$\bar{q}^{n+1} = q^n + n_{S/F}\frac{\Delta t}{2}(\dot{q}^n + \dot{\bar{q}}^{n+1}); \quad \dot{\bar{q}}^{n+1} = \dot{q}^n + n_{S/F}\frac{\Delta t}{2}(\ddot{q}^n + \ddot{\bar{q}}^{n+1})$$

$$\bar{p}_{\Gamma_{F/S}} = 0$$

$$For \ s = \frac{n_{S/F}}{2}, \ ..., \ n_{S/F}$$

$$\{$$

$$Solve \quad \widetilde{\mathbf{M}}\ddot{\mathbf{x}}^{n+1^{(s)}} + \widetilde{\mathbf{D}}\dot{\mathbf{x}}^{n+1^{(s)}} + \widetilde{\mathbf{K}}\mathbf{x}^{n+1^{(s)}} = \widetilde{\mathbf{K}}_c \ \bar{q}^{n+1^{(s)}}$$

$$\mathbf{x}^{n+\frac{1}{2}^{(s)}} = \frac{\mathbf{x}^{n^{(s)}} + \mathbf{x}^{n+1^{(s)}}}{2}; \quad \dot{\mathbf{x}}^{n+\frac{1}{2}^{(s)}} = \frac{\mathbf{x}^{n+1^{(s)}} - \mathbf{x}^{n^{(s)}}}{\Delta t}$$

$$W_i^{n+1^{(0)^{(s)}}} = W_i^{n^{(s)}}$$

$$W_i^{n+1^{(k)^{(s)}}} = \frac{V_i(\mathbf{x}^{n^{(s)}})}{V_i(\mathbf{x}^{n+1^{(s)}})}W_i^{n+1^{(0)^{(s)}}} + \frac{\Delta t^{(s)}F_i^c(W^{(k-1)^{(s)}}, \ \mathbf{x}^{n+\frac{1}{2}^{(s)}}, \ \dot{\mathbf{x}}^{n+\frac{1}{2}^{(s)}})}{(4-k)V_i(\mathbf{x}^{n+1^{(s)}})} \qquad k = 1, 2,$$

$$W_i^{n+1^{(s)}} = W_i^{n+1^{(3)^{(s)}}}; \quad \bar{p}_{\Gamma_{F/S}} = \bar{p}_{\Gamma_{F/S}} + \Delta t^{(s)}p_{\Gamma_{F/S}}^{n+1^{(s)}}$$

$$\}$$

$$\bar{p}_{\Gamma_{F/S}}^{n+1} = \bar{p}_{\Gamma_{F/S}}/\sum \Delta t^{(s)}$$

$$\mathbf{M}\ddot{q}^{n+1} + \mathbf{f}^{int}(q^{n+1}) = \mathbf{f}^{ext}(\bar{p}_{\Gamma_{F/S}}^{n+1})$$

$$q^{n+1} = q^n + n_{S/F}\frac{\Delta t}{2}(\dot{q}^n + \dot{q}^{n+1}); \quad \dot{q}^{n+1} = \dot{q}^n + n_{S/F}\frac{\Delta t}{2}(\ddot{q}^n + \ddot{q}^{n+1})$$

$$(7.58)$$

Algorithm ALG3 is illustrated in Fig. 7.12. The first-half of the computations is identical to that of ALG2, except that the fluid system is subcycled only up to $t^{n+\frac{n_{S/F}}{2}}$, while the structure is advanced in one shot up to $t^{n+n_{S/F}}$. At $t^{n+\frac{n_{S/F}}{2}}$, the fluid and structure kernels exchange pressure, displacement and velocity information. In the second-half of the computations, the fluid system is subcycled from $t^{n+\frac{n_{S/F}}{2}}$ to $t^{n+n_{S/F}}$ using the new structural information, and the structural behavior is re-computed in parallel using the newly received pressure distribution. Note that the first evaluation of the structural state vector can be interpreted as a prediction step, and the second as a correction step.

It can be shown that when applied to the linearized Eqs. (7.54), ALG3 is first-order accurate and reduces the errors of ALG2 by the factor $n_{S/F}$, at the expense of one additional communication step or I/O transfer during each coupled cycle (see Piperno et al. (1995) for a detailed error analysis).

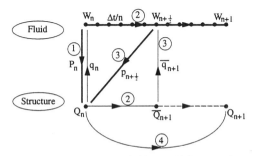

**Figure 7.12** ALG3: subcycling, inter-field parallelism and improved accuracy

## IMPLICIT/IMPLICIT STAGGERED ALGORITHMS

Clearly, the partitioned analysis procedures ALG0, ALG1, ALG2 and ALG3 can be equally employed with an implicit flow solver. However, it is shown in Piperno et al. (1995) that in order for these partitioned procedures to be unconditionally stable, not only an unconditionally stable implicit flow solver must be used, but also an interface coupling operator must be exchanged between the structure and fluid field analyzers. For further details on this topic, we refer the reader to Piperno et al. (1995).

## 7.4 THE FLOW SOLVER (Farhat, Lantéri and Fezoui (1992), Farhat, Fezoui and Lantéri (1993), Lantéri and Farhat (1993), and Farhat and Lantéri (1994))

So far, no restriction has been imposed on the nonlinear flow solver technology, except for the requirement of satisfying the GCL. Hence, all flow solvers based on an ALE finite volume/element discretization can be equally employed within the computational framework presented in this chapter for solving nonlinear transient aeroelastic problems.

In our case, we have opted for a mixed finite element/volume ALE formulation based on unstructured triangular meshes in two-dimensional problems, and unstructured tetrahedra in three-dimensional ones. This approach combines a Galerkin centered approximation for the viscous terms, and a Roe upwind scheme (Roe 1981) for the computation of the convective fluxes. Higher order accuracy is achieved through the use of a piecewise linear interpolation method that follows the principle of the MUSCL (Monotonic Upwind Scheme for Conservative Laws) procedure (Van Leer 1979, Dervieux 1985, and Fezoui and Stoufflet 1989).

## SPATIAL DISCRETIZATION

The conservative form of the equations describing viscous flows can be written in ALE form as

$$\frac{\partial(JW)}{\partial t}\Big|_\xi + J\nabla_x.\mathcal{F}^c(W,\dot{x}) = \frac{1}{Re}\nabla_x\mathcal{R}(W) \tag{7.59}$$

$$\mathcal{F}^c(W,\dot{x}) = \mathcal{F}(W) - \dot{x}W$$

where, $\mathcal{R}$ denotes the diffusive fluxes, $Re$ is the Reynolds number, and as for the case of Euler flows and Eqs. (7.4), $J = det(dx/d\xi)$ is the jacobian of the frame transformation $\xi \to x$, $W$ denotes the fluid conservative variables, $\mathcal{F}^c$ denotes the convective ALE fluxes, and $\dot{x} = \frac{\partial x}{\partial \theta}\big|_\xi$ is the ALE grid velocity that may be different from the fluid velocity and from zero.

The boundary $\Gamma(t)$ of the flow domain is partitioned into a wall boundary $\Gamma_w(t)$ corresponding to the fluid/structure interface boundary $\Gamma_{F/S}$, and an infinity boundary $\Gamma_\infty(t)$

$$\Gamma(t) = \Gamma_w(t) \cup \Gamma_\infty(t) \tag{7.60}$$

Let $U_w$ and $T_w$ denote the wall velocity and temperature. On the wall boundary $\Gamma_w(t)$, a no-slip and a temperature Dirichlet conditions are imposed

$$U = U_w; \quad T = T_w \tag{7.61}$$

and no boundary condition is specified for the density. Hence, the total energy per unit of volume and the pressure on the wall are given by

$$p = (\gamma - 1)\rho C_v T_w; \quad E = \rho C_v T_w + \frac{1}{2}\rho \parallel U_w \parallel^2 \tag{7.62}$$

For external flows around aircraft structures, the viscous effects are assumed to be negligible at infinity, and therefore a uniform free-stream state vector $W_\infty$ is imposed on $\Gamma_\infty(t)$.

Following the procedure described in details in Section 7.2, Eqs. (7.60) can be transformed into

$$\frac{d}{dt}\int_{C_i(t)} W \, d\Omega_x + \int_{\partial C_i(t)} \nabla\mathcal{F}^c(W,\dot{x}) \, d\Omega_x = \int_{C_i(t)} \frac{1}{Re}\nabla R(W) \, d\Omega_x \tag{7.63}$$

Integrating Eq. (7.64) by parts leads to

$$\frac{d}{dt}\int_{C_i(t)} W \, d\Omega_x + \sum_j \int_{\partial C_{i,j}(x)} \mathcal{F}^c(W_i,W_j,\dot{x}).\nu_i \, d\sigma + \int_{\partial C_i(t)\cap\Gamma(t)} \mathcal{F}^c(\overline{W},\dot{x}).\nu_i \, d\sigma$$

$$= -\frac{1}{Re} \sum_{\Delta,S_i\in\Delta} \int_\Delta \mathcal{R}(W).\nabla\phi_i^\Delta \, d\Omega_x \tag{7.64}$$

where $\phi_i^\Delta$ is the finite element shape function at vertex $S_i$ in element $\Delta$ (a triangle in two-dimensional problems, a tetrahedron in three-dimensional ones), and $\overline{W}$ is the specified value of $W$ at the boundaries. The second term of Eq. (7.65) corresponds exactly to the term $F_i^c(\mathbf{W}, \mathbf{x}, \dot{\mathbf{x}})$ introduced in Section 7.2 Each component of this term can be written as

$$\Phi_{ij} = \int_{\partial C_{i,j}(x)} \mathcal{F}^c(W_i, W_j, \dot{x}).\nu_i \, d\sigma = \Phi(W_i, W_j, x, \dot{x}) \tag{7.65}$$

While various upwind algorithms can be used for computing $\Phi_{ij}$, Roe's scheme (Roe 1981) is chosen here. Following the MUSCL approach introduced by Van Leer (Van Leer 1979), second-order accuracy is achieved by computing the numerical fluxes at interpolated values of the fluid state vector on the interface between cells $C_i$ and $C_j$ as follows

$$\Phi_{ij} = \Phi(W_{ij}, W_{ji}, x, \dot{x})$$
$$W_{ij} = W_i + \frac{1}{2}(\nabla W)_i.vector(S_i S_j) \tag{7.66}$$
$$W_{ji} = W_j - \frac{1}{2}(\nabla W)_j.vector(S_i S_j)$$

For three-dimensional problems, the gradient of $W$ at a vertex $S_i$ is computed from

$$\cdot V_i(\nabla W)_i = \sum_{\Delta, S_i \in \Delta} \frac{volume(\Delta)}{4} \sum_{k=1}^{k=4} W_k.\nabla\phi_k^\Delta \tag{7.67}$$

In practice, the interpolation implied by Eqs. (7.67) is performed on the physical instead of the the conservative variables. Optional limiters are also implemented following the approaches discussed in Dervieux (1985) and Fezoui and Stoufflet (1989).

The numerical viscous fluxes are computed using a classical Galerkin method.

*TEMPORAL SOLUTION*

The explicit kernel of our two-dimensional flow solver uses the 3-step Runge-Kutta algorithm discussed in Section 7.3. On the other hand, the explicit module of our three-dimensional flow solver employs the predictor-corrector scheme suggested by Hancock and presented by Van Leer. This scheme has a lower stability limit than the 3-step Runge-Kutta algorithm but is significantly more economical.

The implicit versions of our two- and three-dimensional flow solvers employ a first-order accurate backward Euler time-integration scheme. To solve the system of linearized equations arising at each time-step, we have recently developed a multilevel, overlapping domain decomposition preconditioned Krylov-Schwarz iterative method (Cai et al. 1995). Numerical experiments have shown that this and other members of the family of Krylov-Schwarz algorithms are highly scalable and highly parallelizable. More importantly, the convergence of these methods does not degenerate when the linearized system becomes highly nonsymmetric and possibly indefinite, which occurs, for example, in the case of high-angle of attack and/or high Mach number.

*PARALLELIZATION*

The mesh partitioning paradigm is used for parallelizing both two-dimensional and three-dimensional flow solvers. This paradigm is briefly discussed in Section 7.8.

## 7.5 THE STRUCTURAL DYNAMICS ANALYZER (Farhat, Chen and Mandel 1995, Farhat, Crivelli and Roux 1994a, Farhat, Crivelli and Roux 1994b, and Vanderstraeten et al. 1995)

There is no question that the finite element method is the most popular method for solving arbitrary structural problems such as those governed by the second of Eqs. (7.1). However, with the advent of parallel processing, many of the computational modules of this powerful method are being constantly revisited for improvement in performance.

Nonlinear transient finite element problems in structural mechanics are characterized by the semi-discrete equations of dynamic equilibrium

$$\mathbf{M}\ddot{\mathbf{q}} + \mathbf{f}^{int}(\mathbf{q}) = \mathbf{f}^{ext} \tag{7.68}$$

where, as before, $\mathbf{M}$ is the mass matrix, $\mathbf{q}$ is the vector of nodal displacements, a dot superscript indicates a time derivative, $\mathbf{f}^{int}$ is the vector of internal nodal forces, and $\mathbf{f}^{ext}$ is the vector of external nodal forces. In many low and medium frequency dynamics applications such as transient aeroelasticity, Eq. (7.69) is most efficiently solved using an implicit time-integration scheme. In that case, a nonlinear algebraic system of equations is generated at each time-step. The Newton-Raphson method and its numerous variants collectively known as "Newton-like" methods are the most popular strategies for solving these nonlinear algebraic problems. All of these algorithms require the solution of a linear algebraic system of equations of the form

$$\mathbf{K}^*(\mathbf{q}_{n+1}^{(k)}) \, \Delta \mathbf{q}_{n+1}^{(k+1)} = \mathbf{r}^*(\mathbf{q}_{n+1}^{(k)})$$
$$\Delta \mathbf{q}_{n+1}^{(k+1)} = \mathbf{q}_{n+1}^{(k+1)} - \mathbf{q}_{n+1}^{(k)} \tag{7.69}$$

where the subscript $n$ refers to the $n$-th time step, the superscript $k$ refers to the $k$-th nonlinear iteration within the current time step, $\mathbf{K}^*$ is a time-dependent symmetric positive approximate tangent matrix that includes both mass and stiffness contributions, and $\Delta \mathbf{q}_{n+1}^{(k+1)}$ and $\mathbf{r}^*(\mathbf{q}_{n+1}^{(k)})$ are respectively the vector of nodal displacement increments and the vector of out-of-balance nodal forces (dynamic residuals).

With the advent of parallel processing, domain decomposition (or substructure) based direct and iterative algorithms have become increasingly popular for the solution of finite element systems of equations of the form given in Eq. (7.69). Indeed, domain decomposition provides a higher level of concurrency than parallel global algebraic paradigms, and is simpler to implement on most parallel computational platforms (Farhat and Roux 1994). In general, the subdomain (or substructure) equations are solved using a direct skyline or sparse factorization based algorithm, while both direct and iterative schemes have been proposed for the solution of the interface problem (Farhat and Wilson 1987, Bjordstad and Widlund 1986, Farhat 1991, Farhat and Roux 1991, and Bramble et al. 1986). When the reduced system of equations is solved directly, the overall domain decomposition algorithm becomes a direct frontal or multifrontal method (Duff 1986 and Liu 1992), and its success becomes contingent on finding a good mesh partition and/or reordered system that can achieve an optimal balance between minimizing fill-in and increasing the degree of parallelism (Benner et al. 1987, Lesoinne et al. 1991, Pothen and Sun 1993, and Pothen et al. 1994). When the interface problem is solved iteratively — usually, via a preconditioned conjugate gradient (PCG) algorithm— the overall domain decomposition method becomes a genuine iterative solver whose success hinges on two important properties: *numerical* scalability, and *parallel* scalability. A domain decomposition based iterative method is said to be numerically scalable if the condition number $\kappa$ after preconditioning does not grow or grows "weakly" with the ratio of the subdomain size $H$ and the mesh size $h$ (Fig. 7.13), that is

$$\kappa = O\left(1 + log^{\beta}\left(\frac{H}{h}\right)\right) \tag{7.70}$$

with a small constant $\beta$. Numerous authors have proved Eq. (7.71) with $\beta = 2$ for various domain decomposition methods (see, for example, Bramble et al. (1986), Dryja and Widlund (1994), and Mandel and Tezaur (1995) and references therein).

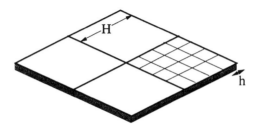

**Figure 7.13** Subdmomain size $H$ and mesh size $h$

It is well known that in order to achieve (7.71), a domain decomposition method must involve a *coarse problem* with a few d.o.f. per subdomain, that must be solved at each iteration to propagate the error globally and accelerate convergence. Parallel scalability characterizes the ability of an algorithm to deliver larger speedups for a larger number of processors. In particular, parallel scalability is necessary for massively parallel processing.

The practical implications of a condition number after preconditioning such as that described in Eq. (7.71) are

- suppose that a given mesh is fixed, one processor is assigned to every subdomain, and the number of subdomains (which varies as $1/H$) is increased in order to increase parallelism. In that case, $h$ is fixed and $H$ is decreased. From Eq. (7.71), it follows that the bound on the condition number decreases and therefore *the number of iterations for convergence is expected to decrease with an increasing number of subdomains.*

- on most distributed memory parallel processors, the total amount of available memory increases with the number of processors. When solving a certain class of problems on such parallel hardware, it is customary to define in each processor a constant subproblem size, and to increase the total problem size with the number of processors. In that case, $h$ and $H$ are decreased, but the ratio $H/h$ is kept constant.

From Eq. (7.71), it follows that a numerically scalable domain decomposition algorithm can solve larger problems with the same number of iterations as smaller ones, simply by increasing the number of subdomains. However, the presence of the coarse problem may limit parallel scalability for a large number of processors.

- When $H/h$ increases, that is, the number of elements assigned to a subdomain increases, the condition number will increase only slightly. Without this property, the condition number may be too large to be practical for subdomains of a size that we wish to work with. If there are only a few substructures, the conjugate gradient algorithm might still converge quickly for some domain decomposition methods because of the presence of gaps in the spectrum of the preconditioned operator; however, for large number of subdomains, the spectrum tends to fill in, and the number of iterations tends to increase (Farhat, Mandel and Roux 1994).

The Finite Element Tearing and Interconnecting (FETI) method developed originally for the solution of self-adjoint elliptic partial differential equations is a numerically scalable domain decomposition method (Farhat 1991, Farhat and Roux 1991, and Farhat, Mandel and Roux 1994). This method was shown to outperform direct skyline solvers and several popular iterative algorithms on both sequential and parallel computing platforms (Farhat 1991, Farhat and Roux 1992). It has recently been extended for dynamics problems (Farhat, Crivelli and Roux 1994b and Farhat, Chen and Mandel 1995) and biharmonic partial differential equations such as those encountered in plate and shell problems (Mandel et al. 1995). For structural mechanics problems, the condition number of the unpreconditioned FETI interface problem is known to grow asymptotically as (Farhat, Mandel and Roux 1994)

$$\kappa = O\left(\frac{H}{h}\right) \tag{7.71}$$

As was observed numerically in Farhat, Mandel and Roux (1994) and Farhat and Roux (1994), and proved mathematically in Mandel and Tezaur (1995) and Mandel et al. (1995), for elasticity problems discretized using plane stress/strain and/or brick elements, the condition number of the FETI interface problem preconditioned with a subdomain based Dirichlet operator (Farhat, Mandel and Roux 1994) varies as

$$\kappa = O\left(1 + log^{\beta}\left(\frac{H}{h}\right)\right), \quad \beta \leq 3. \tag{7.72}$$

For shell and plate problems, this condition number varies as (Mandel et al. 1995)

$$\kappa = O\left(1 + log^2\left(\frac{H}{h}\right)\right) \tag{7.73}$$

The conditioning results (7.72,7.73) highlight the numerical scalability of the FETI method with respect to both the mesh size $h$ and the number of subdomains. The parallel scalability of this domain decomposition method — that is, its ability to achieve larger speedups for larger number of processors —- has also been demonstrated on current massively parallel processors for several realistic structural problems (Farhat and Roux 1994, Farhat 1995, and Farhat, Chen and Stern 1994).

## 7.6 NON-MATCHING INTERFACE BOUNDARIES (Maman and Farhat 1995)

All four partitioned analysis procedures discussed in Section 7.3 require exchanging interface-data only between the field analyzers. More precisely, the structure expects to receive the values of the flow pressure and viscous stresses at the fluid/structure interface boundary $\Gamma_{F/S}$, and convert them into a structural load. Similarly, the fluid expects to receive from the structure the displacement and/or velocity of the interface boundary $\Gamma_{F/S}$, and use them to update the position of the dynamic fluid mesh. This exchange is performed at every time-step, or as required by the subcycling algorithm.

In general, the fluid and structure meshes have two independent representations of the physical fluid/structure interface. When these representations are identical — for example, when every fluid grid point on $\Gamma_{F/S}$ is also a structural node and *vice-versa* — the evaluation of the pressure forces and the transfer of the structural motion to the fluid mesh are trivial operations. However, analysts usually prefer to

- use a fluid mesh and a structural model that have been independently designed and validated.
- refine each mesh independently from the other.

Hence, most realistic aeroelastic simulations will involve handling fluid and structural meshes that are incompatible at their interface boundaries (Fig. 7.14). In Maman and Farhat (1995), we have addressed this issue and proposed a preprocessing "matching" procedure that does not introduce any other approximation than those intrinsic to the fluid and structure solution methods. This procedure can be summarized as follows.

The nodal forces induced by the fluid pressure on the "wet" surface of a structural element $e$ can be written as:

$$f_i^{ext} = \int_{\Omega^{(e)}} N_i(-p\nu + (\tau\nu)\bar{\nu}) \, d\sigma \qquad (7.74)$$

where $\Omega^{(e)}$ denotes the geometrical support of the wet surface of the structural element $e$, $p$ is the pressure field, $\tau$ is the tensor of viscous stresses, $\nu$ is the unit normal to $\Omega^{(e)}$, $\bar{\nu}$ is a tangent to the plane of $\Omega^{(e)}$, and $N_i$ is the shape function associated with node $i$ in element $e$.

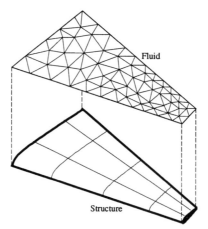

**Figure 7.14** Incompatible fluid and structure meshes

Most if not all finite element structural analysis codes evaluate the integral in Eq. (7.74) via a quadrature rule

$$f_i^{ext} = \sum_{g=1}^{g=n_g} w_g N_i(X_g)(-p(X_g)\nu + (\tau(X_g)\nu)\bar{\nu}) \tag{7.75}$$

where $w_g$ is the weight of the Gauss point $X_g$. Hence, a structural analysis code needs to know only the values of the pressure field at the Gauss points of its wet surface. This information can be easily made available once every Gauss point of a wet structural element is paired with a fluid cell (Fig. 7.15). It should be noted that in Eq. (7.75), $X_g$ are not necessarily the same Gauss points as those used for stiffness evaluation. For example, if a high pressure gradient is anticipated over a certain wet area of the structure, a larger number of Gauss points can be used for the evaluation of the pressure forces $f_i^{ext}$ on that area.

On the other hand, when the structure moves and/or deforms, the motion of the fluid grid points on $\Gamma_{F/S}$ can be prescribed via the regular finite element interpolation

$$x(S_j) = \sum_{k=1}^{k=wne} N_k(X_j)q_k^{(e)} \tag{7.76}$$

where $S_j$, $wne$, $X_j$, and $q_k$ denote respectively a fluid grid point on $\Gamma_{F/S}$, the number of wet nodes in its "nearest" structural element $e$, the natural coordinates of $S_j$ in $\Omega^{(e)}$, and the structural displacement at the $k$-th node of element $e$. From Eq. (7.76), it follows that each fluid grid point on $\Gamma_{F/S}$ must be matched with one wet structural element (Fig. 7.15).

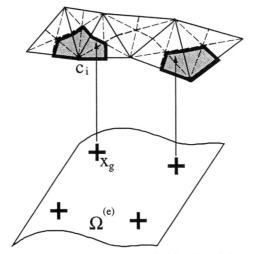

Fig. 7.14. Gauss-point—fluid cell pairing

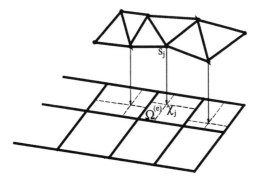

Figure 7.15 Fluid grid point—wet structural element pairing

Given a fluid grid, a structural analysis model, and a discrete description of the fluid/structure interface, the Matcher program described in Maman and Farhat (1995) generates all the data structures needed for evaluating the quantities described in Eqs. (7.75,7.76). If parallel data structures — for example, data structures associated with mesh partitions of the fluid and structure grids — are fed as input, Matcher outputs parallel data structures that allow a painless implementation of the interface-data exchange between the field analyzers and are fully compatible with the intrinsic parallel data structures of the fluid and structural analysis programs. In general, the pairing of fluid and structure entities does not change in time. Therefore, Matcher is run as a preprocessor. The pairing data structures are generated only once, prior to any aeroelastic computation.

Finally, we note that Matcher is written in a message-passing style. Therefore, this software is portable to any parallel computing platform that supports a PVM- or MPI-like communication library. Of course, it also runs on sequential computers. For a complete aircraft configuration where the fluid mesh contained 439272 tetrahedra, 77279 vertices, and was partitioned into 32 subdomains, and the structural model contained 7520 triangular shell elements, 3841 nodes, and was partitioned into 16 subdomains,

Matcher generated the desired fluid/structure pairing data structures in 327 seconds CPU on a 32-processor iPSC-860 system (Maman and Farhat 1995).

## 7.7 THE MESH MOTION SOLVER (Farhat, Lesoinne, Chen and Lantéri 1995)

At the beginning of each step of the chosen staggered solution procedure, the dynamic fluid grid must be updated to conform to the most recently computed configuration of the structure. In general, this is done in two steps

Step 1.  first, the points lying on the interface boundary $\Gamma_{F/S}$ are adjusted to match (in the sense defined in Section 7.6) the newly computed or predicted position of the surface of the structure. This defines a prescribed displacement vector $\mathbf{x}_{\Gamma_{F/S}}$.

Step 2.  next, the remainder of the fluid grid points are repositioned accordingly to the prescribed values of $\mathbf{x}_{\Gamma_{F/S}}$. This completes the computation of the new mesh displacement vector $\mathbf{x}$.

Several procedures have been proposed in the literature for implementing the above two steps. Most of them can be summarized as viewing the fluid domain or its grid as a pseudo-structural continuous or discrete system. For example, in the discrete approach, either or all of the following can be done (see Fig. 7.3)

- lumping a fictitious mass at each vertex of the fluid mesh.

- introducing a fictitious dashpot at each edge connecting two vertices.

- attaching a fictitious spring on each edge connecting two vertices.

Similarly, a pseudo-structural continuous system can be generated with fictitious distributed structural properties. In both cases, the motion of the constructed pseudo-structural system is governed by

$$\widetilde{\mathbf{M}}\ddot{\mathbf{x}} + \widetilde{\mathbf{D}}\dot{\mathbf{x}} + \widetilde{\mathbf{K}}\mathbf{x} = \widetilde{\mathbf{K}}_c \mathbf{x}_{\Gamma_{F/S}} \qquad (7.77)$$

where $\widetilde{\mathbf{M}}$, $\widetilde{\mathbf{D}}$, and $\widetilde{\mathbf{K}}$ are the fictitious mass, damping, and stiffness matrices associated with the dynamic fluid mesh, and $\widetilde{\mathbf{K}}_c$ is the component of the fictitious stiffness matrix that represents the coupling between the fluid points lying on $\Gamma_{F/S}$ and the others. Eq. (7.77) above is integrated in time until the steady-state equilibrium displacement $\mathbf{x}$ is reached. This solution procedure can be speeded up by constructing $\widetilde{\mathbf{M}}$ and $\widetilde{\mathbf{D}}$ as follows

$$\widetilde{\mathbf{M}} = a\widetilde{\mathbf{K}}, \qquad \widetilde{\mathbf{D}} = b\widetilde{\mathbf{K}} \qquad (7.78)$$

and selecting the two scalars $a$ and $b$ so that the governing equations of motion (7.126) are *critically damped*. In that case, the equilibrium solution $\mathbf{x}$ is reached in a few time-steps.

Alternatively, $\widetilde{\mathbf{M}}$ and $\widetilde{\mathbf{D}}$ can be set to zero, and the new position of the dynamic fluid mesh can be computed via the solution of the static problem

$$\widetilde{\mathbf{K}}\mathbf{x} = \widetilde{\mathbf{K}}_c \mathbf{x}_{\Gamma_{F/S}} \qquad (7.79)$$

This approach is often referred to as Batina's network of edge-springs (Farhat and Lin 1990). However, it should be noted that attaching a lineal spring on the edge connecting two vertices of a tetrahedron prevents these two vertices from colliding during the mesh deformation, but does not prevent a vertex from interpenetrating the facet of a tetrahedron. To prevent such a detrimental interpenetration that is more likely to happen when the structure undergoes large motions, torsional springs must also be added at the mesh vertices, and their stiffnesses must be carefully calibrated.

In this work, the pseudo-structural system associated with the unstructured dynamic fluid mesh is constructed with lineal and torsional springs only $(\widetilde{\mathbf{M}} = \widetilde{\mathbf{D}} = 0)$. Each fictitious lineal spring attached to an each edge connecting two fluid grid points $S_i$ and $S_j$ is attributed the following stiffness

$$k_{ij} \;=\; \frac{1}{||S_i S_j||_2} \,\text{`}$$

(7.80)

The grid points located on the upstream and downstream boundaries are held fixed. At each time-step $t^{n+1}$, the new position of the interior grid points is determined from the solution of Eq. (7.79) via a two-step iterative procedure. First, the displacements of the interior grid points are predicted by extrapolating the previous displacements at time-steps $t^n$ and $t^{n-1}$ in the following manner

$$\Delta\mathbf{x} \;=\; 2\Delta^n\mathbf{x} - \Delta^{n-1}\mathbf{x}$$

(7.81)

where $\Delta^n\mathbf{x} = \mathbf{x}^{n+1} - \mathbf{x}^n$. Next, the predicted values are corrected with a few explicit Jacobi relaxations as follows

$$\Delta^{n+1}\mathbf{x_i} \;=\; \frac{\sum\limits_{j} k_{ij}\Delta\mathbf{x_j}}{\sum\limits_{j} k_{ij}}$$

(7.82)

Finally, the position of the fluid grid points at $t^{n+1}$ is computed from

$$\mathbf{x}^{n+1} \;=\; \mathbf{x}^n + \Delta^{n+1}\mathbf{x}$$

(7.83)

## 7.8 A UNIFIED PARALLELIZATION STRATEGY (Vanderstraeten et al. 1995 and Farhat, Lantéri and Simon 1995)

It is now well accepted that the mesh partitioning and message-passing paradigms lead to the most portable software design for parallel computational mechanics. Essentially, the computational mesh is assumed to be partitioned into several submeshes, each defining a subdomain. The same "old" serial code can be executed within every subdomain. The "gluing" or assembly of the subdomain results can be implemented in a separate software module. Therefore, all of our flow solvers, structural analyzers, and mesh motion solvers are designed to work with mesh partitions, and are written in a message-passing style. Consequently, their performance is not only machine dependent, but sometimes also mesh partition dependent.

Research in mesh partitioning has focused so far on the automatic generation of subdomains with minimum interface points. In a recent paper (Vanderstraeten et al. 1995), we have revisited this issue and emphasized other aspects of the partitioning problem including the fast generation of large-scale mesh decompositions on conventional workstations, the optimization of initial decompositions for specific kernels such as parallel frontal solvers and domain decomposition based iterative methods. More specifically, we have overviewed in Vanderstraeten et al. (1995) a two-step partitioning paradigm for tailoring generated mesh partitions to specific applications, and a simple mesh contraction procedure for speeding up the optimization of initial mesh decompositions. We have discussed what defines a good mesh partition for a given problem, and shown that the methodology summarized in Vanderstraeten et al. (1995) can produce better mesh partitions than the well celebrated multilevel Recursive Spectral Bisection algorithm (Barnard and Simon 1994), and yet be an order of magnitude faster. Our two-step decomposition methodology and contraction procedure are available in the TOP/DOMDEC (Farhat, Lantéri and Simon 1995) interactive software package for mesh partitioning and parallel processing.

## 7.9 APPLICATIONS AND PERFORMANCE RESULTS

Here, we demonstrate the aeroelastic computational methodology described in the previous sections with the numerical investigation of the instability of flat panels with infinite aspect ratio in supersonic airstreams, and the solution of three-dimensional wing response problems in the transonic regime. All flow computations are performed using the Euler equations and the explicit solver.

*TWO-DIMENSIONAL AEROELASTIC SUPERSONIC COMPUTATIONS*

### Problem Definition

The flat panel with infinite aspect ratio considered here (Fig. 7.16) is assumed to have a length $L = 0.5m$, a uniform thickness $h = 1.35 \times 10^{-3}$ $m$, a Young modulus $E = 7.728 \times 10^{10}$ $N/m^2$, a Poisson ratio $\mu = 0.33$, a density $\rho = 2710$ $Kg/m^3$, and to be clamped at both ends. Its rectangular cross section is discretized into $1111 \times 3$ plane strain 4-node elements. This fine discretization — which generates 3333 elements with perfect aspect ratios and 4448 nodes — is not needed for accuracy; we have designed this structural mesh only because we are also interested in assessing some computational and I/O performance issues.

The two-dimensional flow domain above the panel is discretized into 32568 triangles and 16512 vertices. A slip condition is imposed at the fluid/structure boundary. Because the fluid and structural meshes are not compatible at their interface (Fig. 7.17), the Matcher software (Maman and Farhat 1995) is used to generate in a single preprocessing step the data structures required for transferring the pressure load to the structure, and the structural deformations at the upper surface of the panel to the fluid.

We consider several supersonic flows at different Mach numbers and discuss the performances of the partitioned analysis procedures ALG0, ALG1, ALG2, and ALG3. Whenever subcycling is used, the $I^1$ interpolation scheme is used to prescribe the motion of the fluid grid points on $\Gamma_{F/S}$.

## Computational Platform

All computations are performed on an iPSC-860 parallel processor using double precision arithmetic. The fluid and structure solvers are implemented as separate programs that communicate via the intercube communication procedures described in Barszcz (1992).

## Assessment of the Partitioned Procedures

In order to illustrate the relative merits of the partitioned procedures ALG0, ALG1, ALG2 and ALG3, we consider first two different series of transient aeroelastic simulations at Mach number $M_\infty = 1.90$ that highlight

- the relative accuracy of these algorithms for a fixed subcycling factor $n_{S/F}$.

- the relative speed of these algorithms for a fixed level of accuracy, on both sequential and parallel computational platforms.

In all cases, 64 processors are allocated to the fluid system, and 2 processors are assigned to the structural solver. Initially, a steady-state flow is computed above the panel at $M_\infty = 1.90$ (Fig. 7.18), speed at which the panel described above is not supposed to flutter. Then, the aeroelastic response of the coupled system is triggered by a displacement perturbation of the panel along its first mode (Fig. 7.19).

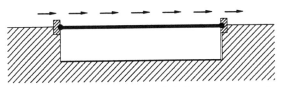

Figure 7.16 A flat panel with infinite aspect ratio

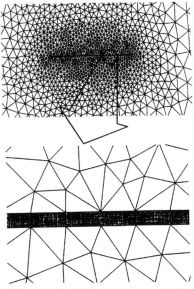

Figure 7.17 Mesh incompatibility

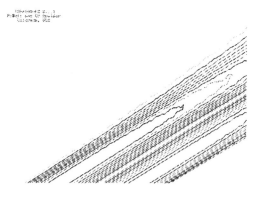

**Figure 7.18** Pressure isovalues for the steady-state flow solution ($M_\infty = 1.90$)

**Figure 7.19** Initial perturbation of the panel displacement field

First, the subcycling factor is fixed to $n_{S/F} = 30$, and the lift coefficient is computed using the time-step $\Delta t = 3.9 \times 10^{-6}$ corresponding to the stability limit of the explicit flow solver in the absence of coupling with the structure. The obtained results are depicted in Fig. 7.20 for the first 4102 time-steps. For $n_{S/F} = 30$, ALG1 and ALG3 exhibit essentially the same accuracy. In the long run, their amplitude and phase errors are less important than those of ALG2. Clearly, this highlights the superiority of ALG3 which, despite its inter-field parallelism and unlike ALG2, is capable of delivering the same accuracy as the sequential algorithm ALG1.

Next, the relative speed of the focus partitioned solution procedures is assessed by comparing their CPU performance for a certain level of accuracy dictated by ALG0. It turns out that in order to meet the accuracy requirements of ALG0, ALG1 and ALG3 can use a subcycling factor as large as $n_{S/F} = 10$, but ALG2 can subcycle only up to $n_{S/F} = 5$ (Fig. 7.21).

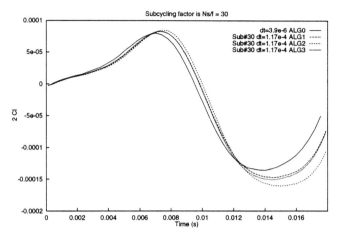

**Figure 7.20** Lift coefficient history for $n_{S/F} = 30$

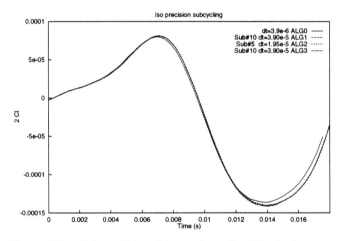

**Figure 7.21** Lift coefficient history for a fixed level of accuracy

The performance results measured on the iPSC-860 are reported in Table 1 where ICC denotes the intercube communication time. Note that ICC is measured in the fluid kernel and includes idle time when the flow and structural communications do not overlap.

**Table 1** Performance results on the iPSC-860

| Fluid: 64 processors | Structure: 2 processors |
| --- | --- |

Elapsed time for 4102 fluid time-steps

| Algorithm | Fluid | Structure | Fluid-Wait+ICC | Total CPU |
| --- | --- | --- | --- | --- |
| ALG0 | 2617.23 s. | 1267.93 s. | 1283.10 s. | 3900.33 s. |
| ALG1 $(n_{S/F} = 10)$ | 2625.11 s. | 126.67 s. | 127.90 s. | 2753.01 s. |
| ALG2 $(n_{S/F} = 5)$ | 2643.57 s. | 253.34 s. | 1.67 s. | 2645.24 s. |
| ALG3 $(n_{S/F} = 10)$ | 2603.56 s. | 253.23 s. | 1.37 s. | 2604.93 s. |

From the results reported in Table 1, the following observations can be made

- the fluid computations dominate the simulation time. This is partly because the structural model is simple in this case, and a linear elastic behavior is assumed for the panel.
- considering that the iPSC-860 has 128 processors and that only clusters of $2^n$ processors can be defined on this machine, allocating 64 processors to the fluid and 2 processors to the structure achieves the minimum possible inter-field load imbalance for this coupled problem.
- the effect of subcycling on intercube communication costs is clearly demonstrated. Because the flow solution time is dominating, the effect of subcycling on the total CPU time is less important for ALG2 and ALG3 which feature inter-field parallelism in addition to intra-field multiprocessing, than for ALG1 which features intra-field parallelism only (note that ALG1 with $n_{S/F} = 1$ is identical to ALG0).
- ALG2 and ALG3 allow a perfect overlap of inter-field communications, which reduces intercube communication and idle time to less than 0.3% of the amount corresponding to ALG0.
- The superiority of ALG3 over ALG2 is not clearly demonstrated for this problem because of the simplicity of the structural model and the subsequent load imbalance between the fluid and structure computations.

## Panel Flutter

The classical and analytical solution of the instability problem of flat panels with infinite aspect ratio in supersonic airstreams assumes a shallow shell theory for the structure and a linearized formulation for the flow problem (piston theory). Within this analytical approach, the dynamics of the focus coupled fluid/structure system are governed by a fourth-order partial differential equation [2, page 419], and the flutter condition is obtained by analyzing the roots of the corresponding characteristic equation. For the panel described the beginning of this section, the classical linear theory predicts flutter at the critical Mach number $M_\infty^{cr} = 1.98$. The objective of this section is to validate the aeroelastic simulation capability presented in this chapter by reproducing the theoretical critical Mach number for the given panel. Note that in order to compare the analytical and finite element approaches, the coefficients of the shallow shell equation described in [2, page 419] must be computed to represent the same equation as that corresponding to the finite element model used in this chapter.

Four different runs at $M_\infty = 2.0$, $M_\infty = 2.05$, $M_\infty = 2.095$, and $M_\infty = 2.13$ are performed using ALG3. For each run, a steady-state flow is first computed. Then, a displacement perturbation of the panel along its first mode (Fig. 7.19) is imposed, and the aeroelastic response of the coupled system is computed. The predicted time histories of the lift coefficient are depicted in Fig. 7.22 for all four cases.

From the results reported in Fig. 7.22, it follows that the flutter speed predicted by our formulation verifies $2.05 \leq M_\infty^{cr} \leq 2.095$. Hence, this flutter speed is 4.5 % higher than that predicted by the piston theory. This is a rather good agreement, given that the piston theory and the computational approach presented herein do not share exactly the same approximations.

Finally, we report in Fig. 7.23 the history of the accumulated external energy at $M_\infty = 2.095$ for both the fluid and structural systems. At this speed, the panel is clearly shown to extract energy from the fluid, and therefore to flutter. Note that Fig. 7.23 also highlights the quality of the matching performed by Matcher: the amount of external energy extracted by the structure is shown to be equal to that lost by the fluid, as it should be.

## THREE-DIMENSIONAL AEROELASTIC TRANSONIC COMPUTATIONS

### Problem Definition

Next, we consider transient aeroelastic reponse problems associated with a simple structural model of the ONERA M6 wing.

The wing is represented by an equivalent plate model discretized in 1071 triangular plate elements, 582 nodes, and 6426 degrees of freedom (Fig. 7.24). Four meshes $M1 - -M4$ are designed for the discretization of the three-dimensional flow domain around the wing. The characteristics of theses meshes are given in Table 2 where $N_{ver}$, $N_{tet}$, $N_{fac}$, and $N_{var}$ denote respectively the number of vertices, tetrahedra, facets (edges), and fluid variables. A partial view of the discretization of the flow domain is shown in Fig. 7.24.

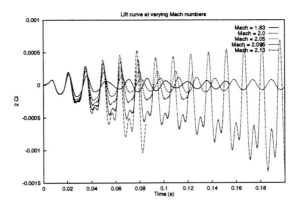

**Figure 7.22** Flutter analysis

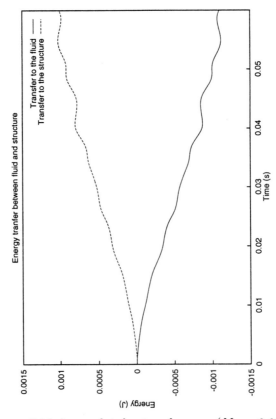

**Figure 7.23** Accumulated external energy $(M_\infty = 2.095)$

**Table 2** Characteristics of meshes $M1--M4$

| Mesh | $N_{ver}$ | $N_{tet}$ | $N_{fac}$ | $N_{var}$ |
|------|-----------|-----------|-----------|-----------|
| $M1$ | 15460 | 80424 | 99891 | 77300 |
| $M2$ | 31513 | 161830 | 201479 | 157565 |
| $M3$ | 63917 | 337604 | 415266 | 319585 |
| $M4$ | 115351 | 643392 | 774774 | 576755 |

**Figure 7.23** Finite element plate model of the wing

**Figure 7.24** Partial view of the fluid mesh $M1$ on the skin of the ONERA M6 wing

The sizes of the fluid meshes $M1--M4$ have been tailored for parallel computations on respectively 16 ($M1$), 32 ($M2$), 64 ($M3$), and 128 processors ($M4$) of a Paragon XP/S and a Cray T3D systems. In particular, the sizes of these meshes are such that

the processors of a Paragon XP/S machine with 32 Mbytes per node would not swap
when solving the corresponding flow problems.

Here again, the fluid and structural meshes are not compatible at their interface.
Matcher (Maman and Farhat 1995) is used to generate in a single preprocessing step
the data structures required for transferring the pressure load to the structure, and the
structural deformations to the fluid.

## Computational Platforms

All computations are performed on an iPSC - 860, and/or a Paragon XP/S, and/or a
Cray T3D, and/or an IBM SP2 computers using double precision arithmetic. Message
passing is carried out via NX on the Paragon XP/S multiprocessor, PVM T3D on the
Cray T3D system, and MPI on the IBM SP2 parallel processor. The fluid and structure
solvers are implemented as separate programs that communicate via the intercube
communication procedures described in Barszcz (1992).

## Parallel Performance of the Flow Solver

The performance of the parallel flow solver is assessed with the computation of the
steady state of a flow around the given wing at a Mach number $M_\infty = 0.84$ and an
angle of attack $\beta = 3.06$ degrees (Fig. 7.25) . The CFL number is set to 0.9. The
four meshes $M1 - -M4$ are decomposed in respectively 16, 32, 64, and 128 *overlapping*
subdomains using TOP/DOMDEC (Farhat, Lantéri, and Simon 1995). The motivations
for employing overlapping subdomains and the impact of this computational strategy
on parallel performance are discussed in Farhat and Lantéri 1994. The CPU timings
in seconds are reported in Tables 3-5 for the first 100 iterations on a Paragon XP/S
machine (128 processors), a Cray T3D system (128 processors), and an IBM SP2
computer (32 processors), respectively. In these tables, $N_p$, $N_{var}$, $T_{comm}^{loc}$, $T_{comm}^{glo}$,
$T_{comp}$, $T_{tot}$ and $mflops$ denote respectively the number of processors, the number
of variables (unknowns) to be solved, the time elapsed in short range interprocessor
communication between neighboring subdomains, the time elapsed in long range global
interprocessor communication, the computational time, the total simulation time, and
the computational speed in millions of floating point operations per second. Typically,
short range communication is needed for assembling various subdomain results such
as fluxes at the subdomain interfaces, and long range interprocessor communication
is required for reduction operations such as those occurring in the the evaluation of
the stability time-steps and the norms of the nonlinear residuals. *Because message-
passing is also used for synchronization, the reported communication timings include
any idle-time due to load imbalance.* We also note that we use the same fluid code for
steady state and aeroelastic computations. Hence, even though we are benchmarking in
Tables 3-5 a steady state computation with a local time stepping strategy, we are still
timing the kernel that evaluates the global time-step in order to reflect its impact on the
unsteady computations that we perform in aeroelastic simulations such as those that
are discussed next. The $mflop$ rates reported in Tables 3-5 are computed in a strict
manner: they exclude all the redundant computations associated with the overlapping
subdomain regions.

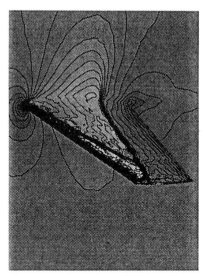

**Figure 7.25** Steady-state Mach lines ONERA M6 wing — mesh $M4$)

**Table 3** Performance of the parallel flow solver
on the Paragon XP/S system (16—128 processors)

| 100 iterations—CFL = 0.9 | | | | | | |
| Mesh | $N_p$ | $T_{comm}^{loc}$ | $T_{comm}^{glo}$ | $T_{comp}$ | $T_{tot}$ | $mflops$ |
| --- | --- | --- | --- | --- | --- | --- |
| $M1$ | 16 | 2.0 s. | 40.0 s. | 96.0 s. | 138.0 s. | 84 |
| $M2$ | 32 | 4.5 s. | 57.0 s. | 98.5 s. | 160.0 s. | 145 |
| $M3$ | 64 | 7.0 s. | 90.0 s. | 103.0 s. | 200.0 s. | 240 |
| $M4$ | 128 | 6.0 s. | 105.0 s. | 114.0 s. | 225.0 s. | 401 |

**Table 4** Performance of the parallel flow solver
the Cray T3D system (16—128 processors)

| 100 iterations—CFL = 0.9 | | | | | | |
| Mesh | $N_p$ | $T_{comm}^{loc}$ | $T_{comm}^{glo}$ | $T_{comp}$ | $T_{tot}$ | $mflops$ |
| --- | --- | --- | --- | --- | --- | --- |
| $M1$ | 16 | 1.6 s. | 2.1 s. | 87.3 s. | 91.0 s. | 127 |
| $M2$ | 32 | 2.5 s. | 4.1 s. | 101.4 s. | 108.0 s. | 215 |
| $M3$ | 64 | 3.5 s. | 7.2 s. | 100.3 s. | 111.0 s. | 433 |
| $M4$ | 128 | 3.0 s. | 7.2 s. | 85.3 s. | 95.5 s. | 945 |

**Table 5** Performance of the parallel flow solver
on the IBM SP2 system (4—32 processors)

| 100 iterations—CFL = 0.9 | | | | | | |
|---|---|---|---|---|---|---|
| Mesh | $N_p$ | $T_{comm}^{loc}$ | $T_{comm}^{glo}$ | $T_{comp}$ | $T_{tot}$ | $mflops$ |
| $M1$ | 4 | 0.8 s. | 0.4 s. | 70.8 s. | 72.0 s. | 160 |
| $M2$ | 8 | 1.1 s. | 0.6 s. | 73.3 s. | 75.0 s. | 308 |
| $M3$ | 16 | 1.4 s. | 0.7 s. | 78.9 s. | 81.0 s. | 594 |
| $M4$ | 32 | 2.0 s. | 1.0 s. | 79.0 s. | 82.0 s. | 1102 |

The reader can easily verify that the number of processors assigned to each mesh is such that $N_{var}/N_p$ is almost constant. This means that larger numbers of processors are attributed to larger meshes in order to keep each local problem within a processor at an almost constant size. For such a benchmarking strategy, parallel scalability of the flow solver on a target parallel processor implies that the total solution CPU time should be constant for all meshes and their corresponding number of processors. This is clearly not the case for the Paragon XP/S system. On this machine, short range communication is shown to be inexpensive, but long range communication costs are reported to be important. While some of these global commmunication costs are due to the latency of the Paragon XP/S parallel processor which is an order of magnitude slower than that of the Cray T3D system, most of them are induced by the load imbalance of the mesh partitions. Indeed, global communication occurs at the beginning and end of each large chunk of parallel computations. Since communication is also used for synchronization, load imbalance between the processors infaltes the timing of global message passing. Surprisingly, the effect of load imbalance on global communication appears to be insignificant on the T3D and SP2 parallel processors.

On the other hand, parallel scalability is well demonstrated for the Cray T3D and IBM SP2 systems. The results reported in Tables 4 and 5 show that all computations using meshes $M1--M4$ and the corresponding number of processors consume almost the same total amount of CPU time. For 128 processors, the Cray T3D system is shown to be more than twice faster than the Paragon XP/S machine. The difference appears to be strictly in long range communication as the computational time is reported to be almost the same on both machines. However, most impressive is the fact that an IBM SP2 with 32 processors only is shown to be three times faster than a 128 - processor Paragon XP/S, and faster than a Cray T3D with 128 processors.

## Performance of the Partitioned Analysis Procedures

As in the two-dimensional application, we consider first two different series of transient aeroelastic simulations at Mach number $M_\infty = 0.84$ that highlight

- the relative accuracy of these coupled solution algorithms for a fixed subcycling factor $n_{S/F}$.
- the relative speed of these coupled solution algorithms for a fixed level of accuracy.

In all cases, mesh $M2$ is used for the flow computations, 32 processors of an iPSC-860 system are allocated to the fluid solver, and 4 processors of the same machine are assigned to the structural code. Initially, a steady-state flow is computed around the wing at $M_\infty = 0.84$, Mach number at which the wing described above is not supposed to flutter. Then, the aeroelastic response of the coupled system is triggered by a displacement perturbation of the wing along its first mode (Fig. 7.26).

First, the subcycling factor is fixed to $n_{S/F} = 10$ then to $n_{S/F} = 30$, and the lift is computed using a time-step corresponding to the stability limit of the explicit flow solver in the absence of coupling with the structure. The obtained results are depicted in Fig. 7.27 and Fig. 7.28 for the first half cycle.

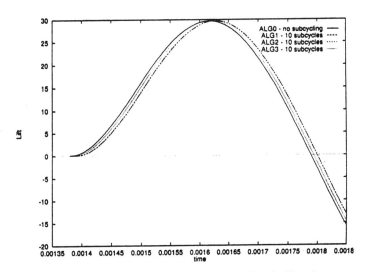

Figure 7.26 Initial perturbation of the structure

Figure 7.27 Lift history for the first half cycle
$$(n_{S/F} = 10)$$

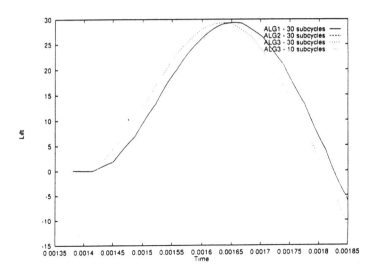

**Figure 7.28** Lift history for the first half cycle
$$(n_{S/F} = 30)$$

The superiority of the parallel fluid-subcycled ALG3 solution procedure is clearly demonstrated in Fig. 7.27 and Fig. 7.28. For $n_{S/F} = 10$, ALG3 is shown to be the closest to ALG0, which is supposed to be the most accurate since it is sequential and non-subcycled. ALG1 and ALG2 have comparable accuracies. However, both of these algorithms exhibit a significantly more important phase error than ALG3, especially for $n_{S/F} = 30$.

Next, the relative speed of the partitioned solution procedures is assessed by comparing their CPU time for a certain level of accuracy dictated by ALG0. For this problem, it turned out that in order to meet the accuracy requirements of ALG0, the solution algorithms ALG1 and ALG2 can subcycle only up to $n_{S/F} = 5$, while ALG3 can easily use a subcycling factor as large as $n_{S/F} = 10$. The performance results measured on an iPSC-860 system are reported in Table 6 for the first 50 coupled time-steps. In this table, ICWF and ICWS denote the inter-code communication timings measured respectively in the fluid and structural kernels; these timings include idle and synchronization (wait) time when the fluid and structural communications do not completely overlap. For programming reasons, ICWS is monitored together with the evaluation of the pressure load.

**Table 6** Performance results on the iPSC-860

Fluid: 32 processors        Structure: 4 processors

Elapsed time for 50 fluid time-steps

| Alg. | Fluid Solver | Fluid Motion | Struc. Solver | ICWS | ICWF | Total CPU |
|------|------|------|------|------|------|------|
| ALG0 | 177.4 s. | 71.2 s. | 33.4 s. | 219.0 s. | 384.1 s. | 632.7 s. |
| ALG1 | 180.0 s. | 71.2 s. | 16.9 s. | 216.9 s. | 89.3 s. | 340.5 s. |
| ALG2 | 184.8 s. | 71.2 s. | 16.6 s. | 114.0 s. | 0.4 s. | 256.4 s. |
| ALG3 | 176.1 s. | 71.2 s. | 10.4 s. | 112.3 s. | 0.4 s. | 247.7 s. |

From the results reported in Table 6, the following observations can be made:

- the fluid computations dominate the simulation time. This is partly because the structural model is again simple in this case, and a linear elastic behavior is assumed. However, by allocating 32 processors to the fluid kernel and 4 processors to the structure code, a reasonable load balance is shown to be achieved for ALG0.

- during the first 50 fluid time-steps, the CPU time corresponding to the structural solver does not decrease linearly with the subcycling factor $n_{S/F}$ because of the initial costs of the FETI reorthogonalization procedure designed for the efficient iterative solution of implicit systems with repeated right hand sides.

- the effect of subcycling on intercube communication costs is clearly demonstrated. The impact of this effect on the total CPU time is less important for ALG2 and ALG3 which feature inter-field parallelism in addition to intra-field multiprocessing, than for ALG1 which features intra-field parallelism only (note that ALG1 with $n_{S/F} = 1$ is identical to ALG0), because the flow solution time is dominating.

- ALG2 and ALG3 allow a certain amount of overlap between inter-field communications, which reduces intercube communication and idle time on the fluid side to less than 0.001% of the amount corresponding to ALG0.

Most importantly, the performance results reported in Table 6 demonstrate that subcycling and inter-field parallelism are desirable for aeroelastic simulations even when the flow computations dominate the structural ones, because these features can significantly reduce the total simulation time by minimizing the amount of inter-field communications and overlapping them. For the simple problem described herein, the parallel fluid-subcycled ALG2 and ALG3 algorithms are more than twice faster than the conventional staggered procedure ALG0.

## 7.10 CONCLUSIONS

In this chapter, we have highlighted some key elements of the solution of large-scale three-dimensional nonlinear aeroelastic problems on high performance computational platforms. We have described a three-field arbitrary Lagrangian-Eulerian (ALE) finite volume/element formulation for the coupled fluid/structure problem, presented geometric conservation laws for three-dimensional flow problems with moving boundaries and unstructured and deformable meshes, and discussed the solution of the corresponding coupled semi-discrete equations with partitioned analysis procedures. In particular, we have presented a family of mixed explicit/implicit staggered solution algorithms, and discussed them with particular reference to accuracy, stability, subcycling, and parallel processing. We have described a general framework for the solution of coupled aeroelastic problems on heterogeneous and/or parallel computational platforms, and illustrated it with two- and three-dimensional applications on an iPSC-860, a Paragon XP/S, and a Cray T3D massively parallel systems. We have shown that even when the flow computations dominate the total CPU time of a coupled aeroelastic simulation, subcycling and inter-field parallelism are desirable as they can significantly speedup the total solution time.

## ACKNOWLEDGEMENTS

The author acknowledges partial support by the National Science Foundation under Grant ASC-9217394, partial support by RNR NAS at NASA Ames Research Center under Grant NAG 2-827, and partial support by CMB at the NASA Langley Research Center under Grant NAG-1536427. He wishes to thank Po-Shu Chen, Stéphane Lantéri, Michel Lesoinne, Serge Piperno, and Paul Stern for their help in this research effort.

## REFERENCES

Albano E. and Rodden W.P. 1969. A Doublet-Lattice Method for Calculating Lift Distribution on Oscillating Surfaces in Subsonic Flow. *AIAA J.*, 7: pp. 279–285.

Anderson W.K., Thomas J.L. and Rumsey C.L. 1987. *Extension and Application of Flux-Vector Splitting to Unsteady Calculations on Dynamic Meshes.* AIAA Paper No 87-1152-CP.

Barnard S.T. and Simon H.D. 1994. A Fast Multilevel Implementation of Recursive Spectral Bisection for Partitioning Unstructured Problems. *Concurrency: Practice and Experience*, 6: pp. 101–107.

Barszcz E. 1992. Intercube Communication on the iPSC/860. In: *Proceedings of the Scalable High Performance Computing Conference*, Williamsburg, April 26-29, 1992.

Batina J.T. 1989. *Unsteady Euler Airfoil Solutions Using Unstructured Dynamic Meshes* AIAA Paper No. 89-0115, AIAA 27th Aerospace Sciences Meeting, Reno, Nevada, January 9–12, 1989.

Belytschko T., Smolenski P. and Liu W.K. 1985. Stability of Multi-Time Step Partitioned Integrators for First-Order Finite Element Systems. *Comput. Meths. Appl. Mech. Engrg.*, 49: pp. 281–297.

Belytschko T. and Kennedy J.M. 1978. Computer Models for Subassembly Simulation. *Nucl. Eng. Design*, 49: pp. 17–38.

Benner R.E., Montry G.R. and Weigand G.G. 1987. Concurrent Multifrontal Methods: Shared Memory, Cache, and Frontwidth Issues. *Int. J. Supercomp. Appl.*, 1: pp. 26–44.

Bisplinghoff R.L. and Ashley H. 1962. *Principles of Aeroelasticity*, Dover Publications, Inc.

Bjordstad P.E. and Widlund O.B. 1986. Iterative Methods for Solving Elliptic Problems on Regions Partitioned into Substructures. *SIAM J. of Num. Anal.*, 23: pp. 1097–1120.

Blair M., Williams M.H. and Weisshaar T.A. 1995. *Time Domain Simulations of a Flexible Wing in Subsonic Compressible Flow.* ICASE Report No. 95-3, NASA Langley Research Center, February 1995.

Borland C.J. and Rizzetta D.P. 1982. Nonlinear Transonic Flutter Analysis. *AIAA Journal*, 1606–1615.

Bramble J.H., Pasciak J.E., and Schatz A.H. 1986. The Construction of Preconditioners for Elliptic Problems by Substructuring. *I, Math. Comp.*, 47: pp. 103–134.

Cai X.-C., Farhat C. and Sarkis M. 1995. Schwarz Preconditioners and Implicit Methods for Compressible Flows Problems on Unstructured Meshes. *Eighth International Conference on Domain Decomposition Methods for Partial Differential Equations*, AMS, in press.

Chen P.C. and Liu D.D. 1983. A Harmonic Gradient Method for Unsteady Supersonic Flow Calculations, *AIAA 830887 CP*

Dervieux A. 1985. Steady Euler Simulations Using Unstructured Meshes. *Von Kármán Institute Lecture Series.*

Donea J. 1982. An Arbitrary Lagrangian-Eulerian Finite Element Method for Transient Fluid-Structure Interactions. *Comput. Meths. Appl. Mech. Engrg.*, 33: pp. 689–723.

Dowell E.H. and Llgamov M. 1988. *Studies in Nonlinear Aeroelasticity*, Springer-Verlag.

Dryja M. and Widlund O.B. 1994. Domain Decomposition Algorithms with Small Overlap. *SIAM J. Sci. Comput.*, 15: pp. 604–620.

Duff I.S. 1986. Parallel Implementation of Multifrontal Schemes. Parallel Computing, 3: pp. 193–204.

Edwards J.W. and Malone J.B. 1992. Current Status of Computational Methods for Transonic Unsteady Aerodynamics and Aeroelastic Applications. *Comput. Sys. Engrg.*, 3: pp. 545–569.

Farhat C. 1995. Optimizing Substructuring Methods for Repeated Right Hand Sides, Scalable Parallel Coarse Solvers, and Global/Local Analysis, In: D. Keyes, Y. Saad and D. G. Truhlar, eds., *Domain-Based Parallelism and Problem Decomposition Methods in Computational Science and Engineering*, D. Keyes, Y. Saad and D. G. Truhlar, eds., SIAM, pp. 141–160.

Farhat C., Lantéri S. and Simon H.D. 1995. TOP/DOMDEC, A Software Tool for Mesh Partitioning and Parallel Processing. *J. Comput. Sys. Engrg.*, 6: pp. 13–26.

Farhat C., Chen P.S. and Mandel J. 1995. A Scalable Lagrange Multiplier Based Domain Decomposition Method for Implicit Time-Dependent Problems. *Internat. J. Numer. Meths. Engrg.*, 38: pp. 3831–3854.

Farhat C., Lesoinne M. and Maman N. 1995. Mixed Explicit/Implicit Time Integration of Coupled Aeroelastic Problems: Three-Field Formulation, Geometric Conservation and Distributed Solution. *Internat. J. Numer. Meths. Fluids*, 21: pp. 807-835.

Farhat C., Lesoinne M., Chen P.S. and Lantéri S. 1995. *Parallel Heterogeneous Algorithms for the Solution of Three-Dimensional Transient Coupled Aeroelastic Problems*. AIAA Paper 95-1290, AIAA 36th Structural Dynamics Meeting, New Orleans, Louisiana, April 10-13, 1995.

Farhat C., Mandel J. and Roux F.X. 1994. Optimal Convergence Properties of the FETI Domain Decomposition Method. *Comput. Meths. Appl. Mech. Engrg.*, 115: pp. 367–388.

Farhat C., Crivelli L. and Roux F.X. 1994a. Extending Substructure Based Iterative Solvers to Multiple Load and Repeated Analyses. *Comput. Meths. Appl. Mech. Engrg.*, 117: pp. 195–209.

Farhat C., Crivelli L. and Roux F.X. 1994b. A Transient FETI Methodology for Large-Scale Parallel Implicit Computations in Structural Mechanics. *Internat. J. Numer. Meths. Engrg.*, 37: pp. 1945–1975.

Farhat C. and Lantéri S. 1994. Simulation of Compressible Viscous Flows on a Variety of MPPs: Computational Algorithms for Unstructured Dynamic Meshes and Performance Results. *Comput. Meths. Appl. Mech. Engrg.*, 119: pp. 35–60.

Farhat C. and Roux F.X. 1994. Implicit Parallel Processing in Structural Mechanics. *Computational Mechanics Advances*, 2: pp. 1–124.

Farhat C., Chen P.S. and Stern P. 1994. Towards the Ultimate Iterative Substructuring Method: Combined Numerical and Parallel Scalability, and Multiple Load Cases. *J. Comput. Sys. Engrg.*, 117: pp. 195–209.

Farhat C., Fezoui L., and Lantéri S. 1993. Two-Dimensional Viscous Flow Computations on the Connection Machine: Unstructured Meshes, Upwind Schemes, and Massively Parallel

Computations. *Comput. Meths. Appl. Mech. Engrg.*, 102: pp. 61–88.

Farhat C., Lantéri S. and Fezoui L. 1992. Mixed Finite Volume/Finite Element Massively Parallel Computations: Euler Flows, Unstructured Grids, and Upwind Approximations, In: *Unstructured Scientific Computation on Scalable Multiprocessors*, P. Mehrotra, J. Saltz, and R. Voigt, eds., MIT Press, pp. 253–283.

Farhat C. and Roux F.X. 1992. An Unconventional Domain Decomposition Method for an Efficient Parallel Solution of Large-Scale Finite Element Systems. *SIAM J. Sc. Stat. Comp.*, 13: pp. 379–396.

Farhat C. 1991. A Lagrange Multiplier Based Divide and Conquer Finite Element Algorithm. *J. Comput. Sys. Engrg.*, 2: pp. 149–156.

Farhat C. and Roux F.X. 1991. A Method of Finite Element Tearing and Interconnecting and its Parallel Solution Algorithm. *Internat. J. Numer. Meths. Engrg.*, 32: pp. 1205–1227.

Farhat C., Park K.C. and Pelerin Y.D. 1991. An Unconditionally Stable Staggered Algorithm for Transient Finite Element Analysis of Coupled Thermoelastic Problems. *Comput. Meths. Appl. Mech. Engrg.*, 85: pp. 349–365.

Farhat C. and Lin T.Y. 1990. *Transient Aeroelastic Computations using Multiple Moving Frames of Reference.* AIAA Paper No. 90-3053, AIAA 8th Applied Aerodynamics Conference, Portland, Oregon, August 20–22, 1990.

Farhat C. and Wilson E. 1987. A New Finite Element Concurrent Computer Program Architecture. *Internat. J. Numer. Meths. Engrg.*, 24: pp. 1771–1792.

Fezoui L. and Stoufflet B. 1989. A Class of Implicit Upwind Schemes for Euler Simulations with Unstructured Meshes. *J. Comp. Phys.*, 84: pp. 174–206.

Fung Y.C. 1969. *An Introduction to the Theory of Aeroelasticity.* Dover Publications, Inc.

Guruswamy G.P. 1988. *Time-Accurate Unsteady Aerodynamic and Aeroelastic Calculations of Wings Using Euler Equations* AIAA Paper No. 88 2281, AIAA 29th Structures, Structural Dynamics and Materials Conference, Williamsburg, Virginia, April, 18–20, 1988.

Hughes T.J.R., Liu W.K. and Zimmermann T.K. 1978. Lagrangian-Eulerian Finite Element Formulation for Incompressible Viscous Flows. In: *U.S.-Japan Seminar on Interdisciplinary Finite Element Analysis*, Cornell Univ., Ithaca, New York.

Jones W.P. and Appa K. 1977. Unsteady Supersonic Aerodynamic Theory for Interfering Surfaces by the Method of Potential Gradient, *NASA CR 2898*.

Kandil O.A. and Chuang H.A. 1988. *Unsteady Vortex-Dominated Flows Around Maneuvering Wings Over a Wide Range of Mach Numbers.* AIAA Paper No. 88-0317, AIAA 26th Aerospace Sciences Meeting, Reno, Nevada, January 11–14, 1988.

Lantéri S. and Farhat C. 1993. Viscous Flow Computations on MPP Systems: Implementational Issues and Performance Results for Unstructured Grids, In: *Parallel Processing for Scientific Computing*, R. F. Sincovec, *et. al.*, ed., SIAM, pp. 65–70.

Lesoinne M. and Farhat C. 1995. *Geometric Conservation Laws for Aeroelastic Computations Using Unstructured Dynamic Meshes.* AIAA Paper 95-1709, 12th AIAA Computational Fluid Dynamics Conference, San Diego, California, June 19-22, 1995.

Lesoinne M. 1994. *Mathematical Analysis of Three-Field Numerical Methods for Aeroelastic Problems.* Ph. D. Thesis, The University of Colorado at Boulder.

Lesoinne M. and Farhat C. 1993. *Stability Analysis of Dynamic Meshes for Transient Aeroelastic Computations.* AIAA Paper No. 93-3325, 11th AIAA Computational Fluid Dynamics Conference, Orlando, Florida, July 6–9, 1993.

Lesoinne M., Farhat C. and Géradin M. 1991. ParallelVvector Improvements of the Frontal Method. *Internat. J. Numer. Meths. Engrg.*, 32: pp. 1267–1282.

Liu J.W.H. 1992. The Multifrontal Method for Sparse Matrix Solution: Theory and Practice. *SIAM Review*, 34: pp. 82–109.

Maman N. and Farhat C. 1995. Matching Fluid and Structure Meshes for Aeroelastic Computations: a Parallel Approach. *Comput. & Struc.*, 54: pp. 779–785.

Mandel J., Tezaur R. and Farhat C. 1995. An Optimal Lagrange Multiplier Based Domain Decomposition Method for Plate Bending Problems. submitted to *SIAM J. Sci. Stat. Computing*.

Mandel J. and Tezaur R. 1995. Convergence of a Substructuring Method with Lagrange Multipliers. *Numerische Mathematik*, in press.

Masud A. 1993. *A Space-Time Finite Element Method for Fluid Structure Interaction*. Ph. D. Thesis, Stanford University.

NKonga B. and Guillard H. 1994. Godunov Type Method on Non-Structured Meshes for Three-Dimensional Moving Boundary Problems. *Comput. Meths. Appl. Mech. Engrg.* 113: pp. 183–204.

Park K.C. and Felippa C.A. 1983. Partitioned Analysis of Coupled Systems, In: *Computational Methods for Transient Analysis*. T. Belytschko and T. J. R. Hughes, eds., North-Holland Pub. Co. pp. 157–219.

Piperno S., Farhat C. and Larrouturou B. 1995. Partitioned Procedures for the Transient Solution of Coupled Aeroelastic Problems. *Comput. Meths. Appl. Mech. Engrg.* 124: pp. 79–112.

Pothen A., Rothberg E., Simon H. and Wang L. 1994. Parallel Sparse Cholesky Factorization with Spectral Nested Dissection Ordering. *RNR-094-011*, NASA Ames Research Center, May 1994.

Pothen A. and Sun C. 1993. A Mapping Algorithm for Parallel. Sparse Matrix Factorization. *SIAM J. Sci. Comput.* 4: pp. 1253–1257.

Pramono E. and Weeratunga S.K. 1994. *Aeroelastic Computations for Wings Through Direct Coupling on Distributed-Memory MIMD Parallel Computers*. AIAA Paper No. 94-0095, 32nd Aerospace Sciences Meeting & Exhibit, Reno, January 10-13, 1994.

Rausch R.D., Batina J.T. and Yang T.Y. 1989. *Euler Flutter Analysis of Airfoils Using Unstructured Dynamic Meshes* AIAA Paper No. 89-13834, 30th Structures, Structural Dynamics and Materials Conference, Mobile, Alabama, April 3-5, 1989.

Roe P.L. 1981. Approximate Riemann Solvers, Parameters Vectors and Difference Schemes. *J. Comp. Phys.*, 43: pp. 357–371.

Schlichting H. 1960. *Boundary Layer Theory*. Fourth edition, McGraw-Hill, New York.

Shankar V. and Ide H. 1988. Aeroelastic Computations of Flexible Configurations. *Comput. & Struc.* 30: pp. 15–28.

Steger J. and Warming R.F. 1981. Flux Vector Splitting for the Inviscid Gas Dynamic with Applications to Finite-Difference Methods. *Journ. of Comp. Phys.*, 40: pp. 263–293.

Strganac T.W. and Mook D.T. 1990. Numerical Model of Unsteady Subsonic Aeroelastic Behavior. *AIAA Journal* 28: pp. 903–909.

Thomas P.D. and Lombard C.K. 1979. Geometric Conservation Law and its Application to Flow Computations on Moving Grids. *AIAA J.* 17: pp. 1030–1037.

Tijdeman H. and Seebass R. 1980. Transonic Flow Past Oscillating Airfoils. *Ann. Rev. Fluid Mech.*, 12: pp. 181–222.

Vanderstraeten D., Farhat C., Chen P.S., Keunings R., and Zone O. 1995. A Retrofit and Contraction Based Methodology for the Fast Generation and Optimization of Mesh Partitions: Beyond the Minimum Interface Size Criterion. *Comput. Meths. Appl. Mech. Engrg.*, in press.

Van Leer B. 1979. Towards the Ultimate Conservative Difference Scheme V: a Second-Order Sequel to Goudonov's Method. *J. Comp. Phys.*, 32: pp. 361–370.

Venkatakrishnan V. 1995. *A Perspective of Unstructured Grid Flow Solvers*. ICASE Report No. 95-3, NASA Langley Research Center, February 1995.

Zhang H., Reggio M., Trépanier J.Y. and Camarero R. 1993. Discrete Form of the GCL for Moving Meshes and its Implementation in CFD Schemes. *Computers in Fluids*, 22: pp. 9–23.

# 8

# Finite Element And Active Element Placement Techniques For Control-Structure Interaction Applications

Rong C. Shieh[1]

## 8.1 GENERAL

Control-structure interaction (CSI) problems of flexible space structures (including components such as appendages) involve both static shape control (for precision pointing) and dynamic (vibration) jitter control problems. Space-borne structures, particularly the precision-oriented space structures, are usually flexible and may be very large. Because of this, they will have vibrations with extremely low frequencies, densely spaced and lightly damped modes, and long decay times unless some vibration control measures are in place. The vibrations (dynamic responses) can be induced by environmental excitations (such as thermally induced vibrations) and by operational excitations (such as pointing and retargeting maneuvers, onboard motors, etc.). But space structures supporting weapons or antennas must accommodate retargeting maneuvers without detrimental jitter from vibrations and thermo-mechanical flutter. Solely relying on small inherent damping of the structures in most cases is not sufficient to damp out in time or limit excessive vibration jitter to meet the stringent pointing and retargeting requirements. Therefore, to meet such requirements, some active technique, preferably augmented by a passive technique (such as a damping enhancement) is needed to control the structural dynamic responses induced by environmental and operational excitations. Similarly, active shape control is needed to augment the conventional passive means of increasing structural stiffness to meet the stringent pointing/targeting requirement so as to have effective shape control without adding excessive structural weight to the system.

An active vibration jitter control is traditionally approached by using centralized or decentralized discrete controllers. However, a major problem in using these active control techniques in large flexible space structures is the actuator/observer spillover. Properly designed distributed controllers (actuators/sensors), such as those using

[1]   MRJ, Inc., 10560 Arrowhead Drive Fairfax, VA 22030

*Parallel Solution Methods in Computational Mechanics* edited by M. Papadrakakis

piezoceramics (PZT), appear to provide mechanisms of controlling all the modes of the system thus avoiding the problem of spillover of the uncontrolled modes. This relatively new technology has been an active research area in recent years (e.g., Anderson *et al.* 1992; Doesh *et al.* 1992; Lee 1987, 1993; Tzou *et al.* 1989, 1994; etc.). Applications of distributed control techniques using smart materials, such as piezoelectrics and shape memory alloys, have also been extended to vibration, divergence and flutter control problems of flexible aircraft/helicopter components (panels, rotor blades, wings, etc.) (cf. Zhou *et al.* 1994; Suleman *et al.* 1994; Xue *et al.* 1994; Nam *et al.* 1994; Giurgiutiu *et al.* 1994; etc.). In an earlier study, the writer (Shieh, 1992a, 1994a) developed new active beam elements, including theory, design, and finite element (FE) models for partial or complete control of 3-D vibrational components (induced by axial extension, bendings in two axis directions, and torsional twisting).

Spatially distributed point actuators/sensors are also used in the efficient active control of flexible space structures in the high modal kinetic energy region(s). Considerable research efforts have been expended in recent years to maximize the robustness of such CSI systems, as well as for increasing the stability margin to reduce potential spillover of higher frequency modes of vibration (Haftka 1990; Sepulveda *et al.* 1992; Thomas *et al.* 1989; etc.). Therefore, a general CSI system will contain both distributed and point sensors/actuators. Such active and/or passive damping enhanced systems form a unique class of dynamic systems, known as nonproportionally damped elastic systems (and also nonconservatively loaded for the noncollocated control case). Dynamic response analyses and design optimization of such active damping controlled structures or CSI systems are typically computationally intense because the equations of motion cannot be decoupled (or assumed to be decoupled) by simply using the undamped free vibration modes as the modal coordinates in a decoupling process.

Also in greater demand are computationally efficient integrated design optimization methods of CSI systems because a large number of cycles of reanalyses and, thus, enormous computational time is required in such design optimization calculations unless the problem involved is relatively small. A particularly important and active subarea of research is the placement optimization of active elements for shape or vibration control. Such placement optimization problems have been approached mostly by using stochastic optimization methods, such as the Genetic Algorithms (Goldberg 1989; Anderson *et al.* 1994; Onoda *et al.* 1992; Rao *et al.* 1991, 1992) and Simulated Annealing (Otten *et al.* 1989; Anderson *et al.* 1994; Chen *et al.* 1991; Kincaid *et al.* 1990; Onoda *et al.* 1992; Seeley *et al.* 1994) in a sequential environment. A design optimization based on a stochastic method is usually much more computationally intense than that based on a gradient-based optimization method. The latter is typically used in structural sizing design optimization problems and cannot be directly applied in such placement optimization problems. Therefore, whether it is for design, design analysis, and/or optimization, efficient and reliable methods are needed to predict the structural response and dynamic performance of structures with active piezoelectric elements, or of those acting in concert with onboard distributed and/or discrete controllers for maneuvering, pointing, vibration/noise suppression, or shape control.

Recent progress in computer architecture, particularly the multiple-instruction-multiple-data (MIMD) massively parallel architecture, enables one to solve these classes of analysis and design problems by drastically reducing the computational time required in the calculations. This also enables one to solve new, unprecedented, large-scale problems. However, new parallel or parallelized algorithms need to be formulated to take full advantage of the enormous computational power of MPP computers. The main objective of this chapter is to present such FE method-based MPP computational algorithms/solution procedures for transient response analysis and sensor/actuator placement optimization for shape control of CSI systems, largely based on writer's previous work (Shieh 1993a, 1994c). The study emphasis is placed on distributed piezoelectrically actuator/sensor (active element) controlled CSI systems of the multiaxially active, collocated piezoelectric sensor/actuator structural element type. The structure model used is of space (three-dimensional) frame and truss types. Due to space and resource limitations, the design synthesis computational problems based on linear optimal control laws of the linear quadratic regulator (LQR)/linear quadratic Gaussian (LQG) type (Haftka 1989; Park *et al.* 1989; Belvin 1990) are not dealt with here. Formulation of the CSI transient response analysis problems, presentation of several numerical algorithms/solution procedures, and numerical demonstration example problems are given in Section 8.2. A similar parallel presentation of a class of placement optimization problems of active elements for static structural control is given in Section 8.3.

## 8.2 FORMULATIONS OF TRANSIENT RESPONSE PROBLEMS AND FE METHOD-BASED MPP COMPUTATIONAL PROCEDURES

*INITIAL-VALUE PROBLEM FOR TRANSIENT RESPONSES OF SMART FRAME STRUCTURES*

Consider an adaptive space (3-D) frame (or truss) structure comprised of active and passive beam (and/or rod) elements, containing also discrete actuators/sensors, and undergoing 3-D motions. Through the use of the displacement finite element (FE) method, the structure is spatially discretized into an FE model, which is an assemblage of $N_E$ beam (rod) elements that are joined together at the $N_N$ nodal points. Let $\{u\}_i$ be the $(12 \times 1)$ end displacement vector of a typical ($i$-th) beam or rod element and $\{q\}$ be the $(N_q \times 1)(N_q = 6 \times N_N)$ nodal point displacement vector, referred to as the global coordinate system. These displacement vectors are related to each other by

$$\{u\}_i = [B]_i\{q\}(i = 1, 2, \ldots, N_E) \tag{8.1}$$

where $[B]_i$ is a $(12 \times N_q)$ Boolean matrix.

### Finite Element Equations of Motion for Active Beam Elements

The governing equations of motion (dynamic end force ($\{f\}_i$)–displacement ($\{u\}_i$) relationships) for a spatially discretized $i$-th active beam element can be written as follows:

$$\{f\}_i = [m]_i\{u''\}_i + [c]_i\{u'\}_i + [k]_i\{u\}_i - \{p\}_i \tag{8.2a}$$

$$([c]_i = [c_v]_i + [c_p]_i, \quad [k]_i = [k_e]_i + [k_p]_i) \tag{8.2b}$$

where $\{p\}_i$ is the force vector associated with the distributed forces; $[m]_i$, $[c]_i$, and $[k]_i$ denote the $i$-th element's mass matrix and velocity- and position-dependent force matrices, respectively; and $[c_v]_i$ and $[k_e]_i$ are the viscous damping and elastic stiffness matrices, respectively. The $[c_p]_i$ and $[k_p]_i$ matrices are generally nonsymmetric matrices associated with the negative velocity and position feedback control forces of the $i$-th active beam element and become symmetric and can be called active damping and stiffness matrices, respectively, for the collocated sensor/actuator case. These active element force matrices are given below for a multiaxially active, piezoelectric beam element based on a "four-sectored" sensor/actuator design concept.

## Feedback Piezoelectric Actuator Force Matrices, $[k_p]$ and $[c_p]$, for Multiaxially Active Beam Elements

These matrices for a multiaxially active, piezoelectric beam element based on "four-sectored" sensor/actuator design concept are given by (Shieh 1994a)

$$[k_p]_i = [b_a]_i [g_k]_i [b_s]_i, \quad [c_p]_i = [b_a]_i [g_c]_i [b_s]_i \tag{8.2c}$$

Here, $[b_a]$ and $[b_s]$ are the $(12 \times 4)$ element actuator force ($\{f_a\}$) and $(4 \times 12)$ sensor signal matrices, respectively, and $[g_k]$ and $[g_c]$ the $(4 \times 4)$ diagonal element position and velocity feedback control gain matrices, respectively. They are interrelated to the $(4 \times 1)$ sensor and actuator voltage vectors, $\{v_s\}$ and $\{v_a\}$, by

$$\{v_s\} = -([C^*]^{-1}[b_s]\{u\} + [R][b_s]\{u'\}) \tag{8.2d}$$

$$[v_a] = -[g_{sa}]\{v_s\} = [g_k][b_s]\{u\} + [g_c][b_s]\{u'\} \tag{8.2e}$$

$$([g_k] = [g_{sa}][C^*]^{-1}, \quad [g_c] = [g_{sa}][R]) \tag{8.2f}$$

$$\{f_a\} = [b_a]\{v_a\} = [k_p]\{u\} + [c_p]\{u'\} \tag{8.2g}$$

where $[C^*]$ and $[R]$ are the $(4 \times 4)$ element diagonal electric capacitance and resistance matrices, respectively, and $[g_{sa}]$ is the diagonal amplification matrix, and where, for brevity, all subscripts $i$ standing for the $i$-th structural element have been omitted.

For the collocated actuator/sensor case, such as the self-sensing actuator design, $[b_a] = [b_s]^T$ so that both $[k_p]$ and $[c_p]$ are symmetric as in the case of $[k_e]$. Also, for the adjacent layered sensor/actuator pair design case, this matrix is usually nearly symmetric because the thickness of these layers is usually thin compared with the host beam thicknesses or depths.

The $[b_a]$ and $[b_s]$ (and, thus, $[k_p]$ and $[c_p]$) matrices have been derived by Shieh (1994a) for the hollow beam rectangular cross section case with PVFD piezoelectric sensor/actuator pair layers bonded to the four outer beam surfaces (sectors). The elements of these matrices under the cubic interpolation displacement functions are as follows:

$$b_{sj,1} = -b_{sj,7} = e_{s31}^j h_{sz} \quad (j = 1, 3); \quad = -b_{sj,7} = e_{s31}^j h_{sy} (j = 2, 4)$$

$$b_{sj,4} = -b_{sj,10} = -e_{s36}^j h_{sy} h_{sz}/2 \quad (j = 1-4); \quad b_{s2,5} = -b_{s2,11} = e_{s31}^2 h_{sy} h_{sz}/2$$

$$b_{s4,5} = -b_{s4,11} = -e_{s31}^2 h_{sy} h_{sz}/2; \quad b_{s1,6} = -b_{s1,12} = -e_{s31}^1 h_{sy} h_{sz}/2$$

$$b_{s3,6} = -b_{s3,12} = e_{s31}^3 h_{sy} h_{sz}/2; \quad \text{all other } b_{sj,k} = 0 \tag{8.2h}$$

$$b_{aj,k} = b_{sk,j}(s \to a) \tag{8.2i}$$

with $j, k = 1 - 12$. In these expressions, $h_i$'s $(i = y, z)$ are the beam cross-sectional heights in the $i$-th coordinate (transverse) direction; subscripts a and s stand for the "actuator layer" and "sensor layer," respectively.

$$e_{31}(\theta_1) = E(d_{3'1'} \cos^2 \theta_1 + d_{3'2'} \sin^2 \theta_1) - G d_{3'6'} \sin 2\theta_1$$

$$e_{36}(\theta_1) = e_{35} = G[(d_{3'1'} - d_{3'2'})\sin 2\theta_1 + d_{3'6'} \cos 2\theta_1] \qquad (8.2j)$$

where $E$ and $G$ are Young's and shear moduli of the piezoelectric layer in question; $d$'s are the piezoelectric strain constants in the material coordinate systems (1'2'3' with 3' in the thickness direction); $\theta_1$ is the material skew angle; and subscripts a and s have been omitted here for brevity.

A special "uncoupled" active beam element design case in which there is no cross-coupling between various internal force and deformation components is also obtained for the uniformly distributed surface electrode and polarization strength case using the same PVFD material for all sensors and actuator layers. The skew angle $(\theta_1)$ relationships among four sectors' piezoelectric sensor/actuator layers for this special design are shown in Figure 8.1. The nonzero elements for the piezoelectric matrix, $[\kappa_p]$, $(\kappa = k \text{ or } c)$ for this case are:

$$\kappa_{1,1} = -\kappa_{1,7} = -\kappa_{7,1} = -\kappa_{7,7} = 2g_\kappa (e_{31})^2 h_{az}(h_{ay} + h_{az})$$

$$\kappa_{4,4} = -\kappa_{4,10} = -\kappa_{10,4} = -\kappa_{10,10} = g_\kappa (e_{36})^2 h_{sy}(h_z^*)^2 (h_{ay} + h_{az})/2$$

$$\kappa_{5,5} = -\kappa_{5,11} = -\kappa_{11,5} = -\kappa_{11,11} = g_\kappa (e_{31})^2 h_{ay}(h_z^*)^3/2$$

$$\kappa_{6,6} = -\kappa_{6,12} = -\kappa_{12,6} = -\kappa_{12,12} = g_\kappa (e_{31})^2 (h_y^*)^2 (h_z^*)^2/2 \qquad (8.2k)$$

with

$$g_\kappa = g_{\kappa 1}; h_j^* = (h_{aj} + h_{sj})/2 \qquad (\kappa = \text{k,c}; j = y, z)$$

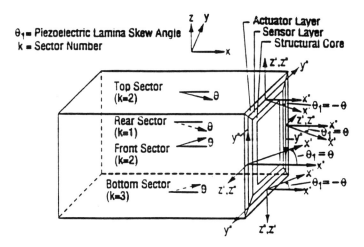

**Figure 8.1** Lamina Skew Angle Relationship among Four Sectored Sensor/Actuator Pairs for Special "Uncoupled" Active Beam Element

where subscripts $y$ and $z$ stand for the beam $y$- and $z$-coordinate (transverse) directions, respectively. In arriving at these expressions, the design conditions of

$$g_{\kappa i+2} = g_{\kappa i}(i = 1, 2); \quad g_{\kappa 2}/g_{\kappa 1} = (h_{az}/h_{ay}) = (h_{sz}/h_{sy}) \tag{8.21}$$

have been used. In the self-sensing actuator design case, the sensor and actuator share the same piezoelectric layer so that $h_{ai} = h_{si} = h_i^* \; (i = y, z)$ and the approximations contained in the above expressions become exact.

## Global (Nodal) Equations of Motion in Vector Form

The nodal equations of motion can be expressed as

$$\sum_{i=1,N_E} [B]_i^T \{f\}_i = \{P(t)\} + [H_a]\{U\} \tag{8.3a}$$

where $\{f\}_i$ is given by (8.2); $\{P(t)\}$ and $\{U\}$ are the applied concentrated mechanical and control force vectors, respectively; [Ha] the actuator location matrix; and superscript T stands for "transpose". The actuator force is related to the sensor measurements of displacements $\{y\}$ and velocities $\{y'\}$ by

$$\{U\} = -([G_d]\{y\} + [G_v]\{y'\}) \tag{8.3b}$$

while $\{y\}$ is related to the displacement vector $\{q\}$ by

$$\{y\} = [H_s]\{q\} \tag{8.3c}$$

Hence, in view of (8.3b,c), (8.3a) now becomes

$$\{F_I(\{q\}, \{q'\}, \{q''\})\} + \{F_C(\{q\}, \{q'\}, \{q''\})\} = \{0\} \tag{8.3d}$$

with

$$\{F_I\} = \sum_{i=1,N_E} [B]_i^T \{f\}_i$$

$$\{F_C\} = [M_C]\{q''\} + [C_U]\{q'\} + [K_U]\{q\} - \{P(t)\} \tag{8.3e}$$

$$([K_U] = [H_a][G_d][H_s], \quad [C_U] = [H_a][G_v][H_s])$$

where $[M_C]$ is the diagonal concentrated mass matrix. For the collocated sensor/actuator case, as in the distributed piezoelectric sensor/actuator case, the gain matrices, $[G_d]$ and $[G_v]$, are diagonal and $[H_a] = [H_s]^T$ so that the discrete actuator stiffness and damping matrices, $[K_U]$ and $[C_U]$, are symmetric.

Equations (8.2) and (8.3d), together with the initial conditions

$$\{q(t_o)\} = \{q_o\}, \quad \{q'(t_o)\} = \{q'_o\}, \tag{8.4}$$

constitute an initial-value problem that governs the transient response of a control-augmented structure.

## FULLY IMPLICIT ALGORITHM-BASED SOLUTION PROCEDURES

The spatially discretized element equations of motion can be further and fully discretized by using a numerical integration scheme to obtain a system of (linear algebraic) recurrence equations (of motion) of the form

$$\{F_{\mathrm{I}}\}_n + \{F_{\mathrm{C}}\}_n = \{0\} \tag{8.5a}$$

with

$$\{F_{\mathrm{I}}\}_n = \sum_{i=1,N_{\mathrm{E}}} [B]_i^{\mathrm{T}}\{f\}_{in} \tag{8.5b}$$

$$\{f\}_{in} \rightarrow \{f^*\}_{in} = [a]_i\{u\}_{in} - \{b\}_{in} \tag{8.5c}$$

$$\{F_{\mathrm{C}}\}_n \rightarrow \{F_{\mathrm{C}}^*\}_n = [A_{\mathrm{C}}]\{q\}_n - \{B_{\mathrm{C}}\}_n \tag{8.5d}$$

where subscripts $n$ are used to denote the quantities being evaluated at the $n$-th integration step at which time $t_n = t_{n-1} + \Delta t$, with $\Delta t$ denoting the incremental time step size. Depending on the type of numerical integration scheme used, the $[a]_i$ matrix and $\{b\}_i$ vector have different expressions as given below.

### Fixed-Step Newmark Beta (NB) Scheme-Based Element Recurrence Equations

In the NB scheme (Newmark 1959, Burnett 1987), the velocity $\{x'\} = \{u'\}_n$ or $\{q'\}_n$ and acceleration $\{x''\} = \{u''\}_n$ or $\{q''\}_n$ at the $n$-th step of integration time, $t_n$ are expressed as

$$\{x'\}_{in} = \mu(\{x\}_{in} - \{x\}_{i,n-1})/\Delta t + (1-\mu)\{x'\}_{i,n-1} + (1-\mu/2)\Delta t\{x''\}_{i,n-1} \tag{8.6a}$$

$$\{x''\}_{in} = (\{x\}_{in} - \{x\}_{i,n-1} - \{x'\}_{i,n-1}\Delta t)/(\beta\Delta t^2) + [1 - 1/(2\beta)]\{x''\}_{i,n-1}$$
$$(\{x\} = \{u\} \text{ or } \{q\}) \tag{8.6b}$$

where $\mu = \gamma/\beta$ and $\beta$ and $\gamma$ are two parameters defined below. Substitution of (8.6a,b) into (8.2) and (8.3e) at time $t = t_n$ yields the recurrence in form of (8.5a–d), where

$$[a]_i \equiv [a^*]_i + [k]_i \tag{8.7a}$$

$$\{b\}_{in} = \{p(t_n)\}_{in} + [a^*]\{u^*\}_{in} - [c]\{u'^*\}_{in} - [m]\{u''\}_{i,n-1} \tag{8.7b}$$

$$[A_{\mathrm{C}}] \equiv [A_{\mathrm{C}}^*] + [K_{\mathrm{C}}] \tag{8.7c}$$

$$\{B_{\mathrm{C}}\}_n = \{P(t_n)\}_n + [A_{\mathrm{C}}^*]\{q^*\}_n - [C_{\mathrm{U}}]\{q'^*\}_n - [M_{\mathrm{U}}]\{q''\}_{n-1} \tag{8.7d}$$

with

$$[a^*]_i = [m]_i/(\beta\Delta t^2) + \mu[c]_i/\Delta t \quad (\mu = \gamma/\beta) \tag{8.7e}$$

$$[A_{\mathrm{C}}^*] = [M_{\mathrm{U}}]/(\beta\Delta t^2) + \mu[C_{\mathrm{U}}]/\Delta t \tag{8.7f}$$

$$\{u^*\}_{in} = [B]_i\{q^*\}_n, \quad \{u^{*2/3}\}_{in} = [B]_i\{q^{*2/3}\}_n \tag{8.7g}$$

$$\{q^*\}_n = \{q\}_{n-1} + \{q'\}_{n-1}\Delta t + \{q''\}_{n-1}\Delta t^2/2 \tag{8.7h}$$

$$\{q'^*\}_n = \{q'\}_{n-1} + \{q''\}_{n-1}\Delta t \tag{8.7i}$$

The $\gamma$ and $\beta$ in these expressions are parameters to be assigned. Two sets of $\beta$ and $\gamma$ that are especially popular are (1) $\beta = 1/4$ and $\gamma = 1/2$ and (2) $\beta = 1/6$ and

$\gamma = 1/2$, corresponding to constant acceleration and linear acceleration assumption cases in formulating the integration scheme.

## Fixed-Step Implicit Central Difference (ICD) Scheme-Based Element Recurrence Equations

In this scheme (cf. Burnett 1987),

$$\{x'\}_{n-1} = (\{x\}_n - \{x\}_{n-2})/(2\Delta t) \qquad (\{x\} = \{u\}_i \text{ or } \{q\}) \tag{8.8a}$$

$$\{x''\}_{n-1} = (\{x\}_n - 2\{x\}_{n-1} + \{x\}_{n-2})/(\Delta t)^2 \tag{8.8b}$$

Again, substitution of (8.8a,b) into (8.2) and (8.3e) yields recurrence equations in the form of (8.5a–d), with

$$[a]_i = [m]_i/\Delta t^2 + [c]_i/(2\Delta t) \tag{8.9a}$$

$$\{b\}_{in} = \{p\}_{in} - [k]\{u\}_{i,n-1} + [c]\{u\}_{i,n-2}/(2\Delta t)$$

$$+ [m]_i(2\{u\}_{i,n-1} - \{u\}_{i,n-2})/\Delta t^2 \tag{8.9b}$$

$$[A_C] = [M_U]/\Delta t^2 + [C_U]/(2\Delta t) \tag{8.9c}$$

$$\{B_C\}_n = \{P\}_n - [K_U]\{q\}_{n-1} + [C_U]\{q\}_{i,n-2}/(2\Delta t)$$

$$+ [M_U](2\{q\}_{n-1} - \{q\}_{n-2})/\Delta t^2 \tag{8.9d}$$

## Implicit Nodal Recurrence Equations of Motion

From (8.3d) and (8.5a–e), the nodal recurrence equations are obtained as

$$\{F(\{q\})\}_n = \{Q\}_n \tag{8.10a}$$

where

$$\{F(\{q\})\}_n = \sum_{i=1,N_E} [B]_i^T \{f_a(\{u\})\}_{in} + \{F_a(\{q\})\}_n \equiv [A]\{q\}_n$$

$$\{Q\}_n = \sum_{i=1,N_E} [B]_i^T \{b\}_{in} + \{B_C\}_n \tag{8.10b}$$

with

$$\{f_a(u)\}_{in} = [a]_i\{u\}_{in}, \quad \{F_a\}_n = [A_C]\{q\}_n$$

$$[A] = \sum_{i=1,N_E} [B]_i^T [a]_i [B]_i + [A_C] \tag{8.10c}$$

and $u_{in} = \{u\}_{in} = [B]_i\{q\}_n$ (cf. (8.1)). It should be noted that, although the expression for the assembled global master structural coefficient matrix, $[A]$, is given here as (8.10c), it is not actually assembled. Rather, only the diagonal elements of this matrix are assembled nodal-wise to be used as the diagonal preconditioner when using the diagonally preconditioned conjugate gradient (DPCG) method (cf. Golub Van Loan 1989 for PCG method).

For a given set of initial velocity and displacement vectors, (8.4), the initial acceleration vector, $\{q_o''\}$, is calculated from the ordinary differential equations of motion, (8.3) with $\{f\}_i$ given by (8.2), i.e.,

$$\sum_{i=1,N_E} [B]_i^T \{f_m(u_n'')\}_i + \{F_{C_m}\} = \{B_C^*\}_n + \sum_{i=1,N_E} [B]_i^T \{b^*\}_{in} \qquad (8.11a)$$

or

$$[M]\{q''\}_n = \{P^*\}_n \qquad (8.11b)$$

where

$$\{f_m(u_n'')\}_{in} = [m]_i \{u''\}_{in} = [m]_i [B]_i \{q''\}_n$$

$$\{F_{C_m}\}_n = [M_C]\{q\}_n$$

$$\{b^*\}_{in} = \{p(t_n)\}_i - ([c]_i \{u'\}_{in} + [k]_i \{u\}_{in})$$

$$\{B_C^*\}_n = \{P(t_n)\} - ([C_U]_i \{q'\}_n + [K_U]\{q\}_n)$$

$$[M] = \sum_{i=1,N_E} [B]_i^T [m]_i [B]_i + [M_C]$$

$$\{P^*\}_n = \{B_C^*\}_n + \sum_{i=1,N_E} [B]_{in}^T \{b^*\}_{in} \qquad (8.11c)$$

## SOLUTION PROCEDURES IN AN MPP ENVIRONMENT

### Virtual Processor (Logical Unit) Assignment and MPP Computational Strategies

The following strategies can be used:

• Element-wise Processor Assignment/Computational Strategy — In this strategy, one allocates the task of computing and storing the element quantities (matrices, forces, etc.) for each element of the finite element mesh to a virtual processor (VP) of element VP (E-VP) set and nodal quantities (nodal displacements, forces, solution of equations of motion, etc.) for each nodal point to a VP of nodal (N-) VP set with $N_E$ and $N_N$ denoting the numbers of beam elements and nodal points in the FE structural model. Thus, there are total numbers of $N_E$ and $N_N$ VPs in the E-VP and N-VP sets, respectively. The N-VP set and the E-VP set use the same group of $N_G$ VPs, but otherwise are independent, where $N_G = \max(N_E, N_N)$. This processor assignment strategy permits a maximum number of multiple sets of independent response analyses to be performed simultaneously in parallel. Such a multiple analysis strategy can be used in structural (or structural-related interdisciplinary) optimization, design sensitivity analysis, frequency response analysis, etc., to greatly accelerate the numerical computations. This strategy is particularly efficient for a data parallel computational paradigm on an MPP computer for the case in which the VP ratios are small and is used in the CM–DYNASTRAN code used below in generating numerical data. Other logical unit assignment strategies used elsewhere by other researchers are the unassembled (element) nodal point-wise VP assignment strategy by Mathur et al. (1989), and the purely element-wise logical unit assignment strategy of Farhat et al. (1990) and Belytschko et al. (1987), in which the all nodal quantities adjacent to a structural element are calculated and stored in that element (E)-VP. The former is efficient in a single structural analysis if physical processors are at least twice as many as the number of structural elements (which is usually the case for use of the CM-2 where the number of processors, $N_p$, may be up to 65 536, but may not be for the CM-5 with $N_p < 1024$). A disadvantage in the latter strategy, compared with the independent assignment of E-VP and N-VP set strategy, is that it requires additional processor memory to store redundant $N_{en} - 1$ duplicated nodal data, where $N_{en}$ is the number of element (unassembled) nodal points.

• Substructure-wise Processor Assignment/Domain Decomposition Computational Strategy — In this strategy, a structure is subdivided into an $N_S$ number of equal or nearly equal size substructures (sections of structure) to have more-or-less equal computational loads for each substructure. The data calculational/storage task for each substructure is assigned to one processor. A number of different numerical procedures can then be used to assemble and solve (for transient response) the equations of motion using $N_S$ processors. The most direct implicit procedure similar to that outlined above is as follows: after calculation of substructure element forces, they are properly summed up to obtain the internal force portion nodal forces, which are needed to be balanced with the applied nodal forces via an iterative method. In this procedure, only those element forces associated with the substructure boundary nodal points need to be sent to different processors, i.e., all those intra-nodal quantities are calculated/treated in the same processor as those of substructure element quantities. An alternative procedure is the one based on a substructuring method (e.g., Craig 1985) in which the full-order equations of motion are first reduced to those of substructural interface displacements. These reduced equations of motion can then be solved by using an iterative method by using multiple processors.

Domain decomposition methods for general parallel computation and substructure methods for parallel structural analysis have been active areas of research. Some relevant recent studies on these subjects can be found in the work of Farhat (1993); Keyes et al. (1995); Khan et al. (1993); and Storaasli et al. (1987).

## Numerical Solution Steps

In view of the sparsity of the FE method-based semidiscretized equations and, thus, the fully discretized linear system of equations of motion, (8.10b), it is advantageous to solve the latter via an iterative (implicit) method in a parallel, distributed memory computational environment. If the coefficient matrix $[A]$ given in (8.10c) is positive definite (e.g., in the case of collocated control case), it is particularly efficient to use the preconditioned conjugate gradient (PCG) method. Otherwise, other iterative methods, such as a generalized PCG method (Barrett et al. 1994), can be used. The computational procedures using the fully implicit algorithms are as follows:

*Step 1.* At the beginning of numerical integration at $t = t_o$, calculate the $[a]_i$ matrices concurrently by using the $N_E$ element VPs. Also calculate the diagonal preconditioner

$$[A_d] \equiv \text{diag}([A]) = \sum_{i=1,N_E} [B]_i^T \text{diag}([a]_i)[B]_i + \text{diag}([A_C]) \qquad (8.11d)$$

and store only the diagonal elements in the N-VP set node-wise (i.e., 6 and 3 diagonal elements per N-VP set processor for the 3-D frame and truss analysis cases, respectively) if the diagonal PCG (DPCG) method is used. Also calculate the initial acceleration vector $\{q_0''\}$ by using (8.4) and (8.11a) or (8.11b) by directly solving them if a lumped mass (LM) approach is used and by using the DPCG method if the mass matrix $[M]$ is nondiagonal.

*Step 2.* At time $t_n$, the unknown $\{q\}_n$ is fairly accurately estimated by one of the following two formulas:

- The second-order Taylor series expansion formula (designated as A-type trial vector)

$$\{\tilde{q}\}_n = \{q^*\}_n = \{q\}_{n-1} + \{q'\}_{n-1}\Delta t + \{q''\}_{n-1}\Delta t^2/2 \quad \text{(cf. (8.7e))} \quad (8.12a)$$

- The Explicit Central Difference (ECD) formula (designated as B-type trial vector)

$$\{\tilde{q}\}_n = \{q\}_{n-2} + 2\Delta t\{q'\}_{n-1} \quad (8.12b)$$

Because of dual use of $\{q^*\}_n$ in calculating the $\{b\}_n$ vector in (8.7b) (via the first of (8.7g)) and (8.7d) as the initial trial vector given by (8.12a), the NB scheme-based algorithm set based on the A-type trial vector should be an efficient algorithm set. The B-type trial vector, on the other hand, involves fewer computational operations than the A-type and, thus, is more suitable in conjunction with using the ICD scheme-based algorithm set. Also, because a sufficiently accurate estimate of the nodal displacement vector can be obtained by using (8.12a) or (8.12b) by appropriately choosing the size of $\Delta t$, (8.12a) or (8.12b) can be viewed as a predictor of $\{q\}_n$.

*Step 3.* This estimated (predicted) displacement vector, $\{\tilde{q}\}_n$, is then used as the initial trial vector in an iterative method (e.g., the DPCG method given below) to solve the recurrence equations, (8.10a), implicitly (by directly calculating force vectors) to obtain an improved solution $\{q\}_n$. Because $\{\tilde{q}\}_n$ is fairly accurate to begin with, only a small number of iteration cycles are usually needed in the use of an iterative method, such as the DPCG method, to obtain a convergent solution to a (or a set of) specified error norm value(s), $E_r$, particularly if a sufficiently small integration step size is used. Thus, the iterative solution method of recurrence equations, such as the DPCG method, can be viewed as a corrector method. The initial trial vector $\{\tilde{q}\}_n$ is used initially to concurrently calculate the element force vectors $\{f_a\}_i$ and $\{b\}_i$ in (8.10a) in the E-VP set. These force vectors are then sent to appropriate nodal VPs to assemble (properly sum up) the equations of motion in vector form as shown in (8.10a). The force residuals

$$\{r\}_n \equiv \{F(q_n)\} - \{B^*\}_n \quad (8.13a)$$

(i.e., the difference between the left- and right-hand-side quantities of (8.10a)) are then computed, and the iterative solution (e.g., DPCG) algorithm is used to obtain a new trial vector of $\{q\}_n$. The above computational steps are repeated until an error norm (or a set of) is (or are) satisfied. The following solution convergence test criterion is usually used:

$$(e_r)_n \equiv (\{r\}_n^{\mathrm{T}}\{r\}_n/\{B^*\}_n^{\mathrm{T}}\{B^*\}_n)^{1/2} \leq E_r \quad (8.13b)$$

*Step 4.* After displacement vector, $\{q\}_n$, is computed, the velocity and acceleration vectors are computed by the following expressions:

- For the NB Scheme-Based Algorithm Case

$$\{q'\}_n = \mu(\{q\}_n - \{q\}_{n-1})/\Delta t + (1 - \mu)\{q'\}_{n-1} + (1 - \mu/2)\Delta t\{q''\}_{n-1} \quad (8.14a)$$

$$\{q''\}_n = (\{q\}_n - \{q\}_{n-1} - \{q'\}_{n-1}\Delta t)/(\beta\Delta t^2) + [1 - 1/(2\beta)]\{q''\}_{n-1} \quad (8.14b)$$

- For the ICD Scheme-Based Algorithm Case

$$\{q'\}_{n-1} = (\{q\}_n - \{q\}_{n-2})/(2\Delta t) \quad (8.15a)$$

$$\{q''\}_{n-1} = (\{q\}_n - 2\{q\}_{n-1} + \{q\}_{n-2})/(\Delta t)^2 \quad (8.15b)$$

*Step 5.* Then, the next integration step computations (i.e., increase $n$ by 1) are performed by repeating steps 2 through 4 until $t = t_f$. At $t = t_f$ (last time point), for the ICD-scheme based algorithm case only, the final velocity $\{q'\}_f$ is calculated by

$$\{q'\}_f = \{q'\}_{f-2} + 2\Delta t\{q''\}_{f-1} \quad (8.16)$$

while the final $\{q''\}_f$ is calculated in the same manner as that of $\{q''_0\}$ in step 1.

It should be noted that the NB numerical integration scheme is numerically unconditionally stable, while the ICD numerical integration scheme is only numerically conditionally stable. The stable integration step size for an undamped system is

$$\Delta t \leq 2/\omega_{\max} \quad (8.17)$$

where $\omega_{\max}$ is the maximum circular eigenfrequency of the undamped system. Belytschko (1983) formulated a method based on structural element properties to calculate an upper bound frequency that can be used in setting stable integration step size according to (8.17).

## DPCG Method-Based Recurrence Equation Solution Procedure

At the typical $n$-th integration step at time $t = t_n$, the DPCG procedure, with the omission of brackets and braces for vector and matrix quantities for brevity, is as follows:

(1) Initialization $k = 0$ ($k$ = No. of iterations)
- Assume $q_n^{(o)} = q_n$
- Calculate $r^{(o)} = Q_n - F_n^{(0)}$ ($F_n^{(k)} = F(q_n^{(k)})$) (cf. (8.10a))

(2) Iteration $k \geq 1$
   Calculate $e_r^{(k)} = [r^{(k)\mathrm{T}}r^{(k)}/Q_n^{\mathrm{T}}Q_n]^{1/2}$
   while $e_r^{(k)} > E_r$ ($E_r$ = given error limit)
      solve $M^{*(k)}z^{(k)} = r^{(k)}$ ($M^{*(k)}$ = preconditioner = $[A_d]$)
      $\rho^{(k)} = r^{(k)\mathrm{T}}z^{(k)}$
      $k = k + 1$
      if $k = 1$
         $p^{(1)} = z^{(o)}$
      else
         $\beta^{(k)} = \rho^{(k-1)}/\rho^{(k-2)}$
         $p^{(k)} = z^{(k-1)} + \beta^{(k)}p_n^{(k-1)}$

```
end if
```
$$\gamma^{(k)} = F(p^{(k)}) \ (F \text{ given by the first of } (8.10b) \text{ with } q \text{ substituted with } p^{(k)})$$
$$\alpha^{(k)} = \rho^{(k-1)}/p^{(k)\mathrm{T}}\gamma^{(k)}$$
$$q_n^{(k)} = q_n^{(k-1)} + \alpha^{(k)}p^{(k)}$$
$$r^{(k)} = r^{(k-1)} - \alpha^{(k)}\gamma^{(k)}$$
```
end
```
$$q_n = q_n^{(k)}$$

All inner products of various vectors in the above procedure are performed node-wise in nodal VP set processors by first performing the subvector inner products in each of $N_N$ processors and then summing up the subvector inner product values intra- and inter-processor wise via a data reduction procedure.

## SEMI-IMPLICIT ALGORITHM SET-BASED SOLUTION PROCEDURES

If an explicit numerical integration scheme, such as the explicit central difference (ECD) or the fourth-order Runge-Kutta (RK4) scheme, is used in conjunction with the lumped mass approach, the recurrence equations are completely uncoupled and, thus, can be solved for the unknown displacement vector explicitly, i.e., there is no need to use a linear algebraic equation solver, such as the DPCG method for the unknown displacements. Two such semi-implicit algorithm-based solution procedures are given below. As for the processor assignment and MPP computational strategies, similar strategies to those using the fully implicit computational algorithms can be used.

### Numerical Solution Procedure Based on the ECD/Lumped Mass (LM) Method-Based Semi-Implicit Algorithm Set

An alternative set of recurrence equations, based on the ECD scheme, is as follows (cf. Cheney and Kincaid 1985):

$$\{q'\}_{n-1/2} = \{q'\}_{n-3/2} + \Delta t\{q''\}_{n-1} \tag{8.18a}$$

$$\{q\}_n = \{q\}_{n-1} + \Delta t\{q'\}_{n-1/2} \tag{8.18b}$$

$$\{q'\}_n = \{q'\}_{n-2} + 2\Delta t\{q''\}_{n-1} \tag{8.18c}$$

$$\{q''\}_n = [M]^{-1}(\{P\}_n - \{F_{\mathrm{V}}(q'_n)\} - \{F_{\mathrm{D}}(q_n)\}) \tag{8.18d}$$

where $[M]$ is the diagonal mass matrix and $\{F_{\mathrm{V}}(q'_n)\}$ and $\{F_{\mathrm{D}}(q_n)\}$ are the velocity- and displacement-dependent nodal forces, which are obtained from their adjacent element end forces (calculated in the E-VP set) by properly summing up these forces. At the typical time point $t = t_n$, the mid-point velocity vector, $\{q'\}_{n-1/2}$ is calculated first by (8.18a). The displacement and velocity vectors at step n are then calculated from (8.18b,c). The acceleration vector, $\{q''\}_n$, is then calculated from (8.18d). In particular, if the mass matrix is lumped, then the $[M]$ matrix is diagonal, and (8.18d) can be easily and very efficiently solved for $\{q''\}_n$ even for the slightly damped case. In an MPP environment, the element lumped mass matrices, as well as the various element forces, are calculated in the E-VPs. Then the diagonal elements of the system diagonal mass matrix are assembled in the N-VPs (where each N-VP, for the 3-D motion case, contains six diagonal matrix elements associated with the nodal point assigned to the N-VP) in a manner similar to assembling the diagonal preconditioner matrix (cf. (8.11d)). The other element forces are concurrently computed in the E-VPs in parallel, and the nodal forces are assembled in the N-VPs; i.e., via use of the implicit force calculation/equations-of-motion assembly procedure described previously. The numerically stable $\Delta t$ is that given by (8.17). This set of semi-implicit algorithms is very efficient for the undamped or slightly damped system case. In general, it will sooner or later run into a numerical instability problem in the finitely damped system case due to exclusion of the damping force term on the left-hand side of recurrence equations given by (8.5d), (8.9a–d), and (8.10a–c).

## Numerical Solution Procedure Using the Fourth-Order Runge Kutta (RK4)/Lumped Mass Method-Based Semi-Explicit Algorithm Set

It is more convenient to put the equations of motion in the following state vector form in order to describe the algorithms:

$$\{x'\} = \{f(x), t\} \tag{8.19a}$$

where

$$\{x\} = \{x_1^T, x_2^T\}^T = \{q^T, q'^T\}^T \tag{8.19b}$$

$$\{f\}\left\{\begin{matrix} f_1 \\ f_2 \end{matrix}\right\} = \left\{ \frac{x_2}{[M]^{-1}(\{P\}_n - \{F_V(q'_n)\} - \{F_D(q_n)\})} \right\} \tag{8.19c}$$

The RK4 scheme-based recurrence equations can then be written as (Cheney *et al.* 1985):

$$\{x(t + \Delta t)\} = \{x(t)\} + (\Delta t/6)(\{r_1\} + 2\{r_2\} + 2\{r_3\} + \{r_4\}) \tag{8.20a}$$

where

$$\{r_1\} = \{f(t, x)\}, \quad \{r_2\} = \{f(t + \Delta t/2, x + r_1 \Delta t/2)\}$$

$$\{r_3\} = \{f(t + \Delta t/2, x + r_2 \Delta t/2)\}$$

$$\{r_4\} = \{f(t + \Delta t, x + r_3 \Delta t)\} \tag{8.20b}$$

The calculation procedure for the present set of algorithms is essentially the same as the preceding set and, thus, will not be repeated here.

The RK4 integration scheme appears to be particularly popular with applied mathematicians and control analysts due to its simplicity. The advantage of the RK4 scheme-based set of algorithms is that, unlike the ICD scheme-based algorithm set whose recurrence equation is given by (8.5d), (8.9a–d) and (8.10a–c), the consistent damping matrix can be used in conjunction with the lumped mass (LM) approach in a very efficient manner without actually needing to solve a set of coupled linear algebraic equations. This is particularly important in the slightly to moderately damped system solution case in which the semi-implicit ECD scheme-based algorithms are numerically unstable. Its major disadvantage is that it generally requires increasingly smaller numerically stable integration time-step size as the system damping magnitude increases, so that the overall computational time required for the response solution in a given response time duration may become very large if the system damping level is finite or large.

For such systems (e.g., most of CSI systems), use of a set (particularly the NB scheme-based one) of fully implicit algorithms presented above is much more efficient than the RK4 scheme-based one.

## MASS LUMPING TECHNIQUES

Mass lumping is usually performed on the element mass matrix $[m]_i$ in (8.2) prior to assembly. Given below are the four lumped element mass (LEM) matrices, of which the second is newly formulated and the other three are existing. The diagonal elements of these LEM matrices can be written as

$$\text{diag}\,[m]_e = (\rho A/2, \rho A/2, \rho A/2, \rho I_x L/2, k\rho AL^3, k\rho L^3,$$
$$\rho A/2, \rho A/2, \rho A/2, \rho I_x L/2, k\rho AL^3, k\rho L^3) \qquad (8.21)$$

These four LEM matrices differ only in the bending rotational mass inertia coefficient $k$.

- LEM matrix 1 (Burnett 1987): $k = 1/420$
- LEM matrix 2 (newly formulated): $k = 1/60$
- LEM matrix 3 (Belytschko 1983): $k = 1/24$
- LEM matrix 4 (classical one): $k = 0$

The LEM matrix 2 is obtained from the consistent mass matrix by imposing the condition of symmetric beam element translational and rotational displacements, which is equivalent to the first rotational bending mode of vibration for a symmetrically constrained beam element. As the mesh size becomes smaller and smaller, the element end rotational angles in the local coordinate system would approach the above condition and, thus, the solution is guaranteed to converge to the exact one. Numerical study results of several example problems show that this LEM matrix (No. 2) is superior to the other LEM matrices in yielding satisfactory results for both transient response and eigenvalue solutions for a simply-supported beam, as shown in Figure 8.2.

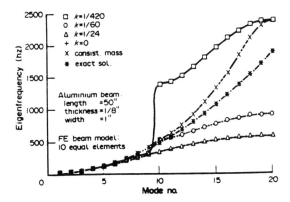

**Figure 8.2** Comparison of Eigenfrequency Solutions for a Simply-Supported Beam Based on Exact, Consistent, and Various Lumped Mass Models

## NUMERICAL DEMONSTRATION EXAMPLE PROBLEMS ON THE CONNECTION MACHINE (CM) COMPUTERS

The foregoing MPP algorithms/procedures have been implemented in both SIMD-type CM-2 and MIMD-type CM-5 model Connection Machine computers as CM-DYNASTRAN (DYNamic Adaptive STRucture ANalyis code). In this implementation, the element-wise processor (logical unit) assignment strategy is used. Before proceeding further, let's pause for a moment to briefly introduce various Connection Machine computer models in use.

### CM-Computers

The CM computer models currently in use include the CM-2, the CM-200, and the CM-5 models (cf. Table 8.1). The first two (older) models are massively parallel, SIMD (Single-Instruction-Multiple-Data) machines, while the last model is a mixed SIMD/MIMD (Multiple-Instruction-Multiple-Data) machine manufactured by Thinking Machines Corporation. The CM computers are available in modular form. That is, the CM-2 can have the configurations of 8K (8192), 16K, 32K, and 64K (65536) one-bit processors, while the CM-5 modules can be configured to yield $16 \times 2^n$ $(n = 1, \ldots, 5)$, i.e, 32 to 1024 64-bit processors. The active structural elements implemented are of multiaxially active, collocated, laminated piezoelectric beam and rod sensor/actuator element types whose element equations of motion are given by (8.2a) with the negative feedback control matrices (piezoelectric damping and stiffness matrices in this case) given by (8.2k). The iterative method used in solving the recurrence algebraic equations of motion is the DPCG (diagonally preconditioned conjugate gradient) method. This code was used in the transient response analysis of the two example problems given below. In all analyses, the lumped mass approach with $k = 1/60$ in (8.21) was used.

### Example 1: Transient Response Analysis of a Piezo Antenna Frame

Consider an adaptive antenna structure composed of both regular elastic beam elements (i.e., element Nos. 5–18) and special multiaxially active piezoelectric beam sensor/actuator pair elements (element Nos. 1–4) and subjected to time-dependent forces (a transverse force, $P_{5y}$, and a torsional moment, $M_{5x}$, at node 5) as shown in Figure 8.3a. The active elements are, by special design, free from cross-coupling among various force/deformation components of the negative velocity feedback type

**Table 8.1**  Some CM-2, CM-200, and CM-5 Characteristics.

| Connection Machine Model | Computer Architect. | Available Config. (no. processors) | Peak Performance (Giga-FLOPS) | |
|---|---|---|---|---|
| | | | Theoretical | Actually realized |
| CM-2 | SIMD* | 8K to 64K 1-bit | 20 | reg.: 2–4 record: 14 |
| CM-200 | SIMD* | Ditto | 40 | reg.: 5–10 record: 21 |
| CM-5 | SIMD/MIMD* | 32 to 1024 64-bit | 128 | — |

* SIMD, MIMD = single- and multiple-instruction-multiple-data, respectively

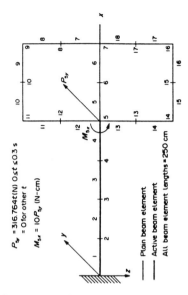

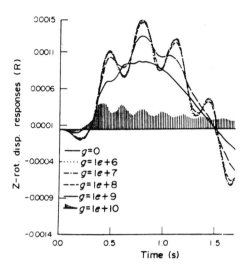

**Figure 8.3a** An Antenna Structure with Some Active Piezo Beam Elements

**Figure 8.3b** Z Rotational Displacement Responses for Node 9 Under Various Gain Values, $g = g_c$

(cf. (8.2k)). Presence of such special active beam elements induces a piezoelectric damping force term $[c_P]_i\{u'\}$ in (8.2b). In what follows, all numerical results are obtained by using the 18-element finite element model of Figure 8.3a. This problem was studied by Shieh (1993a, 1994b) to test various semi- and fully implicit algorithms sets on a (serial) Sun Workstation and the MPP CM-2. Here, the example problem is used in generating transient response results and performance data (wall-clock time and iteration cycles) on the CM-2 and CM-5 computers as functions of piezoelectric control gain parameter of sensor current amplifier, $g_c$ (V/A = volt/ampere), which is an indicator of the active damping level.

Some such performance data and response results are shown in Figure 8.3b and Table 8.2, respectively. In all analyses, a lumped mass matrix is used. The $\beta$ and $\gamma$ values used in the Newmark Beta (NB) scheme are: $\beta = 1/4$ and $\gamma = 1/2$, while the

**Table 8.2** Computational Performance Timing Data for Transient Response Analysis of a Piezo-Antenna Frame Using the NB/DPCG Algorithm Sets ($\Delta t = 1$ ms; $E_r = 1.\text{E-5}$ for All Cases).

| $g_c$ (V/A) | Duration Time (s) | Perf. Time Ratio | Avg. Iter. Cycles |
|---|---|---|---|
| 0 | 1 | 1 | 1.799 |
| 1.0E + 06 | 1 | 0.887 | 1.596 |
| 1.0E + 08 | 1 | 0.950 | 1.661 |
| 1.0E + 10 | 1 | 1.095 | 2.306 |
| 1.0E + 12 | 1 | 1.542 | 4.559 |
| 1.0E + 12 | 4 | 6.568 | 5.255 |

predictor used is the second-order Taylor formula given by (8.12a). Viscous damping is assumed to be absent, i.e., $[c_p] \neq [0]$, $[c_v] \equiv [k_p] \equiv [0]$ in (8.2b). Figure 8.3b shows the $z$-rotational displacement responses for Node 9 of the adaptive antenna structure under various g-values. This displacement response component is induced mainly by the bending deformations in beam element Nos. 1–4 resulting from applying the transverse force at node 5. As the control effort is increased upward from $g = 0+$, the corresponding active damping forces of the active beam elements increase. No appreciable damping effects on the transient response behavior are seen until $g \geq 1E + 10$ V/A.

The average iteration cycle values shown in Table 8.2 for obtaining the response solution to the error norm of $E_r = 1.e - 5$ (cf. (8.12b)) via the NB/DPCG methods and, thus, the wall-clock (WC) time, are seen to decrease initially and then to increase as g is increased from 0+ upward. Except for the two larger g-value cases shown, average cycle values are seen to vary between 1 and 2, even though an integration step size ($\Delta t$) as large as 1 ms is used. Use of a smaller $\Delta t$ value will result in a smaller average iteration cycle value, but the overall computational performance time may not decrease.

### Example 2: Transient Response Analysis of Open-Loop 3-D Building Frames

Consider a class of large, 3-D rectangular building frames (Figure 8.4a) subjected to suddenly applied constant nodal forces of 100 lbs at all nodal points on the $x = 0$ and $y = 0$ planes in the 30° polar angle direction for 0.01 s time duration. Figure 8.4b and Table 8.3 show the performance times for transient response analyses of different size structures using the NB/DPCG fully implicit algorithm set of the CM-DYNASTRAN code. Also shown are the corresponding timing data using the Newmark Beta integration scheme/LU matrix decomposition with the forward-backward substitution (NB/LU) method of MSC/NASTRAN code on the serial IBM RS-6000/560 model

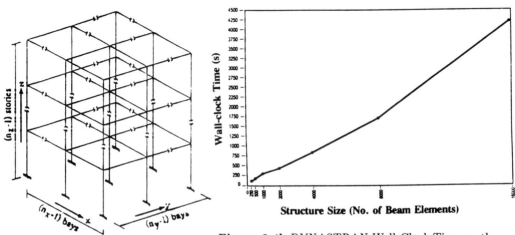

Figure 8.4a A Typical 3-D Frame Structure

Figure 8.4b DYNASTRAN Wall-Clock Time on the 32 Node CM-5 for Response Analysis (1000 Integration Steps)

**Table 8.3** Performance Timing Results for Transient Response Analyses of Large 3-D Building Frames Using the Newmark Beta (NB) Integration Scheme of CM-DYNASTRAN and RS6000-MSC/NASTRAN ($\Delta t = 1$ ms; $E_r = 1.e - 5$).

| | | | CM-DYNASTRAN code case (Using the NB/DPCG algorithm set) | | | | | | | RS-NAST.[a] | CM-DYNAST. speed-up factor vs. RS-NAST. | |
| Computer code and speedup factor (SF) → | | | av. no. of iters. | | | $N_P$-node cm2 wc time[b] | | 32-node cm5 wc time[b] | | | scaleup or mult. anal[d] | |
| No. of elems | No. of nodes | dyn. dof | 1k step case | 10k step case | $N_P$ | 1k step case (s) | 10k step case (hr) | 1k step case (s) | 10k step case (hr) | cpu hour (10k steps) | 1024-cm5 | 64k-cm2 |
|---|---|---|---|---|---|---|---|---|---|---|---|---|
| 255 | 100 | 480 | 1.587 | — | — | — | — | 110.7 | 0.34 | 0.21 | | |
| 502 | 270 | 1080 | 2.894 | — | — | — | — | 172.9 | 0.53 | 0.42 | | |
| 995 | 432 | 2160 | 1.71 | — | 8k | 1616 | 4.85 | 259.4 | 0.79 | 1.00 | 40.5 | 12.3 |
| 1990 | 792 | 4320 | 1.24 | — | 8k | 1548 | 4.58 | 422.3 | 1.29 | 2.59 | 64.2 | 18.1 |
| 3980 | 1512 | 8640 | 1.13 | — | 8k | 1649 | 4.73 | 829.5 | 2.53 | 7.06 | 89.3 | 23.8 |
| 7960 | 2952 | 17280 | 1.29 | — | 8k | 1530 | 4.74 | 1719 | 5.26 | 20.6[c] | 118 | 34.8 |
| 15920 | 5832 | 34560 | 1.62 | 1.97 | 16k | 2232 | 5.21 | 4202 | 12.53 | 69.1[c] | 176 | 54.3 |
| 32560 | 11664 | 69120 | 1.71 | 1.95 | 32k | 3553 | 5.95 | ← estimated | — | 318[c] | — | 106 |
| 63680 | 23112 | 138240 | — | — | 64k | | 7.03 | | — | 2233[c] | — | 318 |

a. The structural analysis timing results for 10K integration steps based on the serial IBM RS-X6000/560 model computer version of the MSC/NASTRAN code via NB/LU methods.
b. The elapsed time shown is the double precision run wall-clock (wc) time.
c. Estimated via least-square curve fitting techniques.
d. Speedup factor for the assumed case of linearly scaling-up to the 1024 node CM-5 or of machine full of concurrent multiple analyses using the 64K-processor CM-2.

computer. In the CM-2 run cases, for the structures with beam elements not exceeding 8K, for which the 8K processor CM-2 was used, the WC (wall-clock) times are seen to be virtually independent of the structural size. For the 15920 and 32560 beam element structure cases (which used 16K and 32K processors, respectively) only small increases in WC times from those of the smaller size structure cases are seen. As a result of this, the CM-2 is seen to be particularly efficient in single analysis of large-size and concurrent multiple analyses (required in structural optimization, etc.) of any size structural problems. Its speedup factor vs. MSC/NASTRAN on the RS-6000 is estimated to be 318 for the 63680 beam element case based on extrapolated timing results via the linear least-square-curve fitting technique.

For the 32 node CM-5 run cases, the WC times are seen from Figure 8.4b to be approximately a linear function of the number of beam elements, with a favorable rate of increase in WC time that is approximately 80% that of beam element. The estimated CM-DYNASTRAN speedup factors on the commercially available 1024 node CM-5 over the RS6000 computer-based MSC/NASTRAN are obtained by using the 32-node CM-5 run results under the assumption that the 1024 node CM-5 is 32 times (= processor ratio) as fast as the 32 node CM-5 and are also listed in Table 8.3. For the 7960 beam element structure case, the speedup factor is seen to be 118 compared with 35 for the case of DYNASTRAN on the 64K node CM-2.

Table 8.4 shows the efficiency of using the type-A predictor given by (8.12a). The speedup factor versus the no predictor case is seen to be slightly over 1.6 for the structure sizes shown. Thus, it is efficient to use the predictor.

The timing results for the transient response analysis are as follows for 1000 steps of integration for the undamped 255 element structure case, based on the fully implicit NB/DPCG and semi-implicit ECD/LM (lumped mass) and RK4/LM algorithm sets:

$$\text{NB/DPCG: 110.7 s} \quad \text{ECD/LM: 38.5 s} \quad \text{RK4/LM: 153.5 s}$$

Thus, the ECM/LM algorithm set is seen to be three and four times as fast as those of NB/DPCG and RK4/LM, but its maximum stable integration time step size is only one-third that of the RK4/LM set. Therefore, the computational efficiency for these two semi-implicit algorithms sets in performing transient response analysis is comparable, assuming that the numerical integration stability condition dictates the size of the integration step. Because such a numerical stability condition usually dictates in

**Table 8.4** Transient Response Analysis Timing Results Based on the NB/DPCG Method-Based Algorithm Set with and without Type-A Predictor, (8.12a) (1000 Integration Steps; $\Delta t = 1$ ms).

| case | 15920 beam elems. on 16k-cm2 | | | | 995 beam elems. on 32-cm5 | | | |
|---|---|---|---|---|---|---|---|---|
| predictor $(q_n^{\sim})$ type | tot. iter. cycs | process time (s) | | | tot. iter. cycs | process time (s) | | |
| | | pre-integ. | integ. | total | | pre-integ. | integ. | total |
| $q_n^{\sim} = 0$ | 6002 | 614 | 3055 | 3669 | 5175 | 1.856 | 428.6 | 430.4 |
| $q_n^{\sim} = q_n$ | — | — | — | — | 4056 | 1.653 | 365.9 | 367.5 |
| 8.12a | 1640 | 636 | 1602 | 2234 | 1708 | 1.640 | 257.8 | 259.4 |
| speedup | 3.66 | — | 1.90 | 1.64 | 3.03 | 1.13 | 1.66 | 1.66 |
| factor | — | — | — | — | 2.37 | 1.01 | 1.42 | 1.42 |

setting the integration time step size in use of the semi-implicit algorithm set (while such step size is determined solely on the computational accuracy basis in use of the NB integration scheme), the NB/DPCG scheme is usually preferable over the other algorithm sets presented in this chapter. This is particularly true in a finitely damped system for which the ECD or even the RK4 explicit integration scheme is virtually useless, due to the requirement for an excessively small integration time step size to ensure numerical stability. For such a system, use of a fully implicit algorithm set (such as that based on the NB/PCG or ICD/PCG methods, particularly the former (cf. Shieh 1993a)) should be used.

## 8.3 PLACEMENT OPTIMIZATION TECHNIQUES OF ACTIVE ELEMENTS (SENSORS/ACTUATORS) FOR STATIC STRUCTURAL SHAPE CONTROL

*PROBLEM FORMULATION*

Most of regular and adaptive static structural optimization problems can be stated as follows: Optimization is to be performed for the objective function $F(\{x\})$ such that

$$Z = F(\{X\}) \rightarrow \text{Minimum} \tag{8.22a}$$

with respect to design variables $\{X\}$ subject to the constraints

$$\{X_\mathrm{L}\} \leq \{X\} \leq \{X_\mathrm{U}\} \tag{8.22b}$$

$$\{h(\{X\})\} \leq \{0\} \tag{8.22c}$$

where $\{h\}$ is the constraint function vector, and subscripts L and U stand for "lower" and "upper" bounds, respectively. Many subclasses of regular and adaptive structural optimization problems exist. Only the following active element placement/ specification design problems for static structural shape control will be considered here.

The problem is to minimize the structural strain energy or maximum displacement with respect to the number, location, and gain value of active elements under stress and gain constraints with the sum of absolute gain values specified to be a constant value. In mathematical terms, the optimal design problem can be expressed as:

$$Z = F(\{X\}) = \text{Strain energy or maximum displacement} \rightarrow \text{Min.} \tag{8.23a}$$

$$\{X\} = \{\{G\}|N_\mathrm{G}\} \tag{8.23b}$$

$$G_{j\mathrm{L}} \leq G_j \leq G_{j\mathrm{U}} \quad (j = 1, \ldots, N_\mathrm{E}) \tag{8.23c}$$

$$\sum_{j=1,N_\mathrm{E}} |G_j| = G^* \tag{8.23d}$$

$$1 \leq N_G \leq N_\mathrm{E} \tag{8.23e}$$

$$\text{Max.}|s_i| \leq s_M (i = 1, \ldots, N_\mathrm{E}) \quad (\{s\} = \text{stress vector}) \tag{8.23f}$$

where $N_\mathrm{E}$ is the number of finite structural elements in the structure, $\{G\} = \text{diag}([g_k])([g_c] = \{0\}, \text{cf. } (8.23f))$ is the $(N_\mathrm{E} \times 1)$ control gain vector of active elements with the zero-value denoting that the beam element is a regular (passive) element, $G^*$

is a prescribed total gain (voltage) value, $N_G$ is the number of active elements, and $s_M$ is the allowable stress.

## A STOCHASTIC GLOBAL OPTIMIZATION ALGORITHM, IIGO

The nonlinear optimizer, IIGO or IGO/Koosh (Shieh et al. 1991), is a vastly improved and massively parallelized version of the Integral Global Optimization (IGO) algorithm (Chew and Zheng, 1988). The unconstrained IGO algorithm requires three quantities: a search region, a current average, and a list of the points currently under consideration. To begin with, one first specifies an initial search region and establishes an initial average. The search region is then sampled at random. Points at which the objective function are below the current average are added to the list of points, called "good" points. When the length of this list grows above some threshold, the average value of the objective function is computed for points on the list. Points at which the value of the objective function is above this new average are removed from the list. The new search region is chosen to be an $n$-dimensional rectangle slightly larger than required to hold all points on the current list of points. Now the iteration is ready to begin again. When many points have been examined but few points have been added to the list of points under consideration, one goes to another mode that is called Koosh search, which constitutes an improvement to the original IGO algorithm. In this mode, small neighborhoods of the best points on the current list are searched. This is done only long enough to get enough good points on the list to be able to make meaningful updates to the average and search region. The search is then returned to the top level (IGO) algorithm. This improvement greatly decreases the number of points and, thus, run-times required for obtaining good optimization results (cf. the second example problem given below).

The IGO/Koosh algorithm was implemented on the CM-2 and CM-5 Connection Machine as an OPTIM code module. The large numbers of processing elements make the CMs a natural environment for a stochastic method like the IGO/Koosh search. In the implementation, all or fractions of available processors simultaneously generate a random point in the search region and evaluate the objective function at that point. Processors for which the value of the objective function is below the current upper bound, $B$, then send their points to a buffer which represents the set of good points, $G$. If the buffer $G$ becomes full, or if the number of points sampled exceeds a threshold, a new $B$ is computed and the points in $G$ with $F(\{X\})$ less than the new $B$ are compressed to the beginning of the buffer. When many random points are sampled but few good points are found, the code switch to the Koosh search. This means that the search region is generated in the neighborhood of the best points in $G$. Otherwise, the algorithm was the same.

## OPTIMAL PLACEMENT/SPECIFICATION PROCEDURE FOR COLLOCATED, ACTIVE PIEZOELECTRIC ELEMENTS

The optimization procedure for the active element placement/specification problem formulated consists of the following steps:

  *Step 1.* Assign the data calculational/storage tasks of or associated with each structural element of FE mesh to each processor; then estimate the number ($N_M$) of machines full of $N_A = N_P/N_E$ numbers of simultaneous independent structural analyses required for the optimization, where $N_P$ is the number of physical or virtual processors. The total number of structural analyses are then $N_{AT} = N_M \times N_A$.

*Step 2.* Randomly select a single integer number $N_{Gj}$ for each ($j$-th) analysis ($j = 1, \ldots, N_E$) among integer number set $\{n\}^T = \{1, \ldots, N_E\}$. Then, randomly generate $N_{AT}$ sets of $G_{ij}(G_L \leq G_{ij} \leq G_U)(i = 1, \ldots, N_E; j = 1, \ldots, N_{AT})$. Rank these gain values for each analysis in the descending order and set to zero those gain values whose ranks are greater than $N_{Gj}$. The resulting gain value sets are designated as $G'_{ij}(i = 1, \ldots, N_E; j = 1, \ldots, N_{Gj})$, of which only $N_{Gj}$ are nonzero for the $j$-th analysis.

*Step 3.* At each cycle of iteration, satisfy the equality constraint of (8.23d) by multiplying the randomly chosen $G_i$ with $k_j = G^*/G'_j$, where $G'_j$ is the sum of $G'_{ij}$ over index $i$ from $i = 1$ to $i = N_E$, i.e., $G''_{ij} \equiv$ new $G'_{ij} = k_j G'_{ij}$. Use them in the structural analysis/reanalysis and optimization.

*Step 4.* Perform $N_M$ machines full of $N_A$-simultaneous structural analyses by using the $G''_{ij}$ values and PP multiple concurrent FE structural analysis code module for multiaxially active piezoelectric-type adaptive structures, CM-ASTRAND.

*Step 5.* Satisfy the stress constraint (and other constraints, if any) for each analysis by using a linear scaling design method or any other constraint treatment algorithm.

*Step 6.* Calculate $N_{AT}$ number of objective functions (strain energies, $E_j$, or maximum displacements), one each for each analysis.

*Step 7.* Use the IIGO search algorithm to find the average of the $N_{AT}$ number of objective functions (points); retain the good points, which are below the average value; and repeat steps 2–7 to narrow down the search region. During the optimization, if the number of good points are below a certain threshold value, one switches the random search mode from the global to the neighborhood regions of the good points. The search mode is then returned to the global mode when the number of good points are equal to or above the threshold number. The optimization procedure is terminated when the search region size measurement quantity satisfies certain small search region size tolerance criterion. The final set of objective functions and the associated design variables for the analysis that give minimum objective function value is then accepted as the optimal design set. The number, location, and values of nonzero gain values for this optimal design set are then the desired optimal active element placement/specification set.

## A CM-COMPUTER CODE CAPABILITY AND NUMERICAL EXAMPLES FOR ACTIVE ELEMENT PLACEMENT OPTIMIZATION

The foregoing computational procedure has been implemented into a previously developed, CM static FE method-based analysis code module for multiaxially active, piezoelectric type adaptive structures, the CM-ASTRAND (*A*daptive *STR*uctural *AN*alysis and *D*esign) code.

### Example 3: Piezo Antenna Shape Control Problem

The frame structural configuration, as well as the applied forces and a torsional moment, is shown in Figure 8.3a. The structure is symmetric with respect to the

$x$-axis so that the active beam elements also need to be symmetrically placed to maintain the structural symmetry. In the minimization of the total strain energy or the maximum displacement $(u_M)$ under the total gain value equality constraint of 4 Giga units, the following two cases of design variables are considered:

- Case 1 — Minimization with respect to active beam element locations and gain values under a given number of active beam elements, $N_G = 2, 3, 4, 6, 7,$ or 18.
- Case 2 — Minimization of $F$ with respect to the active beam element number, locations, and gain values.

The CM-5 computer/ASTRAND code-based optimal design results for these two subcases are shown in Table 8.5a for the strain energy minimization case and in Table 8.5b for the maximum displacement (max. u) minimization case. Comparison of the final displacements at these maximum displacement-based (MDB) and energy functional-based (EFB) optimum states are shown in Table 8.5c. From these tables, one can observe or conclude the following:

- As might be expected, for the cases of small, preassigned numbers $(N_G \leq 3)$ of active beam elements (with $G > 0$), the optimal active element locations are adjacent to the cantilever frame support, with the optimal gain value decreasing rapidly as the degree of proximity of active element location decreases because strains are largest in the immediate vicinity of the support. Note that $G = 0$ corresponds to passive element.
- For the case of $N_G = 4$, the optimal active beam elements selected are Nos. 1–3 and 6, for which the last element is impossible to be intuitively predicted as the fourth active element.
- For the case of $N_G = 6$, the six beam elements (Nos. 1–6) along the frame symmetry line are selected as the six optimal active elements, but the ranks of active element gain values in descending order are nonconsecutive, that is, 1, 2, 3, 6, 5, 4 for active elements 1–6, respectively.
- For the case of $N_G = 7$, for which the $N_G$ value actually used in the computer run is 8 but only seven active elements are selected due to imposition of adaptive structural symmetry condition, a somewhat surprising active element selection result is seen. The sixth beam element, which is selected in the $N_G = 4$ and 6 cases and is ranked as the 4th largest in gain values, now is not selected at all as an active element. Instead, element Nos. 12 and 13 are selected as the 6th and 7th ranked optimal active elements.
- For the fully active element placement/specification case of $N_G = 18$, the ranking order of $G$-values is seen to be different from the fewer active element number cases, except for the first three ranks. Therefore, for the preassigned active element number of $N_G > 3$, the method of ranking and elimination from the upper end number of active elements toward the fewer, preassigned number of active elements may not give true, optimal active element placement results.
- As mentioned previously, shown in Case 2 of Table 8.5a are the optimal design results of active element placement/specification in which the number of active elements, as well as their gain values and element locations (identification numbers), are treated as an optimization design variable. The present optimization methodology yields optimal $N_G = 3$ and other optimal results which are virtually in

**Table 8.5a** Strain Energy-Based Optimal Active Element Placement/Specification Design Results—Design Variables: Element Locations and Gains (G) for Case 1 and Number ($N_G$), Locations, and Gain for Case 2.

| Case→ | 1. Preassigned $N_G$ (no. of active element) case (G in Giga unit) | | | | | | | | | | 2. optimally obtained $N_G$ case | | 3 (part of sol.) | |
|---|---|---|---|---|---|---|---|---|---|---|---|---|---|---|
| $N_G^\dagger →$ | 2 | | 3 | | 4 | | 6 | | 7 | | 18 | | 3 | |
| Elem # | $G^\dagger$ | R | G | R | G | R | G | R | G | R | G | $R^\dagger$ | G | R |
| 1 | 2.46 | 1 | 2.252 | 1 | 1.933 | 1 | 2.065 | 1 | 1.875 | 1 | 1.924 | 1 | 2.231 | 1 |
| 2 | 1.54 | 2 | 1.322 | 2 | 1.406 | 2 | 1.206 | 2 | 1.256 | 2 | 0.961 | 2 | 1.333 | 2 |
| 3 | 0 | — | 0.426 | 3 | 0.634 | 3 | 0.583 | 3 | 0.434 | 3 | 0.698 | 3 | 0.436 | 3 |
| 4 | 0 | — | 0 | — | 0 | — | 0.010 | 6 | 0.173 | 4 | 0.088 | 5 | 0 | — |
| 5 | 0 | — | 0 | — | 0 | — | 0.023 | 5 | 0.100 | 5 | 0.216 | 4 | 0 | — |
| 6 | 0 | — | 0 | — | 0.027 | 4 | 0.113 | 4 | 0 | — | 0.010 | 10 | 0 | — |
| 7,18 | 0 | — | 0 | — | 0 | — | 0 | — | 0 | — | 0.006 | 11,12 | 0 | — |
| 8,17 | 0 | — | 0 | — | 0 | — | 0 | — | 0 | — | 0.003 | 15,16 | 0 | — |
| 9,16 | 0 | — | 0 | — | 0 | — | 0 | — | 0 | — | 0.019 | 6,7 | 0 | — |
| 10,15 | 0 | — | 0 | — | 0 | — | 0 | — | 0 | — | 0.003 | 17,18 | 0 | — |
| 11,14 | 0 | — | 0 | — | 0 | — | 0 | — | 0 | — | 0.004 | 13,14 | 0 | — |
| 12,13 | 0 | — | 0 | — | 0 | — | 0 | — | 0.081 | 6,7 | 0.017 | 8,9 | 0 | — |
| Energy | 36.63 | | 36.17 | | 36.38 | | 36.67 | | 38.36 | | 38.31 | | 36.01 | |
| $u_M$ (cm) | 3.142 | | 3.043 | | 3.044 | | 3.060 | | 3.139 | | 3.162 | | 3.042 | |

† $N_G$ = No. of active elements; $G$ = Gain; $R$ = Rank of gain values in descending order.

**Table 8.5b**  Maximum Displacement-Based Optimal Design Results for the Three Active Beam Element Cases.

| Quantity | Element No. | | | | Max. disp (cm) |
|---|---|---|---|---|---|
| | 1 | 2 | 3 | 4–18 | |
| Gain | 2.052 | 1.332 | 0.616 | 0 | 3.034 |
| Rank of Gain | 1 | 2 | 3 | — | |

**Table 8.5c**  Comparison of Final $y$-Displacements $(u_y)$ at the Maximum Displacement-Based (MDB) and Energy Function-Based (EFB) Optimization States.

| Node # | 1 | 2 | 3 | 4 | 5 | 6 | 7 | 8 | 9 |
|---|---|---|---|---|---|---|---|---|---|
| MDB $u_y$ | 0 | .1085 | .4093 | .8913 | 1.526 | 2.238 | 2.973 | 2.968 | 2.974 |
| EFB $u_y$ | 0 | .1121 | .4175 | .9008 | 1.536 | 2.248 | 2.984 | 2.979 | 2.985 |

| Node # | 10 | 11 | 12 | 13 | 14 | 15 | 16 | 17 | — |
|---|---|---|---|---|---|---|---|---|---|
| MDB $u_y$ | 2.241 | 1.513 | 1.516 | 1.546 | 1.572 | 2.300 | 3.034 | 2.998 | — |
| EFB $u_y$ | 2.251 | 1.523 | 1.526 | 1.556 | 1.582 | 2.310 | 3.044 | 3.009 | — |

complete agreement with those of the pre-assigned active element number of $N_G = 3$ in Case 1 of the same table. (Slight discrepancies in optimal gains and energy values of these two solution set results are due to the iteration nature of the solution and can be made arbitrarily small by appropriately making the iteration error or tolerance criterion (or criteria) smaller.) The maximum displacements $(u_M)$ for the open- and closed-loop cases are 5.77 cm and 3.04 cm, respectively.

- The optimal active element placement results for the preassigned active element number cases of $N_G = 3$ in Table 8.5a and 5b that are based on the energy and maximum displacement objective functions, respectively, are seen to be identical for the element locations and similar for the gain distributions. However, the gains for the former case are seen to be more concentrated in the active element immediately adjacent to the support than in those for the latter case.

- From Table 8.5c, one observes that the final $y$-directional displacements for the energy and max. displacement objective function-based optimization cases differ only slightly, i.e., they completely agree up to the first two digits.

- The CM-5 elapsed time required for obtaining the optimal results of active beam number, locations, and gain distribution values, i.e., Case 2 in Table 8.5a, is $2\frac{1}{2}$ minutes.

## 72-BAR Piezo-Truss Shape Control Problem

Figure 8.5 depicts a minimum weight-wise designed 72-bar passive truss structure with respect to member cross sections under two loading sets and the maximum stress and displacement constraints of 25 ksi and 0.25 inches, respectively. This passive truss design problem is used by Shieh (1993b) to evaluate the numerical performance of the IIGO algorithm in the MPP environment of the CM-2. In the present study, an optimal replacement design is performed of some passive members with active members of this

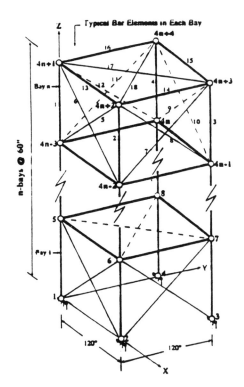

| Load set | Node | X | Y | Z |
|---|---|---|---|---|
| 1 | 4n+1 | 5000 | 5000 | -5000 |
| 2 | 4n+1 | 0 | 0 | -5000 |
| 2 | 4n+2 | 0 | 0 | -5000 |
| 2 | 4n+3 | 0 | 0 | -5000 |
| 2 | 4n+4 | 0 | 0 | -5000 |

| Design/material data | Value |
|---|---|
| Displacement limits | ± 0.25 in |
| Young modulus (Al) | $10^6$ ksi |
| Specific mass | 0.1 lbm/in³ |
| Allowable stresses | ± 25 ksi |
| Minimum c.-s. area | 0.1 in² |

| Bay No. | Areas (in²) for member # | | | |
|---|---|---|---|---|
| | 1-4 | 5-12 | 13-16 | 17-18 |
| 1 | 1.897 | .5158 | 0.1 | 0.1 |
| 2 | 1.280 | .5148 | 0.1 | 0.1 |
| 3 | .5067 | .5200 | 0.1 | 0.1 |
| 4 | .1571 | .5356 | .4099 | .5690 |
| Total Mass (lbm): 382.48 | | | | |

**Figure 8.5** 72-Member Truss Structure $(n = 4)$

optimally designed passive truss structure under the allocated total (summed-up) gain value of 50 Giga units to further reduce the structural distortions for which the maximum displacement is 0.25 inch (0.635 cm). The active structure, as in the case of its passive counterpart, is to maintain symmetry conditions with respect to both $xz$- and $yz$-central planes.

Shown in Table 8.6a are the CM-2 computer run results of the CM-ASTRAND code for the optimal number, gain, and location design results of active truss members. The left (Case 1) and right (Case 2) values of each pair of values listed correspond to those obtained by minimizing the energy and maximum-displacement objective functions, respectively. The CM-ASTRAND code automatically selected 54 and 56 active members out of 72 truss members (as shown in Table 8.6a) for the optimal

**Table 8.6a**  72-Bar Truss Optimal Design Results — General Case.

| Bay No. | Gains (Giga unit) for Intra-Bay truss Member Nos. | | | |
|---|---|---|---|---|
| | 1–4 | 5–12 | 13–16 | 17–18 |
| 1 | 1.192/1.232 | 0.3807/0.5331 | 0/0 | 0/0 |
| 2 | 1.146/0.9899 | 0.3840/0.5805 | 0/0 | 0/0 |
| 3 | 2.195/1.791 | 0.4768/0.7375 | 0/0 | 0/0.8753 |
| 4 | 3.921/2.064 | 0.4418/0.6152 | 0.3183/0.5807 | 0.7229/0.9425 |

Opt. act. elem no. = 54/56; total energy = 2513/2737 N-cm; maximum disp = .235/.190 cm

**Table 8.6b**   72-Bar Truss Optimal Active Member Placement/ Specification Results (Case in Which the Active Element Number Is Pre-assigned to be $N_G = 18$).

| Bay No. | Gains (Giga unit) for Intra-Bay truss Member Nos. | | | |
|---------|-----------|-------|-------|-------|
|         | 1–4       | 5–12  | 13–16 | 17–18 |
| 1       | 1.602/3.653 | 0/0 | 0/0 | 0/0 |
| 2       | 1.406/2.493 | 0/0 | 0/0 | 0/0 |
| 3       | 2.634/0   | 0/0   | 0/0   | 0/3.566 |
| 4       | 6.328/0   | 0/2.286 | 0/0 | 1.061/0 |

Total strain energy = 3362/5469 N-cm; maximum disp = 0.497/0.391

designs cases 1 and 2, respectively. All vertical and diagonal elements parallel to either of the two vertical coordinate planes, as well as the rest of the elements in the top bay (i.e., those parallel to the horizontal coordinate plane), are automatically selected (by CM-ASTRAND) as the optimal active elements for the Case 1 design. The same is true for the Case 2 design, provided that two additional horizontal–diagonal elements of the third bay are also selected as active members. Some significant differences between these two optimal design results are seen. The total final strain energy (which is the sum of two strain energies contributed by two loading sets) for the energy-based design is significantly smaller than that of the maximum displacement-based design (as it should be), while the opposite is true for the final maximum displacement results, which are 0.235 cm and 0.190 cm, respectively, for design cases 1 and 2. In comparison, the maximum displacement for the passive structure is 0.635 cm.

Table 8.6b shows the optimal placement results of the 72-bar truss under the condition of 18 preassigned numbers of active members for both the strain energy and the maximum displacement minimization design cases. As before, the first and second numbers in each pair of data given correspond to the former and latter cases, respectively. Markedly different active member selection results, as well as gain distribution results, are seen between these two cases. In the first (energy-based) optimization case, all (4) vertical members of all (4) bays and all (2) horizontal-diagonal members of the top (4-th) bay are identified as the optimal active members (locations) with the optimal gain values as shown in Table 8.6b. In the second case (maximum displacement-based optimization), the vertical members of only the first two bays, 8 lateral diagonal members of the top bay, and two horizontal diagonal members of the third bay constitute the optimal active member set. The final total strain energy for the second case is seen to be considerably higher (63%) than that of the first case, while the final maximum displacement for the second case is seen to be significantly smaller (21%) than that of the first case. Such significant differences in optimal results based on the two different objective functions are not totally unexpected because the energy objective function is to do with the overall distortions while the maximum displacement objective function is to do with a single component displacement.

Under the same total gain equality constraint (gain sum) of 50 Giga units, the final total strain energy ratios and final maximum displacement ratios between the corresponding subcases are as follows for the optimally selected and preassigned numbers of active members: strain energy ratios = 0.75/0.50; maximum displacement ratios = 0.47/0.49. Again, the first and second numbers for each pair of data correspond to strain energy and maximum displacement objective function-based optimal design subcases, respectively. These ratios show that by optimally determining the number, locations, and gains, the energy and maximum displacement can be reduced by 25% and 53%, respectively, for the subcase 1, and 53% and 51%, respectively, for the subcase 2, from those of the Case 2 in which the active member number is preassigned to be 16, whose locations and gain values are also optimally determined. This shows the importance of determining the optimal number of active members as well as the locations and gain distribution values. However, in reality, the cost of active members is not only a function of gain, but also a function of the number of active members (for which the basic cost of electronics may be dominant). Therefore, in a follow-on study, it is desirable to include such cost factor in the optimal design for achieving a balanced economical and control-wise efficient design goal.

## 8.4 CONCLUDING REMARKS

Finite element transient response analysis and active element placement optimization techniques for control-structure interaction applications in a massively data parallel computational environment have been presented in this Chapter. Although the structural models used are of adaptive frame and truss types, the techniques/algorithms presented are equally applicable to other types of structures. In the transient response analysis portion, two fully implicit algorithm sets (based on NB/PCG and ICD/NB methods) and two semi-implicit algorithm sets (based on the RK4/LM and ECD/LM methods) have been presented. These algorithm sets were numerically demonstrated to be efficient in varying degrees via example analysis problems of both regular and adaptive structure (distributed, active piezoelastic element controlled) in the massively data parallel environments of the Connection Machine CM-2 and CM-5 computers. (For an active and/or passive damping controlled CSI system, the use of an implicit integration scheme, such as NB and ICD, is almost a must.)

In the active element placement optimization portion of the study for static structural shape control, a vastly improved and parallelized stochastic-type Integral Global Algorithm (IGO), IIGO, was introduced as a viable and efficient alternative optimization algorithm to those of Genetic Algorithms and Simulated Annealing. The IIGO algorithm was used in performing optimal placement example problems of active piezoelectric elements for static frame and truss shape control on the CM-2 and CM-5 by minimizing the structural strain energy or the maximum displacement objective function with respect to active element location, number, and gain values. It was numerically shown that such an algorithm works excellently both time-wise and solution-wise.

A more complete set of CSI analysis techniques normally would also include frequency response and stability analysis and eigensolution elements, as well as a design synthesis element based on the LQR/LQG (linear quadratic regulator/linear quadratic Gaussian) control laws (e.g., cf. studies in a sequential environment by Park (1989) Belvin (1990) Haftka (1990)). However, these are not presented here due to resource, time, and space constraints. A discussion of related, potentially efficient numerical solution techniques/algorithms in a parallel environment can be found in a technical report by Shieh (1992b) and, more recently, by Menon et al. (1995).

In addition to these CSI analysis and design synthesis subareas, additional studies are desirable in the MPP design optimization subarea of CSI systems. This will include: (a) development of more efficient optimization algorithms for CSI systems with large numbers of design variables and constraints with combined continuous, integer and/or discrete design variables (present in the actuator/sensor placement optimization problems of composite structures); (b) integrated design optimization with multiple or combined control (efforts or performance indices) and structural (weight, etc.) design objective function(s) with respect to structural cross-sectional sizing parameters and actuator/sensor (or active element) placement/gain paraments, etc., under design and certain mechanical and electrical behavior constraints. In the placement optimization subarea, the basic electronic costs of active structural elements must also be factored into the cost (objective) function for determining the optimal number of active elements and allowable voltage constraints for the actuator/ sensor materials. An extension study for developing efficient, optimization algorithms on the IBM SP2 computer platform is currently being done by the author by properly combining the IGO or IIGO algorithm with a gradient-based optimization algorithm.

# REFERENCES

Anderson, E., Hagwood, N. and Goodliffe, J., 1992, Self-Sensing Piezoelectric Actuation: Analysis and Application to Controlled Structures, *Proc. of 33rd AIAA/ASME/ASCE/ AHS/ASC Structures, Structural Dynamics and Materials (SDM) Conference*, pp. 2141–2155.

Anderson, E. and Hagood, N., 1994, A Comparison of Algorithms for Placement of Passive and Active Dampers, *35th AIAA/ASME/ASCE/ AHS/ASC Structures, Structural Dynamics and Materials (SDM) Conference*, pp. 2701–2714.

Barrett, R., et al. 1994, *Templates for the Solution of Linear Systems: Building Blocks for Iterative Methods*, templates@cs.utk.edu.

Belytschko, T., 1983, Overview of Semidiscretization in *Computational Methods for Transient Analysis*, Ed. by T. Belytschko and T. Hughes, North-Holland, pp. 1–63.

Belytschko, T., Plaskacz, E. J., Kennedy, J. M., and Greenwell, D. L., 1987, Finite Element Analysis on the Connection Machine, Computer Methods in Applied Mechanics and Engineering 81, 229–254.

Belvin, W. K. and Park, K. C., 1990, Computer Implementation of Analysis and Optimization Procedure for Control Interaction Problems, *Proceedings of AIAA 1990 Dynamics Specialist Conference*, pp. 32–41.

Burnett, D. S. 1987, *Finite Element Analysis*, Addison-Wesley Publishing Co., Reading, MA, p. 463.

Chen, C., Bruno, R., and Salama, M., 1991, Optimal Placement of Active/Passive Members in Truss Structures Using Simulated Annealing, *AIAA Journal 29* (8), pp. 1327–1334.

Cheney, W. and Kincaid, D., 1985, *Numerical Mathematics and Computing (2nd Ed.)*, Brooks/Cole Publishing Company, Monterey, CA.

Chew, S. H. and Zheng, Q., 1988, *Integral Global Optimization*, Springer-Verlag, N.Y., 1988.

Craig, R. 1985, A Review of Time-Domain and Frequency-Domain Component Mode Synthesis Method, in *Combined Experimental/ Analytical Modeling of Dynamic Structural Systems*, Matinez, D. & Miller, K. Editors, AMR 67, pp. 1–30.

Doesh, J., Inman, D., and Garcia, E., 1992, A Self-Sensing Piezoelectric Actuator for Collocated Control, *J. of Intelligent Material Systems and Structures 3*, pp. 166–185.

Farhat C., Sobh, N., and Park, K. C., 1990, Transient Finite Element Computations on 65536 Processors: the Connection Machine, *International J. for Numerical Methods in Engineering 30*, pp. 27–55.

Farhat, C., 1993, Fast Structural Design and Analysis via Hybrid Domain Decomposition on Massively Parallel Processors, *Computing System in Engineering 4(4-6)*, 1993, pp. 453–472.

Giurgiutiu, V., Chaundry, Z., and Rogers, C., 1994, Active Control of Helicopter Rotor Blades with Induced Strain Actuators, *AIAA/ASME Adaptive Structures Forum*, pp. 288–297.

Goldberg, D. E., 1989, *Genetic Algorithms in Search, Optimization & Machine Learning*, Addison-Wesley.

Golub, G. and Van Loan, C., 1989, *Matrix Computation*, Second Edition, The Johns Hopkins University Press, Baltimore.

Haftka, R. T., 1990, Integrated Structure-Control Optimization of Space Structures, *Proceedings of AIAA Dynamics Specialist Conference*, pp. 1–9.

Kahn, A. and Topping, B., 1993, Subdomain Generation for Parallel Finite Element Analysis, *Computing Systems in Engineering 4(4-6)*, pp. 473–488.

Keyes, D., Saad, Y. and Truhlar, D. (Editors), 1995, *Domain-Based Paralleism and Problem Decomposition Methods in Computational Science and Engineering*, SIAM Publication, Philadelphia, PA.

Kincaid, R. and Padula, S., 1990, Minimizing Distortion and Internal Forces in Truss Structures by Simulated Annealing, *Proc. of the 31st AIAA/ASME/ASCE/AHS/ASC Structures, Structural Dynamics, and Materials Conference*, pp. 327–333.

Lee, C.-K., 1987, Piezoelectric Laminates for Torsional and Bending Modal Control: Theory and Experiment, *Ph.D Thesis*, Cornell University, Ithaca NY.

Lee, C.-K., 1993, Piezoelectric Laminates: Theory and Experiments for Distributed Sensors and Actuators, *Engr. Technology*.

Mathur, K. K. and Johnsson, S. L., 1989, The Finite Element Method on a Data Parallel Computer System, *Int. J. of High Speed Computing 1(1)*, pp. 29–44.

Menon, R. and Park, K., 1995, Parallel Simulation and Control of Massively Actuated Structures, *Proc. of the 36th AIAA/ASME/ASCE/AHS/ASC Structures, Structural Dynamics, and Materials Conference*, pp. 749–758.

Nam, C., Kim, W., and Oh, S., 1994, Active Flutter Suppression of Composite Plate with Piezoelectric Actuators, *AIAA/ASME Adaptive Structures Forum*, pp. 127–134.

Newmark, N. M., 1959, A Method of Computation for Structural Dynamics, *1959 ASCE Journal of Engineering Mechanics Division 85 (EM3)*, pp. 67–94.

Onoda, J. and Hanawa, Y., 1992, Optimal Locations of Actuators for Statistical Static Shape Control of Large Space Structure: A Comparison of Approaches, *Proc. of the 33rd AIAA/ASME/ASCE/AHS/ASC Structures, Structural Dynamics and Materials Conference*, pp. 2788–2795.

Otten, R. and van Ginneken, L., 1989, *The Annealing Algorithm*, Kluwer Academic Publishers, Boston.

Park, K. C. and Belvin, W. K., 1989, Stability and Implementation of Partitioned CSI Solution Procedures, *Proceedings of AIAA 30th SDM Conference*, pp. 701–709.

Rao, S., Pan, T., and Venkayya, V., 1991, Optimal Placement of Actuators in Actively Controlled Structures Using Genetic Algorithms, *AIAA Journal 29(6)*, pp. 942–943.

Rao, S. and Venkayya, V., 1992, Mixed-Integer Programming and Multi-Objective Optimization Techniques for the Design of Control Augmented Structures, *Proc. of the Fourth AIAA/USAF/ NASA/OAI Symposium on Multidisciplinary Analysis and Optimization*, pp. 389–396.

Seeley, C. E., and Chattopadhyay, A., 1994, Development of Intelligent Structures Using MultiObjective Optimization and Simulated Annealing, *Proc. of AIAA/ASME Adaptive Structures Forum*, pp. 411–421.

Sepulveda, A. E. and Schmidt, L. A., Jr., 1992, Optimal Placement of Active Elements in Control Augmented Structural Synthesis, *Proc. 33rd AIAA/ASME/ASCE/AHS/ASC Structures, Structural Dynamics and Materials Conference*, pp. 2768–2787.

Shieh, R. C. and Wilson, G. V., 1991, A Massively Parallel Nonlinear Optimization Code Capability for Structural Design, *Proc. of the 32nd AIAA/ASME/ASCE/AHS/ASC Structures, Structural Dynamics and Materials Conference*, pp. 636–643.

Shieh, R. C. 1992a, Multiaxial Piezoelectric Beam Sensor/Actuator Theory and Design for 3-D Structural Vibration Control and Analysis, *Proc. of the Third International Conference on Adaptive Structures*, pp. 647–661.

Shieh, R. C., 1992b, Massively Parallel Computational Control/ Structure Interaction Dynamics Including Development of Multiaxial Laminated Piezoelectric Beam Sensor/Actuator Elements, *Final Report*, Strategic Defense Initiative Organization Contract No. F29601-91-C-0062.

Shieh, R. C., 1993a, Massively Parallel Computational Methods for Finite Element Analysis of Transient Structural Responses, *Journal of Computing Systems in Engineering 4(4–6)*, pp. 421–434.

Shieh, R. C., 1993b, Stochastic Method-Based Optimization for Static Shape Control of Adaptive Structures in a Massively Parallel Environment, *Proc. of the Fourth International Conference on Adaptive Structures*, pp. 482–496.

Shieh, R. C., 1994a, Governing Equations and Finite Element Models for Multiaxial Piezoelectric Beam Sensors/Actuators, *AIAA Journal 32 (6)*, pp. 1250–1258.

Shieh, R. C., 1994b, Fully Implicit Massively Parallel Computational Methods for Transient Structural Response Analysis, *Proc. of 5th International Structural Dynamics Conference*, July 18–21, pp. 900–910.

Shieh, R. C., 1994c, Massively Parallel Optimal Placement/ Specification Design Methodology of Active Elements for Structural Shape Control, *Proc. of the 5th International Conference on Adaptive Structures*, pp. 95–106.

Storaasli, O. O., and Bergan, P., 1987, Nonlinear Substructuring Methods for Concurrent Processing Computers, *AIAA Journal 25*, pp. 96–118.

Suleman, A. and Venkayya, V., 1994, Flutter Control of an Adaptive Composite Panel, *Proc. of AIAA/ASME Adaptive Structures Forum*, pp. 118–126.

Thomas, H. L. and Schmidt, L. A., Jr., 1989, Control Augmented Structural Synthesis With Dynamic Stability Constraints, *Proceedings of AIAA 30th SDM Conference*, pp. 521–531.

Tzou, H. S. and Gadre, M., 1989, Theoretical Analysis of a Multi-Layered Thin Shell Coupled with Piezoelectric Shell Actuators for Distributed Vibration Controls, *J. of Sound and Vibration 132(3)*, pp. 433–450.

Tzou, H. S. and R. Ye, 1994, Piezothermoelasticity and Precision Control of Piezoelectric Systems: Theory and Finite Element Analysis, *ASME Transactions, Journal of Vibration and Acoustics 116(4)*, pp. 489–495.

Xue, D and Mei, C., 1994, A Study of the Application of Shape Memory Alloy in Panel Flutter Control, *Proc. of the 5th International Structural Dynamics Conference*, pp. 412–422.

Zhou, R.,Hsiao, M., Xue, D., Lai, Z., Mei, C., and Huang, J.-K., 1994, Nonlinear Flutter Control of Composite Panels with Embedded Piezoelectric Materials Using Finite Element Method, *AIAA/ASME Adaptive Structures Forum*, pp. 107–117.

# 9

# Structural Optimization Based on Evolution Strategies

Georg Thierauf and Jianbo Cai [1]

## 9.1 INTRODUCTION

Structural Optimization may be defined as the rational establishment of a structural design that is the best of all possible designs within a prescribed objective and a given set of geometrical and/or behavioral limitations (Olhoff & Taylor 1983). This topic has a quite long history. The first analytical works in structural optimization could be by Maxwell in 1869 and Michell in 1904 (Vanderplaats 1982). In his pioneering work Schmit (1960) offers a comprehensive statement of the use of mathematical programming techniques to solve the nonlinear-inequality-constrained problem of designing elastic structures under a multiplicity of loading conditions. Since then many methods and techniques for solving structural optimization problems have been developed.

Review on research, techniques and applications of structural optimization has been made by many researchers. Schmit (1981) provides a historical review of 20 years "structural synthesis" and a perspective of the various developments of structural optimization. Belegundu and Arora (1985) present a comprehensive study of various mathematical programming methods for structural optimization. Vanderplaats (1982) points out the contribution of numerical techniques to the development of structural optimization and potential areas for future research and development. Olhoff and Taylor (1983) present a survey of the field of optimal structural design and discuss the mathematical formulation and the characteristic properties and features of optimization problems for both discrete and continuum structures. Levy and Lev (1987) provide a review of structural optimization over the period 1980-1984, emphasizing developments and applications of optimization techniques to structural design problems. Approximation concepts and techniques are reviewed by Barthelemy and Haftka (1993). Several algorithms that have been applied successfully to engineering design problems were discussed by Arora (1990). Status and promise of structural optimization are presented by some of the most prominent experts in this field (Kamat 1993).

A special class of structural optimization are mixed-discrete optimization problems

---

[1] Department of Civil Engineering, University of Essen, 45117 Essen, Germany

*Parallel Solution Methods in Computational Mechanics* edited by M. Papadrakakis
© 1997 John Wiley & Sons Ltd

where both continuous and discrete variables are involved. Bremicker *et al.* (1990) present a overview of the methods for discrete problems. In their review of the methods for solving mixed-discrete optimization problems Arora *et al.* (1994) classify the methods into six categories, like branch and bound, simulated annealing, sequential linearization, penalty functions, Lagrangian relaxation and other methods, and describe the basic ideas of each method and details of some of the methods. Thanedar and Vanderplaats (1995) review three categories of methods for mixed-discrete optimization, like branch and bound, approximation using branch and bound and ad-hoc methods such as simulated annealing and genetic algorithms, and compare the methods with a numerical example.

In recent years, research and application of several stochastic search methods for both continuous and mixed-discrete optimization problems have increased and gained wide acceptance. These methods are genetic algorithms (GAs) (Holland 1975, Goldberg 1989), evolution strategies (ESs) (Rechenberg 1973, Schwefel 1981), evolutionary programming (EP) (L. Fogel 1966, D. Fogel 1992) and simulated annealing (SA) (Kirkpatrick 1983). These strategies have their basis in processes found in natural evolution and in thermodynamics. The mechanism of the search with these strategies offers a much higher probability of locating the global optimum in the design space than the conventional methods (Kamat 1993). The strategies are widely used in structural optimization (Fogel 1991, Lee 1992, Rajeev 1992, Zhang 1993, Wu 1995).

In this chapter, the theory, parallelization and application of the evolution strategies for continuous and mixed-discrete optimization problems are introduced.

The evolution strategies were developed in the seventies by Rechenberg (1973) and Schwefel (1981). Similar to GAs the ESs imitate biological evolution in nature and have two characteristics that differ from other conventional optimization algorithms:

- in place of the usual deterministic operators, ESs use randomized operators: *mutation*, *selection* and *recombination*;

- instead of a single design point, ESs work simultaneously with a population of design points in the space of variables.

The second characteristic allows for an implementation of the evolution strategies on a parallel computing environment. The ESs developed by Rechenberg and Schwefel were commonly applied for continuous optimization problems. For solving discrete and mixed-discrete optimization problems some modifications and parallelization techniques have been introduced by the authors (Cai 1993, 1995, Thierauf 1994, 1995). In the following, the original and the modified evolution strategies for continuous and discrete optimization problems are described.

## 9.2 THE ORIGINAL EVOLUTION STRATEGIES

The ESs can be divided into the two-membered evolution strategy (2-ES) and the multi-membered evolution strategy (M-ES).

## THE 2-ES

The following optimization problem is considered:

$$
\begin{aligned}
&\min && f(X) && X = \{x_1, x_2, \ldots, x_n\}^T, \\
&\text{subject to :} && g_i(X) \geq 0, && i = 1, 2, \ldots, m.
\end{aligned}
\tag{9.1}
$$

The 2-ES works in two steps:

*Step 1* (mutation):

In the $g$−th generation a new design vector is computed from

$$
X_O^{(g)} = X_P^{(g)} + Z^{(g)},
\tag{9.2}
$$

where $X_P^{(g)}$ is the known design vector, called parent vector, and $X_O^{(g)}$ is the new design vector, called offspring vector. $Z^{(g)} = [z_1^{(g)}, z_2^{(g)}, \ldots, z_n^{(g)}]^T$ is a vector of random change.

For continuous problems the components $z_i^{(g)}$ are random numbers from a normal distribution (Schwefel 1981):

$$
p(z_i) = \frac{1}{\sqrt{(2\pi)}\,\sigma_i} \exp\left(-\frac{(z_i - \xi_i)^2}{2\sigma_i^2}\right),
\tag{9.3}
$$

where $\xi_i$ is the expectation, which should have the value zero, and $\sigma_i^2$ is the variance, which should be small. The standard deviation $\sigma$ can be considered as a step length wich can be adjusted during the search according to Rechenberg's 1/5 success rule (Hoffmeister 1992):

> "*The ratio of successful mutations to all mutations should be 1/5. If it is greater, increase; if it is less, decrease the standard deviation $\sigma$*".

According to Schwefel (1981), the frequency of successful mutations should be measured over intervals of $10n$ trials and the increase and decrease factors should be $(1/0.85)$ and $0.85$, respectively. The adjustment should take place every $n$ mutations.

*Step 2* (selection):

The selection chooses the best individual from the parent and the offspring to survive:

$$
X_P^{(g+1)} = \begin{cases} X_O^{(g)} & \text{if } g_i(X_O^{(g)}) \geq 0, \ i = 1, \ldots, m \\ & \text{and } f(X_O^{(g)}) \leq f(X_P^{(g)}); \\ X_P^{(g)} & \text{otherwise .} \end{cases}
\tag{9.4}
$$

According to Schwefel (1981) the search is terminated if the value of the objective functions has been improved by less than a given value $\varepsilon$ in the course of $20n$ trials.

## THE M-ES

In the case of multi-membered ES a population of $\mu$ parent vectors and $\lambda$ offspring vectors is considered simultaneously.

*Step* 1 (recombination and mutation):

In the $g$−th generation $\mu$ vectors, called parent vectors, are given. $\lambda$ offspring vectors are created from the $\mu$ parent vectors by means of recombination and mutation. The operator recombination is described in the following.

*Step* 2 (selection):

There are two variants of M-ES, the $(\mu + \lambda)$−ES and the $(\mu, \lambda)$−ES.

$(\mu + \lambda)$−ES

The best $\mu$ individuals are selected from parents and offsprings to form the parents of the next generation.

$(\mu, \lambda)$−ES (with $\lambda > \mu$)

The best $\mu$ individuals are selected only from the $\lambda$ offsprings. The $\mu$ parents in the $g$−th generation die out.

In step 1, for every offspring vector a temporary parent vector $\tilde{X} = [\tilde{x}_1, \tilde{x}_2, \ldots, \tilde{x}_n]^T$ is first built by means of recombination. For a continuous problem five recombination cases which can be used selectively are given by Hoffmeister and Baeck (1992):

$$\tilde{x}_i = \begin{cases} x_{a,i} \text{ or } x_{b,i} \text{ randomly} & \text{(A)} \\ 1/2(x_{a,i} + x_{b,i}) & \text{(B)} \\ x_{bj,i} & \text{(C)} \\ x_{a,i} \text{ or } x_{bj,i} \text{ randomly} & \text{(D)} \\ 1/2(x_{a,i} + x_{bj,i}) & \text{(E)} \end{cases} \tag{9.5}$$

where $\tilde{x}_i$ is the $i$-th component of the temporary parent vector $\tilde{X}$, $x_{a,i}$ and $x_{b,i}$ are the $i$-th components of the vectors $X_a$ and $X_b$ which are two parent vectors randomly chosen from the population. In case C of Eq. (9.5), $\tilde{x}_i = x_{bj,i}$ means that the $i$-th component of $\tilde{X}$ is chosen randomly from the $i$-th components of all $\mu$ parent vectors (Figure 9.1). From the temporary parent $\tilde{X}$ an offspring can be created in the same way as in the two-membered ES (Eq. 9.2).

During the search, not only the design variables $x_i$, but also the parameters, such as the deviations $\sigma_i$, will be modified by the random operator *mutation* which replaces the 1/5 success rule (Schwefel 1981).

For a continuous problem the search will be terminated (Schwefel 1981),

1) if the absolute or relative difference between the best and the worst objective function values is less than a given value $\varepsilon_1$, or

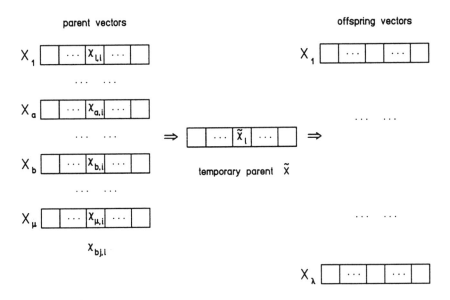

Figure 9.1: Recombination

2) if the mean value of the objective function values of all parent vectors in the last $K$ ($\geq 2n$) generations has been improved by less than a given value $\varepsilon_2$.

For the examples in this chapter the values $\varepsilon_1$ and $\varepsilon_2$ are usually set to 0.01.

## 9.3 THE MODIFIED EVOLUTION STRATEGIES

A discrete structural optimization problem can be formulated in the following form:

$$
\begin{aligned}
&\min && f(X), && X = [x_1, x_2, \cdots, x_n]^T, \\
&\text{subject to}: && g_i(X) \geq 0, && i = 1, \ldots, m \\
&&& x_j \in R^d, && j = 1, \ldots, n
\end{aligned}
\tag{9.6}
$$

where $R^d$ is a given set of discrete values. The design variables $x_i$ ($i = 1, 2, \ldots, n$) can only take the discrete values of this given set.

### MUTATION

It is assumed that in biological evolution small mutations occur frequently but large ones only rarely. For continuous optimization problems such mutations of the objective function will be ensured in such a way that every component of the normally distributed random change vector $Z^{(g)}$ (Eq. 9.2) has small standard deviations $\sigma_i$ and an expectation of zero. This means that all components of a parent vector can be changed at the same time, but the changes are usually small.

For discrete problems, the design variables, i. e. the components of the parent or offspring vectors, can only take their values from a given discrete set, hence a random change can only lead from one discrete value to another adjacent one. The differences between any two adjacent values are usually not small; it is therefore not possible that the objective function gets a small mutation when all components of a parent vector are changed at the same time.

For this reason, we suggest that not all of the $n$ components of a parent vector, but only a few (say $l$), will be randomly changed every time (Cai 1993, 1995). This means that $(n - l)$ components of the random change vector $Z^{(g)}$ in Eq. (9.2) have the value zero. The components of the random change vector $Z^{(g)}$ then have the form

$$z_i^{(g)} = \begin{cases} (\kappa + 1)\delta x_i & l\,(l < n) \quad \text{randomly chosen} \\ & \qquad\qquad \text{components,} \\ 0 & n - l \quad\ \text{other components,} \end{cases} \tag{9.7}$$

where $\delta x_i$ is the current difference between two adjacent values in the discrete set and $\kappa$ is a Poisson-distributed integer random number with the distribution

$$p(\kappa) = \frac{(\gamma)^\kappa}{\kappa!}\, e^{-\gamma}, \tag{9.8}$$

where $\gamma$ is the deviation of the random number $\kappa$; for a very small $\gamma$ (say 0.001), the number $\kappa$ has the value zero with a probability greater than 99%; for $\gamma = 0.05$ the number $\kappa$ can have the value zero with a probability of 95% and the value one with a probability of 5%. This means that the random change $z_i^{(g)}$ is controlled by the parameter $\gamma$.

A uniformly distributed random choice decides which $l$ components should be changed for a mutation according to Eq. (9.7). For structural optimization problems, according to our research, a suitable $l$ value ranges from 8 to 12 (Cai 1995).

## RECOMBINATION

For discrete optimization problems the recombination cases (B) and (E) in Eq. (9.5) are not suitable, because the mean value of two discrete values is usually not a discrete value from the given set. For discrete optimization problems the following recombinations are employed (Cai 1993, 1995):

$$\tilde{x}_i = \begin{cases} x_{a,i}\ or\ x_{b,i}\ \text{randomly} & \text{(A)} \\ x_{bj,i} & \text{(B)} \\ x_{a,i}\ or\ x_{bj,i}\ \text{randomly} & \text{(C)} \\ x_{m,i}\ or\ x_{b,i}\ \text{randomly} & \text{(D)} \\ x_{m,i}\ or\ x_{bj,i}\ \text{randomly} & \text{(E)} \end{cases} \tag{9.9}$$

The recombination cases (A), (B) and (C) are as in Eq. (9.5). The last two cases (D) and (E) in Eq. (9.9) are suggested by the authors (Cai 1993, 1995). The vector $X_m$ is not chosen at random but as the best of the $\mu$ parent vectors in the $g$-th generation. In cases (D) and (E) the information from the best parent can be used which results in a better convergence for many problems. From the temporary parent $\tilde{X}$ an offspring can be created as in the two-membered ES (Eq. (9.2)).

For discrete problems the following termination criteria are suggested (Cai 1995, Thierauf 1994): The search will be terminated,

a) if the best value of the objective function in the last $K_I$ ($\geq 4n\mu/\lambda$) generations has not been improved, or

b) if the mean value of the objective function values from all parent vectors in the last $K_{II}$ ($\geq 2n\mu/\lambda$) generations has been improved by less than a given value $\varepsilon_b$, or

c) if the relative difference between the best objective function value and the mean value of objective function values from all parent vectors in the current generation is less than a given value $\varepsilon_c$, or

d) if the ratio $\mu_b/\mu$ has reached a given value $\varepsilon_d$, where $\mu_b$ is the number of the parent vectors in the current generation with the best objective function value.

For discrete problem the value $\varepsilon_b$ can be set to $\varepsilon_b = 0$. For the examples in Section 9.6 the values $\varepsilon_c$ and $\varepsilon_d$ are usually set to $\varepsilon_c = 0.0001$ and $0.5 \leq \varepsilon_d \leq 0.8$.

## 9.4 PARALLELIZATION OF THE M-ES

The M-ES works simultaneously with a population of design points in the space of variables. This inherent parallelism allows for an implementation in a parallel computing environment. In the following two parallel versions of the M-ES are presented: a directly parallelized evolution strategy (DPES) and a parallel sub-evolution-strategy (PSES) (Thierauf 1994).

### DIRECTLY PARALLELIZED EVOLUTION STRATEGY (DPES)

In every generation $\lambda$ offspring vectors are generated and the objective function and constraint functions are computed. In structural optimization these computations are based on FE-analyses in most cases. For the DPES the $\lambda$ FE-analyses are performed in parallel on $(n-1)$ processors. In the implementation on a $n$-processor computer, the $n$ processors are divided into one master-processor and $(n-1)$ slave-processors. A master job on the master-processor runs the M-ES, generates offsprings and sends them to the slave-processors. The slave jobs on slave-processors check the feasibility of the offsprings and send the information back to the master job. In the master job $\mu$ best individuals are selected as the parent vectors of next generation from the complete population. A parallel environment works for the communication between the processors and has a role of information transferrer.

If possible, the ratio $\lambda/(n-1)$ should be integer so that the computational load on the processors is well balanced. The program flowchart for DPES is shown in Figure 9.2.

The DPES can be implemented on any parallel computing environment. Two variants are described by Thierauf and Cai (1994). The first implementation was performed on an architecture with 7 HP 700 workstations under PVM (Parallel Virtual

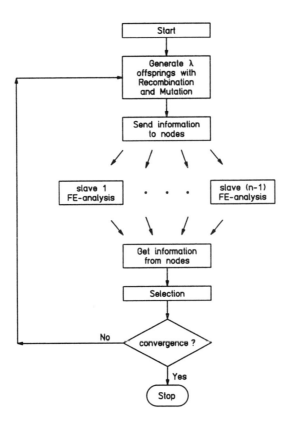

Figure 9.2: Program flowchart with DPES

Machine) environment. Every workstation has its own CPU and all workstations share only one hard disk. A general purpose finite element program B&B (Thierauf 1993) was employed for the FE-analysis on the slave-processors. The test example showed a low speed-up of real time. Two factors might be considered. First, the frequency of information exchange in this implementation is high, because information must be exchanged before and after every FE-analysis. Second, there is no direct connection between the workstations, all information must be transferred under PVM through the local network.

The second implementation was performed on a Power-Xplorer system based on PowerPC 601 processors and transputer communication technology. The parallel operating environment is PARIX (1994).

In this implementation two examples, a 10-bar truss problem and a 200-bar truss problem, were considered. For the FE-analysis a special program for truss was employed. For the first example, a small truss-structures, the FE-analysis requires not more computing time than the ES operations and the communications, only a low speed-up has been reached. In the second example, a much larger truss-structure, the FE-analysis requires more computing operations and a quasi-linear speed-up is obtained

## PARALLEL SUB-EVOLUTION-STRATEGY (PSES)

With an increasing size of the population, the probability to obtain a global optimum by using the M-ES increases almost proportionally. However, large scale problems with an increasing size of population also require much computing time. Following an idea of natural evolution, the PSES subdivides the population into several smaller subpopulations which can undergo their evolution separately and in parallel. In order to prevent the development of evolutionary niches, migration between the subpopulations must be allowed.

In the parallel implementation on a $n$-processor computer, the whole population is divided into $n$ subpopulations, each processor runs the ES on its own subpopulation (called sub-evolution-strategy or sub-ES). Periodically some good individuals will be selected and copies of them will be sent to one of its neighbors (migration). Every subpopulation also receives copies from its neighbors, which replace its own "bad" individuals. This implementation is called parallel sub-evolution-strategy (PSES) (Thierauf 1994). A similar parallelization called "islands model" is discussed by Surry and Radcliffe (1995).

In the realization of the PSES, for convenience of management and control, a further processor can be introduced. A master job on this extra processor plays the role of manager and organizes the information exchange between the other $n$ slave-processors. It starts $n$ optimization processes with sub-ES (slave jobs) on the slave-processors and gets and sends information from or to them.

The exchange is carried out here in a deterministic cyclic form, although a random exchange might be better suited for the probabilistic process. However, a random exchange would produce results which would not be reproducible, even for a pseudo random process when program system restarts occurred.

The exchange of individuals between the subpopulation is the main advantage of PSES over the sequential M-ES and makes it possible that the $n$ subpopulations work with a similar probability to achieve the global optimum as the original large population, but within a much shorter computing time.

The period of exchange and the number of exchanged individuals are two adjustable parameters. The period should not be too long and the number should not be too small, otherwise the evolutionary process in every subpopulation will be isolated and the advantage of a large population to locate a global optimum will be lost. If the period is too short and the number is too large, the individuals in a subpopulation will be quickly replaced by the ones from other subpopulations. After several exchanges the $n$ subpopulations become the same or similar and a few of the good individuals dominate over the whole population, which may lead to a premature termination of the search.

The period of exchange $P_{exch}$ could be selected as the number of generations. After a certain number of generations (say $m$) the exchange can be performed. In many cases this selection of period could lead to long waiting states during the run of PSES. In a generation not all of the $\lambda$ offsprings should be checked for their feasibility by a FE-analysis, some of them might be eliminated only because of their bad value of the objective function. After $m$ generations the unevenness of the numbers of FE-analyses in different slave jobs causes waiting states, because the FE-analyses take much more computing operations than the operators of ES. For this reason, instead of the number

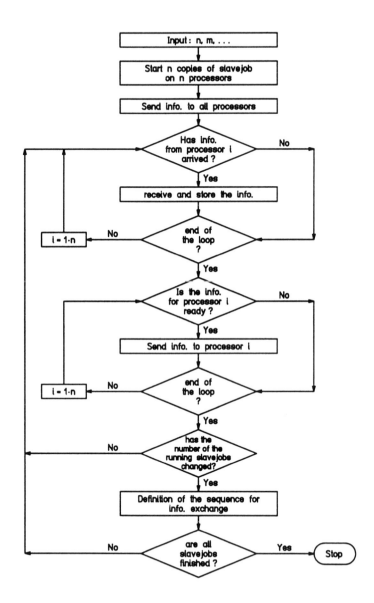

Figure 9.3: PSES - flowchart of the master program

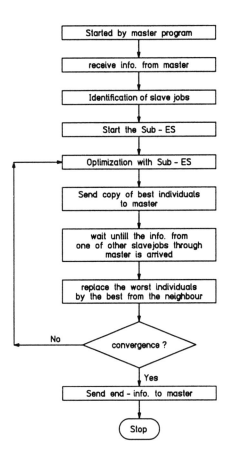

Figure 9.4: PSES - flowchart of the slave program

of generations, we select the number of FE-analyses as the period of exchange, namely after a certain numberof FE-analyses the exchange can be performed once. According to our experience the period $P_{exch}$ can be chosen as $(0.7 \sim 0.9)\lambda_s K_I$ or $(0.7 \sim 0.9)\lambda_s K_{II}$, where $\lambda_s$ is the number of the offsprings of a subpopulation.

The number of exchanged individuals $N_{exch}$ depends on the number of parent vectors in a subpopulation. As result of our study, $N_{exch}$ should be 20% to 30% of the number of parents of a subpopulation.

The implementation with PSES has a lower frequency of exchange than the DPES and is suitable for the parallelization of M-ES on a workstation-cluster. The flowcharts of the master and the slave program for PSES are shown in Figure 9.3 and 9.4, respectively.

## 9.5 THE TWO LEVEL PARALLELIZATION

The main modification of the evolution strategies for solving discrete and continuous optimization problems is the application of different mutation operators. One way of solving mixed-discrete optimization problems could be to handle the discrete and continuous variables with different mutation operators within the same optimization process. Based on a parallel computing technique, another way to solve mixed-discrete problem, namely a two level parallelization method, was suggested by the authors (Thierauf 1995). For a parallel solution of a mixed-discrete optimization problem the discrete and continuous optimization variables are treated separately in different processes. On the first level, a complete optimization problem is divided into two subproblems with discrete and continuous design variables, respectively. The two subproblems are solved simultaneously on a parallel computer. On the second level, each subproblem is further parallelized by means of the parallel sub-evolution-strategy as described before.

*TWO SUBPROBLEMS*

A mixed-discrete optimization problem can be formulated as a nonlinear mathematical programming problem as follows:

$$
\begin{aligned}
&\min && f(X), \\
&\text{subject to}: && g_i(X) \geq 0 && i = 1, \ldots, m, \\
& && X = [x_1, \cdots, x_n, x_{n+1}, \cdots, x_{n+q}]^T, \\
& && x_i^l \leq x_i \leq x_i^u && (i = 1, \cdots, n + q),
\end{aligned}
\tag{9.10}
$$

where

$$
\begin{aligned}
&x_i \in R^d \quad (i = 1, \cdots, n): && \text{discrete variables,} \\
&x_{n+i} \quad (i = 1, \cdots, q): && \text{continuous variables,} \\
&x_i^l, x_i^u: && \text{lower and upper bounds} \\
& && \text{of the variables,} \\
&R^d: && \text{given set of discrete values.}
\end{aligned}
$$

For a parallel solution, problem (9.10) can be divided into two subproblems.
**Subproblem 1 – Optimization problem with discrete variables:**

$$
\begin{aligned}
&\min && f(X_d, X_c^p), \\
&\text{subject to}: && g_i(X_d, X_c^p) \geq 0 && i = 1, \ldots, m, \\
& && X_d = [x_1, x_2, \cdots, x_n,]^T, \\
& && X_c^p = [x_{n+1}^p, x_{n+2}^p, \cdots, x_{n+q}^p]^T, \\
& && x_i^l \leq x_i \leq x_i^u && (i = 1, \cdots, n + q)
\end{aligned}
\tag{9.11}
$$

with

$x_i \in R^d \quad (i = 1, \cdots, n) :$      discrete variables,
optimization variables
of subproblem 1,

$x_{n+i}^p \quad (i = 1, \cdots q) :$      continuous variables,
unchanged during the
optimization process
of subproblem 1,

$x_i^l, x_i^u :$      lower and upper bounds
of the variables,

$R^d :$      given set of discrete values.

**Subproblem 2 – Optimization problem with continuous variables:**

$$
\begin{aligned}
\text{min} \quad & f(X_d^p, X_c), \\
\text{subject to :} \quad & g_i(X_d^p, X_c) \geq 0 \qquad i = 1, \ldots, m, \\
& X_d^p = [x_1^p, x_2^p, \cdots, x_n^p]^T, \\
& X_c = [x_{n+1}, x_{n+2}, \cdots, x_{n+q}]^T, \\
& x_i^l \leq x_i \leq x_i^u \qquad (i = 1, \cdots, n+q)
\end{aligned}
\qquad (9.12)
$$

with

$x_i^p \in R^d \quad (i = 1, \cdots, n) :$      discrete variables,
unchanged during the
optimization process
of subproblem 2,

$x_{n+i} \quad (i = 1, \cdots q) :$      continuous variables,
optimization variables
of subproblem 2,

$x_i^l, x_i^u :$      lower and upper bounds
of the variables.

These two subproblems are solved simultaneously by the modified and the original multi-membered evolution strategies. During the optimization process of a subproblem, say subproblem 1, only the discrete variables vary and the continuous variables are considered as constants.

For the solution of the subproblems formulated in Eq. (9.11) and (9.12) the parallel sub-evolution-strategy (PSES) as described before is employed.

*PARALLEL SOLUTION OF THE MIXED PROBLEM*

For a parallel solution of the mixed-discrete optimization problem (9.10) a parallel computing architecture with $2p$ processors has to be employed. The $2p$ processors are divided into two groups. For example, in case of a parallel computer with 8 processors, the processors with identification number (ID-number) 0, 2, 4, 6 are the first group and are used to solve subproblem 1 in Eq. (9.11), the processors with ID-number 1, 3, 5, 7 belong to the second group and are used to solve the subproblem 2 in Eq. (9.12). An executable code written for solving both subproblems will be started eight times and every processor will get a task according to the ID-number.

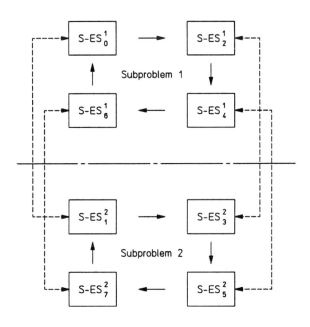

Figure 9.5: two level parallel ES

In each group the local information exchange is carried out in a cyclic form. For example, in the first group the exchange sequence will be $0 \rightarrow 2 \rightarrow 4 \rightarrow 6 \rightarrow 0$ and in the second group $1 \rightarrow 3 \rightarrow 5 \rightarrow 7 \rightarrow 1$. For the information exchange between two groups an one to one partnership like $0-1$, $2-3$, $4-5$, $6-7$ can be set up (Figure 9.5).

The parameters for exchange between the two groups are considered as global parameters, and the parameters for exchange between the subpopulations within a group (s. Section 9.4) are considered as local parameters. The period of global exchange $P_{exch}^{g}$ and the number of globally exchanged individuals $N_{exch}^{g}$ can be defined in a similar way as for the local parameters $P_{exch}^{l}$ and $N_{exch}^{l}$ as described before. According to our experience, to avoid search in a subspace, the number of globally exchanged individuals $N_{exch}^{g}$ can be selected as $N_{exch}^{g} = 2N_{exch}^{l}$. The period of global exchange $P_{exch}^{g}$ can be selected as $P_{exch}^{g} = P_{exch}^{l}$.

Because the original and modified evolution strategies have different convergence behaviors, two subproblems could be terminated at different times. To avoid this, different exchange parameters, for example different exchange periods, for the two subproblems can be selected.

Except for the information exchange, the sub-ES on every processor runs independently and a local search can be stopped according to its own termination criterion. If a sub-ES on a processor terminated normally, it sends a signal to its partner in the other group and the exchange between these two partners is interrupted. To keep the cyclic exchange working, exchange between the processors of the same group is carried out until all computations for the subproblem are terminated.

## 9.6 EXAMPLES

In the following several examples of discrete and mixed-discrete optimization problems are considered for demonstrating the application of the ESs in structural design problems.

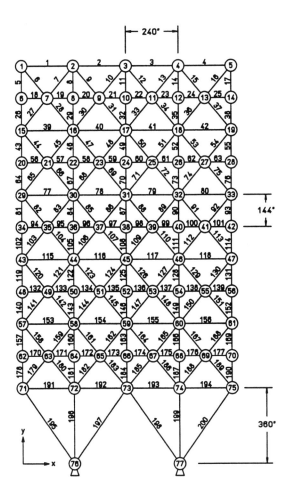

Figure 9.6: 200-bar truss

**Two hundred-bar plane truss**

As a first example the 200-bar plane truss shown in Figure 9.6 (Arora 1977) is solved as a discrete optimization problem. The design parameters are: $E = 30000$ ksi, $\varrho = 0.283$ lb/in$^3$ (for reference, ksi-units are chosen). The objective function of the problem is the weight of the structure. The constraints are the member stresses and the vertical and horizontal displacements of all nodes. The allowable displacement is limited to 0.5

in. and the allowable stress to ±30 ksi. The members of the truss are divided into 96 groups and the cross-sectional areas of every group are the design variables (Table 9.1). There are three loading cases:

a. One kip acting in positive $x$ direction at nodes 1, 6, 15, 20, 29, 34, 43, 48, 57, 62, 71;

b. 10 kips acting in negative $y$ direction at nodes 1, 2, 3, 4, 5, 6, 8, 10, 12, 14, 15, 16, 17, 18, 19, 20, 22, 24, ..., 71, 72, 73, 74, 75;

c. cases $a$ and $b$ combined.

The following discretization of the design variables is considered (according to DIN 1028, double angle profiles).
$$R^{d1} = ((0.100,) \quad 0.347, \quad 0.440, \quad 0.539, \quad 0.954, \quad 1.081, \quad 1.174, \quad 1.333,$$
$$1.488, \quad 1.764, \quad 2.142, \quad 2.697, \quad 2.800, \quad 3.131, \quad 3.565, \quad 3.813,$$
$$4.805, \quad 5.952, \quad 6.572, \quad 7.192, \quad 8.525, \quad 9.300, 10.850, 13.330,$$
$$14.290, 17.170, 19.180, 23.680, 28.080, 33.700) \text{ (sq in.)}$$

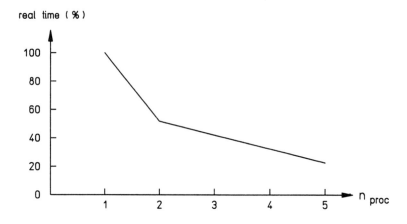

Figure 9.7: 200-bar truss, speed-up of DPES on a Power-Xplorer system

An optimal solution for a continuous problem was given by Arora et al. (1977): optimal volume 102343 in$^3$ and weight 28963 lb. To solve this problem by using DPES on a Power-Xplorer system and by using PSES on a workstation-cluster two tests were made. By the implementation of DPES a (10+10)-ES was considered. Three tests with different numbers of slave-processors were carried out. Because the FE-analyses for this problem take much more computing operations than the ES-operators and the communications, a quasi-linear speed-up has been achieved (Table 9.2 and Figure 9.7). By the implementation of PSES a population of (60+60) was divided into 6 subpopulations. The number of exchanged individuals was set to $n_{exch} = 2$ and the frequency of exchange was $m = 150$. The results are given in Table 9.3.

Table 9.1: Design variables of 200-bar truss

| No. | member numbers | No. | member numbers | No. | member numbers |
|---|---|---|---|---|---|
| 1 | 1,4 | 32 | 65,75 | 65 | 133,134,137,138 |
| 2 | 2,3 | 33 | 66,74 | 66 | 135,136 |
| 3 | 5,17 | 34 | 67,73 | 67 | 140,152 |
| 4 | 6,16 | 35 | 68,72 | 68 | 141,151 |
| 5 | 7,15 | 36 | 69,71 | 69 | 142,150 |
| 6 | 8,14 | 37 | 70 | 70 | 143,149 |
| 7 | 9,13 | 38 | 77,80 | 71 | 144,148 |
| 8 | 10,12 | 39 | 78,79 | 72 | 145,147 |
| 9 | 11 | 40 | 81,93 | 73 | 146 |
| 10 | 18,25,56,63, | 41 | 82,92 | 74 | 153,156 |
|  | 94,101,132,139 | 42 | 83,91 | 75 | 154,155 |
|  | 170,177 | 43 | 84,90 | 76 | 157,169 |
| 11 | 19,20,23,24 | 44 | 85,89 | 77 | 158,168 |
| 12 | 21,22 | 45 | 86,88 | 78 | 159,167 |
| 13 | 26,38 | 46 | 87 | 79 | 160,166 |
| 14 | 27,37 | 47 | 95,96,99,100 | 80 | 161,165 |
| 15 | 28,36 | 48 | 97,98 | 81 | 162,164 |
| 16 | 29,35 | 49 | 102,114 | 82 | 163 |
| 17 | 30,34 | 50 | 103,113 | 83 | 171,172,175,176 |
| 18 | 31,33 | 51 | 104,112 | 84 | 173,174 |
| 19 | 32 | 52 | 105,111 | 85 | 178,190 |
| 20 | 39,42 | 53 | 106,110 | 86 | 179,189 |
| 21 | 40,41 | 54 | 107,109 | 87 | 180,188 |
| 22 | 43,55 | 55 | 108 | 88 | 181,187 |
| 23 | 44,54 | 56 | 115,118 | 89 | 182,186 |
| 24 | 45,53 | 57 | 116,117 | 90 | 183,185 |
| 25 | 46,52 | 58 | 119,131 | 91 | 184 |
| 26 | 47,51 | 59 | 120,130 | 92 | 191,194 |
| 27 | 48,50 | 60 | 121,129 | 93 | 192,193 |
| 28 | 49 | 61 | 122,128 | 94 | 195,200 |
| 29 | 57,58,61,62 | 62 | 123,127 | 95 | 196,199 |
| 30 | 59,60 | 63 | 124,126 | 96 | 197,198 |
| 31 | 64,76 | 64 | 125 |  |  |

## 110 KV transmission tower

This transmission tower with a height of 31 m is a recent application of structural optimization in practice (Figure 9.8). For finite-element-analysis the system is discretisized into a FE-model with 872 beam and truss elements divided into 30 groups, 406 Nodes and 2 material groups. 20 loading combinations including dead weight of the system and wind loads from various directions are considered.

This example is solved as a discrete optimization problem by the modified M-ES. The objective function is the weight of the system. The design variables are the cross-sectional areas of the 30 element groups, each variable has 21 allowable discrete values. The constraints are the member stresses, nodal displacements and the local stability of the elements according to DIN 4114. The maximal nodal displacements and member stresses of the starting design of the system are considered as the allowable values for the stress- and displacement-constraints.

To solve this problem a (10+10) population was taken. The original real design of

Table 9.2: 200-bar truss, speed-up of DPES - with Power-Xplorer

| $n_{proc}$ | real time (s) | speed up | discrete solution volume (in$^3$) | weight (lb) |
|---|---|---|---|---|
| 1 | 18766.87 | 1 | 104730 | 29641 |
| 2 | 9711.30 | 1.93 | 104730 | 29641 |
| 5 | 4215.44 | 4.45 | 104730 | 29641 |

Table 9.3: Results for 200-bar truss, PSES

| slave job | sub-population | recom.-case | number of FE-analyses | number of generations | discrete solution volume (in$^3$) | weight (lb) |
|---|---|---|---|---|---|---|
| 1 | 10+10 | E | 11261 | 2299 | 105105 | 29745 |
| 2 | 10+10 | E | 11856 | 2438 | 105097 | 29742 |
| 3 | 10+10 | E | 12521 | 2632 | 105076 | 29737 |
| 4 | 10+10 | E | 11451 | 2306 | 105097 | 29742 |
| 5 | 10+10 | E | 12454 | 2552 | 105076 | 29737 |
| 6 | 10+10 | E | 10773 | 2182 | 105105 | 29745 |

the system with a weight of 2886 kg was selected as one of the starting points. After 323 generations and 2025 FE-analyses an improved design with a weight of 2354 kg and an improvement of 18.4% was obtained.

**Design of a pressure vessel**

This example is a design optimization problem of a pressure vessel, shown in Figure 9.9. This problem was solved both by Fu *et al.* (1991) and Li *et al.* (1994), with different penalty methods. The objective function is the combined costs of material, forming and welding of the pressure vessel. The constraints are set in accordance with the respective ASME codes. The design optimization problem is formulated as follows:

$$
\begin{aligned}
\min \\
f(X) = {} & 0.6224x_1x_3x_4 + 1.7781x_2x_3^2 + \\
& 3.1661x_1^2x_4 + 19.84x_1^2x_3, \\
\text{subject to:} \\
g_1(X) = {} & -x_1 + 0.0193x_3 \le 0 \\
g_2(X) = {} & -x_2 + 0.00954x_3 \le 0 \\
g_3(X) = {} & -\pi x_3^2 x_4 - \tfrac{4}{3}\pi x_3^3 + 750.0 \cdot 1728.0 \le 0 \\
g_4(X) = {} & -240.0 + x_4 \le 0 \\
& 1.000 \le x_1 \le 1.375 \\
& 0.625 \le x_2 \le 1.000
\end{aligned}
\tag{9.13}
$$

The Design variables $x_3$ and $x_4$ are continuous and the design variables $x_1$ and $x_2$ are discrete values with integer multiples of 0.0625 inches.

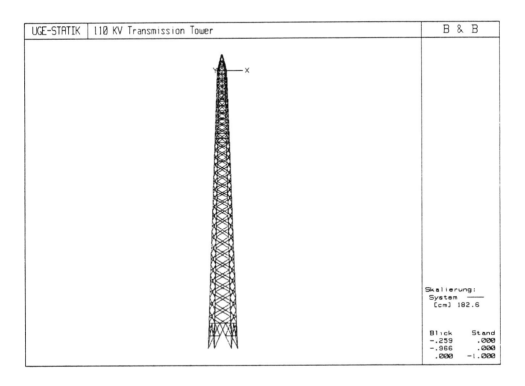

Figure 9.8: 110 KV transmission tower

On the Power-Xplorer system, this problem is solved by using the TLPES as described before. The size of the subpopulations was (10+10) and the information exchange parameters were chosen as $P^l_{exch-1} = P^g_{exch-1} = 20$ (for subproblem 1), $P^l_{exch-2} = P^g_{exch-2} = 100$ (for subproblem 2), $N^l_{exch} = 2$ and $N^g_{exch} = 2$. Because each discrete variable has only 7 feasible values, the subproblem 1 converges much faster than the subproblem 2. The solution obtained by TLPES is $X^* = [1.0, 0.625, 51.812, 84.591]^T$ with the objective function value $f(X^*) = 7006.931$. In Table 9.4 this solution and the solutions of Fu *et al.* (1991) and Li *et al.* (1994) are listed.

## Three-dimensional 39-bar tower

A three-dimensional 39-bar truss (Bremicker 1990) which is shown in Figure 9.10 is considered. The objective function is the weight of the structure subject to stress constraints for the bars and deflection constraints for three horizontal loads with P=25 kN applied at nodes 13, 14 and 15 in the positive $y$-direction. The coordinates of the three bottom nodes 1, 2 and 3 and the three top nodes 13, 14 and 15 are given in Table 9.5.

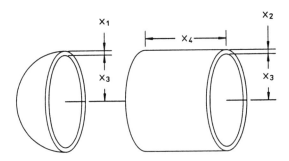

Figure 9.9: pressure vessel

Table 9.4: A comparison of solutions for the pressure vessel design

|           | Solution of Fu (1991) | Solution of Li (1994) | Solution by TLPES |
|-----------|-----------------------|-----------------------|-------------------|
| $f_{min}$ | 8048.6                | 7127.3                | 7006.9            |
| $x_1$     | 1.125                 | 1.000                 | 1.000             |
| $x_2$     | 0.625                 | 0.625                 | 0.625             |
| $x_3$     | 48.380                | 51.250                | 51.812            |
| $x_4$     | 111.745               | 90.991                | 84.591            |
| $g_1$     | $-0.191$              | $-1.011$              | 0.000             |
| $g_2$     | $-0.163$              | $-0.136$              | $-0.131$          |
| $g_3$     | $-75.875$             | $-18759.754$          | $-15.000$         |
| $g_4$     | $-128.255$            | $-149.009$            | $-155.409$        |

There are 6 geometrical variables: $x_1$, $x_2$ and $x_3$ define the $z$-coordinates of three levels as shown in Figure 9.10; $x_4$ is the distance AS, BS and CS in the $x$-$y$-plane at the level of the nodes 4, 5, and 6; similarly $x_5$ and $x_6$ are the distances at the level of 7-8-9 and 10-11-12, respectively. Moreover, five sizing variables are defined: $x_7$ is the cross-sectional area of the elements defined by the the pairs of nodes (1, 4), (2, 5) and (3, 6); $x_8$ is the cross-sectional area of the elements defined by (4, 7), (5, 8) and (6, 9); $x_9$ is the cross-sectional area of the elements defined by (7, 10), (8, 11) and (9, 12); $x_{10}$ is the cross-sectional area of the elements defined by (10, 13), (11, 14) and (12, 15); $x_{11}$ is the cross-sectional area of the remaining elements.

In this problem, the geometric variables $x_1, \cdots, x_6$ are considered as continuous variables and the sizing variables $x_7, \cdots, x_{11}$ are discrete variables. Bounds for the geometric variables are: 500 mm $\leq x_1 \leq$ 4000 mm; 1000 mm $\leq x_2 \leq$ 5000 mm; 2000 mm $\leq x_3 \leq$ 6000 mm; 1000 mm $\leq x_i \leq$ 4000 mm for $i = 4, 5, 6$. The design parameters are $E = 2.1 \cdot 10^6$ N/mm$^2$, $\varrho = 7.85 \cdot 10^{-6}$ kg/mm$^3$. The displacement constraint limits the deflection of top nodes 14 and 15 in the positive $y$-direction to a maximum of 3 mm and the allowable stress is $\pm$ 150 N/mm$^2$. The given discrete values are taken as

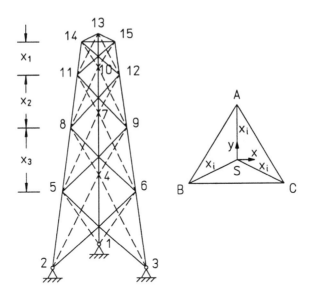

Figure 9.10: 39-bar tower

Table 9.5: Coordinates of the top and bottom nodes of the 39-bar tower

| nodes | coordinates (m) | nodes | coordinates (m) |
|-------|-----------------|-------|-----------------|
| 1 | $(0, 4, 0)$ | 13 | $(0, 1.12, 16)$ |
| 2 | $(-2\sqrt{3}, -2, 0)$ | 14 | $(-0.56\sqrt{3}, -0.56, 16)$ |
| 3 | $(2\sqrt{3}, -2, 0)$ | 15 | $(0.56\sqrt{3}, -0.56, 16)$ |

follows (form DIN 1028, angle sections):
$R^{d2}$ = (112.0,   142.0,   174.0,   185.0,   227.0,   267.0,   308.0,   328.0,
          349.0,   379.0,   430.0,   480.0,   569.0,   582.0,   656.0,   691.0,
          870.0,   903.0,   935.0,   940.0, 1010.0, 1150.0, 1190.0, 1220.0,
          1230.0, 1510.0, 1550.0, 1920.0, 2120.0, 2270.0, 2320.0) $(mm^2)$

Bremicker *et al.* (1990) solved this problem by means of a code named MINSLIP based on a combination of branch and bound and sequential linearization. Three solutions were given. For a first approximation (solution 1, Table 9.6), all variables were considered as continuous. In a second approximation, the computed values of the first solution are assigned to the continuous variables and an integer optimization problem was solved for the discrete variables (solution 2, Table 9.6). In the third approximation, the problem was solved as a mixed discrete problem (solution 3, Table 9.6).

By using evolution strategies three solutions are obtained here. All three solutions are shown in Table 9.7. Solution 1 (Figure 9.11*b*) in Table 9.7 is obtained by means of original ES and all 11 variables are treated as continuous variables. As starting geometry an initial design shown in Figure 9.11*a* is considered.

Table 9.6: Solutions by MINSLIP for the 39-bar tower (Bremicker 1990)

|                        | solution 1 | solution 2 | Solution 3 |
|------------------------|------------|------------|------------|
| $f_{min}$(kg)          | (718.56)   | (776.30)   | (740.92)   |
| $x_1$(mm)              | 1558.60    | 1558.60    | 1567.00    |
| $x_2$(mm)              | 3563.50    | 3563.50    | 3756.10    |
| $x_3$(mm)              | 4532.60    | 4532.60    | 6000.00    |
| $x_4$(mm)              | 3386.40    | 3386.40    | 3738.80    |
| $x_5$(mm)              | 2726.50    | 2726.50    | 3004.30    |
| $x_6$(mm)              | 1720.30    | 1720.30    | 1994.60    |
| $x_7$(mm$^2$)          | 1719.90    | 1920.00    | 1920.00    |
| $x_8$(mm$^2$)          | 1215.30    | 1230.00    | 1220.00    |
| $x_9$(mm$^2$)          | 736.98     | 870.00     | 691.00     |
| $x_{10}$(mm$^2$)       | 239.34     | 267.0      | 227.00     |
| $x_{11}$(mm$^2$)       | 215.67     | 227.0      | 227.00     |

The second solution in Table 9.7 is a discrete solution obtained by the modified ES with the same approximation of Bremicker *et al.* (1990): the geometric variables are fixed at the given continuous solution of Bremicker *et al.* (1990) and a discrete optimization problem with only 5 variables is solved. Solution 3 in Table 9.7 is obtained by using the TLPES presented in this chapter. The same starting geometry as for the solution 1 is used. For solution 3 the original mixed-discrete optimization problem has to be divided into two subproblems: subproblem 1 − the discrete problem with 5 size variables; subproblem 2 − the continuous problem with 6 geometric variables. A parallel Power-Xplorer system with 8 PowerPC 601 processors was employed. The 8 processors were divided into 2 groups, one for subproblem 1 and another one for subproblem 2. On every processor a sub-ES with a (10+10)-subpopulation was started. The information exchange parameters were chosen as $P^l_{exch-1} = P^g_{exch-1} = 30$ (for subproblem 1), $P^l_{exch-2} = P^g_{exch-2} = 40$ (for subproblem 2), $N^l_{exch} = 2$ and $N^g_{exch} = 4$. After 110 − 130 generations in group 1 and 200 generations in group 2 the solution 3 shown in Table 9.7 was reached (Figure 9.11c).

**72 m radio tower**

The radio tower as shown in Figure 9.12a, a steel construction with a height of 72 m, was optimized by TLPES. The objective function is the minimal volume. The FE-model has 1316 beam- and truss-elements divided into 17 groups and 765 nodes. Loading cases including dead weight of the system, ice loads, loads on the platforms and wind loads from various directions were considered. According to DIN 18800 all loading cases were multiplied by a safety factor of 1.35.

This problem was solved as a mixed-discrete optimization problem with 5 continuous geometric variables (Figure 9.12b) and 17 discrete sizing variables. Each discrete variable has 18 allowable discrete values. The constraints are the member stresses, rotation of the antenna and local stability of all elements according to DIN. The maximal

Table 9.7: Solutions by ES for the 39-bar tower

|  | solution 1 (continuous) | solution 2 (mixed) | solution 3 (mixed) |
|---|---|---|---|
| $f_{min}$(kg) | 713.17 | 734.97 | 725.72 |
| $x_1$(mm) | 1854.9 | 1558.6 | 1705.71 |
| $x_2$(mm) | 3336.7 | 3563.5 | 3563.70 |
| $x_3$(mm) | 4460.1 | 4532.6 | 4524.76 |
| $x_4$(mm) | 3265.6 | 3386.4 | 3361.76 |
| $x_5$(mm) | 2572.7 | 2726.5 | 2727.62 |
| $x_6$(mm) | 1580.6 | 1720.3 | 1689.01 |
| $x_7$(mm$^2$) | 1705.3 | 1550.0 | 1920.00 |
| $x_8$(mm$^2$) | 1254.4 | 1150.0 | 1010.00 |
| $x_9$(mm$^2$) | 838.18 | 1150.0 | 656.00 |
| $x_{10}$(mm$^2$) | 318.44 | 227.0 | 227.00 |
| $x_{11}$(mm$^2$) | 207.32 | 227.0 | 227.00 |

allowable stress is $\pm$ 36 kN/cm$^2$, maximal allowable antenna rotation is $\pm$ 0.5$^0$.

The problem was solved by TLPES on a Power-Xplorer parallel computer with 24 PowerPC 601 processors. From a very poor starting design point (all discrete variables taking their upper bound values) with a volume of 24.5 $m^3$, an improved design with a volume of 5.47 $m^3$ is obtained after an iteration of 200 generations. The geometry of the initial design and the improved design are shown in Figure 9.12b. The iteration history is shown in Figure 9.13.

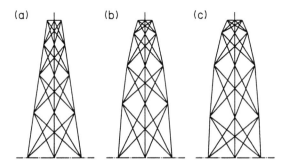

Figure 9.11: Tower geometry for initial design (a), continuous solution (b) and finial mixed-discrete solution (c)

(a) radio tower                    (b) geometry and variables

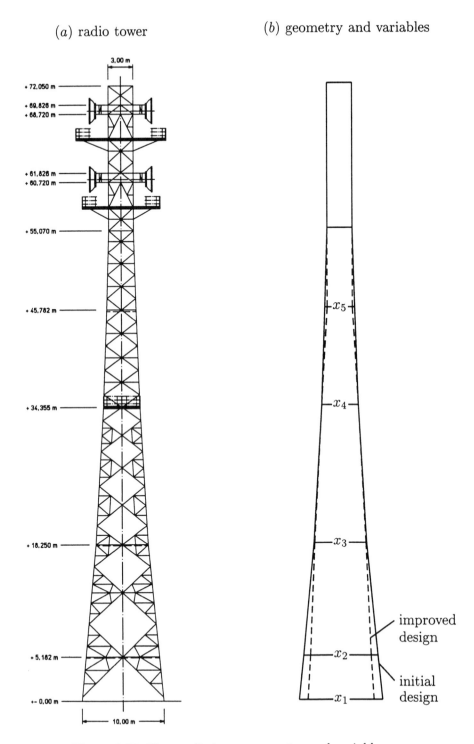

Figure 9.12: 72 m radio tower, geometry and variables

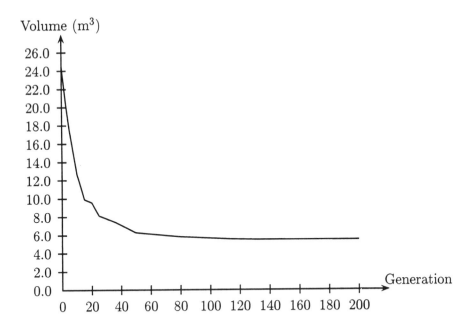

Figure 9.13: Iteration history

## SUMMARY

Compared to mathematical programming, the stochastic search methods offer some distinguished advantages: they are relatively simple and easy to adjust to almost any optimization problem, they are well suited for parallelization and can locate a global optimum with a certain probability. With regard to genetic algorithms, evolution strategies are better suited for the solution of mixed-discrete problems, and do not require the coding and decoding of the variables. However, a detailed comparison for the class of engineering optimization is not known.

## ACKNOWLEDGEMENTS

The research described in this chapter was partially supported by German Research Foundation (DFG) under grant TH-218/12-1 and by Ministry of Science and Research of state North Rhine-Westphalia, Germany. The authors would like to thank DFG and the Ministry for their support.

# REFERENCES

Arora, J. S. and Govil, A. K. 1977. An efficient method for optimal structural design by substructuring. *Comp. & Struct.*, Vol. 7, pp. 507-515.

Arora, J. S. 1990. Computational design optimization: a review and future directions. *Structural Safety*, Vol. 7, pp. 131-148.

Arora, J. S., Huang M. W. and Hsieh, C. C. 1994. Methods for optimization of nonlinear problems with discrete variables: a review. *Struct. Opt.*, Vol. 8, pp. 69-85.

Barthelemy, J. F. M. and Haftka, R. T. 1993. Approximation concepts for optimum structural design – a review. *Struct. Opt.*, Vol. 5, pp. 129-144.

Belegundu, A. D. and Arora, J. S. 1985. A study of mathematical programming methods for structural optimization, Part I and Part II. *Int. J. Num. Meth. Eng.*, Vol. 21, pp. 1583-1623.

Bremicker, M.; Papalambros, P. Y. and Loh, H. T. 1990. Solution of mixed-discrete structural optimization problems with a new sequential linearization algorithm. *Comp. & Struct.*, Vol. 37, No. 4, pp. 451-461.

Cai, J. and Thierauf, G. 1993. Discrete structural optimization using evolution strategies. In: *Neural networks and combinatorial optimization in civil and structural engineering*, edited by B.H.V. Topping and A.I. Khan, Edinburgh, Civil-Comp Limited, pp. 95-100.

Cai, J. 1995. Discrete optimization of structures under dynamic loading by using sequential and parallel evolution strategies (in German). Doctoral Dissertation, Department of Civil Engineering, University of Essen, Germany.

Fogel, L. J.; Owens, A. J. and Walsh, M. J. 1966. Artificial intelligence through simulated evolution. Wiley, New York.

Fogel, D. B. 1991. System identification through simulated evolution: a machine learning approach to modeling. Ginn Press, Needham Heights.

Fogel, D. B. 1992. Evolving artificial intelligence. PhD thesis, University of California, San Diego.

Fu, J.; Fenton, R. G. and Cleghorn, W. L. 1991. A mixed integer-discrete-continuous programming method and its application to engineering design optimization. *Eng. Opt.*, Vol. 17, pp. 263-280.

Goldberg, D. E. 1989. Genetic algorithms in search, optimization and machine learning. Addison-Wesley Publishing Co., Inc., Reading, Massachusetts.

Hoffmeister, F. and Baeck, T. 1992. Genetic algorithms and evolution strategies: similarities and differences. Technical Report No. SYS-1/92, University of Dortmund.

Holland, J. 1975. Adaptation in natural and artificial systems. University of Michigan Press, Ann Arbor, Mich.

Kamat, M. P. (Editor) 1993. Structural optimization: status and promise. (Progress in Astronautic and Aeronautic Series, Vol. 150). AIAA, Washington DC.

Kirkpatrick, S.; Gelatt Jr., C. D. and Vecchi, M. P. 1983. Optimization by simulated annealing. *Science*, Vol. 220, pp. 671-680.

Lee, S. and Wang, H. P. 1992. Modified simulated annealing for multiple-objective engineering design optimization. *J. Intellig. Manufac.*, Vol. 3, No. 2, pp. 101-108.

Levy, R. and Lev, O. E. 1987. Recent developments in structural optimization. *J. Struct. Eng.*, ASCE, Vol. 113, No. 9, pp. 1939-1962.

Li, H.-L. and Chou, C.-T. 1994. A global approach for nonlinear mixed discrete programming in design optimization. *Eng. Opt.*, Vol. 22, pp. 109-122.

Olhoff, N. and Taylor, J. E. 1983. On structural optimization. *J. Appl. Mech.*, Vol. 50, pp. 1139-1151.

PARIX 1.3-PPC, Reference Manual. PARSYTEC Computer GmbH, Aachen, Germany, 1994.

Rajeev, S. and Krishnamoorthy, C. S. 1992. Discrete optimization of structures using genetic algorithms. *J. Struct. Eng.*, ASCE, Vol. 118, No. 5, pp. 1233-1250.

Rechenberg, I. 1973. Evolution strategy: optimization of technical systems according to the principles of biological evolution (in German). Frommann-Holzboog, Stuttgart.

Schmit, L. A. 1960. Structural design by systematic synthesis. Proceedings, 2nd Conference on Electronic Computation, ASCE, New York, pp. 105-122.

Schmit, L. A. 1981. Structural synthesis − its genesis and development. *AIAA J.*, Vol. 19, No.10, pp. 1249-1263.

Schwefel, H.-P. 1981. Numerical optimization of computer models. Wiley & Sons, Chichester.

Surry, P. D. and Radcliffe, N. J. 1995. RPL2: A language and parallel framework for evolutionary computing. Proceedings of the Structural Engineering Computational Technology Seminar, Heriot-Watt University, Edinburgh, 1st-2nd May, 1995.

Thanedar, P. B. and Vanderplaats, G. N. 1995. Survey of discrete variable optimization for structural design. *J. Struct. Eng.*, ASCE, Vol. 121, No. 2, pp. 301-306.

Thierauf, G. 1993. B&B − A program-system for analysis and design of general structures (in German), Universität Essen.

Thierauf, G. and Cai, J. 1994. Evolution strategy, its parallelization and application to discrete optimization problems. NATO Advanced Research Workshop: Emergent Computing Methods in Engineering Design, Nafplio, Greece, August 25-27, 1994.

Thierauf, G. and Cai, J. 1995. A two level parallel evolution strategy for solving mixed-discrete structural optimization problems. The 21st ASME Design Automation Conference, Boston MA, September 17-21, 1995.

Vanderplaats, G. N. 1982. Structural optimization − past, present, and future. *AIAA J.*, Vol. 20, No. 7, pp. 992-1000.

Wu, S.-J. and Chow, P.-T. 1995. Genetic algorithms for nonlinear mixed discrete-integer optimization problems via meta-genetic parameter optimization. *Eng. Opt.*, Vol. 24, pp. 137-159.

Zhang, C. and Wang, H.-P. 1993. Mixed-discrete nonlinear optimization with simulated annealing. *Eng. Opt.*, Vol. 21, pp. 277-291.

# 10

# High Performance Computing Techniques for Flow Simulations

T. Tezduyar, S. Aliabadi, M. Behr, A. Johnson, V. Kalro and M. Litke[1]

## 10.1   INTRODUCTION

High performance computing (HPC), with recent introduction of advanced algorithms capable of accurately simulating complex flow problems, and with advanced computer hardware and networking with sufficient power, memory and bandwidth, has brought flow simulation capabilities to a point of being practical tools in many fields of engineering and applied sciences, in our case, in aerospace engineering and applied fluid mechanics (Tezduyar et al., 1992c; Tezduyar et al., 1993; Behr and Tezduyar, 1994; Mittal and Tezduyar, 1994; Tezduyar et al., 1994; Aliabadi and Tezduyar, 1995; Mittal and Tezduyar, 1995; Tezduyar et al., 1995; Johnson and Tezduyar, 1996a; Kalro et al., 1996). In this chapter, we provide a review, mostly in the context of the research carried out by the Aerospace Engineering and Mechanics group at the Army HPC Research Center (AHPCRC), of how the HPC capabilities in fluid dynamics evolved in recent years, and how all this now is enabling the simulation of many real-world flow problems. This chapter includes material extracted from articles, which appeared or are about to appear, written by the authors and their coworkers.

The flow simulation capabilities developed by our group cover a large class of flow regimes and patterns: flows involving complex geometries (Tezduyar et al., 1994; Aliabadi and Tezduyar, 1995; Mittal and Tezduyar, 1995; Tezduyar et al., 1995; Johnson and Tezduyar, 1996a); flows with moving boundaries and interfaces, including those with free surfaces (Behr and Tezduyar, 1994), two-liquid interfaces (Johnson and Tezduyar, 1994), fluid-particle interactions (Johnson and Tezduyar, 1995a; Johnson and Tezduyar, 1995b; Johnson and Tezduyar, 1996b), fluid-structure interactions (Tezduyar et al., 1994; Mittal and Tezduyar, 1995), and moving mechanical components (Aliabadi and Tezduyar, 1995). The incompressible flow simulations include flows with high Reynolds numbers and sharp boundary layers. The simulations of compressible

---

[1]   Aerospace Engineering and Mechanics, Army HPC Research Center, University of Minnesota, 1100 Washington Avenue South, Minneapolis, MN 55415, USA

*Parallel Solution Methods in Computational Mechanics* edited by M. Papadrakakis

flows include, in addition to flows with high Reynolds numbers and sharp boundary layers, supersonic flows and strong shocks.

All simulations are carried out in 3D and on parallel supercomputing platforms. All flow simulation capabilities developed are based on finite element formulations, applicable to unstructured meshes. The finite element formulations developed are either semi-discrete formulations to solve problems with fixed spatial domains, or space-time formulations (Tezduyar et al., 1992b; Tezduyar et al., 1992d; Aliabadi and Tezduyar, 1993) when we need to solve problems with changing spatial domains, such as those encountered in flows with moving boundaries and interfaces. These formulations are based on stabilization techniques such as the streamline- upwind/Petrov-Galerkin (SUPG) (Brooks and Hughes, 1982; Tezduyar and Hughes, 1983; Le Beau and Tezduyar, 1991; Le Beau et al., 1993; Aliabadi et al., 1993), pressure-stabilizing/Petrov-Galerkin (PSPG) (Tezduyar, 1991; Tezduyar et al., 1992e), and Galerkin/least-squares (GLS) formulations (Hughes et al., 1989; Hansbo and Szepessy, 1990; Tezduyar et al., 1992b; Tezduyar et al., 1992d). These stabilization techniques prevent numerical oscillations and instabilities when the flow involves high Reynolds and/or Mach numbers and strong shocks and boundary layers. The stabilized finite element formulations developed also allow the use of equal-order interpolation functions for velocity and pressure and other unknowns. In all cases, the stabilization is accomplished without introducing excessive numerical dissipation.

The 3D finite element meshes used are generated by using either special mesh generators designed for specific problems, or an automatic mesh generator developed by our group which allows structured layers of elements near solid surfaces and unstructured meshes elsewhere in the domain (Johnson and Tezduyar, 1996a).

In flow problems with moving boundaries and interfaces, the changes in the shape of the spatial domain is handled by updating the mesh by using the combination of moving the mesh and remeshing (i.e., generating a new set of elements and grid points) when the mesh distortion becomes too high (Johnson and Tezduyar, 1994; Johnson and Tezduyar, 1995a; Johnson and Tezduyar, 1996b). Moving the mesh is accomplished by special mesh moving techniques designed for specific problems and by an automatic mesh moving algorithm for more general cases.

The large coupled, nonlinear equation systems (sometimes up to 38 million equations) that need to be solved at every time step (or every pseudo-time step) are solved using iterative strategies (Behr et al., 1993a; Behr et al., 1993b; Kalro and Tezduyar, 1995). The main components of these iterative strategies are: formation of the residual vector of the linear equation system (that needs to be solved at every iteration of a Newton-Raphson sequence used in solving the coupled, nonlinear equations); designing a preconditioning matrix which reasonably approximates the Jacobian matrix in each Newton-Raphson iteration; and updating the solution vector in an optimal way. We form the residual vector of the linear equation systems by using, depending on the size of the problem, the element level matrices or a matrix-free approach. For preconditioning matrices, we are currently using simple ones, such as diagonal or nodal-block- diagonal preconditioners, in production runs, and are experimenting with more sophisticated ones, such as the clustered element-by-element (CEBE) and mixed CEBE and cluster companion (CC) preconditioners (Liou and Tezduyar, 1992;

Tezduyar et al., 1992a). To update the solution vector, we typically use the GMRES technique (Saad and Schultz, 1986).

Currently, the majority of our parallel computations are carried out on the AH-PCRC's 896-node Thinking Machines CM- 5 and the Minnesota Supercomputer Center's 512-node CRAY T3D. Some of our smaller scale parallel computations are performed on the AHPCRC's 20-processor SGI Onyx. The parallel implementations used in these computations are based on the data-parallel, message-passing, and shared memory paradigms.

The applications that will be presented in this chapter are: steady-state decent of a large ram air parachute (Aliabadi et al., 1995; Garrard et al., 1995; Kalro et al., 1996); longitudinal dynamics and flare maneuver of a large ram air parachute; flow past the spillway of the Olmsted Dam; contaminant dispersion around an M1 battle tank; contaminant dispersion in a model subway station; natural convection process; airflow past an automobile; multiple spheres falling in a liquid-filled tube (Johnson and Tezduyar, 1995a; Johnson and Tezduyar, 1995b; Johnson and Tezduyar, 1996b); and dynamics of a paratrooper jumping from a cargo plane.

## 10.2 GOVERNING EQUATIONS OF COMPRESSIBLE AND INCOMPRESSIBLE FLOWS

The flow simulations are based on the solution of time-dependent Navier-Stokes equations of compressible and incompressible flows. In stating those equations here, $\Omega_t$ and $(0, T)$ will denote the space and time domains, where $\Gamma_t$ is the boundary of $\Omega_t$. In general, the spatial domain may change with respect to time, and the subscript $t$ indicates such time-dependence. The symbols $\rho(\mathbf{x}, t)$, $\mathbf{u}(\mathbf{x}, t)$, $p(\mathbf{x}, t)$ and $e(\mathbf{x}, t)$ represent the density, velocity, pressure and the total energy, respectively. The external forces (e.g., the gravity) are represented by $\mathbf{f}(\mathbf{x}, t)$.

*COMPRESSIBLE FLOWS*

The Navier-Stokes equations of compressible flows can be written in the following vector form:

$$\frac{\partial \mathbf{U}}{\partial t} + \frac{\partial \mathbf{F}_i}{\partial x_i} - \frac{\partial \mathbf{E}_i}{\partial x_i} = \mathbf{0} \quad \text{on } \Omega_t \quad \forall t \in (0, T), \tag{10.1}$$

where $\mathbf{U} = (\rho, \rho u_1, \rho u_2, \rho u_3, \rho e)$ is the vector of conservation variables, and $\mathbf{F}_i$ and $\mathbf{E}_i$ are, respectively, the Euler and viscous flux vectors defined as

$$\mathbf{F}_i = \begin{pmatrix} u_i \rho \\ u_i \rho u_1 + \delta_{i1} p \\ u_i \rho u_2 + \delta_{i2} p \\ u_i \rho u_3 + \delta_{i3} p \\ u_i (\rho e + p) \end{pmatrix}, \tag{10.2}$$

$$E_i = \begin{pmatrix} 0 \\ [\mathbf{T}]_{i1} \\ [\mathbf{T}]_{i2} \\ [\mathbf{T}]_{i3} \\ -q_i + [\mathbf{T}]_{ik} u_k \end{pmatrix}. \tag{10.3}$$

Here $[\mathbf{T}]_{ij}$ are the components of the Newtonian viscous stress tensor:

$$\mathbf{T} = 2\mu\boldsymbol{\varepsilon}(\mathbf{u}), \tag{10.4}$$

where $\mu$ is the dynamic viscosity and $\boldsymbol{\varepsilon}$ is the strain rate tensor, and $q_i$ are the components of the heat flux vector. The equation of state is modeled with the ideal gas assumption.

Equation (10.1) can also be written in the following form:

$$\frac{\partial \mathbf{U}}{\partial t} + \mathbf{A}_i \frac{\partial \mathbf{U}}{\partial x_i} - \frac{\partial}{\partial x_i}\left(\mathbf{K}_{ij}\frac{\partial \mathbf{U}}{\partial x_j}\right) = 0 \quad \text{on } \Omega_t \quad \forall t \in (0, T), \tag{10.5}$$

where

$$\mathbf{A}_i = \frac{\partial \mathbf{F}_i}{\partial \mathbf{U}}, \tag{10.6}$$

$$\mathbf{K}_{ij}\frac{\partial \mathbf{U}}{\partial x_j} = \mathbf{E}_i. \tag{10.7}$$

Appropriate sets of boundary and initial conditions are assumed to accompany Equation (10.5).

## INCOMPRESSIBLE FLOWS

The Navier-Stokes equations of incompressible flows can be written in the following vector form:

$$\rho\left(\frac{\partial \mathbf{u}}{\partial t} + \mathbf{u}\cdot\nabla\mathbf{u} - \mathbf{f}\right) - \nabla\cdot\boldsymbol{\sigma} = 0 \quad \text{on } \Omega_t \quad \forall t \in (0, T), \tag{10.8}$$

$$\nabla\cdot\mathbf{u} = 0 \quad \text{on } \Omega_t \quad \forall t \in (0, T), \tag{10.9}$$

where $\rho$ is assumed to be constant, and

$$\boldsymbol{\sigma} = -p\mathbf{I} + \mathbf{T}. \tag{10.10}$$

This equation set is completed by an appropriate set of boundary conditions and an initial condition consisting of a divergence-free velocity field specified over the entire domain:

$$\mathbf{u}(\mathbf{x}, 0) = \mathbf{u}_0, \quad \nabla\cdot\mathbf{u}_0 = 0 \quad \text{on } \Omega_0. \tag{10.11}$$

## 10.3 STABILIZED FINITE ELEMENT FORMULATIONS

In the space-time formulation, the weak form of the governing equations is written over the associated space-time domain of the problem, by dividing this domain into a sequence of space-time slabs $Q_n$, where $Q_n$ is the slice of the space-time domain between the time levels $t_n$ and $t_{n+1}$. The integrations involved in the weak form are then performed over $Q_n$. The finite element interpolation functions used are continuous in space but discontinuous across time levels. To reflect this situation, we use the notation $(\cdot)_n^-$ and $(\cdot)_n^+$ to denote the function values at $t_n$ as approached from below and above respectively. Each space-time slab $Q_n$ is decomposed into space-time elements $Q_n^e$, where $e = 1, 2, \ldots, (n_{el})_n$. The subscript $n$ used with $n_{el}$ is to account for the general case in which the number of space-time elements may change from one space-time slab to other. In our computations, we use first-order polynomials as interpolation functions.

### COMPRESSIBLE FLOWS

In the finite element formulation of compressible flows, for each slab $Q_n$, we first define appropriate finite-dimensional function spaces $\mathcal{S}_n^h$ and $\mathcal{V}_n^h$ corresponding to the trial solutions and weighting functions, respectively. The stabilized space-time formulation of (10.5) can then be written as follows: given $(\mathbf{U}^h)_n^-$, find $\mathbf{U}^h \in \mathcal{S}_n^h$ such that $\forall \mathbf{W}^h \in \mathcal{V}_n^h$:

$$\int_{Q_n} \mathbf{W}^h \cdot \left( \frac{\partial \mathbf{U}^h}{\partial t} + \mathbf{A}_i^h \frac{\mathbf{U}^h}{\partial x_i} \right) dQ + \int_{Q_n} \left( \frac{\partial \mathbf{W}^h}{\partial x_i} \right) \cdot \left( \mathbf{K}_{ij}^h \frac{\partial \mathbf{U}^h}{\partial x_j} \right) dQ$$

$$+ \int_{\Omega_n} (\mathbf{W}^h)_n^+ \cdot ((\mathbf{U}^h)_n^+ - (\mathbf{U}^h)_n^-) \, d\Omega$$

$$+ \sum_{e=1}^{(n_{el})_n} \int_{Q_n^e} \boldsymbol{\tau} \, (\mathbf{A}_k^h)^T \left( \frac{\partial \mathbf{W}^h}{\partial x_k} \right) \cdot \left[ \frac{\partial \mathbf{U}^h}{\partial t} + \mathbf{A}_i^h \frac{\mathbf{U}^h}{\partial x_i} - \frac{\partial}{\partial x_i} \left( \mathbf{K}_{ij}^h \frac{\partial \mathbf{U}^h}{\partial x_j} \right) \right] dQ$$

$$+ \sum_{e=1}^{(n_{el})_n} \int_{Q_n^e} \delta \left( \frac{\partial \mathbf{W}^h}{\partial x_i} \right) \cdot \left( \frac{\partial \mathbf{U}^h}{\partial x_i} \right) dQ = \int_{(P_n)_{\boldsymbol{H}}} \mathbf{W}^h \cdot \boldsymbol{H}^h dP. \tag{10.12}$$

Here $\boldsymbol{H}^h$ represents the Neumann-type boundary condition, $(P_n)_{\boldsymbol{H}}$ is the part of the slab boundary with such conditions, and $\boldsymbol{\tau}$ and $\delta$ are the stabilization parameters.

The solution to (10.12) is obtained sequentially for $Q_1, Q_2, \ldots, Q_{N-1}$, starting with

$$(\mathbf{U}^h)_0^- = \mathbf{U}_0^h, \tag{10.13}$$

where $\mathbf{U}_0$ is the specified initial value of the vector $\mathbf{U}$.

In the formulation given by the Equation (10.12), the first three integrals, together with the right-hand-side, represent the time-discontinuous Galerkin formulation of (10.5). The third integral enforces, weakly, the continuity of the conservation variables in time. The first series of element-level integrals are the SUPG stabilization terms, and the second series are the shock-capturing terms added to the formulation. The details regarding the stabilization and the space-time formulation can be found in (Aliabadi and Tezduyar, 1993). For problems not involving moving boundaries and interfaces, Equation (10.12) can be reduced to a semi-discrete formulation by dropping the third integral, and by converting all space-time integrations to spatial integrations.

## INCOMPRESSIBLE FLOWS

In the finite element formulation of incompressible flows, for each slab $Q_n$, we first define appropriate finite-dimensional trial solution $((S_{\mathbf{u}}^h)_n$ and $(S_p^h)_n)$ and weighting function $((\mathcal{V}_{\mathbf{u}}^h)_n$ and $(\mathcal{V}_p^h)_n = (S_p^h)_n)$ spaces for the velocity and pressure.

The stabilized space-time formulation of Equations (10.8) and (10.9) can then be written as follows: given $(\mathbf{u}^h)_n^-$, find $\mathbf{u}^h \in (S_{\mathbf{u}}^h)_n$ and $p^h \in (S_p^h)_n$ such that $\forall \mathbf{w}^h \in (\mathcal{V}_{\mathbf{u}}^h)_n$ and $\forall q^h \in (\mathcal{V}_p^h)_n$:

$$\int_{Q_n} \mathbf{w}^h \cdot \rho \left( \frac{\partial \mathbf{u}^h}{\partial t} + \mathbf{u}^h \cdot \nabla \mathbf{u}^h - \mathbf{f} \right) dQ + \int_{Q_n} \boldsymbol{\varepsilon}(\mathbf{w}^h) : \boldsymbol{\sigma}(p^h, \mathbf{u}^h) dQ$$

$$+ \int_{Q_n} q^h \nabla \cdot \mathbf{u}^h dQ + \int_{\Omega_n} (\mathbf{w}^h)_n^+ \cdot \rho \left( (\mathbf{u}^h)_n^+ - (\mathbf{u}^h)_n^- \right) d\Omega$$

$$+ \sum_{e=1}^{(n_{el})_n} \int_{Q_n^e} \tau_{\text{MOM}} \frac{1}{\rho} \left[ \rho \left( \frac{\partial \mathbf{w}^h}{\partial t} + \mathbf{u}^h \cdot \nabla \mathbf{w}^h \right) - \nabla \cdot \boldsymbol{\sigma}(q^h, \mathbf{w}^h) \right]$$

$$\cdot \left[ \rho \left( \frac{\partial \mathbf{u}^h}{\partial t} + \mathbf{u}^h \cdot \nabla \mathbf{u}^h - \mathbf{f} \right) - \nabla \cdot \boldsymbol{\sigma}(p^h, \mathbf{u}^h) \right] dQ$$

$$+ \sum_{e=1}^{(n_{el})_n} \int_{Q_n^e} \tau_{\text{CONT}} \nabla \cdot \mathbf{w}^h \, \rho \nabla \cdot \mathbf{u}^h dQ = \int_{(P_n)_h} \mathbf{w}^h \cdot \boldsymbol{h}^h dP. \qquad (10.14)$$

Here $\boldsymbol{h}^h$ represents the Neumann-type boundary condition imposed, $(P_n)_h$ is the part of the slab boundary with such conditions, and $\tau_{\text{MOM}}$ and $\tau_{\text{CONT}}$ are the stabilization parameters.

The solution to (10.14) is obtained sequentially for all of the space-time slabs $Q_1, Q_2, \ldots, Q_{N-1}$, and the computations start with

$$(\mathbf{u}^h)_0^- = \mathbf{u}_0^h. \qquad (10.15)$$

In the formulation given by Equation (10.14), the first four integrals, together with the right-hand-side, represent the time-discontinuous Galerkin formulation of (10.8)-(10.9). The fourth integral enforces, weakly, the continuity of the velocity field in time. The two series of element-level integrals in the formulation are the least-squares stabilization terms. The reader can refer to Tezduyar $et$ $al.$ (1992b,1992c) and Behr and Tezduyar (Behr and Tezduyar, 1994) for further details regarding the space-time formulation for incompressible flows, including definitions of the stabilization parameters. For problems not involving moving boundaries and interfaces, Equation (10.14) can be reduced to a semi-discrete formulation by dropping the fourth integral and the term $\frac{\partial \mathbf{w}^h}{\partial t}$, and by converting all space-time integrations to spatial integrations.

## 10.4   3D MESH GENERATION FOR FLOW SIMULATIONS WITH COMPLEX GEOMETRIES

As numerical simulations become larger and are applied to problems with complex geometries, the mesh generation and management phases play more and more important role in the simulation process. To discretize a domain representing a given geometry, there are two different approaches we can take. We could create a special-purpose mesh generator which is designed for a specific application, or we could use an automatic mesh generator which can be applied to many applications. The two approaches are very different and both have advantages and disadvantages, but usually the type of problem under consideration will determine which approach is taken.

### SPECIAL-PURPOSE MESH GENERATION

A special-purpose mesh generator (Kalro et al., 1996) is one that is developed for a specific geometry and can only create meshes for the class of geometries it was originally designed for. An example of such a mesh generator might be one developed for flow past a flat plate where the geometry would be the discretization of a box. Another example might be a mesh generator developed for flow past a wing where only certain parameters about the wing like airfoil type, pitch angle, and wing aspect ratio can be changed. This approach has several advantages:

- Once the special-purpose mesh generator has been written, the cost of generating a finite element mesh is very minimal in terms of computing time.

- The user has direct control over the structure of the mesh and its refinement.

- In problems with moving boundaries and interfaces, such as free surfaces, moving mechanical components, and fluid-structure and fluid-particle interactions, the nodal coordinates need to be updated every time step, and usually a special-purpose mesh moving scheme can be incorporated within the mesh generator. By using such a mesh moving scheme, the deformation of the mesh is user controlled and requires very little computer time to generate a new set of nodes every time the mesh moves.

The disadvantages of this approach to mesh generation is the often time-consuming task of creating the special-purpose mesh generator for a specific application. If the application involves a geometry of any complexity, this method is very difficult to apply due to this difficulty.

*AUTOMATIC MESH GENERATION*

Automatic mesh generators, on the other hand, provide much more flexibility in discretizing complex domains. They involve little or no assumptions on the shape of the domain, and because of that, a mesh can be generated for almost any geometry. Also, the mesh generator is general and can be used in many applications. The disadvantage of this approach is that the user cannot specify exactly the type of mesh structure and refinement they may wish to have in all areas of the domain. Our automatic mesh generation package is based on Delaunay methods which are probably the most general type and yield high quality meshes in 3D.

Our automatic mesh generation package has three components (Johnson and Tezduyar, 1996a). The first component is an interactive 3D geometric modeler based on Bézier surfaces. By using this modeling program, geometries of almost any complex shape can be created. The second component is an automatic surface mesh generator which takes the geometric model as input and discretizes it into a surface mesh composed of triangular elements. The refinement level within this surface mesh is controlled by user-specified parameters within the geometric model. The third component is the actual 3D volumetric mesh generator which takes the surface mesh as input and generates the 3D mesh composed of tetrahedral elements. The level of refinement within the 3D mesh is determined by the refinement of the surface mesh. We also have the capability of creating thin, structured layers of elements around solid surfaces while using a totally unstructured mesh in the rest of the domain. By creating these thin layers, the boundary layer features of the flow can be modeled more accurately which is important when trying to represent viscous flows using unstructured meshes.

## 10.5   3D MESH UPDATE TECHNIQUES FOR FLOW PROBLEMS INVOLVING MOVING BOUNDARIES AND INTERFACES

A large class of fluid dynamics applications involve moving boundaries and interfaces, and in such numerical simulations, methods are needed to handle the changes in the shape of the domain. We employ two strategies, which may be used in combination, to handle these changes in the domain shape. These include techniques based on moving the nodal coordinates of the mesh, and those based on remeshing (i.e. generating an entirely new set of nodes and elements).

*SPECIAL-PURPOSE MESH MOVING SCHEME*

The space-time finite element method gives us the ability to move the mesh during a simulation. The movement of the boundary is known (prescribed or determined at every nonlinear iteration) and we have two methods to specify the displacement of the internal nodes. One approach is to use a special-purpose mesh moving scheme (usually associated with a special-purpose mesh generator) to specify the movement of the mesh based on some pre-defined motion (Kalro et al., 1996). In problems where the motion is pre-determined or constrained in some fashion, this motion can be programmed within the mesh generator, and every time the nodes are required to move a new mesh is generated in the new configuration. The connectivity of the new mesh remains the same as the old one so we are, in effect, just moving the nodes. This method is desirable due to the fact that the deformation of the mesh is user controlled and bounded, but its applicability is limited to known or a constrained motion for simple geometries. An example of such an application is flow past a wing that is allowed to translate and change its pitch angle. The translation can be handled by translating the entire mesh, and the change in pitch angle can be incorporated within the mesh generator (i.e. the pitch angle becomes an input parameter to the mesh generator).

## AUTOMATIC MESH MOVING SCHEME

For more arbitrary or unknown motions, we employ an automatic mesh moving scheme (Johnson and Tezduyar, 1994; Johnson and Tezduyar, 1996b). In this method, we solve the modified equations of linear elasticity to determine the internal nodal displacements with the specified movement of the boundary as a boundary condition. This system is solved using the finite element method every time the mesh is required to move. The modification we introduce to the equations of linear elasticity relies on eliminating from the formulation the Jacobian of the transformation between the element and physical domains. By doing this, smaller elements have a higher effective stiffness coefficient and retain their shape better during mesh motion. This method can be applied to any type of mesh with arbitrary motion, but this comes at the cost of solving a finite element formulation, possibly at every nonlinear iteration. Also, this method is usually the only option for moving the mesh if an unstructured mesh, typically coming from an automatic mesh generator, is being used.

## REMESHING

If, while using the automatic mesh moving scheme, the quality of the mesh is degraded by the mesh motion, we can remesh by generating an entirely new set of nodes and elements and by projection of the solution from the old mesh on to the new one (Johnson and Tezduyar, 1996b). This remeshing is usually carried out in conjunction with an automatic mesh generator. We can, of course, use remeshing to handle all mesh motion, but this is undesirable due to the cost of using the automatic mesh generator, projection of the solution, and the projection errors introduced. By combining remeshing with the automatic mesh moving scheme, we can reduce the remeshing cost as much as possible. We can also reduce the projection errors by using advanced projection schemes such as incorporating this projection within the jump term of the space-time finite element formulation.

## HYBRID MESH MOVING SCHEME

During a typical numerical simulation involving mesh motion, we use the automatic mesh moving scheme and remeshing in combination with each other. The automatic mesh moving scheme is our main mechanism of mesh motion, but when we find that this mesh has become too distorted, we remesh. During a simulation, we typically track some measures of mesh distortion based on element volume and aspect ratio, and after each computer run, we check these measures to determine if remeshing should take place. If so, we generate a new mesh using an automatic mesh generator (often on a different computer than the one used to carry out the simulation) and project the solution onto the new mesh. This remeshing procedure generates a new restart information which is then used in the next computer run. By using these techniques in combination, we can handle an almost unlimited range of motion while still minimizing the cost and errors associated with remeshing.

## 10.6   ITERATIVE SOLUTION STRATEGIES FOR LARGE-SCALE COMPUTATIONS

As outlined in Section 10.3, the finite element formulations we developed are either semi-discrete formulations, which are used to solve problems with fixed spatial domains, or space-time formulations, used to solve problems with changing spatial domains, such as those encountered in flows with moving boundaries and interfaces. Regardless of this distinction, the formulations like the ones outlined in Section 10.3 will lead to a nonlinear coupled system of equations:

$$N\left(\mathbf{d}_n\right) = \mathbf{F}, \tag{10.16}$$

where $\mathbf{d}_n$ is the vector of unknowns associated with marching from time step $n-1$ to $n$ in a semi-discrete formulation, or associated with space-time slab $n$ in a space-time formulation. This system is typically large, involving up to 38 million equations in our computations. It needs to be solved at every time step, in the case of transient simulations, or at every pseudo-time step, in the case of steady-state simulations.

For the nonlinear system of equations (10.16), the Newton-Raphson iterations

$$\left.\frac{\partial N}{\partial \mathbf{d}}\right|_{\mathbf{d}_n^k} \left(\Delta \mathbf{d}_n^k\right) = \mathbf{F} - N\left(\mathbf{d}_n^k\right), \tag{10.17}$$

each require the solution of a linear equation system

$$\mathbf{A}_n^k \mathbf{x}_n^k = \mathbf{R}_n^k, \tag{10.18}$$

where $k$ is the nonlinear iteration counter, $\mathbf{A}_n^k = \partial N/\partial \mathbf{d}|_{\mathbf{d}_n^k}$ is the nonsymmetric Jacobian matrix, $\mathbf{x}_n^k = \Delta \mathbf{d}_n^k$ is the vector of increments for unknown solution values, and $\mathbf{R}_n^k = \mathbf{F} - N\left(\mathbf{d}_n^k\right)$ is the vector of residuals. When discussing the process of the solution of the linear equation system (10.18), the sub- and superscripts identifying the time step and nonlinear iteration will be dropped, as only one such system is solved at a given time.

The size of the linear equation systems under consideration precludes the use of direct solutions techniques. We focus instead on iterative strategies. Such strategies are typically based on selection of a preconditioning matrix (which approximates the original matrix) and an update technique which governs how the solution vector is updated as corrections are computed based on the preconditioning matrix selected.

## PRECONDITIONING

The choice of the preconditioning matrix for the iterative solvers on parallel machines is becoming more varied in recent years. For production runs, the diagonal and nodal-block-diagonal preconditioning (Shakib, 1988) are still extensively used, and work well for the compressible flow problems, and adequately for the incompressible flow problems. We have also developed more complex preconditioners, among them are the clustered-element-by-element (CEBE) preconditioner (Liou and Tezduyar, 1992), and the mixed CEBE and cluster companion (CC) preconditioner (Tezduyar et al., 1992a). We later implemented some of these complex preconditioners on parallel platforms.

## ITERATIVE UPDATE TECHNIQUES

As an update technique, we employ the Generalized Minimum Residual (GMRES) (Saad and Schultz, 1986), which is a popular method for nonsymmetric systems stemming from Navier-Stokes or Euler equations. In the GMRES procedure, the full system is projected onto a much smaller Krylov subspace, and the residual is minimized on that subspace. The size of the Krylov space required for convergence is usually smaller than 15 for compressible flow problems (Shakib et al., 1989). This requirement increases for incompressible flow problems and finer meshes.

## MATRIX-BASED SCHEMES

A related issue is the interaction of the linear equation system matrix with residual-type vectors, which occurs at various stages of the solution update process. For moderately sized problems, the matrix-vector products are computed directly from the element-level matrices, without the global sparse matrix ever being assembled. Employing the element-by-element storage, the matrix-vector product $\mathbf{A}\mathbf{x}$ is actually computed as $\mathsf{A}\,(\mathbf{a}^e \mathbf{x}^e)$, using the fact that $\mathbf{A} = \mathsf{A}\,\mathbf{a}^e$, where $\mathbf{a}^e$ are the element-level matrices, $\mathbf{x}^e$ is the vector $\mathbf{x}$ distributed (gathered) to the element level, and $\mathsf{A}$ represents the finite element assembly operator.

## MATRIX-FREE SCHEMES

For larger problems, a matrix-free scheme is typically used, which results in significantly lower memory demands, but can involve more computations than its matrix-based counterpart described earlier. In this scheme, the matrix-vector product is approximated by two residual evaluations, eliminating the need to store the element-level matrices (Johan et al., 1991). The matrix-vector product $\mathbf{A}(\mathbf{d})\mathbf{x}$ can be written as $(\mathbf{R}(\mathbf{d} + \varepsilon \mathbf{x}) - \mathbf{R}(\mathbf{d}))/\varepsilon$, where $\varepsilon$ is a suitably small number. In matrix-free computations, the complex analytical task of finding the Jacobian matrix $\mathbf{A}(\mathbf{d})$ of the nonlinear system is also avoided.

## 10.7  PARALLEL COMPUTING

For a number of years, the computation speed delivered by a single processing unit has been growing only moderately, and so did the performance of the traditional scalar and vector supercomputers. In contrast, parallel computers compensate for the limited speed of a single processing node by linking tens, hundreds or thousands of nodes together with fast communication channels. Using this approach, computational speeds on the order of a trillion floating point operations per second (TeraFLOPS) are within grasp, with sustained rates of 10–50 GigaFLOPS achievable as of now. More often than not, however, the parallel algorithm design and programming techniques have to take the specific hardware features into consideration, and thus diverge from the methods which were used on scalar machines. For the time being, the task of using a parallel machine efficiently is largely the burden of the numerical analyst, as the parallel compilers and operating systems are yet unable to hide various characteristic features of the parallel machine, especially the memory hierarchy. This burden is smaller on moderately parallel machines which use a shared memory architecture, but quite acute on highly parallel engines employing distributed memory.

All the numerical examples presented in Section 10.8 were obtained on either Connection Machine CM-5, CRAY T3D, or 20-processor Silicon Graphics Onyx computers. Our implementations on the distributed memory CM-5 use a data-parallel computing paradigm, exemplified by the Connection Machine Fortran (CMF) programming language, which hides somewhat the distinction between on-processor and off-processor memory from the user. However, for efficient computation, every effort must still be taken to minimize the amount of the compiler-generated interprocessor communication. The implementations on the distributed memory T3D use a message-passing computing paradigm, making the interprocessor communication explicit. The said communication is accomplished using the Parallel Virtual Machine (PVM) library. Onyx implementations take advantage of the true shared memory architecture of this machine. Consequently, the issue of the interprocessor communication gives way to scalar optimization as the main approach for increasing the efficiency of the implementation.

### DISTRIBUTED MEMORY IMPLEMENTATIONS

The first two implementations, based on the CM-5 and the T3D, share a number of features. The chief among them is the explicit partitioning of the finite element mesh into contiguous subdomains, to be assigned later to individual processors. The corresponding nodes which are completely interior to the subdomain are also assigned to that processor. The nodes which belong to the subdomain interfaces are randomly assigned to one of the subdomains sharing that node, and to that subdomain's processor. An example of the partitioning of an unstructured finite element mesh around an M1 tank is shown in Figure 10.1. The main form of the communication encountered in the "parallel-ready" solution techniques, such as those discussed in Section 10.6, takes the form of exchange between element-level (stiffness matrices, local residuals) and the node-level (global residuals and increments) data structures (Kennedy et al., 1994). The transfer of data from the element-level to the node-level takes the form of a scatter, while the movement in the opposite direction takes the form of a gather. In the presence of partitioning, both these operations can be performed in two stages. In one stage, constituting the bulk of data motion, only local gather or scatter are performed, with no interprocessor communication. In the second stage, the subdomain surface data is communicated to appropriate remote processors. On the CM-5, this technique is built into the Connection Machine Scientific Software Library (CMSSL) partitioned gather and scatter subroutines. On the T3D, PVM-based counterparts of these routines were written by the authors.

## SHARED MEMORY IMPLEMENTATIONS

Shared memory machines, such as the Onyx, are multiprocessor architectures with a central memory, and as a result, interprocessor communication does not figure significantly in the overall cost. This model of parallelism focuses on distributing segments of DO loops across processors. This is achieved by embedding compiler directives, as is illustrated below in a code fragment for a dot product:

```
      dot = 0.0
C$DOACROSS REDUCTION(dot)
      do i=1,n
            dot = dot + a(i) * b(i)
      end do
```

The DOACROSS directive tells the compiler to distribute n/npes iterations of the loop to each of npes processors. Here dot is treated as a *reduction* type variable, i.e., each processor computes its share of the dot product in its private copy of dot, these contributions are at the end summed together. In our implementation, the DOACROSS directive is applied to the loop over the elements, thus the formation of element level quantities in parallel is easily achieved. The gather and scatter operations are embedded within the element loop. All quantities are stored at the nodal level, and are gathered to element level, when needed, using the element connectivity data. Implementation of the scatter is not straightforward. As mentioned earlier, some nodes are shared by elements, which may be allocated to different processors. Direct assembly of the global data components associated with such nodes would result in data collisions, leading to erroneous results. Thus, as a preprocessing step, such nodes are isolated using the element connectivity. Each processor assembles its share for these nodes, these are then summed outside the element loop, similar to the two-step scatter algorithm used on the CM-5 and T3D. The use of mesh partitioning reduces the number of nodes requiring this special treatment. Matrix-free solvers are used to minimize memory requirements. With the current implementation, it was possible to simulate, for example, the flow around a ram-air parachute with 1,139,145 unknowns with less than 280 Mbytes of memory.

## PERFORMANCE MEASUREMENTS

There are many criteria which may be used in evaluating the performance of a numerical method. The one which is perhaps the most relevant to a computational scientist who attempts to solve a given problem, is the time-to-solution. This benchmark takes into account all the diverse factors, such as the efficiency of the chosen algorithm, utilization of the particular hardware platform as a percentage of its peak speed, and even the effort required to include additional capability into the numerical code. However, the multitude of the quantities rolled into this benchmark makes it quite difficult to compare between different researchers. The more universal, and more widespread, performance benchmark is the raw computational speed, typically expressed in FLoating-point Operations Per Second (FLOPS). While acknowledging the limited value of such isolated performance figures, we still believe that they give at least an approximate measure of the capability of a given algorithm-architecture combination.

We provide therefore, for each of the three architectures mentioned, the approximate speed rating of our finite element codes. For the distributed-memory architectures, the ratio of communication to computation is also a telling figure. The CM-5 performance of the class of codes used by the authors has been discussed extensively in (Kennedy et al., 1994). We reiterate, that the communication-free portions of the code, such as the formation of element-level residuals, are being executed at approximately $24 \times 10^6$ FLOPS (24 MFLOPS) per processing node. The ratio of communication to computation depends greatly on the size of the Krylov space in the GMRES solver; typically it ranges from 18% for a matrix-free implementation, to 31% for a matrix-based implementation. As a matrix-free computation is dominated by the formation of the element-level residuals, the overall speed is close to that of the communication-free portion of the code. In a matrix-based implementation, as the other parts of the computation assume more important role, they reduce the overall speed to as low as 10 MFLOPS per node.

On the CRAY T3D, the speed of the communication-free portions of the implementation is measured at around 24 MFLOPS per node. The low-latency communication network of that machine results in an improved communication-to-computation ratio in a 9% range, in the case of a matrix-free code. The overall speed for such code is observed to be 18 MFLOPS per node. For both CM-5 and T3D, these figures are consistent for all partition sizes, and they are taken on a minimum partition which is able to accommodate given problem.

In the SGI ONYX shared memory implementation, the element-level residual formation proceeds at 22 MFLOPS per node. In the absence of performance penalties associated with a distributed memory architecture, the overall speed is quite similar, at 19 MFLOPS per node. The maximum number of processors which was found to sustain this performance is 16.

## 10.8  EXAMPLES

All computations are carried out in 3D and on parallel platforms.

### *STEADY-STATE DESCENT OF A LARGE RAM-AIR PARACHUTE*

Ram-air parachutes are finding increasing use in military operations for accurate delivery of large payloads such as supply crates and vehicles. HPC tools are currently being developed (Aliabadi et al., 1995; Garrard et al., 1995; Kalro et al., 1996) to predict the performance of these parachutes, from inflation to full deployment. The aerodynamics of such parachutes involves very complex phenomena, thus very fine meshes are required for an accurate representation of the flow features. The top image in Figure 10.2 shows the steady-state pressure distribution on a ram-air parachute with no flaps. The angle of attack is 2 degrees and Reynolds number 10 million. This computation, carried out on a CRAY T3D, required the solution of over 38 million coupled, nonlinear equations at every pseudo-time step. This is the largest problem we have solved with our current implementations. The bottom image in Figure 10.2 shows the flow past a parachute with flaps. The pressure distribution is indicated on the parachute surface, together with stream tubes colored with pressure. This simulation was carried out on the CM-5 and required the solution of 2,363,887 equations at every pseudo-time step. This simulation, and the next two, are part of a collaborative project with W. Garrard (University of Minnesota), and K. Stein and E. Steeves (Natick Research, Development, and Engineering Center).

### *LONGITUDINAL DYNAMICS OF A LARGE RAM-AIR PARACHUTE*

It is essential to evaluate the capability of a parachute to settle into steady glide, and indicate conditions where dynamic transients are lightly damped (Lingard, 1995). The desired steady glide configuration for the parachute is achieved by rigging the canopy, i.e., adjusting the center of gravity (c.g.) of the system such that equilibrium is obtained in the desired configuration. Mittal (Mittal and Tezduyar, 1994) has investigated the effect of location of the c.g. on the dynamic behavior of 2D airfoils. In this preliminary simulation, the longitudinal characteristics of ram-air parachutes are assessed. A hexahedral mesh with 291,437 nodes and 279,888 elements is used in this space-time computation. The parachute has a Clark Y airfoil cross section and a rigging angle of 9 degrees. The aspect ratio is 3.0 and the ratio of line length to span is 0.6. At every time step the solution of 2,258,496 coupled, nonlinear equations is required. A mesh-moving scheme similar to the one used in (Mittal and Tezduyar, 1994) accommodates the pitching motion of the parachute. Further, the motion of the parachute is confined to a vertical plane. The effect of line drag is modeled as a single force acting on the midpoint of the line joining the quarter chord of the canopy to the payload (Lingard, 1995). The parachute is released at 45 degrees with 0 pitch angle. It exhibits a tendency to rapidly pitch forward, accelerate and settle into steady glide. This steady glide configuration is used as an initial condition for flare maneuvers. Figure 10.3 shows the parachute at two instants. The pressure field and mesh in corresponding section are also shown. This computation was carried out on the CRAY T3D.

## FLARE MANEUVER OF A LARGE RAM-AIR PARACHUTE

One of the favorable characteristics of ram-air parachutes is the capability to deliver loads with reduced landing impact. This maneuver is achieved by pulling on the flaps at either end, and is termed as flaring. The increase in the effective camber creates large aerodynamic forces, this in turn causes the parachute system to decelerate. A specific mesh generator was developed to represent the parachute geometry together with flaps. This mesh also allows for the motion of the flaps during the flare without overt distortion of elements. As a result, the entire flare maneuver is simulated without the requirement to remesh; thus reducing the overheads in the parallel computation. The mesh used results in 2,258,496 coupled, nonlinear equations which are solved at every time step. The space-time finite element formulation is used in this problem. Here, the mesh moves together with the parachute. The initial condition consists of the steady glide configuration of an unconstrained parachute with no flap deflection. The time for the flare maneuver and total flap deflection is obtained from test data. The parachute is treated as a solid body with changing shape. The shape of the parachute during the maneuver is interpolated from the initial and final flap configurations. At the end of the maneuver there is a significant decrease in the horizontal component of the velocity, and this is consistent with flight data. The Reynolds number for this simulation is approximately 10 million. An algebraic turbulence model is used in the computation. This simulation was carried out on the CRAY T3D. Figure 10.4 shows the pressure distribution on the parachute surface during three instants of the flare maneuver.

## FLOW PAST THE SPILLWAY OF THE OLMSTED DAM

In this problem we investigate the water flow in the spillway of the Olmsted Dam on the Ohio River. The Dam was under study by the U.S. Army Corps of Engineers for possible modifications of the spillway bed. Several experimental models were constructed with the aim of analyzing the candidate designs for the scour protection of the spillway bed. The geometrical model represents a 48 feet wide section of the navigation pass crest and stilling basin, and includes a long upstream channel, the spillway crest, and a set of underwater obstacles designed to dissipate the energy of the flow. This geometry and its surface discretization are shown in Figure 10.5. The mesh consists of 139,352 space-time nodes and 396,682 tetrahedral space-time elements. The top surface of the mesh is free and allowed to move in the vertical direction. A stabilized mesh surface movement mechanism is employed (Soulaimani et al., 1991). In the interior of the domain, the position of the nodes is being updated based on the linear elasticity formulation described in Section 10.5. The 533,172 flow equations and the 168,846 mesh displacement equations are solved using the GMRES update technique. Also shown in Figure 10.5 are the pressure field and streamlines, and the steady shape of the free surface achieved in the final stages of the time-dependent simulation. This problem was computed on the CM-5. This simulation is part of a collaborative project with R. Stockstill and C. Berger (Waterways Experiment Station).

## CONTAMINANT DISPERSION AROUND AN M1 BATTLE TANK

In this problem, contaminant dispersion around a M1 battle tank is simulated on the CRAY T3D. This simulation is carried out in two stages. First, the full Navier-Stokes equations are solved to obtain the velocity field around the tank. This velocity field is used in the second stage in the time-dependent contaminant advection-diffusion equation to obtain the concentration of contaminant around the tank. In this computation, the tank is stationary and is subjected to a wind speed of 20 miles per hour. The contaminant is released in front of the tank from a point source with constant strength. The Reynolds, Prandtl, and Lewis numbers in this computation are 1.6 million, 0.72, and 1.0 respectively. The computations are carried out on an unstructured mesh consisting of 100,379 nodes and 527,586 tetrahedral elements.

Approximately, 0.4 million coupled, nonlinear equations are solved at every pseudo-time step to obtain the steady-state flow field, and 0.1 million equations are solved at every time step to obtain the transient solution of contaminant dispersion. The time-history of the contaminant concentration obtained from such simulations can be used to study visibility during battlefield conditions.

The top picture in Figure 10.6 shows the pressure distribution on the tank surface corresponding to the steady-state. The bottom pictures in Figure 10.6 show the contaminant concentration at two instants, in each case viewed from two different angles. This simulation is part of a collaborative project with C. Nietubicz (Army Research Laboratory).

## CONTAMINANT DISPERSION IN A MODEL SUBWAY STATION

In this problem, the dispersion of a contaminant in a model subway station is simulated. The subway station has two entrances on each side and four vents located on the upper surface, as shown in Figure 10.7. The air is assumed to enter the subway station from the left (inflow boundary) at a rate of 0.75 m/s. The air is also ventilated through each vent at rate of 0.15 m/s. The simulations, carried out on the CRAY T3D, are based on the solution of same set of equations as those solved for the M1 battle tank described earlier. Here, the contaminant is released from a point source with constant strength located close to the inflow boundary. The Prandtl and Lewis numbers in this computation are 0.72, and 1.0 respectively.

The unstructured mesh used in these simulations consists of 187,612 nodes and 1,116,992 tetrahedral elements. The steady-state solution of the incompressible Navier-Stokes equations is obtained by solving over 0.65 million coupled, nonlinear equations at every pseudo-time step. The transient solution of the equations governing contaminant dispersion involves solution of a linear system with more than 0.15 million equations at every time step.

The top image in Figure 10.7 shows the "surface" discretization of the model subway station. The bottom image in Figure 10.7 shows the contaminant concentration at an instant.

## NATURAL CONVECTION PROCESS

These simulations are carried out to study natural convection in a fluid layer with an internal heat source. These types of problems are encountered in a wide variety of astrophysical and geophysical phenomena such as plate motions on the earth. This simulation is carried out on the CRAY T3D for a cubic domain, with unit length dimensions. A structured finite element mesh with $20 \times 20 \times 20$ uniform cells is used in this computation. The Rayleigh and Prandtl numbers are 100,000 and 6.7, respectively. The Reynolds number is equal to the inverse of the Prandtl number and the non-dimensional internal heat source is set to 2. On the top of the cube, the non-dimensional temperature is set to zero, and on the other sides of the cube a zero heat flux condition is imposed. The no slip boundary condition is specified on each side of the cube. As the computation continues, the linear temperature distribution obtained during the early stage of the computation breaks down and the flow-field develops. At steady-state, the temperature gradient is highest on the top side of the cube, and a strong 3D vorticity is developed. The comparison between the computed and experimental Nusselt numbers are quite satisfactory (Kulacki and Emara, 1977).

The top image in Figure 10.8 shows the temperature field together with the mesh on three sides of the cube and an iso-surface of temperature corresponding to the steady-state solution. The bottom image in Figure 10.8 shows the time-history of the Nusselt number.

## AIRFLOW PAST AN AUTOMOBILE

In this example, described in more detail in (Johnson and Tezduyar, 1996a), we simulate airflow past an automobile traveling at a speed of 55 miles per hour. We chose to model the automobile after a Saturn SL2. We first generated a mesh for half of the domain (since the geometry is symmetric) using our automatic mesh generation package. This mesh contains 1,407,579 tetrahedral elements. We then reflected the mesh across the symmetry plane to represent the full automobile. This mesh contains 448,695 nodes and 2,815,158 elements. The automatic mesh generator has created three thin structured layers of elements around the automobile surface and wheels.

The Reynolds number for this flow condition is 6.9 million (based on the automobile length), and the flow is assumed to be incompressible and turbulent. We use a Smagorinsky turbulence model in which the physical viscosity is augmented by an eddy-viscosity to model the unresolved scales in the flow. For efficient use of the computing resources (512 node CM-5), a matrix-free implementation of the flow solver is used. The simulation is carried out under road conditions where the wheels are spinning (rotational velocity field is specified on the wheels), and the free-stream velocity is imposed on the road and inflow boundaries. We also carried out a simulation under wind-tunnel conditions where both the automobile surface and wheels are stationary, and we apply slip boundary conditions along the road. This second simulation is performed so we can compare the computed drag coefficient with those found experimentally in wind tunnels.

The computed drag coefficient under road conditions is 0.46 and under wind-tunnel conditions is 0.35. For comparison, the stated drag coefficient for a Saturn SL2 is 0.34, but the computational model we used is only a rough approximation to a Saturn SL2. Shown in Figure 10.9 is the pressure on the surface of the automobile and a set of streamlines. Velocity vectors at a cutting plane at 1/4 car lengths behind the automobile and at a section cutting the rear wheel are also shown in Figure 10.9. All images in the figures are for road conditions.

## MULTIPLE SPHERES FALLING IN A LIQUID-FILLED TUBE

In this example, described in more detail in (Johnson and Tezduyar, 1996b), we simulate four cases of multiple spheres falling in a liquid-filled tube. The number of spheres ranges from two to five, and the radius of each sphere is 1.0 while the radius of the cylinder is 5.0. The spheres are allowed to translate and rotate freely due to Newton's laws of motion as they react to gravity and the fluid forces. The simulations continue until terminal velocity and a stable configuration is reached.

We use the automatic mesh generation package to create the finite element meshes used in these simulations. The size of these meshes ranges from around 140,000 elements for two spheres to around 320,000 elements for five spheres. Around each sphere we create three structured layers of elements. The motion of the mesh is handled by using the automatic mesh moving scheme in combination with remeshing. Each of the four case differ by the number of spheres and their initial arrangement, but all other parameters are the same for each case. These parameters were set such that one sphere alone would fall at a terminal velocity Reynolds number of 100. In each case, the spheres and the fluid are initially at rest. We used an incompressible flow solver on the CM-5 to perform the calculations.

The first case (see Figure 10.10) involves two spheres initially arranged in a staggered configuration. As the spheres fall, the trailing sphere is attracted to the wake of the leading sphere, the spheres eventually collide and then separate. They fall side by side throughout the rest of the simulation. The Reynolds number at terminal velocity is 93.1. The second case (see Figure 10.10) involves three spheres initially aligned in a row but with uneven spacing. As they fall, they first form a neutrally-stable state where the spacing becomes even, but over a long period of time, eventually form an equilateral triangle. The Reynolds number at terminal velocity is 86.1. The third case (see Figure 10.11) involves five spheres initially in a slightly jumbled pentagon configuration. At the terminal velocity state, the spheres have moved into an exact pentagon shape. The Reynolds number at terminal velocity is 72.1. The fourth case (see Figure 10.11) again involves five spheres but are initially arranged in a pyramid shape. As they fall, the center sphere settles to a level even with the other four to form this second stable state of five spheres. The Reynolds number at terminal velocity for this case is 71.6.

*DYNAMICS OF A PARATROOPER JUMPING FROM A CARGO PLANE*

In this example we are studying the dynamics of a paratrooper jumping from a cargo aircraft which is traveling at 130 Knots. We use the space-time finite element formulation in combination with our mesh moving algorithms to numerically simulate this fluid-body interaction problem on the CRAY T3D.

A finite element mesh has been created for the cargo plane and paratrooper combination using our automatic mesh generation software. This mesh contains approximately 880,000 tetrahedral elements for half of the domain (since the geometry is symmetric). Since the flow conditions are at very high Reynolds numbers, a Smagorinsky turbulence model is used. We use the automatic mesh moving scheme in combination with remeshing to move the mesh in response to the motion of the paratrooper. The paratrooper is allowed to translate and rotate freely governed by Newton's laws of motion.

The initial condition for this time-dependent simulation is the steady-state flow past the cargo plane with the paratrooper located within the open side door of the plane (a portion of the interior of the plane is also modeled). The paratrooper motion is initialized with a small outward velocity. Figure 10.12 shows the steady-state pressure distribution on the cargo aircraft and also shows a close-up-view of the pressure distribution on the paratrooper with streamlines at one instant during the time-dependent simulation. The last image in Figure 10.12 shows the position and orientation of the paratrooper relative to the plane at five instants during the simulation. This simulation is part of a collaborative project with W. Sturek (Army Research Laboratory) and K. Stein (Natick Research, Development, and Engineering Center).

## 10.9   CONCLUDING REMARKS

This chapter focused on demonstrating that flow simulation, with recent, remarkable enhancements in high performance computing (HPC) capabilities in fluid dynamics, has come a long way and is being applied to many complex, real-world problems in aerospace engineering and applied fluid mechanics. The progress in HPC capabilities in fluid dynamics can be seen in both of the two major components of HPC: development of advanced algorithms capable of accurately simulating complex, real-world problems; and availability of new, advanced computer hardware and networking with sufficient power, memory and bandwidth to execute those simulations. It is also the case that while HPC enables flow simulation, flow simulation motivates development of novel HPC techniques. In this chapter, we covered several of our group's HPC tools and efforts, including methods for flow problems with moving boundaries and interfaces; automatic 3D mesh generation with structured meshes near the boundaries; 3D mesh update strategies; and parallel computations on a 512-node CRAY T3D, an 896-node Thinking Machines CM-5, and a 20-processor SGI Onyx. The 3D simulations we presented in this chapter included: flow past large ram-air parachutes; flow past the spillway of a dam; contaminant dispersion around an M1 battle tank; contaminant dispersion in a model subway station; natural convection process: airflow around an automobile; multiple spheres falling in a liquid-filled tube; and dynamics of a paratrooper jumping from a cargo plane.

## ACKNOWLEDGMENT

Sponsored by ARO, ARPA, NASA-JSC, and by the Army High Performance Computing Research Center under the auspices of the Department of the Army, Army Research Laboratory cooperative agreement number DAAH04-95-2-0003/contract number DAAH04-95-C-0008. The content does not necessarily reflect the position or the policy of the Government, and no official endorsement should be inferred. Cray C90 time was provided in part by the Minnesota Supercomputer Institute. The second author was partially supported by the Minnesota Supercomputer Institute.

## REFERENCES

Aliabadi, S., Ray, S., and Tezduyar, T. (1993). SUPG finite element computation of compressible flows with the entropy and conservation variables formulations. *Computational Mechanics*, 11:300–312.

Aliabadi, S. and Tezduyar, T. (1993). Space-time finite element computation of compressible flows involving moving boundaries and interfaces. *Computer Methods in Applied Mechanics and Engineering*, 107(1–2):209–224.

Aliabadi, S. and Tezduyar, T. (1995). Parallel fluid dynamics computations in aerospace applications. *International Journal for Numerical Methods in Fluids*, 21:783–805.

Aliabadi, S., Garrard, W., Kalro, V., Mittal, S., and Tezduyar, T. (1995). Parallel finite element computation of the dynamics of large ram air parachutes. In *Proceedings of AIAA 13th Aerodynamic Decelerator Systems Technology*, AIAA Paper 95-1581, Clearwater Beach, Florida.

Behr, M., Franca, L., and Tezduyar, T. (1993a). Stabilized finite element methods for the velocity-pressure-stress formulation of incompressible flows. *Computer Methods in Applied Mechanics and Engineering*, 104(1):31–48.

Behr, M., Johnson, A., Kennedy, J., Mittal, S., and Tezduyar, T. (1993b). Computation of incompressible flows with implicit finite element implementations on the Connection Machine. *Computer Methods in Applied Mechanics and Engineering*, 108:99–118.

Behr, M. and Tezduyar, T. (1994). Finite element solution strategies for large-scale flow simulations. *Computer Methods in Applied Mechanics and Engineering*, 112:3–24.

Brooks, A. and Hughes, T. (1982). Streamline upwind/Petrov-Galerkin formulations for convection dominated flows with particular emphasis on the incompressible Navier-Stokes equations. *Computer Methods in Applied Mechanics and Engineering*, 32:199–259.

Garrard, W., Tezduyar, T., Aliabadi, S., Kalro, V., Luker, J., and Mittal, S. (1995). Inflation analysis of ram air inflated gliding parachutes. In *Proceedings of AIAA 13th Aerodynamic Decelerator Systems Technology*, AIAA Paper 95-1565, Clearwater Beach, Florida.

Hansbo, P. and Szepessy, A. (1990). A velocity-pressure streamline diffusion finite element method for the incompressible Navier-Stokes equations. *Computer Methods in Applied Mechanics and Engineering*, 84:175–192.

Hughes, T., Franca, L., and Hulbert, G. (1989). A new finite element formulation for computational fluid dynamics: VIII. the Galerkin/least-squares method for advective-diffusive equations. *Computer Methods in Applied Mechanics and Engineering*, 73:173–189.

Johan, Z., Hughes, T., and Shakib, F. (1991). A globally convergent matrix-free algorithm for implicit time-marching schemes arising in finite element analysis in fluids. *Computer Methods in Applied Mechanics and Engineering*, 87:281–304.

Johnson, A. and Tezduyar, T. (1994). Mesh update strategies in parallel finite element computations of flow problems with moving boundaries and interfaces. *Computer Methods in Applied Mechanics and Engineering*, 119:73–94.

Johnson, A. and Tezduyar, T. (1995a). Mesh generation and update strategies for parallel computation of 3D flow problems. In *Computational Mechanics 95, Proceedings of International Conference on Computational Engineering Science*, Mauna Lani, Hawaii.

Johnson, A. and Tezduyar, T. (1995b). Numerical simulation of fluid-particle interactions. In *Proceedings of the Ninth International Conference on Finite Elements in Fluids*, Venice, Italy.

Johnson, A. and Tezduyar, T. (1996a). Parallel computation of incompressible flows with complex geometries. to appear in *International Journal for Numerical Methods in Fluids*.

Johnson, A. and Tezduyar, T. (1996b). Simulation of multiple spheres falling in a liquid-filled tube. to appear in *Computer Methods in Applied Mechanics and Engineering*.

Kalro, V. and Tezduyar, T. (1995). Parallel finite element computation of 3D incompressible flows on MPPs. In Habashi, W., editor, *Solution Techniques for Large-Scale CFD Problems*. John Wiley & Sons.

Kalro, V., Aliabadi, S., Garrard, W., Tezduyar, T., Mittal, S., and Stein, K. (1996). Parallel finite element simulation of large ram-air parachutes. to appear in *International Journal for Numerical Methods in Fluids*.

Kennedy, J., Behr, M., Kalro, V., and Tezduyar, T. (1994). Implementation of implicit finite element methods for incompressible flows on the CM-5. *Computer Methods in Applied Mechanics and Engineering*, 119:95–111.

Kulacki, F. and Emara, A. (1977). Steady and transient thermal convection in a fluid layer with uniform volumetric energy sources. *Journal of Fluid Mechanics*, 55(2):271–278.

Le Beau, G., Ray, S., Aliabadi, S., and Tezduyar, T. (1993). SUPG finite element computation of compressible flows with the entropy and conservation variables formulations. *Computer Methods in Applied Mechanics and Engineering*, 104:397–

422.

Le Beau, G. and Tezduyar, T. (1991). Finite element computation of compressible flows with the SUPG formulation. In Dhaubhadel, M., Engelman, M., and Reddy, J., editors, *Advances in Finite Element Analysis in Fluid Dynamics*, FED-Vol.123, pages 21–27, New York. ASME.

Lingard, J. (1995). Ram-air parachute design. In *AIAA 13th Aerodynamic Decelerator Conference, 2nd ADS Technology Seminar*, Clearwater Beach, Florida.

Liou, J. and Tezduyar, T. (1992). Clustered element-by-element computations for fluid flow. In Simon, H., editor, *Parallel Computational Fluid Dynamics – Implementation and Results*, Scientific and Engineering Computation Series, chapter 9, pages 167–187. The MIT Press, Cambridge, Massachusetts.

Mittal, S. and Tezduyar, T. (1994). Massively parallel finite element computation of incompressible flows involving fluid-body interactions. *Computer Methods in Applied Mechanics and Engineering*, 112:253–282.

Mittal, S. and Tezduyar, T. (1995). Parallel finite element simulation of 3d incompressible flows – Fluid-structure interaction. *International Journal for Numerical Methods in Fluids*, 21:933–953.

Saad, Y. and Schultz, M. (1986). GMRES: A generalized minimal residual algorithm for solving nonsymmetric linear systems. *SIAM Journal of Scientific and Statistical Computing*, 7:856–869.

Shakib, F. (1988). *Finite Element Analysis of the Compressible Euler and Navier-Stokes Equations*. PhD thesis, Department of Mechanical Engineering, Stanford University.

Shakib, S., Hughes, T., and Johan, Z. (1989). A multi-element group preconditionined GMRES algorithm for nonsymmetric systems arising in finite element analysis. *Computer Methods in Applied Mechanics and Engineering*, 75:415–456.

Soulaimani, A., Fortin, M., Dhatt, G., and Ouellet, Y. (1991). Finite element simulation of two- and three-dimensional free surface flows. *Computer Methods in Applied Mechanics and Engineering*, 86:265–296.

Tezduyar, T. and Hughes, T. (1983). Finite element formulations for convection dominated flows with particular emphasis on the compressible Euler equations. In *Proceedings of AIAA 21st Aerospace Sciences Meeting*, AIAA Paper 83-0125, Reno, Nevada.

Tezduyar, T. (1991). Stabilized finite element formulations for incompressible flow computations. *Advances in Applied Mechanics*, 28:1–44.

Tezduyar, T., Behr, M., Aliabadi, S., Mittal, S., and Ray, S. (1992a). A new mixed preconditioning method for finite element computations. *Computer Methods in Applied Mechanics and Engineering*, 99:27–42.

Tezduyar, T., Behr, M., and Liou, J. (1992b). A new strategy for finite element computations involving moving boundaries and interfaces – the deforming-spatial-domain/space-time procedure: I. The concept and the preliminary tests. *Computer Methods in Applied Mechanics and Engineering*, 94(3):339–351.

Tezduyar, T., Behr, M., Mittal, S., and Johnson, A. (1992c). Computation of unsteady incompressible flows with the finite element methods – space-time formulations, iterative strategies and massively parallel implementations. In Smolinski, P., Liu,

W., Hulbert, G., and Tamma, K., editors, *New Methods in Transient Analysis*, AMD-Vol.143, pages 7–24, New York. ASME.

Tezduyar, T., Behr, M., Mittal, S., and Liou, J. (1992d). A new strategy for finite element computations involving moving boundaries and interfaces – the deforming-spatial-domain/space-time procedure: II. Computation of free-surface flows, two-liquid flows, and flows with drifting cylinders. *Computer Methods in Applied Mechanics and Engineering*, 94(3):353–371.

Tezduyar, T., Mittal, S., Ray, S., and Shih, R. (1992e). Incompressible flow computations with stabilized bilinear and linear equal-order-interpolation velocity-pressure elements. *Computer Methods in Applied Mechanics and Engineering*, 95:221–242.

Tezduyar, T., Aliabadi, S., Behr, M., Johnson, A., and Mittal, S. (1993). Parallel finite-element computation of 3D flows. *IEEE Computer*, 26-10(10):27–36.

Tezduyar, T., Aliabadi, S., Behr, M., and Mittal, S. (1994). Massively parallel finite element simulation of compressible and incompressible flows. *Computer Methods in Applied Mechanics and Engineering*, 119:157–177.

Tezduyar, T., Aliabadi, S., Behr, M., Johnson, A., Kalro, V., and Waters, C. (1995). 3D simulation of flow problems with parallel finite element computations on the Cray T3D. In *Computational Mechanics'95, Proceedings of International Conference on Computational Engineering Science*, Mauna Lani, Hawaii.

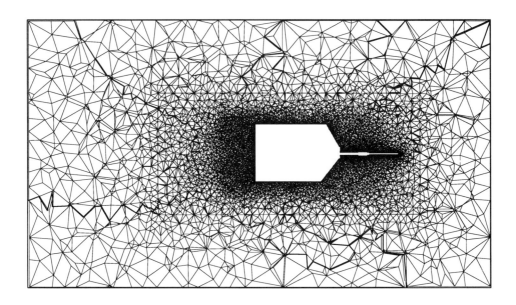

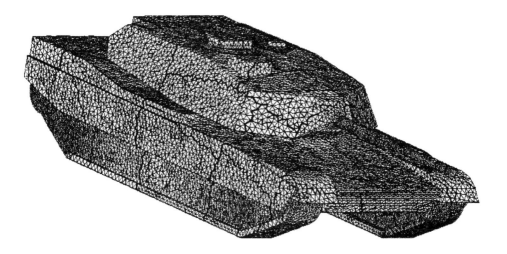

**Figure 10.1** An example of the partitioning of an unstructured finite element mesh around an M1 tank.

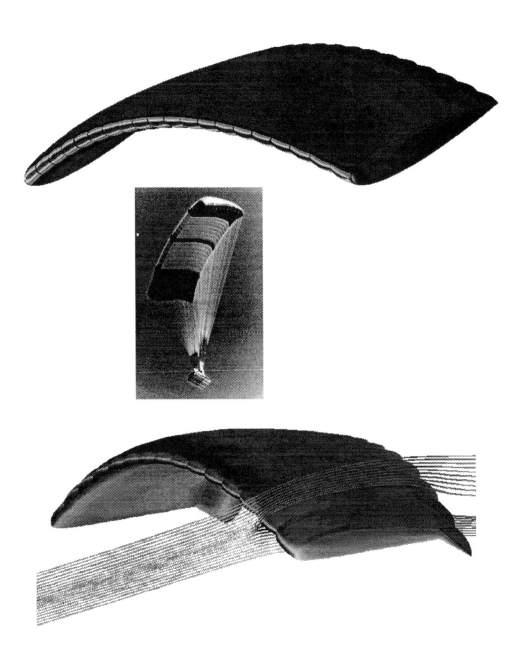

**Figure 10.2** The top image shows the pressure distribution on the surface of a parachute with no flaps. The bottom image shows the pressure distibution together with stream tubes on a parachute with flaps.

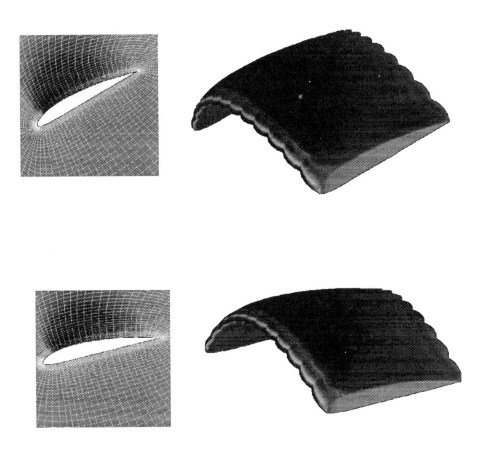

**Figure 10.3** Pressure distribution on the parachute surface and in two cross sections at two instants of pitcing. The mesh can also be seen in these sections.

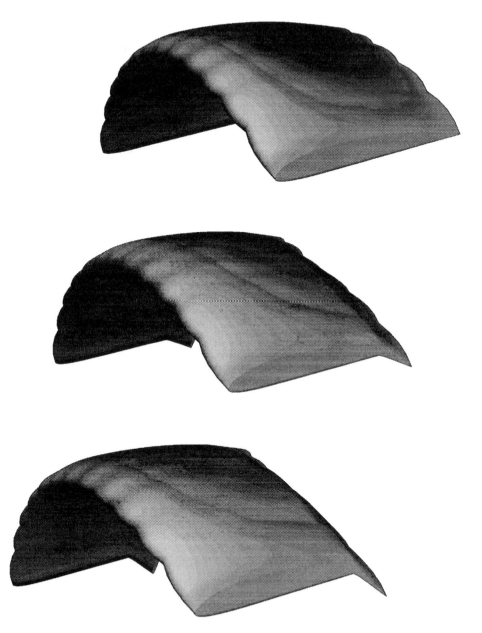

**Figure 10.4** Images show the pressure distribution on the parachute surface at three instants of flare

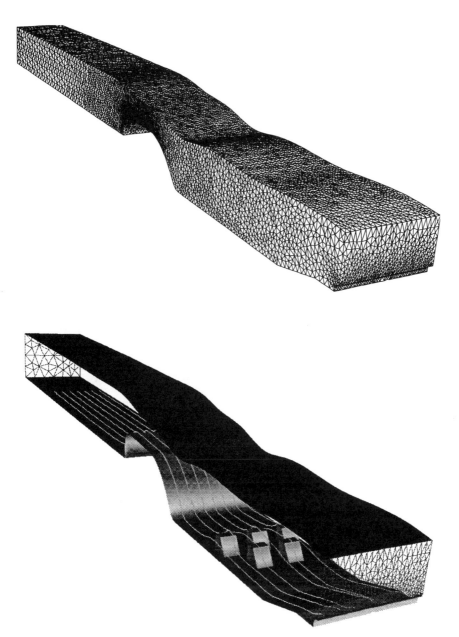

**Figure 10.5** The top figure shows the finite element mesh representing the spillway of the Olmsted Dam on the Ohio Rivier. The bottom figure shows the pressure field on the spillway bed and on the free surface, and a set of streamlines.

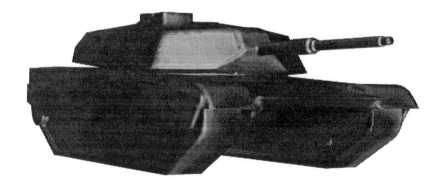

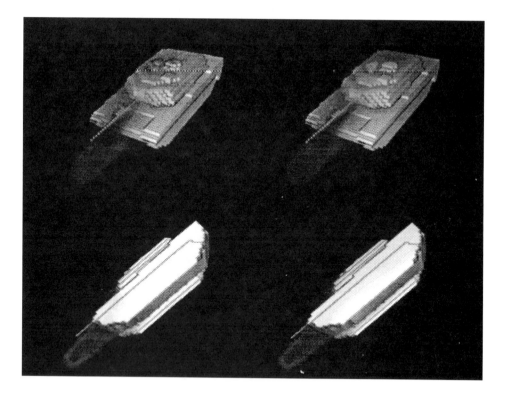

**Figure 10.6** The top picture shows the pressure distribution on the tank surface corresponding to the steady-state. The bottom pictures show the contaminant concentration at two instants from different views.

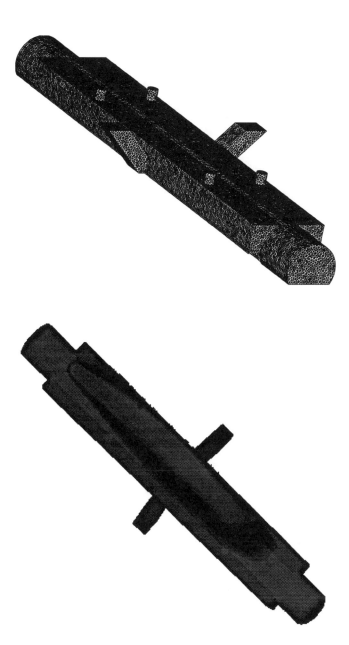

**Figure 10.7** The top image shows the surface discretization of the model subway. The bottom image shows the contaminant concentration field at an instant.

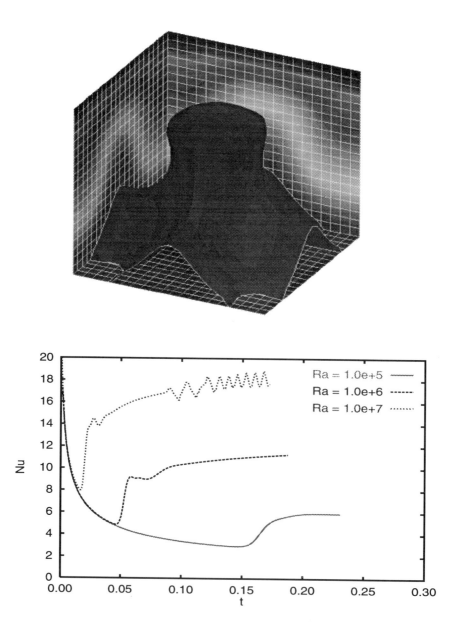

**Figure 10.8** The top image shows the temperature field together with the mesh on three sides of the cube and an iso-surface of temperature corresponding to the steady-state solution. The bottom image shows the time history of the Nusselt number.

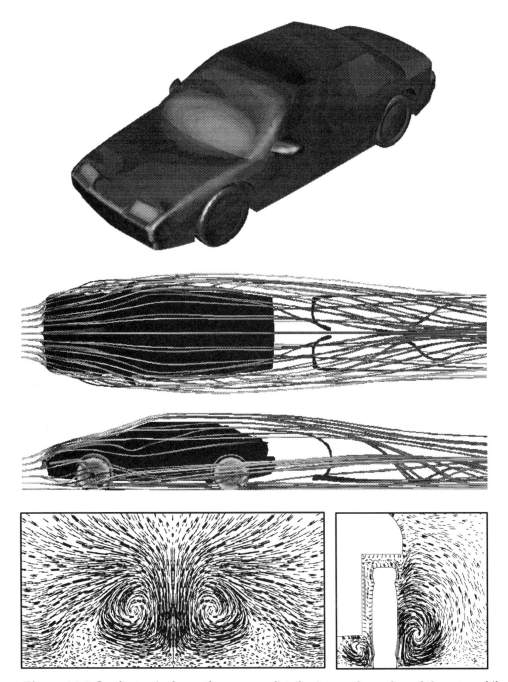

**Figure 10.9** On the top is shown the pressure distribution on the surface of the automobile while in the center is shown streamlines. On the bottom is shown velocity vectors at cross-sectios 1/4 car lengths behind the automobile and at a section cutting the rear wheel.

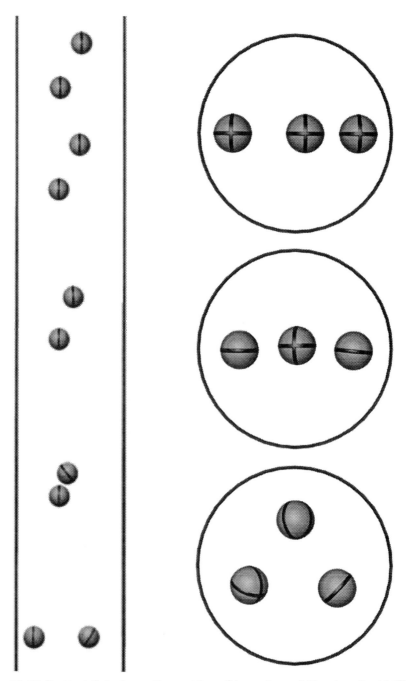

**Figure 10.10** On the left is shown the position of two spheres falling in a liquid-filled tube at five instants during the simulation. On the right is shown the position of three spheres falling in a liquid-filled tube at the initial, neutrally-stable, and stable states.

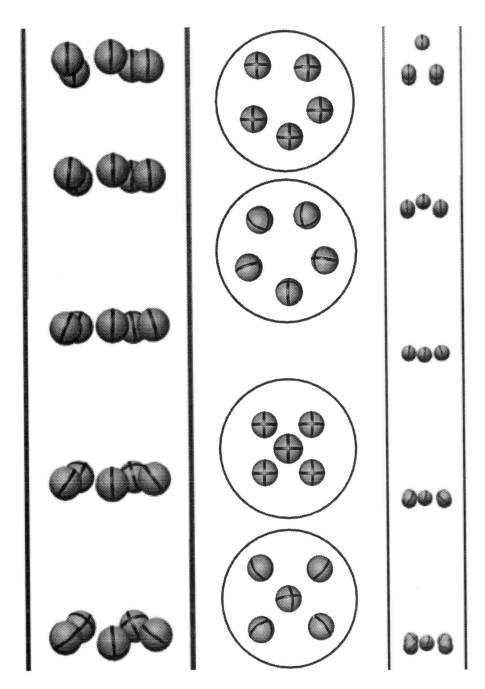

**Figure 10.11** On the left is shown the position of five spheres falling in a liquid-filled tube at five instants. Top views of the initial and final positions are shown at top-center. On the right is shown the position of five spheres falling in a liquid-filled tube at five instants. Top views of the initial and final positions are shown at bottom-center.

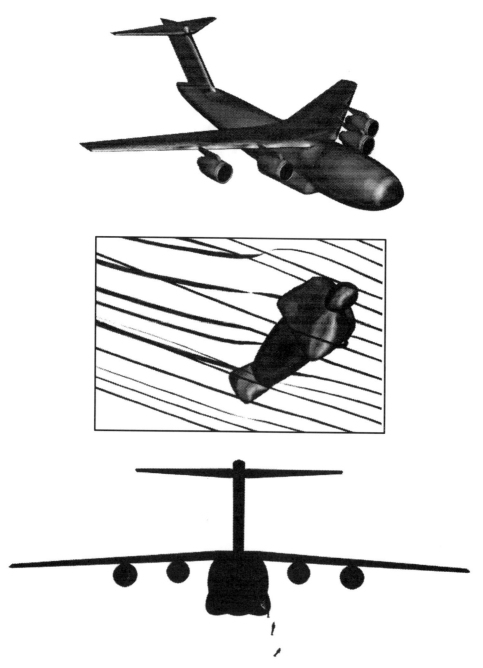

**Figure 10.12** On the top is shown the steady-state pressure distribution on the surface of the cargo plane, and in the center is the pressure distribution on the paratrooper with streamlines. On the botton is shown the position and orientation of the paratrooper at different instants during the simulation.

# 11

# Advanced Computational Technology For Product Design

Ramesh K. Agarwal[1] and Vijaya Shankar[2]

## 11.1 ABSTRACT

The science of physical systems computation, which involves solving appropriate mathematical models using state-of-the-art computer hardware from graphics workstations to supercomputers to massively parallel architectures in simulating various physical processes, has progressed dramatically over the last 30 years. Many aerospace developments from the current Space shuttle to the future reusable launch vehicles (RLV), and other advanced fighters, commercial airplanes and helicopters, have critically benefited from this computational technology in reducing the design-cycle cost and time to market. While this computational technology development has been largely due to major funding from the defense sector, with the recent decline in the aerospace markets as a result of the end of the cold war, increasingly the trend is to apply computational simulations to commercial product development. This paper provides several examples of Computational Fluid Dynamics (CFD) and Computational Electromagnetics (CEM) technology to aerospace products and outlines some examples of the role of computational simulation in the transition from aerospace to commercial products.

## 11.2 INTRODUCTION

The development of modern aerospace vehicles increasingly requires synergism in integrating multidisciplinary technologies such as 1) fluid dynamics for flow management, 2) structures for flexibility effects, 3) propulsion for thrust, 4) controls for stability, and 5) low observables for stealth considerations. Implementing such an integrated approach demands development of a computational environment referred to as 'Computational Sciences' that encompasses many disciplines. The mission of Computational Sciences is to combine the strengths of mathematics and supercomputers to better understand/simulate problems in physical sciences and integrate such multidisciplinary technologies to achieve synergism in the design

[1]   Aerospace Engineering Department, Wichita State University, Wichita, KS 67260, USA
[2]   Rockwell International Science Center, Thousand Oaks, CA 91360, USA

*Parallel Solution Methods in Computational Mechanics* edited by M. Papadrakakis

process. Both government and industry view 'Computational Sciences' as a critical, potentially efficient and cost-effective technology for advanced design.

There are many elements in Computational Sciences that need to be addressed covering issues from research to application stage. At the research end, some of the issues are 1) development of stable numerical algorithms having high order of accuracy addressing not only the physics of the equations but also some of the emerging computer architectures with vector/parallel and massively parallel features, 2) representation of proper physics such as turbulence, shocks, and flow separation in fluid dynamics, material behavior in electromagnetics processes, and interdisciplinary coupling issues, and 3) code validation to ensure proper coupling of numerics and physics. At the applications end, some of the issues are 1) complex geometry representation, both surface modeling and grid setup, addressing the needs of different disciplines, and 2) code documentation and user training (as the computer codes become very complex, addressing increasingly nonlinear physics and general geometry, constant interface with users is crucial to ensure proper usage of the computational tools).

One of the computational disciplines that has been a supercomputing pace setter for the last three decades is CFD. This technology, which started with the development of the transonic small-disturbance theory in the late sixties, has matured both in algorithm and code development to the point that today it is able to solve the time averaged Navier Stokes equations for predicting the flowfield over a complete fighter. Today, CFD is playing a critical role in the development of advanced aerospace configurations and commercial products such as automobiles. Many of the attributes of CFD are also beginning to impact the simulation capabilities in other disciplines such as Computational Electromagnetics (CEM), Computational Aeroacoustics (CAA), Computational Structures and Manufacturing Processes, Semiconductor Device and Process Simulation, etc. Agarwal (1996) describes the application of CFD based approach to a variety of disciplines in computational physics.

## ATTRIBUTES OF CFD AND THEIR SIMILARITY WITH OTHER DISCIPLINES

- The fluid dynamic equations are usually cast in conservation form either as differential or local integral equations conserving mass, momentum, and energy fluxes, thus allowing for numerical capture of flow discontinuities such as shocks and slip surfaces. Equations representing the physics in other disciplines, for example Maxwell's equations in electromagnetics, Shroedinger equation, equations of elasticity, semiconductor hydrodynamic device simulation equations, Einstein's equations of general relativity, etc. can also be cast in similar conservation forms for conservation of appropriate fluxes (Agarwal 1996).
- Recent developments of hyperbolic algorithms for solving the time-domain Euler equations are based on the characteristic theory of signal propagation and are referred to as the 'upwind' schemes. For hyperbolic equations, the upwind based schemes can be constructed to provide the right amount of numerical dissipation to achieve both stability and accuracy. Current state of the art in numerical algorithms address construction of high order of accuracy for arbitrary cell shapes such as hexahedral, triangular prism, and tetrahedral elements.

- For treatment of complex aerospace configurations, CFD methods usually employ either a structured grid based body-fitted coordinate system or an unstructured finite-element grid setup for ease in implementing the various boundary conditions. Similar numerical geometry and grid setup procedures are equally applicable to modeling complex problems in other disciplines.
- For solving large problems as well as to achieve quick-turn-around response, the current state of the art is to employ massively parallel architectures and/or workstation clusters to distribute the problem over many computational nodes with appropriate message passing algorithms.
- Pre and post processing capabilities running on advanced graphics work stations are effectively employed to visualize and animate the geometry/grid and solution.

The goal of 'Computational Sciences' is to effectively employ the many advances in CFD coupled with emerging supercomputer architectures with expected teraflops (trillion floating point operations per second) performance to mature the computational technology in different disciplines and be able to perform multidisciplinary studies critical to advanced design in both aerospace and commercial arena.

## 11.3 COMPUTATIONAL FRAMEWORK

In order to apply conservation principles (for example, in fluid dynamics mass, momentum, and energy are conserved), many of the governing equations representing appropriate physical processes are written in conservation form. The general form of a differential conservation equation can be written as

$$\mathbf{Q}_t + \mathbf{E}_x + \mathbf{F}_y + \mathbf{G}_z = \text{Source } \mathbf{S} \tag{11.1}$$

where $\mathbf{Q}$ is the solution vector and $\mathbf{E}$, $\mathbf{F}$ and $\mathbf{G}$ are the fluxes in $x$, $y$, and $y$ coordinate directions, respectively. The conservation form readily admits weak solutions such as shock waves.

The integral form of the conservation laws which can easily be derived from the differential form by integrating (11.1) with respect to $x, y, z$ over any conservation cell whose volume is $V$ takes the form

$$\iiint_v \left( \frac{\partial \mathbf{Q}}{\partial t} + \frac{\partial \mathbf{E}}{\partial x} + \frac{\partial \mathbf{F}}{\partial y} + \frac{\partial \mathbf{G}}{\partial z} \right) dx \, dy \, dz = \iiint_v \mathbf{S} \, dx \, dy \, dz = \tilde{\mathbf{S}}. \tag{11.2}$$

This can be rewritten in vector notation as

$$\frac{\partial}{\partial t} \iiint_v \mathbf{Q} \, dx \, dy \, dz + \iiint_v (\vec{\nabla} \cdot \vec{\mathbf{F}}) \, dx \, dy \, dz = \tilde{\mathbf{S}} \tag{11.3}$$

In (11.3),

$$\vec{\mathbf{F}} = \mathbf{E}\widehat{\mathbf{j}} + \mathbf{F}\widehat{\mathbf{k}} + \mathbf{G}\widehat{\mathbf{l}} \tag{11.4}$$

Applying the Gauss divergence theorem, we can convert the volume integral into a surface integral and obtain.

$$\frac{\partial}{\partial t} (\tilde{\mathbf{Q}}V) + \iint_s (\vec{\mathbf{F}} \cdot \widehat{\mathbf{n}}) \, ds = \tilde{\mathbf{S}}. \tag{11.5}$$

In (11.5) the cell average of the dependent variables are denoted by $\tilde{Q}$:

$$\tilde{Q} = \frac{\iiint_v Q\,dV}{\iiint_v dV} \tag{11.6}$$

The outward unit normal at any point of the boundary surface of a cell has been denoted by $\hat{n} = n_x \hat{j} + n_y \hat{k} + n_y \hat{l}$ . The integral form of the conservation laws given by (11.5) defines a system of equations for the cell average values of the dependent variables. In order to construct numerical methods to solve the integral form of the conservation laws, one must be able to define cell geometries, approximate the dependent variables, develop spatial discretization procedures, develop time integration procedures to update the cell averages, etc. There are many numerical algorithmic issues that come into play in devising a solution procedure to solve (11.5). Some of them are 1) implicit and explicit schemes, 2) stability and order of accuracy, 3) relaxation and approximate factorization procedures, 4) central difference and upwind schemes, 5) finite-volume and finite difference schemes applied to a structured grid, and 6) finite-element-like finite-volume schemes for an unstructured grid setup.

The Navier–Stokes equations solved in CFD simulations take on the form

$$Q = \begin{pmatrix} \rho \\ \rho u \\ \rho v \\ \rho w \\ \rho e \end{pmatrix}, \quad E = \begin{pmatrix} \rho u \\ \rho u^2 + p - \tau_{xx} \\ \rho v u - \tau_{xy} \\ \rho w u - \tau_{xz} \\ (\rho e + p)u - u\tau_{xx} - v\tau_{xy} - w\tau_{xz} + q_x \end{pmatrix},$$

$$F = \begin{pmatrix} \rho v \\ \rho u v - \tau_{xy} \\ \rho v^2 + p - \tau_{yy} \\ \rho w v - \tau_{xz} \\ (\rho e + p)v - u\tau_{xy} - v\tau_{yy} - w\tau_{yz} + q_y \end{pmatrix}, \tag{11.7}$$

$$G = \begin{pmatrix} \rho w \\ \rho u w - \tau_{xz} \\ \rho v w - \tau_{yz} \\ \rho w^2 + p - \tau_{zz} \\ (\rho e + p)w - u\tau_{xy} - v\tau_{yz} - w\tau_{zz} + q_z \end{pmatrix}, \quad S = 0$$

where

$$p = (\gamma - 1)\left\{\rho e - \tfrac{1}{2}(u^2 + v^2 + w^2)\right\}, \quad \tau_{xx} = \frac{2\mu}{3}\left(2\frac{\partial u}{\partial x} - \frac{\partial v}{\partial y} - \frac{\partial w}{\partial z}\right),$$

$$\tau_{yy} = \frac{2\mu}{3}\left(2\frac{\partial v}{\partial y} - \frac{\partial u}{\partial x} - \frac{\partial w}{\partial z}\right), \quad \tau_{zz} = \frac{2\mu}{3}\left(2\frac{\partial w}{\partial z} - \frac{\partial u}{\partial x} - \frac{\partial v}{\partial y}\right),$$

$$\tau_{xy} = \mu\left(\frac{\partial u}{\partial y} + \frac{\partial v}{\partial x}\right), \quad \tau_{yz} = \mu\left(\frac{\partial v}{\partial z} + \frac{\partial w}{\partial y}\right), \quad \tau_{xz} = \mu\left(\frac{\partial u}{\partial z} + \frac{\partial w}{\partial x}\right),$$

$$q_x = -K\frac{\partial T}{\partial x}, \quad q_y = -K\frac{\partial T}{\partial y}, \quad q_z = -K\frac{\partial T}{\partial z}.$$

In (11.7), $p$ is pressure, $\rho$ is density, $(u, v, w)$ are the Cartesian velocity components, in the $x$, $y$, and $z$ directions, respectively, $e$ is the energy per unit mass and $\mu$ and $K$ are the molecular viscosity and thermal conductivity of the fluid medium. Euler equations are obtained by setting $\mu = 0$ and $K = 0$ in (11.7).

Similarly, the Maxwell equations used in CEM simulations are written in the form

$$\frac{\partial \boldsymbol{B}}{\partial t} = -\nabla x \boldsymbol{E} \tag{11.8}$$

and

$$\frac{\partial \boldsymbol{D}}{\partial t} = \nabla x \boldsymbol{H} - \boldsymbol{J} \tag{11.9}$$

The divergence conditions $\nabla \cdot \boldsymbol{D} = \rho$ and $\nabla \cdot \boldsymbol{B} = 0$ are derived directly from Maxwell's equations, where $\nabla \cdot \boldsymbol{J} = \partial \rho / \partial t$. The vector quantities $\boldsymbol{E} = (E_x, E_y, E_z)$ and $\boldsymbol{H} = (H_x, H_y, H_z)$ are the electric and magnetic field intensities, $\boldsymbol{D} = (D_x, D_y, D_z)$ is the electric displacement, $\boldsymbol{B} = (B_x, B_y, E_z)$ is the magnetic induction, $\boldsymbol{J} = (J_x, J_y, J_z)$ is the current density, and $\rho$ is the charge density. The subscripts $x, y, z$ in the vector representation of $\boldsymbol{E}, \boldsymbol{H}, \boldsymbol{B}$ and $\boldsymbol{D}$ refer to components in respective directions.

$$\mathbf{Q} = \left\{ \begin{array}{c} B_x \\ B_y \\ B_z \\ D_x \\ D_y \\ D_z \end{array} \right\}, \quad \mathbf{E} = \left\{ \begin{array}{c} 0 \\ -D_z/\varepsilon \\ D_y/\varepsilon \\ 0 \\ B_z/\mu \\ -B_y/\mu \end{array} \right\}, \quad \mathbf{F} = \left\{ \begin{array}{c} D_z/\varepsilon \\ 0 \\ -D_z/\varepsilon \\ -B_z/\mu \\ 0 \\ B_z/\mu \end{array} \right\},$$

$$\mathbf{G} = \left\{ \begin{array}{c} -D_y/\varepsilon \\ D_x/\mu \\ 0 \\ B_y/\mu \\ B_x/\mu \\ 0 \end{array} \right\}, \quad \mathbf{S} = \left\{ \begin{array}{c} 0 \\ 0 \\ 0 \\ -J_x \\ -J_y \\ -J_z \end{array} \right\}. \tag{11.10}$$

In what follows, the permittivity coefficient $\varepsilon$ and the permeability coefficient $\mu$ are taken to be isotropic, scalar material properties and satisfy the following relationship, $\boldsymbol{D} = \varepsilon \boldsymbol{E}$ and $\boldsymbol{B} = \mu \boldsymbol{H}$. Generalization to tensor $\varepsilon$ and $\mu$ is rather cumbersome but straightforward. The current density $J$ is usually represented by $\sigma \boldsymbol{E}$, where $\sigma$ is the material electrical conductivity.

(11.5) is usually solved using an appropriate space/time discretization procedure that is applicable to any grid that fills the computational domain with polyhedral cells with the solution vector $\boldsymbol{Q}$ represented by a polynomial within each polyhedron.

If we denote the vector polynomial to be fitted as $\vec{Q}$ and its boundary values as $\vec{Q}^*$, then the quantity to be minimized is

$$e = \int_{(\text{cell boundary})} (\vec{Q} - \vec{Q}^*)^2 \, dS,$$

where

$$\mathbf{Q}_x(\vec{r}) = \langle \mathbf{Q}_x \rangle \alpha + (x - x_\alpha)\frac{\partial \mathbf{Q}_x}{\partial x} + (y - y_\alpha)\frac{\partial \mathbf{Q}_x}{\partial y} + (z - z_\beta)\frac{\partial \mathbf{Q}_x}{\partial z}.$$

The angular brackets denote volume averaging.

From the divergence theorem, the average value of any derivative over the cell volume $V_\alpha$ can be rewritten as a surface integral:

$$\frac{1}{V_\alpha} \int_\alpha \frac{\partial \rho}{\partial x} \, dV = \frac{1}{V_\alpha} \int_{\partial \alpha} n_x \rho \, dS$$

where $\hat{n}$ is the unit outward normal on the boundary $\partial \alpha$.

Therefore

$$\vec{\nabla} \vec{Q} \simeq \frac{1}{V_\alpha} \int_{\partial \alpha} \hat{n} \vec{Q}^* \triangleq K_\alpha.$$

An algorithm that maintains second-order accuracy in both space and time can be constructed from the linear polynomial representation as follows:

$$\langle \mathbf{Q} \rangle_\alpha^{m+1/2} = \langle \mathbf{Q} \rangle_\alpha^m - \frac{\Delta t}{2 V_\alpha} \int_{\partial \alpha} \hat{n} \cdot \vec{F}(\mathbf{Q}_\alpha^{*m}) \, dS$$

$$K_\alpha^m = \frac{1}{V_\alpha} \int_{\partial \alpha} \hat{n} \mathbf{Q}_\alpha^{*m} \, dS$$

$$Q_\alpha^{m+1/2}(\vec{r}) = \langle Q \rangle_\alpha^{m+1/2} + (\vec{r} - \vec{r}_\alpha) \cdot K_\alpha^m \qquad \text{for } \vec{r} \text{ in cell } \alpha$$

$$\langle \mathbf{Q} \rangle_\alpha^{m+1} = \langle \mathbf{Q} \rangle_\alpha^m - \frac{\Delta t}{V_\alpha} \int_{\partial \alpha} \hat{n} \cdot \vec{F}\left(\mathbf{Q}_\alpha^{*(m+1/2)}\right) \, dS$$

Some of the other issues involved in the computational framework are:

## GRIDDING

- geometry modeling using CAD packages
- structured or unstructured surface grids using advancing front technique
- volume grids using hexahedral, prismatic or tetrahedral cells
- optimization routines for achieving grid quality and smoothness
- adaptive gridding

## MASSIVELY PARALLEL COMPUTING

- domain decomposition and load balancing
- internodal message passing with minimum communication delays
- synchronization for time accurate computation
- measure of MFLOP rating
- scalability measures
- pre and post processing parallel environment
- transportability

## USER ISSUES

- problem set up and boundary conditions — how flexible and general is the code?
- geometry and grid set up time, resolution requirements and computational domain size
- internal consistency check for spotting user errors and diagnostic measures

- run environment
  - — selection of number of nodes
  - — load balancing and domain decomposition
  - — automatic termination criteria
  - — complete monostatic runs
  - — post processing routines: FFT, plotting, etc.
- reliability of solution

In addition, proper implementation of boundary conditions appropriate for the physical process being modeled is crucial for validation and general applications of these computational tools. References at the end provide some background material.

## 11.4 IMPLEMENTATION ON MIMD COMPUTERS

The numerical algorithm described in this paper for both the solution of Navier–Stokes equations and Maxwell's equations is a time-explicit finite-volume scheme applicable for both structured and unstructured grids. The algorithm described in Section 11.3 has been the basis of an industrial production level CFD / CEM code ULTRA developed at McDonnell Douglas Research Laboratories and another production level CEM code RCSMPP developed at Rockwell International Science Center. Over the years, ULTRA has been implemented on a variety of vector supercomputers, and SIMD and MIMD computing platforms (Agarwal & Lewis 1992), for example, Cray X-MP/4, Cray X-MP/8, IBM 3090S, BBN Butterfly II, Symult Series 2010, Intel iPSC/2, Connection Machine CM2, and MASPAR. A recent implementation of ULTRA on IBM SP1 is shown in Figure 11.4. RCSMPP has also been implemented on a variety of vector supercomputers, and SIMD and MIMD platforms such as 16-processor Cray C-90, Ncube, Intel Paragon, and IBM SP2 among others (Shankar 1991b). An implementation of RCSMPP on an Ncube is shown in Figure 11.21.

Since the potential teraflop architectures of the future are going to be either distributed or shared memory MIMD machines, here we briefly describe the implementation of ULTRA and RCSMPP on a distributed memory MIMD machine. For implementation details on SIMD and shared memory MIMD machines, the reader is referred to Agarwal & Lewis (1992).

The parallelization procedure can be divided into three main steps — the first is to partition the computational grid and download it to the nodes, then to upload and recombine the results from the nodes. In order to obtain high efficiencies, it is necessary to maintain a good load balance among the nodes, and to minimize the number and length of inter-node communications. Due to the explicit nature of the algorithm, the interactions in the code are primarily local, that is, each grid point is affected most strongly by its nearest neighbors. To take advantage of the local nature of the problem, the overall three-dimensional computational domain is divided into a series of equal rectangular-parallelopiped subdomains, and each processor is given one subdomain. For partitioning an unstructured grid, several strategies exist, namely the coordinate bisection, the graph bisection, and the spectral bisection which have been implemented into a software package both at NASA Ames Research Center and NASA Langley Research Center. The software, designated as PARTI, from NASA Langley Research Center was used in partitioning the unstructured grid in applications described in this paper. Each processor is responsible for updating the grid points in its subdomain, which are called the *active* grid points. This involves calculating second-order differences at each grid point, based on the values of the variables at neighboring grid points. In order to do this for the grid points that lie on the edge of its subdomain, each processor needs values for the grid points which lie in an adjacent subdomain, and this is done by exchanging data for the grid points on the faces between adjacent subdomains. Thus, each processor must store the data for one additional row of grid points on each side of its region of interest. These are referred to as the auxiliary points. The ratio of the number of auxiliary points to active points is proportional to $1/N$, where $N$ is a typical linear dimension of the problem.

In both the CFD and CEM codes, various boundary conditions are applied on the solid boundary surfaces, in the far-field and across branch cuts, for example along the wake of the wing. In implementing these boundary conditions, care must be taken to ensure that only the processors that lie on the boundaries are used to apply the respective boundary conditions. This was achieved by the use of logical flags for each of the edges, and applying the boundary conditions only if the flags were true. The parallelization process for the codes is straightforward, and no special difficulties were experienced in implementation. The code was implemented directly in FORTRAN, and only required the use of the standard messaging subroutine calls which are available through the operating system libraries of various machines.

## 11.5 COMPUTATIONAL AEROSPACE APPLICATIONS

Many of the computational examples presented in this section were done either using a CFD code or a CEM code, both of which were developed primarily for aerospace applications, and more recently adapted to solve problems encountered in commercial product development.

As mentioned earlier, the development of advanced aerospace vehicles increasingly require synergism in integrating multidisciplinary technologies such as fluid dynamics, electromagnetics, structures and other related disciplines.

*COMPUTATIONAL FLUID DYNAMICS*

One of the computational disciplines that has been a supercomputing pace setter for the last three decades is CFD. This technology, which started with the development of the transonic small-disturbance theory in the late 60's, has matured both in algorithm and code development to the point of today being able to solve the time averaged Navier–Stokes equations for predicting the flowfield over a complete fighter. Today, CFD is playing a critical role in the development of next generation fighters, transport airplanes and helicopters, and reusable space launch vehicles. Current CFD applications include a wide range of flow conditions from subsonic to hypersonic Mach numbers simulating different flow effects such as perfect or real gas, inviscid or viscous, laminar or turbulent, compressible or incompressible flows. Depending on the level of physics being modeled, appropriate forms of the fluid dynamic equations and boundary conditions are used in the CFD simulation.

Some CFD aerospace calculations are shown in Figures 11.1–11.17. These calculations illustrate the complexities in geometry and associated flow fields with shocks

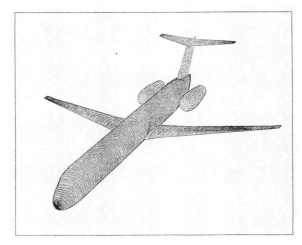

**Figure 11.1** Structured surface grid on an MD-80 aircraft

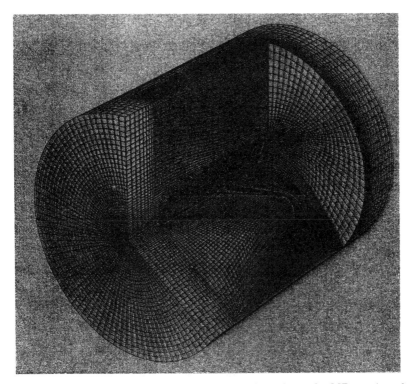

**Figure 11.2** Single block hexahedral structured mesh on the MD-80 aircraft

and vortex effects encountered in the CFD simulations. Figure 11.1 shows the structured surface grid on an MD-80 twin-jet transport airplane and Figure 11.2 shows the single block hexahedral structured mesh around the plane in physical domain. Figure 11.3 shows the pressure distribution on the wing of the aircraft at a cruise mach number of 0.76 and angle of attack of 2° obtained with an Euler code and a

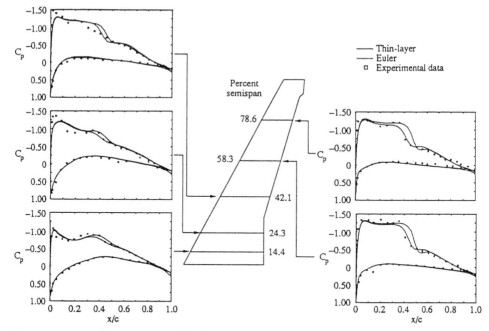

**Figure 11.3** Surface pressure distributions on the MD-80 wing at $M_\infty = 0.76$, $\mathrm{Re}_c = 6.36 \times 10^6$, $\alpha = 2.0°$, $160 \times 34 \times 42$ mesh

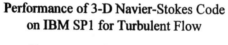

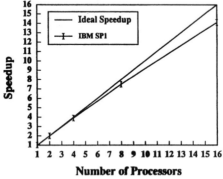

**Figure 11.4** Parallelization of the Navier–Stokes code for flow over a wing on IBM SP1

thin-layer Navier–Stokes code (Deese 1988). The Navier–Stokes computations show good agreement with the experimental data except in the region near the suction peak, where better grid resolution is needed to capture the high flowfield gradients. Figure 11.4 shows the results of parallelization on a 16-processor IBM SP1 (Agarwal & Lewis 1992). The scalability of the code is excellent. Figure 11.5 shows the

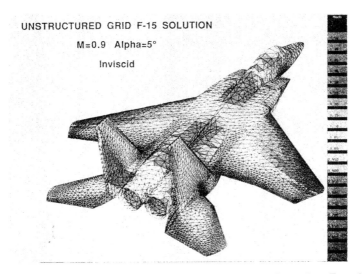

**Figure 11.5** Unstructured triangular elements on the surface of an F-15 fighter

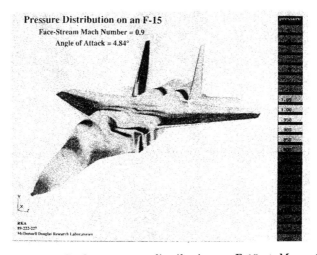

**Figure 11.6** Surface pressure distribution on F-15 at $M_\infty = 0.9$

unstructured triangular grid elements on the surface of an F-15 fighter with fared-in inlets. Figure 11.6 shows the pressure distribution on the aircraft at Mach number of 0.90 and angle of attack of 4.84° obtained with an Euler code. Figure 11.7 shows the comparison of the computed and experimental pressure distributions at various spanwise locations of the fighter wing (Agarwal *et al.* 1989). The Euler solution tends to predict a suction peak that is higher than observed experimentally. Better resolution of the relatively small-radius fighter wing should improve agreement with the data. The shocks on the wing upper surface are predicted to be stronger and slightly downstream of the measured shocks as is typical of inviscid solutions.

Figures 11.8 and 11.9 show the structured zonal and unstructured surface grids on a launch vehicle with nine boosters. Figure 11.10 shows the comparison of the computed

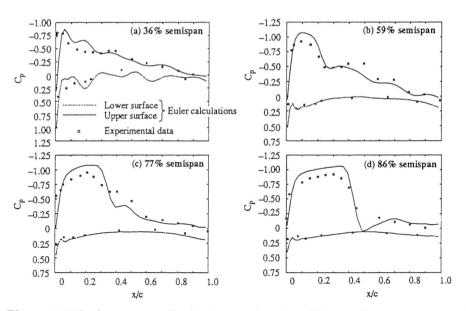

**Figure 11.7** Surface pressure distributions on the wing of F-15 at $M_\infty = 0.9$, $\alpha = 4.89°$

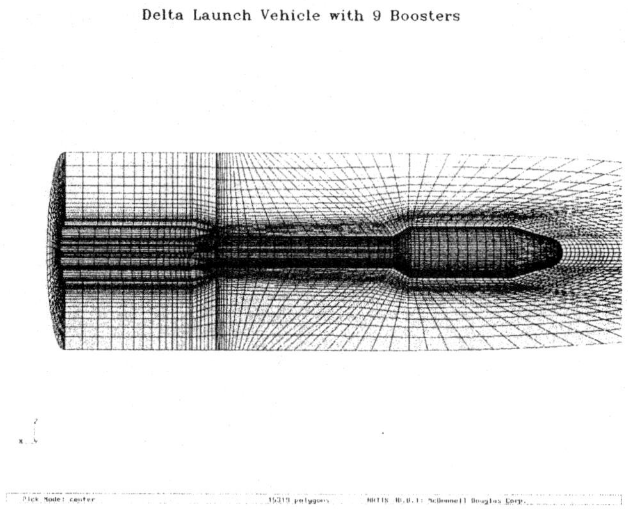

Delta Launch Vehicle with 9 Boosters

**Figure 11.8** Structured zonal grid around a launch vehicle with nine boosters

pressure distributions using Euler and Navier–Stokes codes on the core of the vehicle and different boosters with the experimental data (Deese *et al.* 1992). The viscous predictions are in good agreement with the data. Additional grid resolution may further improve the results. Figure 11.11 shows a chimera grid around a helicopter fuselage with a 2-bladed rotor. Figure 11.12 shows the pressure distribution on an

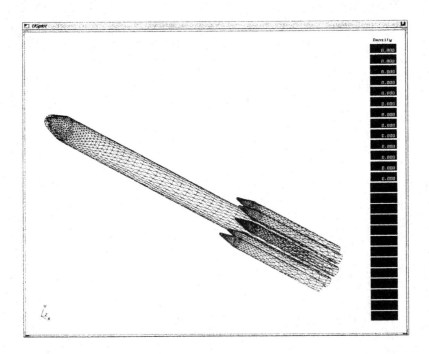

**Figure 11.9** Unstructured surface grid on a launch vehicle with nine boosters

isolated MD-500 helicopter fuselage obtained with the Navier–Stokes code described in Agarwal (1991). Figure 11.13 shows the pressure distribution on an ONERA rotor blade in forward flight obtained with the rotary-wing Euler code described in Agarwal & Deese (1990). Figure 11.14 compares computed and experimental pressure distributions on the ONERA rotor in forward flight.

As an example of bodies in relative motion, Figure 11.15 shows the unstructured grid about a store ejecting from a cavity. Figure 11.16 shows the computed density distribution on the store and the cavity at a given instant of time at Mach 10. These calculations were performed with the Euler code described in Marcum & Agarwal (1992).

Figure 11.17 shows the flowfield simulations for four extremely complex configurations. In the design of a supermaneuverable fighter (X-31), CFD calculations provide the designer with high angle-of-attack flow characteristics that are difficult to simulate both experimentally and computationally. In the case of a mated Shuttle computation, the key is to understand the effects of shock loadings on the wing in the presence of plumes at transonic and supersonic Mach numbers. For the B-1 bomber development and upgrades, CFD simulations play a key role in understanding the safe store separation problem. The design of reusable launch vehicles (RLV), which are the next

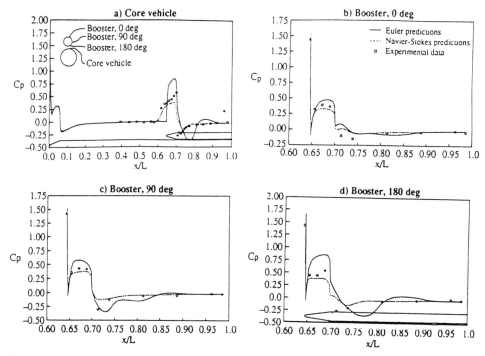

**Figure 11.10** Pressure distributions on the launch vehicle (core and the boosters), $M_\infty = 1.6$, $\mathrm{Re} = 0.8 \times 10^{16}$/inch, $\alpha = 0.0$

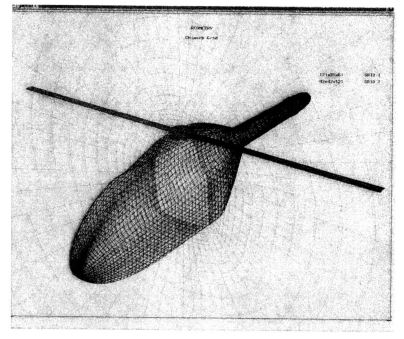

**Figure 11.11** Chimera grid around a helicopter fuselage with a two-bladed rotor

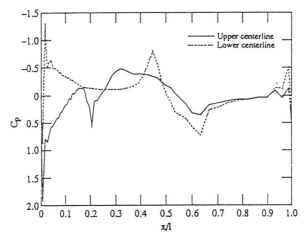

**Figure 11.12** Pressure distributions in the symmetry plane of an MD-500 fuselage, $M_\infty = 0.4$, $\alpha = -5°$

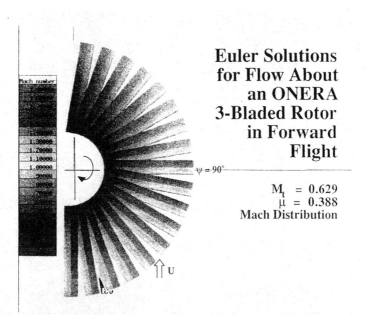

**Figure 11.13** Pressure distributions on an ONERA rotor blade in forward flight, $M_t = 0.629$, $\mu = 0.388$

generation space vehicles, again critically depend on computational simulations to reduce the design cycle cost and time.

Figure 11.18 shows the calculations for a free jet using a large-eddy simulation (LES) code. On the currently available parallel platforms LES simulations cannot be routinely performed for the complex configurations and flow phenomenon described in Figures 11.1–11.17.

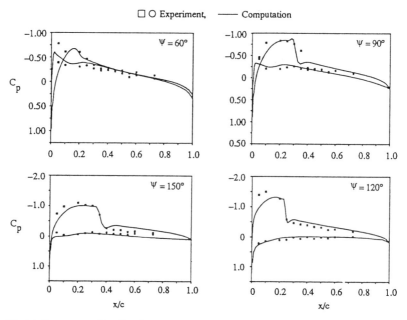

**Figure 11.14** Pressure distributions on an ONERA rotor blade in forward flight, $M_t = 0.629$, $\mu = 0.388$, 90% span location

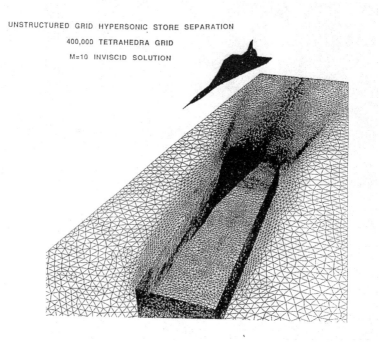

**Figure 11.15** Unstructured grid about a store ejecting from a cavity

**Figure 11.16** Pressure distributions on the store and the cavity of Figure 11.15 at Mach 10

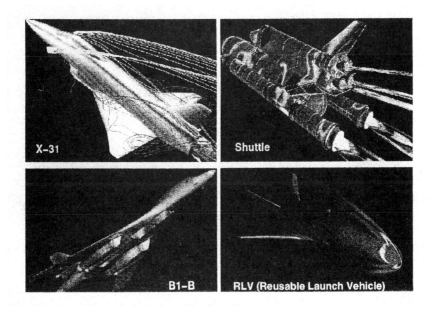

**Figure 11.17** Application of CFD in flow field simulation of aerospace configuration

# Large Eddy Simulation of Free Jets:
# Time-Averaged and Acoustic Quantities

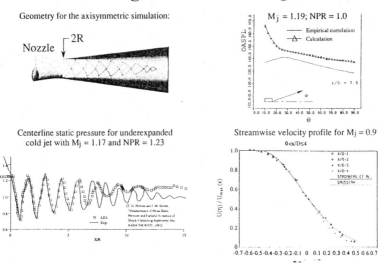

Figure 11.18 Large eddy simulation of a free jet

## COMPUTATIONAL ELECTROMAGNETICS

The ability to predict radar return from complex structures with layered material media over a wide frequency range (100 MHz to 20 GHz) is a critical technology need for the development of stealth aerospace configurations. Traditionally, radar cross section (RCS) calculations have employed one of two methods: high frequency asymptotics, which treats scattering and diffraction as local phenomena; or solution of an integral equation (in the frequency domain) for radiating sources on (or inside) the scattering body, which couples all parts of the body through a multiple scattering process. A third approach is the direct integration of the differential or integral form of Maxwell's equations in the time-domain.

CEM is a critical technology in the advancement of future aerospace development through supercomputing. As we transition from the present gigaflops to the next generation teraflops computing, CEM will become integral to aerospace design not only as a stand alone technology but also as part of the multidisciplinary coupling that leads to well optimized designs.

Toward establishing a computational environment for performing multidisciplinary studies, the initial goal is to advance the state-of-the-art in CEM with the following specific objectives.

1) Apply algorithmic advances in Computational Fluid Dynamics (CFD) to solve/break Maxwell's equations in general form to study scattering (radar cross

section), radiation (antenna), and a variety of electromagnetic environmental (electromagnetic compatibility, shielding, and interference) problems of interest to both the defense and commercial communities.

2) Establish the viability of MIMD massively parallel architectures for tackling large scale problems not amenable to present day supercomputers.

3) Mature the CEM technology to the point of being able to perform coupled CFD/CEM optimization design studies.

Proper development of a CEM capability appropriate for all aspects of aerospace design must consider various issues associated with electomagnetics. Some of them are:

1. *Physics of Maxwell's Equations*
- differential and integral forms of Maxwell's equations suitable for numerically satisfying tangential field flux conservation
- flexibility in implementing total field, scattered field, and diffracted field forms depending on the nature of the problem being solved
- incorporation of various source terms

2. *Material Properties*
- perfectly conducting surfaces
- lossy/lossless $\varepsilon$ and $\mu$
- resistive sheet (conductivity $\sigma$ and thickness $d$)
- impedance layer
- anisotropic media
- chiral media
- dispersive media—$\varepsilon(\omega)$ and $\mu(\omega)$
- nonlinear materials

3. *Boundary Conditions*
- perfectly conducting, $n \times E = 0$ (Electric wall) and $n \times H = 0$ (magnetic wall)
  — accurate evaluation of $n \times H$ on the perfectly conducting (PC) surface is crucial
- material interface, $|n \times E|$ and $|n \times H|$ are zero
  — algorithms must account for any variation in $\varepsilon$ and $\mu$ at interface
- resistive sheet, $n \times E$ and $n \times H$ jump across resistive sheet (RS)
- impedance layer, $n \times n \times E = -\eta n \times H$
  — implementation in time-domain involves convolution integrals
- nonreflecting farfield
  — characteristics based hierarchy of absorbing conditions for total field, scattered field, and diffracted field formulations
- periodic
- zero flux

4. *Algorithmic Issues*
- unstructured grid-based finite-volume algorithms that include structured grids as a special case
- stability and accuracy of schemes using spectral techniques
- construction of higher-order schemes, including boundaries, using polynomial representations for electric and magnetic field variations inside general polyhedral cell

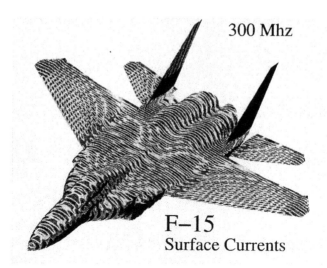

**Figure 11.19** Application of CEM technology to compute the RCS of an F-15

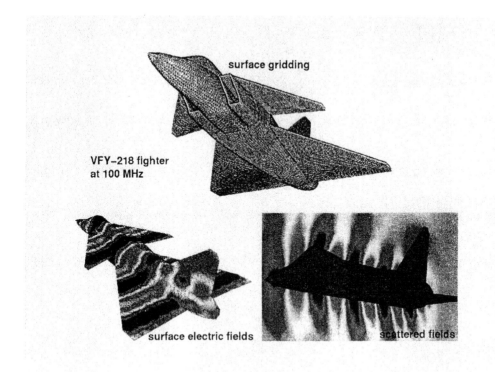

**Figure 11.20** Application of CEM technology to compute the RCS of VFY-218

- implicit and explicit schemes with appropriate space and time discretization
- *Validation and Applications*
- code validation on Electromagnetic Code Consortium test cases and canonical solutions
- radar cross section of low observable platforms
- antenna performance
- Microwave monolithic integrated circuit (MMIC) modeling and photonic band gap periodic structures
- electromagnetic environmental effects ($E^3$), such as EMP, EMI, and compatibility problems
- bioelectromagnetics such as microwave hypothermia cancer treatment of humans and effects of cellular phones

Figures 11.19 and 20 show CEM computations for predicting the RCS of an F-15 (Shu & Agarwal 1994) and a VFY-218 (Shankar 1991b) fighter. Maxwell's equations are solved to get the near field surface currents from which the RCS information is derived. Use of unstructured gridding enables one to model complex topologies including radar absorbing material layers with relative ease.

## PARALLEL COMPUTING
The emergence of massively parallel computing architectures with the potential for teraflops performance requires code development activities to effectively utilize a

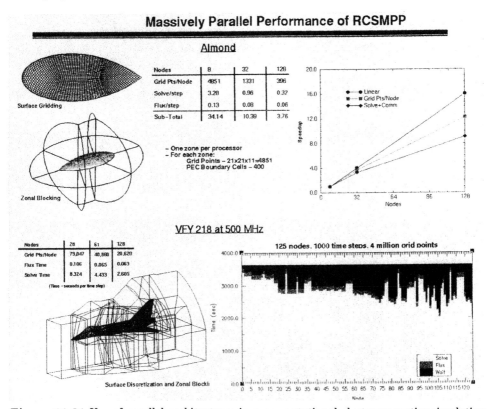

**Figure 11.21** Use of parallel architectures in a computational electromagnetics simulation

computer architecture to achieve proper load balancing with minimum inter-processor data communication.

Figure 11.21 shows some typical parallel performance results for a simple almond geometry and a complex fighter configuration. An 8 zone almond shown in Figure 11.21 was used to verify that the solution time for the same problem could be sped-up by distributing the problem over more processors. The almond grid was run using 8, 32, and 128 processors of an NCUBE machine. The process of distributing the grid over more processors adds grid points to the total problem size. Between the 8 processor and 32 processor runs the number of grid points per processor decreased by 3.64, while the solve time plus communication time decreased by 3.3. Similarly between the 8 and 128 processor runs, the number of grid points decreased by 12.25 while solve time plus communication time decreased by 9.07. As seen in the chart, the speed-up is not linear. Results for the 128 processor run may be poor due to the very small amount of grid points per processor. In this case the solve time is no longer the dominating time factor. Flux communication now takes up a large percentage of total time, around 20% compared to 4% for the 8 processor run. Complex problems such as full scale fighter geometries also show encouraging scaling and speed-up results.

## 11.6 COMPUTATIONAL COMMERCIAL APPLICATIONS

The CFD and CEM tools that play a major role in the design of aerospace systems are also being applied to simulate problems in commercial product development. In this era of defense-conversion, for aerospace computational tools to be effective in the commercial markets, they have to become even more cost effective and efficient and be able to run effectively on workstations and workstation clusters. In general, the complexity of physics being modeled in commercial problems is less severe than those of aerospace problems which usually involve shocks and other high Mach number effects.

### AUTOMOBILE SIMULATIONS

A compressible Navier–Stokes solver at low Mach numbers or an incompressible Naviere–Stokes solver can be used to study a number of flow related problems associated with the development of a new automobile. Figure 11.22 shows a few sample automobile calculations (Ota 1993). Automobile sunroofs are designed to ventilate the passenger compartment. A major problem associated with sunroofs is that for certain speed ranges, the open sunroof creates uncomfortable levels of noise within the passenger compartment. As an alternative to road tests or wind tunnel tests, CFD tools play a role in minimizing noise problems of new sunroofs. One can study the acoustics environment inside the passenger compartment as well as the climatization problems associated with the air-conditioning system. CFD simulations of smoke visualization over a complete automobile in the presence of wheels and ground plane to achieve proper aerodynamic characteristics are routine in automobile design.

Another example of CFD application is the simulation of flow fields around two trains moving in opposite directions inside a tunnel. This transient simulation provides the pressure buildup inside and on the tunnel walls that determines the acoustic levels and the structural integrity of the tunnel. Such calculations involve relative grid motions associated with the moving trains with respect to a stationary grid fixed to

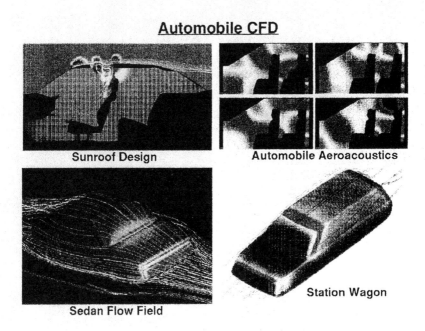

**Figure 11.22** Application of computational fluid dynamics technology in automobile simulation

the tunnel. Figure 11.23 shows results for a transient moving train problem done using an incompressible Navier–Stokes model (Pan & Chakravarthy 1989).

## THERMAL MODELING

CFD simulations play a key role in modeling the thermal environment that is critical to the design of many commercial products such as that encountered in electronic packaging and compact drives for power devices. Figure 11.24 shows a low speed Navier–Stokes computation to study the heat transfer inside a power assembly used in the control of motors (Chen & Ota 1996). A major bottleneck in this process is the proper dissipation of heat produced by various electronic components. CFD simulations that model the forced convection cooling with turbulent recirculating flows help designers better understand the heat dissipation process in the existing designs and develop new thermal management concepts. As the size of these power modules become increasingly compact, the design is dominated by achievable thermal efficiency.

## IMAGE ENHANCEMENT

Partial differential equation (PDE)-based image enhancement tools are beginning to derive benefit from CFD shock capturing algorithms in solving problems associated with noise removal and deblurring of images encountered in image enhancement

## CFD Simulation of Moving Trains

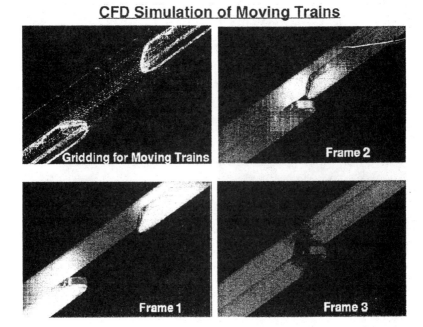

**Figure 11.23** CFD simulation of high-speed moving trains inside a tunnel

## Power Assembly Heat Transfer Simulation
### (Incompressible Navier–Stokes)

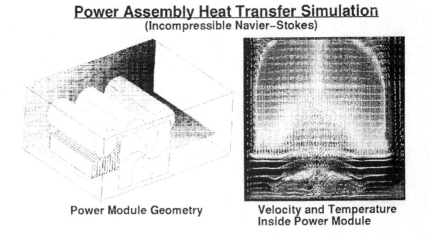

**Figure 11.24** Thermal modeling of heat transfer in power modules

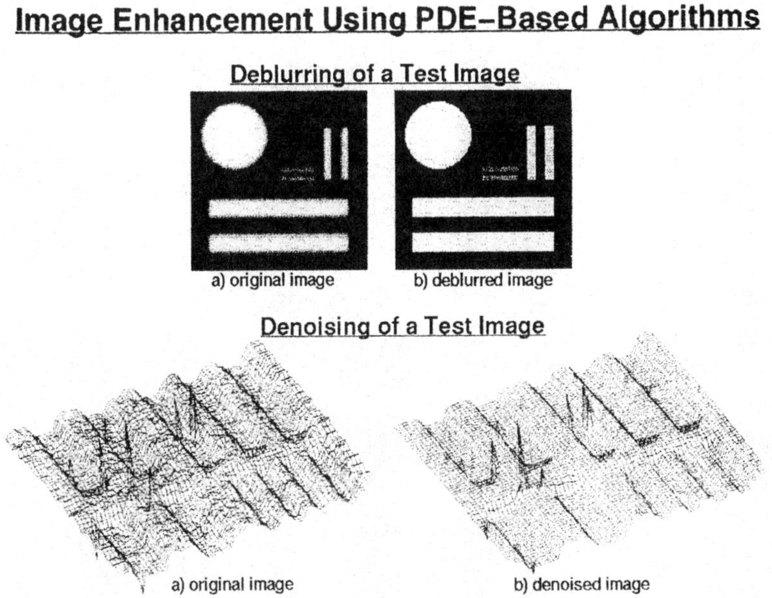

## Image Enhancement Using PDE–Based Algorithms

### Deblurring of a Test Image

a) original image          b) deblurred image

### Denoising of a Test Image

a) original image                          b) denoised image

**Figure 11.25** Image enhancement using PDE-based algorithms

studies. Since blurring often occurs because of some heat-like diffusion process, one idea is to solve the "backward heat equation." This is an ill-posed problem. Using a nonlinear form of a backward heat equation model, results of Figure 11.25 are generated. Beyond its application of enhancement to photographic images, the method has potential application to enhancing infrared images and reconstruction of compressed images. Bihari (1996) provides some details.

## MICRO ELECTRO MECHANICAL SYSTEMS (MEMS)

Micro Electro Mechanical Systems (MEMS) advances provide a powerful tool for miniaturization of mechanical systems into a dimensional domain not accessible by conventional machining. A unique feature of MEMS technology is that it applies equally well to both sensors and actuators, and couples intimately with microoptics and microelectronics. MEMS is an enabling technology that can potentially provide a higher performance than mechanical systems at a reduced size and cost, leveraging the mature semiconductor microfabrication technology already in use by the electronics industry. Fluid dynamics simulation of MEMS devices, which involves the coupling of electrostatic forces with the structural response and air damping effects, is currently being exercised in the design of many MEMS devices such as the micro-gyro sensors for antilock brakes in automobiles, RF switches, flux sensors, current transducers, optical scanners, and other micro sensors and actuators with different applications.

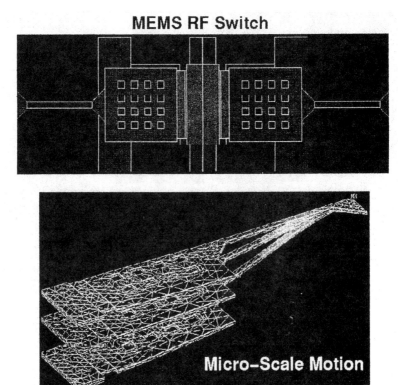

**Figure 11.26** Multidisciplinary coupling simulations in MEMS design

Because the devices are not designed to operate in ultra high vacuum, CFD simulations can play a role in precisely determining the device performance as it relates to mechanical motions at micro-scale levels. For these micro-scale flows, the assumption of a continuum may not be valid. The Knudsen number becomes an important parameter along with the Mach number and Reynolds number. Figure 11.26 shows a model coupled simulation for a MEMS RF switch.

## COMPUTATIONAL MANUFACTURING

Incompressible Naviere–Stokes models play a role in simulating a number of manufacturing processes such as laser welding, laser surface heat treatment, resin transfer molding (RTM), plastic injection molding, metal casting and forging, and so on that basically involve understanding the heat transfer mechanism in determining the cooling rates in the presence of solid, liquid and gaseous phases simultaneously present. Figures 11.27–11.29 show different laser material processing simulations performed using a Navier–Stokes model (Shankar & Gnanamuthu 1986). Figure 11.27 shows results for laser surface hardening of iron-base alloys to improve their wear and tear and corrosion resistance. The simulation basically provides a chart for determining the case depth for a given laser travel speed. Figure 11.28 shows the formation of vortices inside a laser melted molten pool due to surface tension driven forces. One needs to account for these effects in understanding the cooling rates that play a role

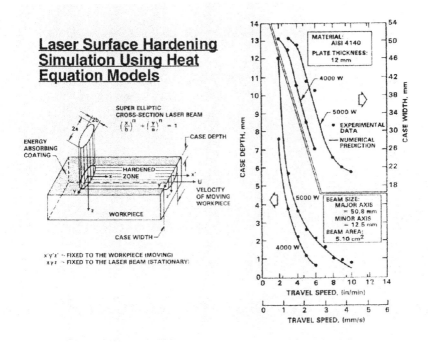

Figure 11.27 Laser surface hardening simulation

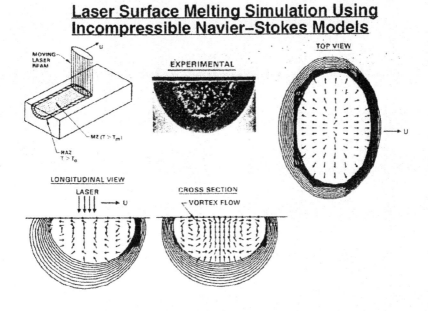

Figure 11.28 Laser surface melting simulation

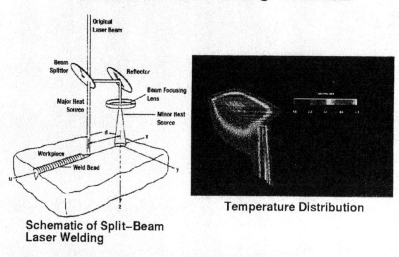

**Figure 11.29** Split-beam laser welding simulation

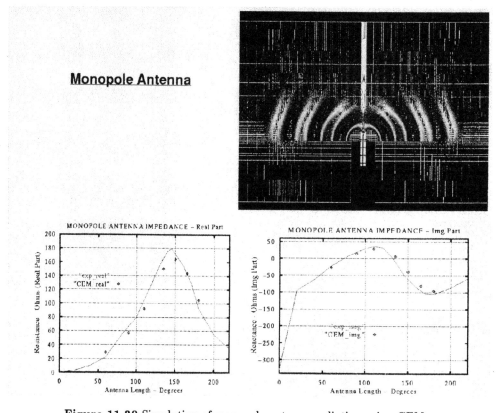

**Figure 11.30** Simulation of monopole antenna radiation using CEM

## Photonic Band Structure Simulation for MMIC

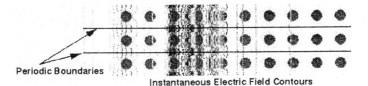

Periodic Boundaries

Instantaneous Electric Field Contours

Transmission Coefficient

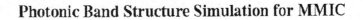

0.125"  0.1875"  0.118"

acrylic  pyrex  air

Geometry of the Photonic Structure

**Figure 11.31** Simulation of photonic band gap filters using CEM

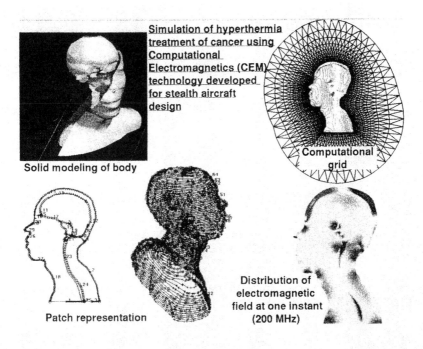

Simulation of hyperthermia treatment of cancer using Computational Electromagnetics (CEM) technology developed for stealth aircraft design

Solid modeling of body

Computational grid

Patch representation

Distribution of electromagnetic field at one instant (200 MHz)

**Figure 11.32** Bioelectromagnetic simulation of microwave radiation effects on human body

in the quality in welding. The cooling rates can be altered using a split beam laser welding where a preheat source is used to heat the work piece prior to actual welding. This is shown in Figure 11.29.

## BIOELECTROMAGNETICS AND FLUIDS

The CEM tools developed for predicting the radar cross section of low observable targets are also applied to study the effects of microwaves on humans. For example, the effects of electromagnetic fields emitted by cellular phones on the human brain is a debatable issue. Figure 11.30 shows a simulation of the electromagnetic radiation fields around a monopole antenna, typical of a cellular phone (Shankar *et al.* 1990). Figure 11.31 shows a photonic band gap device that can filter certain frequencies which can be used inside cellular phones to basically cut off transmission of waves into the brain. Figure 11.32 shows a CEM simulation of electromagnetic waves propagating through a human head (Shankar 1991a). Such simulations play a role in hypothermia treatment of cancer using patient specific anatomical details.

Figure 11.33 shows a biofluids application of an incompressible Navier–Stokes solver to study the problem of detached retina and the seepage of vitreous humor, a jelly-like fluid with fibers, in the eye surgical process of retinal reattachment.

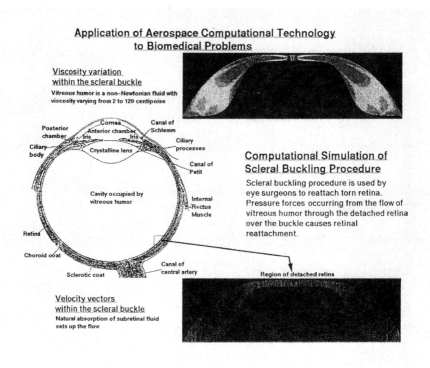

**Figure 11.33** Biofluids simulation of vitreous humor in human eye

## 11.7 CONCLUSIONS

Computational simulation is maturing and becoming an integral part of product development in both aerospace and commercial markets. We now face a real challenge in making these tools cost effective and quick-turn-around to be responsive to the market pressures faced by the product development teams. We will continue to seek advances in the numerical modeling of the physical processes as well as to better utilize the increasing power of workstation clusters and parallel architectures.

## 11.8 ACKNOWLEDGMENTS

The material presented here is a culmination of many years of research, development and application of computational tools developed by members of the CFD and CEM groups at Rockwell Science Center and McDonnell Douglas Research Laboratories. Funding for this work comes from Rockwell and McDonnell Douglas IR&D projects and number of government sponsored contracts.

## REFERENCES

Agarwal R. K., Deese J. E., Johnson J. G., Steinhoff J. 1989. Euler Calculations for Flow Over a Complete Aircraft. *AIAA Paper No. 89-2221*.

Agarwal R. K., Deese J. E. 1990. Euler/Navier–Stokes Computation of the Flowfield of a Helicopter Rotor in Hover and Forward Flight. *Progress in Aeronautics, P. A. Henne, American Institute of Aeronautics and Astronautics*, 125:533–555.

Agarwal R. K., Deese J. E. 1991. Recent Advances in Computational Aerodynamics. *Computer Physics Communications*, 65:8–16.

Agarwal R. K., Lewis J. C. 1992. Computational Fluid Dynamics on Parallel Processors. Computing Systems in Engineering, 3:251-259.

Agarwal R. K. 1996. A Unified CFD Based Approach to a Variety of Problems in Computational Physics. In: *Proceedings of Third World Conference in Applied Computational Fluid Dynamics, Basel World CFD User Days 1996, Freiburg, Germany.*

Bihari B. L. 1996. Application of Nonlinear Noise Removal to Infrared Images. In: *Proceedings of the Signal Processing Applications Conference and Exhibition at DPSX '96, San Jose, California.*

Chakravarthy S. R., Szema K. -Y., Haney J. W. 1988. Unified Nose-to-Tail Computational Method for Hypersonic Vehicle Applications. *AIAA Paper No. 88-2564.*

Chen C. L., Ota D. K. 1996. Heat Transfer Modeling of Power Assemblies. *AIAA Paper No. 96-1827.*

Deese J. E., Agarwal R. K. 1988. Navier–Stokes Calculations of Transonic Viscous Flow About Wing-Body Configurations. *Journal of Aircraft*, 25:1106–1112.

Deese J. E., Pavish D. L., Johnson J. G., Agarwal R. K., Soni B. K. 1992. Flowfield Predictions for Multiple Body Launch Vehicles. *AIAA Paper No. 92-2681.*

Marcum D. L., Agarwal R. K. 1992. Finite Element Navier–Stokes Solver for Unstructured Grids. *AIAA Journal*, 30:648.

Ota D. K. 1993. Computation of External Automobile Body Shapes. *SAE Technical Paper Series, Worldwide Passenger Car Conference and Exhibition, Dearborn, Michigan.*

Pan D., Chakravarthy S. R. 1989. Unified Formulation for Incompressible Flows. *AIAA Paper No. 89-0122.*

Shankar V., Gnanamuthu D. 1986. Computational Simulation of Heat Transfer in Laser Melted Material Flow. *AIAA Paper No. 86-0461*.

Shankar V., Hall W. F., Mohammadian A. H. 1990. A Time Domain, Finite-Volume Treatment for the Maxwell Equations. *Electromagnetics*, 10:127.

Shankar V. 1991a. Research to Application — Supercomputing Trends for the 90's and Opportunities for Interdisciplinary Computations. *AIAA Dryden Research Lecture, AIAA Paper No. 91-0002*.

Shankar V. 1991b. Gigaflop Performance Algorithms in Computational Science — Application to Maxwell's Equations for Electromagnetics Computations. *AIAA 10th Computational Fluid Dynamics Conference, Honolulu, Hawaii*.

Shu M., Agarwal R. K. 1994. A Spatially Compact Solver for Electromagnetic Scattering. In: *Numerical Methods for the Solution of Maxwell Equations, R. Lohner, J. Periaux, and H. Steve, eds., John Wiley and Sons, New York*.

Szema K. -Y., Chakravarthy S. R., Dresser H. 1988. Multizone Euler Marching Technique for Flows over Multibody Configurations. *AIAA Paper No. 88-0276*.

# 12

# Multi-Color Neural Network with Feedback Mechanism for Parallel Finite Element Fluid Analysis

Hiroshi Okuda and Genki Yagawa[1]

## 12.1 INTRODUCTION

Evolutional computer technologies like the fuzzy theory and the neural network seem to play key roles to meet the requests of engineers and designers, who would claim more talented CAE (Computer Aided Engineering) systems to be developed (Bennet & Englemore 1979, Taig 1986). They should satisfy such requirements as (a) high accuracy, (b) faster computation than real time, (c) user-friendliness even for beginners and (d) portability among various computer facilities. This paper focuses on the artificial neural network and its applications to the field of the computational mechanics. Discussing recent research activities of neuro-computations, feasibilities of the neural network as a computational mechanics tool, especially in the area of CFD (Computational Fluid Dynamics), which is one of the most CPU-time-consuming area, are explored.

Figure 12.1 schematically depicts the biological neuron. Each neuron receives signals, which are either positive or negative impulses, from many other neurons. When the accumulated signals reached a certain amount, the signal is transfered to other neurons through the axon. Thus, the signals are successively transfered over the biological neural network, which is constructed by numerous neurons. For exmaple, a human being is estimated to have around $10^{10}$ brain cells. The artificail neural network (simply called 'neural network' in the following descriptions) was developed to mimic the biological brain neural network (Minsky & Papert 1988, Hecht-Nielsen 1989, Wasserman 1993, Fausett 1994). The neural network is characterized as the information processing structure consisting of highly parallel distributed elements. Each element possesses a local memory and carries out localized information processing operations. The neural network model can be classified into the interconnected neural network and the hierarchal neural network. Figure 12.2 schematically depicts these

---

[1] Department of Quantum Engineering and Systems Science, University of Tokyo, 7-3-1 Hongo, Bunkyo-ku, Tokyo 113, Japan

*Parallel Solution Methods in Computational Mechanics* edited by M. Papadrakakis

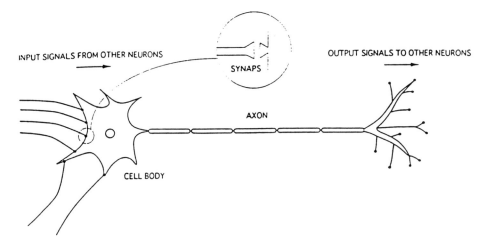

Figure 12.1 Biological neuron

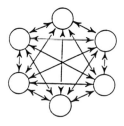

(a) interconnected neural network

Output patterns

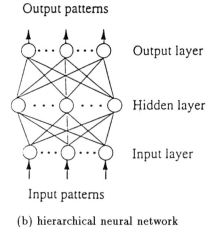

Input patterns

(b) hierarchical neural network

Figure 12.2 Neural network configurations

network structures. They are also called the recurrent neural network and the feed-forward neural network, respectively, which stand for the way of data transmission.

There have been published many papers about the neural network (Minsky & Papert 1988, Hecht-Nielsen 1989, Wasserman 1993, Fausett 1994). Starting from Hopfield's network (Hopfield & Tank 1985, Tank & Hopfield 1986), the interconnected neural network has been applied to a wide variety of optimization problems. On the other hand, the hierarchal neural network (Rumelhalt *et al.* 1986), which transmitts signals in one direction between perceptron layers, has become a powerful tool in the pattern recognition, control, diagnosis and so on. The ability of parameter identification of the hierarchal neural network has also been utilized to solve the inverse problems. Progress in learning performances of the neural network would strengthen the credibility of the neural network aided computation tools, and also would widen the area of applications. Various types of learninig algorithms (Amari 1993, Biegler-Koenig & Baermann 1993, Karayiannis & Venetsanopoulos 1992, Du *et al.* 1992), many of which are for multi-layered error back-propagarion networks, have been developed to improve the learning performance. Since the optimum size and topology of the neural network are usually unknown, users of the neural network would have to resort to a time-consuming heuristic choice before constructing application phase networks. As for the hierarchal neural network, an a priori estimation method of optimum network size and topology is discussed (Weymaere & Martens 1994, Yu *et al.* 1993, Mehrotra 1991). The generalization is an important and attractive feature of the hierarchal neural network. Improvement of the generalization performance is investigated in Wedd (1994), Jean & Wang (1994), Cun (1992) and Yoshimura *et al.* (1994). Other fundamental features of the neural network, which may affect the computational performance and/or the robustnes of the neurological computation strategy, are also discussed in Mukhopadhyay & Narendra (1993), Feuer & Cristi (1993) and Lee & Kil (1994).

The most outstanding characteristics of the neural network aided computation is that neither complicated programmings nor rigid algorithms are needed. Given the teaching data repeatedly, solutions of the system are pursued iteratively until a certain tolelance would be satisfied. Another important point is that the neural network's inherent parallelism, that is, concurrent signal transmissions over numerous information processing elements suits the massively parallel computer architectures, which is a leading faction of current computer developments. Thus, the neural networks have been implemented on various parallel machines (Kung & Chou 1993, Zell et al.1993, Grajski 1993, Majumder & Dandapani 1993, Yagawa & Aoki 1993, Botros & Abdul-Azia 1993, Okawa 1993, Fujimoto *et al.* 1992, Misra & Prasanna 1992, Witbrock & Zagha 1993).

This paper is organized as follows. In Section 12.2, neural network applications, especially to the computational mechanics fields, are reviewed from recent publications. In Section 12.3, we briefly describe the basis of the neural network as preliminaries for Section 12.4, which describes the neural network solver proposed by the authors. The neuro-solver will be utilized for the Poisson's equation, which has to be solved for the pressure at every time step in the incompressible viscous flow analysis. In Section 12.5, feasibilities of the neuro-solver are investigated with respect to the convergence and the parallel efficiency. Concluding remarks are made in Section 12.6.

## 12.2 REVIEW OF NEURAL NETWORK APPLICATION

The neurological computations have been exploited in many applications such as nondestructive evaluation, structural identification, design optimization, reasoning of the numerical analysis procedure and so on. In these applications, the neural networks are utilized through the process of either inverse analysis or direct analysis.

### NEURAL NETWORK AS INVERSE ANALYSIS TOOL

As an inverse analysis tool, the hierarchal neural network is employed in a wide vriety of problems. The neural network's identification ability is effectively used, for instance, in the quantitative nondestructive evaluation (QNDE) (Upda & Upda 1990, Kitahara *et al.* 1992, Brown *et al.* 1992, Yoshimura *et al.* 1993, Yagawa *et al.* 1992, Song & Schmerr 1991, Elshafiey et al. 1992), which detects sizes, shapes and/or locations of the defects hidden in solids. QNDE techniques are very important to assure the structural integrity of operating plants and structures, and to evaluate their residual life time. Various NDE techniques using ultrasonic wave, X-ray, magnetic powder, eddy current and so on have been studied so far. The principle of the neural-network-based inverse analysis is described here giving an example of the defect identification problem by the ultrasonic wave. We note that the inverse analysis approach explained below is applicable not only to QNDE, but also to other various engineering fields such as the structural identification, the material identification and so on.

In the ultrasonic-based defect identification problems, defect parameters such as the size and the location of a defect are determined from the dynamic responses of solid measured at several points on the solid surface. This is one of the inverse problems. On the other hand, if the defect parameters are given, the dynamic responses of solid can be computed through the computational mechanics, e.g. the finite element simulations, which is a direct problem. These direct and inverse problems can be simply regarded as the nonlinear mappings from the multiple data, i.e. the dynamic responses, to the multiple data, i.e. the defect sizes and the locations. Thus, the inverse analysis procedure using the hierarchical neural network is constructed through the following three phases:

Phase 1: The dynamic finite element simulations of solid containing a defect are performed repeatedly with parametrically changing defect parameters. Data pairs of the defect parameters vs. the dynamic responses of solid are collected. (direct analysis phase)

Phase 2: The neural network is trained with the data pairs collected in Phase 1. The defect parameters and the dynamic responses of solid are given to the network as the teaching data and the input data, respectively.

Phase 3: The trained network is utilized to identify defect parameters from the measured dynamic responses of solid in an application process. (inverse analysis phase).

The above inverse analysis procedure can be applied to any inverse problem by replacing the kinds of the computational mechanics simulations. Thus, the present inverse analysis approach using the neural networks and the computational mechanics are widely applicable to various engineering fields. There have been reported such inverse problems as the structural identification of a vibrating plate (Yoshimura & Yagawa 1993), parameter identification included in a viscoplastic constitutive equation (Yoshimura *et al.* 1992), mixture ratio identification of inhomogeneous materials (Yagawa *et al.* 1995).

## NEURAL NETWORK AS DIRECT ANALYSIS TOOL

Many of industrial systems have no first-principles to reproduce their dynamic behaviors or mapping patterns, which often makes it impossible to utilize currently available computational mechanics methodologies. The neural networks are employed to model such nonlinear systems instead of, or otherwise, hybridized with the numerical simulations (Ramamurthy 1993, Shahla 1993, Chang & Chang 1993, Kim & Lee 1993, Saravanan *et al.* 1993, Rico-Martinez 1993). Generally, most of the nonlinear analyses involve several parameters to be tuned, which seriously influence both the numerical results and the stability. Methodologies for estimating multiple parameters involved in the nonlinear and the time-dependent finite element analyses are proposed (Yagawa *et al.* 1994).

In the field of structural design, the structural optimization problems have been solved by the mathematical methods combined with the computational mechanics (Bennett & Botkin 1986), while recently proposed are the use of the neural networks (Berke & Hajela 1992, Chen & Chan 1993, Tsutsumi *et al.* 1993). Another strategy is to use the design window (DW) that indicates an area of satisfactory solutions in a permissible design space. A multilayer neural network is utilized as the DW search engine (Mochizuki *et al.* 1993). The neural network is also used directly to model the inelastic material behaviors in place of the constitutive equations (Ghaboussi *et al.* 1991, Yamamoto 1991, Okuda *et al.* 1994).

Not only the structural optimization problems but also a wide variety of constrained or unconstrained optimization problems have been solved by the neural networks (Maa & Schanblatt 1992, Zhang et al. 1992, Sun & Fu 1993, Francelin & Gomide 1993, Bouzerdoum & Pattison 1993, Ling & Jin 1993). Considering that seeking the solution of equations can be formulated as an optimization problem, an equation solver, which will be detailed in Section 12.4, is also constructed on the neural network (Yagawa 1994, Yagawa & Okuda 1996).

## 12.3 BASIS OF NEURAL NETWORK

As a preliminary step for Section 12.4, which will describe a neural network solver for the Poisson's equation, in this section, we briefly describe the basis of the neuron model, the interconnected neural network and the hierarchal neural network. More details would be obtained in the literature (Minsky & Papert 1988, Hecht-Nielsen 1989, Wasserman 1993, Fausett 1994).

## NEURON MODEL

An artificial model of neuron is shown in Figure 12.3. The neurons are connected with neighboring ones by nerve fibers depicted by the solid lines so that the network system is organized as shown in Figure 12.2. In general, a neuron has multiple inputs and a

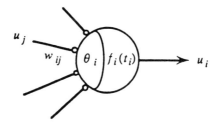

Unit $i$

$u_i$    Output from unit $i$
$u_j$    Input from unit $j$
$w_{ij}$    Weight
$\theta_i$    Offset
$t_i$    State variable
$f()$    Activation function

**Figure 12.3** Artificial model of neuron

single output. The inputs are operated and transformed into the output by the state transition rule as:

$$u_i = f(t_i) \tag{12.1}$$

$$t_i = \sum_{\substack{j \\ (i \neq j)}} w_{ij} u_j + \theta_i \tag{12.2}$$

where $u_j$ in (12.1) and $u_i$ in (12.2) are the input from a neuron $j$ and the output from a neuron $i$, respectively. $w_{ij}$ is the connection weight, $\theta_i$ the offset, $t_i$ the state variable and $f(t_i)$ the activation function. Input signals received from other neurons are multiplied by the connection weights, which imply the connection strength between the neurons. The weighted signals are then summed up and transformed into the output signal through an activation function. Since activations of neurons are expected to occur independently and randomly, the state transition of the neural network is an attractive computational model for a massively parallel platform.

As the activation function, the sigmoid function, which is a continuous and nonlinear function taking the value of 0 through 1 (see Figure 12.4) is often used. A typical sigmoid function is expressed as:

$$f(t_i) = 1/\{1 + \exp(-t_i/T)\} \tag{12.3}$$

where $T$ is called a temperature of the sigmoid function.

## NETWORK CONFIGURATIONS

As is shown in Figure 12.2(a), the interconnected neural network is composed of neurons, which allow mutual communications among themselves. Repetitious state

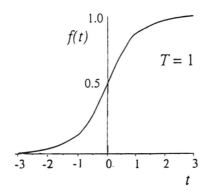

**Figure 12.4** Sigmoid function

transitions according to (12.1) and (12.2) minimize the energy of network, which is defined as:

$$E(u) = -\frac{1}{2} \sum_{\substack{i,j \\ (i \neq j)}} w_{ij} u_i u_j - \sum_i \theta_i u_i \tag{12.4}$$

Therefore, considering the analogy of the equilibrium of the network energy to the minimization of a cost function, the state transition procedure of the interconnected neural network can be utilized for solving various kinds of optimization problems.

A three-layered hierarchical neural network is schemitically shown in Figure 12.2(b). The whole units are formed into multiple layers, i.e. an input layer, a hidden (or intermediate) layer and an output layer. No connections exist among units in the same layer, while every two units in the neighboring layers have connections. Attractive features of the hierarchical neural networks can be summarized as follows:

a. One can automatically construct a nonlinear mapping from multiple input data to multiple output data in the network through a learning process of some or many sample input vs. output relations.
b. The network has a capability of the so-called "generalization", i.e. a kind of interpolation, such that the trained network estimates appropriate output data even for unlearned input data.
c. The trained network operates quickly in an application phase. The CPU power required to operate it may be equivalent only to that of a personal computer.

Although the number of hidden layers can be more than two, a three-layered neural network having sufficient number of units is known to map nearly all kinds of relationship. As a teaching method, the error back-propagation is often employed. The connection weights and the offset values are iteratively adjusted to minimize the square error between the output and the teaching data. During the teaching process, the connection weight and the offset values are iteratively modified until the convergence is attained.

## 12.4 MULTI-COLOR NEURAL NETWORK SOLVER FOR THE POISSON'S EQUATION (YAGAWA 1994, OKUDA & YAGAWA 1995, YAGAWA & OKUDA 1996)

### INTERCONNECTED NEURAL NETWORK WITH FEEDBACK MECHANISM

A concept of neuron called 'feedback neuron' is introduced into the interconnected neural network. To avoid confusion, neurons without the feecback function are called 'base neurons' in the followings. Figure 12.5 schematically shows the structure of the interconnected neural network accompanying feedback neurons. In this model, each base neuron has one feedback neuron. Thus, the number of the feedback neurons is the same as that of the base neurons. The function of the feedback neuron is also formulated by (12.1) and (12.2), except that the weights defined between the feedback neuron and its counterpart are not symmetrical. The state transition rules of the present network structure are summarized as follows:

Base neurons:
$$u_i = f_i(t_i) \tag{12.5}$$

$$t_i = \sum_{\substack{j \\ (i \neq j)}} w_{ij} u_j + w_{ii'} u_{i'} + \theta_i \tag{12.6}$$

Feedback neurons:
$$t_{i'} = w_{i'i} u_i + \theta_{i'} \tag{12.7}$$

$$u_{i'} = f_{i'}(t_{i'}) \tag{12.8}$$

where $i'$ indicates the feedback neuron connected to its counterpart base neuron $i$. $w_{ii'}$ is the weight defined from $i'$ to $i$, while $w_{i'i}$ is defined in the opposite direction. By virtue of the feedback neuron, the output from the base neuron is mixed with its previous value, which would help accelerating the minimization of the network energy.

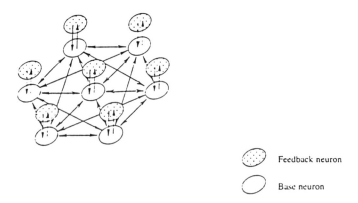

Feedback neuron

Base neuron

**Figure 12.5** Interconnected neural network accompanying feedback neurons

According to (12.4), the network energy is expressed as:

$$E(u) = -\frac{1}{2}\sum_{\substack{i,j \\ (i\neq j)}} w_{ij}u_iu_j - \sum_i \theta_i u_i - \frac{1}{2}\sum_i w_{ii'}u_iu_{i'} - \frac{1}{2}\sum_{i'} w_{i'i}u_{i'}u_i - \sum_{i'}\theta_{i'}u_{i'} \quad (12.9)$$

It is noted here that the third through fifth terms of the right hand side of (12.9) are contributions from the feedback neurons. Weights, offsets and activation functions in (12.5)–(12.8), are determined such that the network energy of (12.9) is equated to a cost function of an optimization problem.

## STATE TRANSITION RULE FOR THE POISSON'S EQUATION

We consider the following Poisson's equation:

$$-\Delta u = b \text{ in } \Omega \quad (12.10)$$

where $\Delta$ is the Laplace operator, $u$ the unknown variable and $b$ the source term. Both Dirichlet and Neumann type boundary conditions are imposed. The variational principle implies that the solution of (12.10) minimizes the following functional:

$$J(u) = \int_\Omega \left\{ \frac{1}{2}(\nabla u)^2 - bu \right\} d\Omega \quad (12.11)$$

Using an approximation procedures, e.g. the finite element method, (12.11) is expressed in a discretized form as:

$$J(u) = \frac{1}{2}\sum_{i,j} k_{ij}u_iu_j - \sum_i b_iu_i \quad (12.12)$$

where $u_i$ and $b_i$ are nodal values of $u$ and $b$, respectively. $k_{ij}$ is the 'stiffness' coefficient, which has meaningful value only if there exists interactions between nodes $i$ and $j$.

It is found that the functional of (12.12) can be expressed as the energy of the interconnected neural network with the feedback mechanism (12.9), by assuming weights, offsets and activation functions as:

Base neurons: $\qquad\qquad\qquad w_{ij} = w_{ji} = -Ak_{ij}$ $\qquad\qquad\qquad\qquad$ (12.13)

$\qquad\qquad\qquad\qquad\qquad \theta_i = Ab_i$ $\qquad\qquad\qquad\qquad\qquad\qquad$ (12.14)

$\qquad\qquad\qquad\qquad f_i(t_i) = t_i/k_{ii}$ $\qquad\qquad\qquad\qquad\qquad$ (12.15)

Feedback neurons: $\qquad\qquad\quad w_{ii'} = (1-A)k_{ii}$ $\qquad\qquad\qquad\qquad$ (12.16(a))

$\qquad\qquad\qquad\qquad\quad w_{i'i} = -k_{ii}$ $\qquad\qquad\qquad\qquad\qquad$ (12.16(b))

$\qquad\qquad\qquad\qquad\qquad \theta_{i'} = 0$ $\qquad\qquad\qquad\qquad\qquad\qquad\quad$ (12.17)

$\qquad\qquad\qquad\qquad f_{i'}(t_{i'}) = -t_{i'}/k_{ii}$ $\qquad\qquad\qquad\qquad\quad$ (12.18) where $A$ is a positive parameter, which controls the magnitude of feedback for accelerating the minimization process. $A = 1$ as an extreme case implies the zero-feedback. From (12.7), (12.8), (12.16(b)), (12.17) and (12.18), the state transition in the feedback neuron is simply written as:

$$u_{i'} = u_i \qquad (12.19)$$

Equation (12.19) implies that the output from the base neuron is copied and memorized as the feedback neuron's output.

Substituting (12.13)–(12.18) into (12.9):

$$E(u) = \frac{1}{2} \sum_{\substack{i,j \\ (i \neq j)}} A k_{ij} u_i u_j - \sum_i A b_i u_i + \frac{1}{2} \sum_i A k_{ii} u_i u_{i'} \qquad (12.20)$$

Considering (12.19), the network energy is finally expressed as:

$$E(u) = A \left[ \frac{1}{2} \sum_{i,j} k_{ij} u_i u_j - \sum_i b_i u_i \right] \qquad (12.21)$$

It is obvious that search for a minimum point of $J$ (12.12) is equivalent to search for that of $E$ (12.21) if a positive constant value of A is employed. It can be shown that the network energy E decreases monotonously by the state transition if $0 < A < 2$.

As is described above, the memory function in the interconnected neural network is realized by the feedback neurons, where the extent of feedback of the previous state is controled by the parameter $A$ ($0 < A < 2$). Because of the difficulty in finding the theoretical optimal value of $A$, the value has to be determined in a heuristic manner. Here, we should note that the equivalence between $E$ and $J$ is degraded due to the variation of $A$. However, since the variation of $A$ is gentle enough compared with that of the outputs of neurons, the equivalence may still hold in the practical computations.

Also, it should be mentioned that the state transition rules formulated with (12.5) through (12.8) lead to the same operations among the outputs as the successive over-relaxation (SOR) iterations. With other choice of weights, offsets and activation functions, different type of iteration solvers could be derived. It is emphasized here that the feedback mechanism is essential to construct an equation solver by the interconnected neural network.

Generally speaking, heuristic determination of parameters, which would influence the accuracy and/or the convergence of the computations, should be avoided from the scheme in order to keep the robustness. Dynamic fuzzy control of A was found to be effective (Yagawa & Okuda 1996), where during the minimization process, the present interconnected neural network is "trained" in a sense that the degree of feedback from the previous state is dynamically optimized by varying the weight and the offset. The stability and the gradient of $E$'s variations are judged by the membership functions, and increase or decrease of $A$ is determined according to the combinations of the smoothness and the gradient level.

## UNSTRUCTURED MULTI-COLOR PARALLELISM (OKUDA & YAGAWA 1995)

Trend in the supercomputer development is towards the distributed-memory type massively parallel architecture, and full potential of the neural networks would be disclosed if they are implemented on (massively) parallel computers, which would realize their inherent parallelism. To make the best use of the massively parallel computers, sophisticated algorithms have been exploited (Carey 1989).

Considering that independent activations of neurons are suitable for a parallel platform, the present neural-net solver is executed here on KSR1, which is a massively parallel processor machine consisting of 64 PEs (Processor Elements). Each PE has 32 Mbytes local cash memory. The full automatic parallelization preprocessor called KAP is available on KSR1. On the other hand, it is also possible to manually direct the parallelization procedure to the compiler. In this study, we selected the manual method to attain high parallel efficiency. Parallel computation was realized by the DO loop decomposition, which has equivalent effect of the domain decomposition. The parallelization by a domain decomposition was mainly directed to DO loops, which have long array lengths and occupies major part of CPU time. As shown below, inserting the directives into a source program enables the parallel processing of the DO loop:

$$c^*ksr^* \text{ tile (i,j, numthreads} = 16)$$

$$\text{DO } 10 \text{ I} = 1, \text{N}$$

$$\ldots\ldots\ldots\ldots$$

$$10 \text{ CONTINUE}$$

$$c^*ksr^* \text{ end tile}$$

To assess the effectiveness of the parallel computation, the following variables are often referred to:

$$S_p(n) = \frac{T_{(1)}}{T_{(n)}} \tag{12.22}$$

$$\varepsilon(n) = \frac{S_p(n)}{n} \tag{12.23}$$

where $S_p(n)$ and $\varepsilon(n)$ are speed-up and parallel efficiency, respectively. $T(n)$ denotes the CPU time when $n$ PEs are used. The parallel efficiency cannot, in general, reach unity due to the existence of the communication and the synchronization. It should be also noted that when employing the iterative methods for equation solvers, number of subdomains or number of PEs may degrade the convergence property, because the optimum iteration order of the solvers cannot be maintained. The odd-even and/or the multi-color method have been developed to tackle this difficulty in the finite difference (structured mesh) community. In the present study, unstructured multi-color method is proposed, and implemented in the neural network solver described above. Referring to the integer array of element-node connectivities of the finite element method, nodes, which are expressed by the neurons, are grouped into several colors such that the nodes having the same color have no interaction with each other. Figures 12.6(a) and 12.6(b) schematically show the multi-colored neurons as to a lid-driven cavity

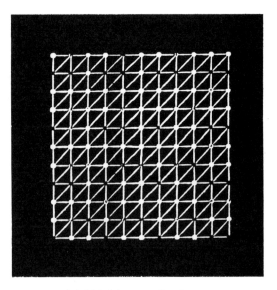

(a) Lid-driven cavity flow

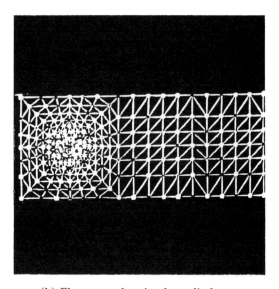

(b) Flow around a circular cylinder

**Figure 12.6** Multi-colored neurons

flow and a flow around a cylinder, respectively. The grouping procedure is done only
once at the beginning of computation. The advantage of the present method is that
when the neurons having the same color are computed in parallel, the operation
orders in the iterative solver are not affected by the number of subdomains nor the
number of PEs. Hence, the convergence property is maintained. Figure 12.7 shows
the parallel computation procedure of the multi-color method. After computing the

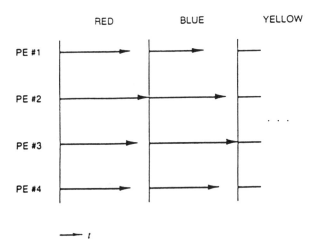

**Figure 12.7** Parallel computation procedure of the multi-color method

same colored neurons in parallel, communications between the PEs are performed. The synchronization is needed before going to the next color.

## 12.5 APPLICATIONS TO INCOMPRESSIBLE FLOW ANALYSIS (YAGAWA 1994, OKUDA & YAGAWA 1995, YAGAWA & OKUDA 1996)

*GOVERNING EQUATIONS AND SOLUTION ALGORITHM OF INCOMPRESSIBLE FLOW*

The interconnected neural network with feedback mechanism proposed here can be utilized in wide range of optimization problems including even an iterative equation solver. In this section, the applicability of the present neural-net approach to the incompressible viscous flow analysis is investigated.

The motion of the incompressible viscous fluid is governed by the Navier–Stokes equations and the incompressible constraint equation as follows:

$$\dot{u}_i + u_j u_{i,j} = -P,i + vu_{i,j,j} + f_i \tag{12.24}$$

$$u_{i,i} = 0 \tag{12.25}$$

where $u_i$, $p$, $f_i$ and $v$ imply the velocity component in the $i$-th direction, the pressure divided by the fluid density, the external force component and the kinematic viscosity, respectively. Subscripts $i$ and $j$ mean the directions in the Cartesian coordinate system running from 1 to 3. The notation $i$ means the partial differentiation with respect to $x_i$. The Einstein's summation convention is utilized. The essential boundary condition (EBC) and the natural boundary condition (NBC) are considered as

$$u_i = \hat{u}_i \quad \text{on } \Gamma_E \tag{12.26}$$

$$\tau_i = \hat{\tau}_i \quad \text{on } \Gamma_N \tag{12.27}$$

where $\Gamma_E$ and $\Gamma_N$ are complimentary parts of the whole boundary on which EBC and NBC are imposed, respectively. Superscript denotes the prescribed value. $\tau_i$ is the traction component defined as:

$$\tau_i = (-p\delta_{ij} + vu_{i \cdot j})n_j \tag{12.28}$$

where $\delta_{ij}$ and $n_i$ are the Kronecker's delta and the $i$-th component of the outward unit vector normal to $\Gamma_N$, respectively. Applying the weighted residual method to (12.22) and (12.23), the spatial discretizations are performed based on the Galerkin finite element method (Zienkiewicz & Taylor 1991). The finite element adopted here assumes quadratic and linear profiles for velocity and pressure, respectively. The Navier–Stokes equations and the incompressibility constraint equation are, respectively, expressed in the matrix form as follows:

$$\boldsymbol{M\dot{u} + Bu - Cp + Du = f} \tag{12.29}$$

$$\boldsymbol{C^t u} = 0 \tag{12.30}$$

where $\boldsymbol{u}$, $\boldsymbol{p}$ and $\boldsymbol{f}$ are the global vectors consisting of the nodal velocity, the nodal pressure and the nodal external force vectors, respectively. $\boldsymbol{M}$, $\boldsymbol{B}$, $\boldsymbol{C}$, $\boldsymbol{C^t}$ and $\boldsymbol{D}$ are the global matrices, which stand for the mass, the advection, the gradient, the divergence and the diffusion, respectively.

One of the computational difficulties in solving the above equation system is in dealing with the incompressibility constraint (Pelletier 1989). The incompressibility implies the occurrence of the infinite speed of pressure propagation and hence needs implicit treatment of pressure. In the so-called decomposition algorithms, for instance, the fractional step method (Donea $et$ $al.$ 1982), this difficulty is overcome by converting the incompressibility constraint equation into the Poisson's equation for pressure, which would yield the pressure field such that the velocity at the next time-step would satisfy the continuity.

Discretizing (12.27) and (12.28), respectively, in the direction of time as:

$$\boldsymbol{u}^{n+1} = \boldsymbol{u}^n + \Delta t \boldsymbol{M}^{-1}(\boldsymbol{Cp}^{n+1} - \boldsymbol{Bu}^n - \boldsymbol{Du}^n) \tag{12.31}$$

$$\boldsymbol{C^t u}^{n+1} = 0 \tag{12.32}$$

where $n$ is the time step. Here, external force is omitted for the sake of simplicity. In the fractional step method, (12.29) is decomposed into two stages as follows:

$$\tilde{\boldsymbol{u}} = \boldsymbol{u}^n + \Delta t \boldsymbol{M}^{-1}(-\boldsymbol{Bu}^n - \boldsymbol{Du}^n) \tag{12.33}$$

$$\boldsymbol{u}^{n+1} = \tilde{\boldsymbol{u}} + \Delta t \boldsymbol{M}^{-1} \boldsymbol{Cp}^{n+1} \tag{12.34}$$

where $\tilde{\boldsymbol{u}}$ is the intermediate velocity. Introducing (12.32) into (12.30), the following Poisson's equation for pressure is obtained:

$$\boldsymbol{C^t M}^{-1} \boldsymbol{Cp}^{n+1} = -\frac{\boldsymbol{C^t}}{\Delta t} \tilde{\boldsymbol{u}} \tag{12.35}$$

where $\boldsymbol{C^t M}^{-1} \boldsymbol{C}$ implies the discretized Laplacian operator.

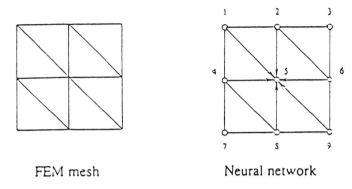

FEM mesh    Neural network

**Figure 12.8** Finite elements vs. neural network

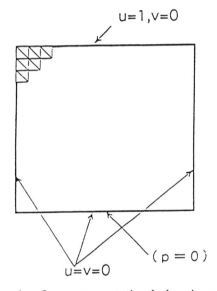

**Figure 12.9** Lid-driven cavity flow : computational domain and boundary conditions (441 neurons for pressure calculation)

The solution algorithm is summarized as follows:

Step 1  Compute $\tilde{u}$ from (12.31)

Step 2  Solve (12.33) for $p^{n+1}$

Step 3  Compute $u^{n+1}$ from (12.32)

Step 4  Increment the time step and return to Step 1

In the above algorithm, the velocity is advanced in time explicitly, while the Poisson's equation for pressure needs to be solved at each time step. Therefore, large part of the computational time is consumed in solving the Poisson's equation. Computational efficiency of the incompressible flow analysis largely depends on how fast the Poisson's equation for pressure is solved at every time step. Regarding nodes of pressure as neurons of network (see Figure 12.8), the neural network solver described before can be straightforwardly applied to the Poisson's equation for pressure.

## VERIFICATION OF NEURAL NETWORK SOLVER WITH FEEDBACK MECHANISM

Comparing with the conventional case that incorporates the conjugated gradient (CG) method as a solver, the efficiency of the present neural-net approach is studied. The minimization procedure is judged to be converged if $|E(t+1) - E(t)| < 1.0 \times 10^{-3}$.

First, a lid-driven cavity flow of Re = 100 is taken as an example among flows, which finally reach steady states. A square domain is regularly subdivided with 800 ($20 \times 20 \times 2$) triangular elements; that is, 441 ($21 \times 21$) neurons for pressure are located at regular intervals as shown in Figure 12.9. The time increment is set to $5.0 \times 10^{-3}$ sec and the computation is performed until 10 sec (2,000 steps). Figure 12.10 compares the pressure distribution. Agreements between the CG-method and the present method (the neural-net approach) is fairly good. Figure 12.11 shows the CPU-time monitored during the computations, which employ $A = 1.0$ and $A = 1.9$ for

(a) Present (neuro) method

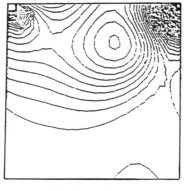

(b) CG method

**Figure 12.10** Lid-driven cavity flow: pressure contours

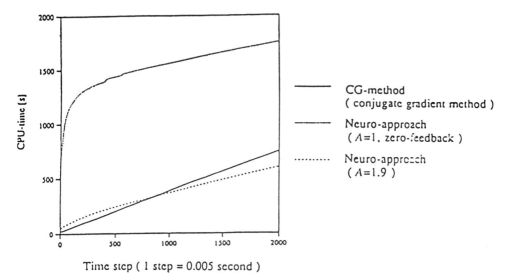

**Figure 12.11** Lid-driven cavity flow: CPU-time versus time step

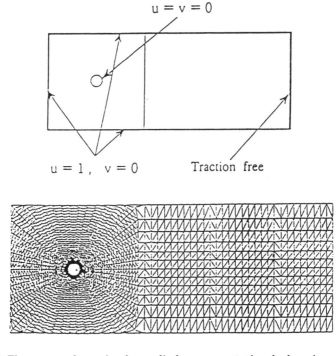

**Figure 12.12** Flow around a circular cylinder: computational domain and boundary conditions (1,878 neurons for pressure calculation)

the present method. We note here again that $A = 1.0$ means the zero-feedback. On the other hand, $A = 1.9$ was heuristically specified when computing with the feedback mechanism. It is observed that the computational burden required at an early stage, i.e., up to around 500 step, is dramatically reduced by adopting the feedback mechanism.

Second example is a flow around a circular cylinder of Re = 100. Different from the previous example, a cyclic time-dependent flow will be obtained at a quasi-steady state. Figure 12.12 shows a finite element mesh composed of 3,600 elements. Number of neurons for pressure calculation is 1,878. The time increment is set to be $5.0 \times 10^{-3}$ sec and the computation is performed until 250 sec or 50,000 steps. The value of $A$ is set to be 1.9. Figure 12.13 shows the pressure distribution obtained by the neural-net approach. Fairly good agreements between the CG-method and the neural-net approach is also observed. The CPU-time is compared in Figure 12.14. It is found from the figure that the present neural network method is more efficient than the CG-method for the time-dependent flow considered.

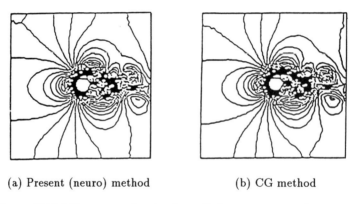

  (a) Present (neuro) method                    (b) CG method

**Figure 12.13** Flow around a circular cylinder: pressure contours

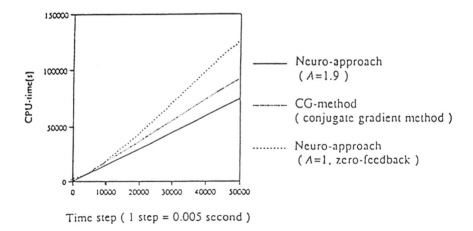

Time step ( 1 step = 0.005 second )

**Figure 12.14** Flow around a circular cylinder: CPU-time versus time step

## MULTI-COLOR PARALLEL COMPUTATIONS

The parallel efficiency of the neuro-solver with the multi-color method was evaluated for a two-dimensional cavity flow of Re = 5000. Figure 12.15 shows the network configuration. In this test, the number of pressure nodes was 6,561, and these nodes were grouped into nine colors. Pressure and stream line are depicted in Figure 12.16. Figure 12.17 compares the obtained velocity profile along the center line with the reference solution by Ghia *et al.* (1982). The CPU time consumed in the neuro-solver for the first time step was measured for assessing the parallel efficiency. First, effect of

**Figure 12.15** Lid-driven cavity flow : computational domain (6,561 neurons for pressure calculation)

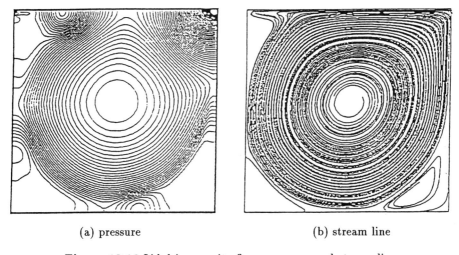

(a) pressure                (b) stream line

**Figure 12.16** Lid-driven cavity flow : pressure and stream line

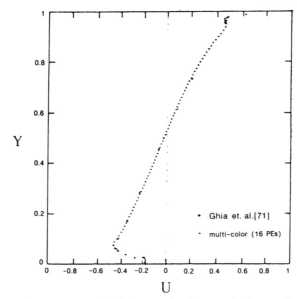

**Figure 12.17** Lid-driven cavity flow : velocity profile

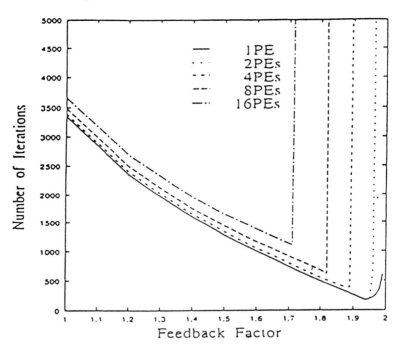

**Figure 12.18** Lid-driven cavity flow : number of iterations of the uni-color method

the number of PEs onto the optimum value of the feedback parameter $A$ is investigated. Figure 12.18 depicts the number of iterations of the conventional uni-color method, which stands for the neuro-solver without employing the multi-color grouping, varing the number of PEs. It can be observed that the optimum value of $A$ decreases according to the increase of the number of PEs. Therefore, even when adopting the optimum

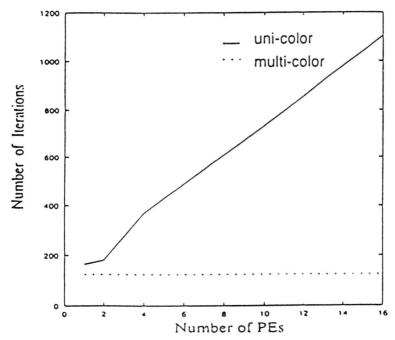

**Figure 12.19** Lid-driven cavity flow : number of iterations of the uni-color method and the multi-color method

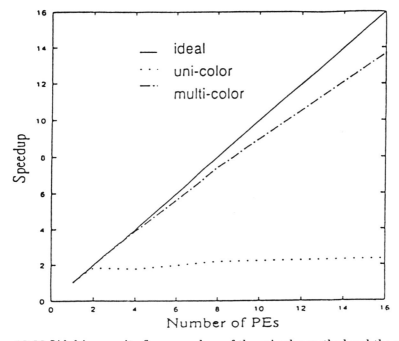

**Figure 12.20** Lid-driven cavity flow : speed-up of the uni-color method and the multi-color method

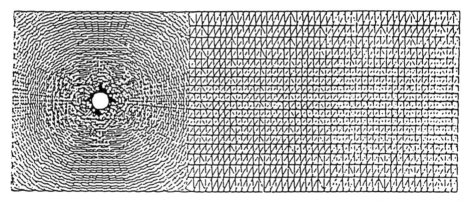

**Figure 12.21** Flow around a circular cylinder: computational domain (4,167 neurons for pressure calculation)

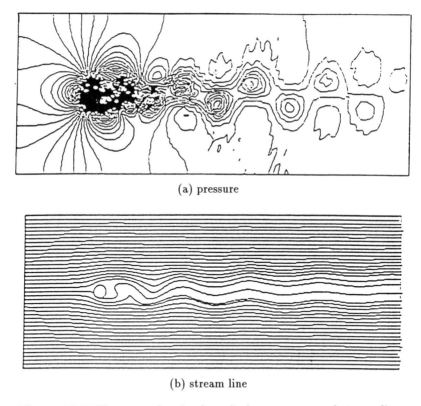

(a) pressure

(b) stream line

**Figure 12.22** Flow around a circular cylinder : pressure and stream line

value of $A$, the number of iterations required in the uni-color method increases as the number of PEs. On the other hand, the multi-color method is never affected by the number of PEs as shown in Figure 12.19. This result implies the favorable convergence nature of the multi-color method. Next, Figure 12.20 shows the speed-up curve of the both methods. The multi-color method attained 82% parallel efficiency when using

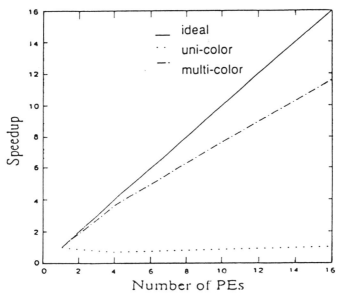

**Figure 12.23** Flow around a circular cylinder: speed-up of the uni-color method and the multi-color method

16 PEs, while the uni-color method showed unsuccessfull results. Deterioration of the multi-color method from the ideal linear speed-up is caused by the synchronization overhead, since the numbers of neurons are not completely equal among the PEs.

Second example is a flow around a circular cylinder of $Re = 100$. Figure 12.21 shows a finite element mesh composed of 8,100 elements. Number of neurons for pressure calculation is 4,167, and they were grouped into 12 colors. Pressure and stream line are depicted in Figure 12.22. Figure 12.23 shows the speed-up curve of the both methods. The multi-color method attained 73% parallel efficiency when using 16 PEs.

## 12.6 CONCLUDING REMARKS

In this chapter, for the purpose of seeking the possibilities of the evolutional computation techniques, firstly recent advances in the neural networks as well as their applications to the computational mechanics fields have been reviewed. Next, considering the analogy between the functional of the Poisson's equation and the energy of the interconnected neural network, the neuro-solver with the feedback mechanism has been developed, and has been successfully applied to the incompressible viscous flow analysis. Inherent parallel nature was maintained by incorporating the multi-color method, which groups the neurons into multi colors such that the neurons having the same color does not have interactions with each other, and hence these neurons can be computed in parallel.

The neurological research as well as its applications to the computational mechanics, applied mechanics and the CAE fields are remarkably developing with the advances of the massively parallel computers. These neural-net approaches would become more promising by virtue of future hardware development like optical neuro-computers.

## ACKNOWLEDGEMENTS

The authors wish to thank Canon Supercomputing SI Inc. for offering us an opportunity to use KSR1-64 supercopmuter. Also, the authors wish to thank Mr. Nagai and Mr. Someya, who were the undergraduate students of the University of Tokyo, for their help in performing the computations.

## REFERENCES

Amari, S. 1993. A universal theorem on learning curves. *Neural Networks*, 6: 161–166.

Bennet, J. S. and Englemore, R. S. 1979. SACON: A Knowledge-Based Consultant for Structural Analysis. *Proc. 6th Int. Conf. Artificial Intelligence*, 47–49.

Bennett, J. A. and Botkin, M. E. (eds.) 1986. *The Optimum Shape (Automated Structural Design)*. Plenum Press.

Berke, L. and Hajela, P. 1992. Applications of artificial neural nets in structural mechanics. *Structural Optimization*, 4: 90–98.

Biegler-Koenig, F. and Baermann, F. 1993. Learning algorithm for multilayerd neural networks based on linear least squares problems. *Neural Networks*, 6: 127–131.

Botros, N. M. and Abdul-Azia, M. 1993. Hardware implementation of an artificial neural network. 1993 IEEE International Conference on Neural Networks IEEE Service Center, San Francisco, California, USA, 1252–1257.

Bouzerdoum, A. and Pattison, T. R. 1993. Neural network for quadratic optimization with bound constraints. *IEEE Trans Neural Networks*, 4: 293–304.

Brown, L. M., Newman, R. W., DeNale, R., Lebowitz, C. A. and Arcella, F. G. 1992. Graphite Epoxy Defect Classification of Ultrasonic Signatures Using Statistical and Neural Network Techniques. *Review of Progress in Quantitative Nondestructive Evaluation*, 11 A: 677–684.

Carey, G. F. (Ed.), 1989. Wiley. *Parallel Supercomputing: Methods, Algorithms and Applications.*

Chang, G. -W. and Chang, P. -R. 1993. Neural plant inverse control approach to color error reduction for scanner and printer. *1993 IEEE International Conference on Neural Networks* IEEE Service Center, San Francisco, California, USA, 1979–1983.

Chen, R. M. M. and Chan, W. W. 1993. An efficient tolerance design procedure for yield maximization using optimization techniques and neural network. *Proc IEEE Int. Symp. Circuits Syst.*, 3: 1793–1796.

Cun, Y. L. 1992. Improving generalization performance using double backpropagation. *Harris Drucker, IEEE Trans Neural Networks* 3: 991–997.

Donea, J., Giuliani, S., Laval, H. and Quartapelle L. 1982. Finite Element Solution of the Unsteady Navior–Stokes Equations by a Fractional Step Method. *Comp. Meth. in Appl. Mech. and Eng.*, 30: 53–73.

Du, L. -M., Hou, Z. -Q. and Li, Q. -H. 1992. Optimum block-adaptive learning algorithm for error back-propagation networks. *IEEE Trans Signal Process*, 40: 3032–3042.

Elshafiey, I., Upda, L. and Upda, S. S. 1992. A Neural Network Approach for Solving Inverse Problems in NDE. *Review of Progress in Quantitative Nondestructive Evaluation*, 11 A: 709–716.

Fausett, L. 1994. *Fundamentals of Neural Networks: Architecture, Algorithms, and Applications*, Prentice-Hall, Inc., New Jersey.

Feuer, A. and Cristi, R. 1993. On the optimal weight vector of a perceptron with Gaussian data and arbitrary nonlinearity. *IEEE Trans Signal Process*, 41: 2257–2259.

Francelin, R. A. and Gomide, F. A. 1993. A Neural Network to Solve Discrete Dynamic Programming Problems. *1993 IEEE International Conference on Neural Networks* IEEE Service Center, San Francisco, California, USA, 1433–1525.

Fujimoto, Y., Fukuda, N. and Akabane, T. 1992. Massively parallel architectures for large scale neural network simulations. *IEEE Trans Neural Networks*, 3: 876–888.

Ghaboussi, J., Garrett, Jr . J. H. and Wu, X. 1991. Knowledge-Based Modeling of Material Behavior with Neural Networks. *Journal of Engineering Mechanics*, 117: 132–153.

Ghia, U., *et al.* 1982. High-Re Solutions for Incopressible Flow Using the Navier–Stokes Equations and a Multigrid Methd. *J. Compt. Physics*, 48: 387–411.

Grajski, K. A. 1993. Neurocomputing Using the Maspar MP-1. *Parallel Digital Implementation of Neural Networks, PTR Prentice Hall*, 51–76.

Hecht-Nielsen, R. 1989. *Neurocomputing*, Addison-Wesley.

Hopfield, J. J. and Tank, D. W. 1985. Neural Computation of Decisions in Optimization Problems. *Biol.Cybern.*, 52: 141–152.

Jean, J. S. N. and Wang, J. 1994. Weight Smoothing to Improve Network Generalization. *IEEE Transactions on Neural Network*, 5: 752–763.

Karayiannis, N. B. and Venetsanopoulos, A. N. 1992. Fast learning algorithms for neural networks. *IEEE Trans Circuits Syst II Analog Digital Signal Process*, 39: 453–474.

Kim, S. -W. and Lee, J. -J. 1993. Unknown Parameter Identification of Parameterized System Using Multi-layered Neural Networks. *1993 IEEE International Conference on Neural Networks* IEEE Service Center, San Francisco, California, USA, 438–443.

Kung, S. Y. and Chou, W. H. 1993. Mapping Neural Networks onto VLSI Array Processors. *Parallel Digital Implementation of Neural Networks*, PTR Prentice Hall, 3–49.

Kitahara, M., Achenbach, J. D., Guo, Q. C., Peterson, M. L., Notake, M. and Takadoya, M. 1992. Neural Network for Crack-Depth Determination from Ultrasonic Backscattering Data. *Review of Progress in Quantitative Nondestructive Evaluation*, 11: 701–708.

Lee, S. and Kil, R. M. 1994. Inverse Mapping of Continuous Functions Using Local and Global Information. *IEEE Transactions on Neural Network*, 5: 409–423.

Ling, P. and Jin, K. 1993. Solving Search Problems with Subgoals Using an Artificial Neural Network. *1993 IEEE International Conference on Neural Networks* IEEE Service Center, San Francisco, California, USA, 81–86.

Maa, C. -Y. and Schanblatt, M. A. 1992. A Two-Phase Optimization Neural Network. *IEEE Trans Neural Networks*, 3: 1003–1009.

Majumder, A. and Dandapani, R. 1993. Neural networks as a massively parallel automatic test pattern generators. *1993 IEEE International Conference on Neural Networks* IEEE Service Center, San Francisco, California, USA, 1724–1730.

Mehrotra, K. G. 1991. Bounds on the number of samples needed for neural learning. *IEEE Trans Neural Networks*, 2: 548–558.

Minsky, M. L. and Papert, S. A. 1988. *Perceptrons*. MIT Press, Cambridge.

Misra, M. and Prasanna, V. K. 1992. Implementation of neural networks on massive memory organizations. *IEEE Trans Circuits Syst II Analog digiral Signal Process*, 39: 476–480.

Mukhopadhyay, S. and Narendra, K. S. 1993. Disturbance rejection in nonlinear systems using neural networks. *IEEE Trans Neural Networks*, 4: 63–72.

Mochizuki, Y., Yoshimura, S. and Yagawa, G. 1993. Automated Structural Design System Based on Design Window Search Approach: Its Application to ITER First Wall Design. *Proc. of SMiRT-12 Post Conf. Seminar Nr.13*: 465–474.

Okawa, Y. 1993. Parallel computation of neural networks in a processor pipeline with partially shared memory. *1993 IEEE International Conference on Neural Networks* IEEE Service Center, San Francisco, California, USA, 1638–1643.

Okuda, H., Miyazaki, H. and Yagawa, G. 1994. A Neural Network Approach for Modelling of Viscoplastic Material Behaviors, Advanced Computer Applications. *ASME/PVP 274*: 141–145.

Okuda, H. and Yagawa, G. 1995. Parallel Finite Element Fluid Analysis on Element-by-Element and Unconstructured Multi-Color Basis. *Proc. ICES'95, Hawaii*, 917–922.

Pelletier, D., Fortin, A. and Camarero, R. 1989. Are FEM solutions of incompressible flows really incompressible? (or how simple flows can cause headaches!). *Int. J. Numer. Meth. in Fluids*, 9: 99–112.

Ramamurthy, A. C. 1993. Stone impact damage to automotive paint finishes — A neural net analysis of electro-chemical impedance data. *1993 IEEE International Conference on Neural Networks* IEEE Service Center, San Francisco, California, USA, 1708–1712.

Rico-Martinez, R. 1993. Continuous time modeling of nonlinear systems: a neural network-base approach. *1993 IEEE International Conference on Neural Networks* IEEE Service Center, San Francisco, California, USA, 1522–1525.

Rumelhalt, D. E., Hinton, G. E. and Williams, R. J. 1986. Learning Representations by Back-Propagating Errors. *Nature*, 329–9: 533–536.

Saravanan, N., Duyar, A., Guo, T.H. and Merrill, W. C. 1993. Modeling of the space shuttle main engine using feed-forward neural networks. *American Control Conference, Proc 1993 Am Control Conf 1993*, 2897–2899.

Shahla, K. 1993. Evaluation of the performance of various artificial neural networks to the signal fault diagnosis in nuclear reactor systems. *1993 IEEE International Conference on Neural Networks* IEEE Service Center, San Francisco, California, USA, 1719–1723.

Song, S. J. and Schmerr, L. W., Jr. 1991. Ultrasonic Flaw Classification in Weldments Using Neural Networks. *Review of Progress in Quantitative Nondestructive Evaluation*, 10: 697–704.

Sun, K. T. and Fu, H. C. 1993. Hybrid neural network model for solving optimization problems. *IEEE Trans Comput*, 42: 218–227.

Taig, I. C. 1986. Experts Aids to Finite Element System Applications. *Proc. 1st. Int. Conf. Applications of Artificial Intelligence in Engineering Problems*, II–2: 795–770.

Tank, D. W. and Hopfield, J. J. 1986. Simple Neural Optimization Networks : An A/D Converter, Signal Decision Circuit, and a Linear Programming Circuit. *IEEE Trans. Circuits Syst.,CAS-33*: 533–541.

Tsutsumi, K., Tani, A., Kawamura, H., Matsubayashi, T., Kataoka, H., Komono, K. and Fukushima, Y. 1993. Study of a Computer System for Evaluation of Bearing Wall Stresses in Structural Planning Reasoning by Neural Network, Computing in Civil and Building Engineering. *Proc 5 Int Conf Comput Civ Build Eng V ICCCBE*, 817–824.

Upda, L. and Upda, S. S. 1990. Eddy current defect characterization using neural networks. *Materials Evaluation*, 48: 342–347.

Wasserman, P. D. 1993. *Advanced Methods in Neural Computing*, Van Nostrand Reinhold.

Wedd, A. R. 1994. Functional Approximation by Feed-Forward Networks: A Least-Squares Approach to Generalization. *IEEE Trans Neural Networks*, 5: 363–371.

Weymaere, N. and Martens, J. P. 1994. On the Initialization and Optimization of Multilayer Perceptrons. *IEEE Trans Neural Networks*, 5: 738–751.

Witbrock, M. and Zagha, M. 1993. Back-Propagation Learning on the IBM GF11. *Parallel Digital Implementation of Neural Networks*, PTR Prentice Hall, 77–104.

Yamamoto, K. 1991. Study on Applications of Neural Networks to Non-Linear Structural Analysis (Part-1) Modeling of Hysteretic Behavior and its Application to Non-Linear Dynamic Response Analysis. *CRIEPI Report*, U91046 (in Japanese).

Yagawa, G.and Aoki, O. 1993. Neural Network-Based Direct FEM on a Massively Parallel Computer AP1000. *Proc. of the Second Parallel Computing Workshop*, Nov. 11, P1-K-1-K.11.

Yagawa, G., Yoshimura, S., Mochizuki, Y. and Oichi, T. 1992. Identification of Crack Shape Hidden in Solid by Means of Neural Network and Computational Mechanics. *Inverse Problems in Engineering Mechanics, Proc. IUTAM Symp. Tokyo*, 213–222.

Yagawa, G., Yoshimura, S. and Okuda, H. 1994. Neural Network Based Parameter Optimization for Nonlinear Finite Element Analyses. *Extended Abstracts, 2nd Japan-US Symposium on Finite Element Methods in Large-Scale Computational Fluid Dynamics*, 83–86.

Yagawa, G., Matsuda, A., Kawate, H. and Yoshimura, S. 1995. Neural Network Approach to Estimate Stable Crack Growth in Welded Specimens. *International Journal of Pressure Vessels and Piping*, 63: 303–313.

Yagawa, G. 1994. Finite Element with Network Mechanism. *Proc. of WCCM III (The Third World Congress on Computational Mechanics)*, II: 1474–1481.

Yagawa, G. and Okuda, H. 1996. Finite Element Solutions with Feedback Network Mechanism Through Direct Minimization of Energy Functionals. *Int. J. for Num. Meth. in Eng.*, 39: 867–883.

Yoshimura, S., Matsuda, A. and Yagawa, G. 1994. A new regularization method for neural-network-based inverse analyses and its application to structural identification. *Inverse Problems in Engineering Mechanics, Proc. 2nd ISIP*, 461–466.

Yoshimura, S., Yagawa, G., Oishi, A. and Yamada, K. 1993. Quantitative Defect Identification By Means of Neural Network and Computational Mechanics. *3rd Japan International SAMPE Symposium*, 2263–2268.

Yoshimura, S. and Yagawa, G. 1993. Inverse Analysis by Means of Neural Network and Computational Mechanics : Its Application to Structural Identification of Vibrating Plate. *Inverse Problems, S. Kubo* (Eds.) Atlanta Technology Publications, 184–193.

Yoshimura, S., Hishida, H. and Yagawa, G. 1992. Parameter Optimization of Viscoplastic Constitutive Equation Using Hierarchical Neural Network. *VII International Congress of Experimental Mechanics*, 1: 296–301.

Yu, X., Loh, N. K. and Miller, W. C. 1993. New acceleration technique for the backpropagation algorithm. *1993 IEEE International Conference on Neural Networks* IEEE Service Center, San Francisco, California, USA, 1157–1161.

Zell, A., Mache, N., Vogt, M. and Huttel, M. 1993. Problems of massive parallelism in neural network simulation, *1993 IEEE International Conference on Neural Networks* IEEE Service Center, San Francisco, California, USA, 1890–1895.

Zhang, S., Zhu, X. and Zou, L. -H. 1992. Second-order neural nets for constrained optimization. *IEEE Trans Neural Networks*, 3: 1021–1024.

Zienkiewicz, O. C. and Taylor, R. L. 1991. *The Finite Element Method*, 4th Ed., McGraw-Hill.

# 13

# Parallel Automatic Mesh Generation and Adaptive Mesh Control

M.S. Shephard, J.E. Flaherty, H.L. de Cougny,
C.L. Bottasso, and C. Ozturan[1]

## 13.1  INTRODUCTION

Automated and adaptive techniques provide the promise of reliably solving complex problems to the desired level of accuracy, while relieving the user of the time consuming and error prone tasks associated with mesh generation. For many problems, the computational requirements of the solution processes can only be met by scalable parallel computers. The development of effective parallel algorithms for adaptive techniques is challenging due to the irregular nature of adaptive discretizations and the constant modification of the discretization. This chapter discusses the techniques required to support automated adaptive analysis on distributed memory MIMD parallel computers.

Three assumptions underlying the techniques presented are (i) the parallel computation algorithms assume a partitioning of the mesh onto the processors, (ii) the meshes are unstructured, and (iii) the mesh generation and enrichment processes interact directly with a geometric definition of the domain being analyzed as it exists in a CAD system. These assumptions have a defining influence on the procedures developed.

A key aspect to supporting calculations on adaptively evolving meshes is the data structure used to describe the mesh and support its evolution during the adaptive analysis process. When the analyses are performed on parallel computers, capabilities must be available to support the communications between the partitions of the mesh assigned to various processors. As the mesh is adapted, partition work load becomes unbalanced, therefore procedures must be available to effectively modify the mesh partitions to regain load balance for the next computational step. Section 13.2 presents a set of data structures and algorithms for the effective parallel control of evolving meshes.

The demand for continuously larger meshes indicates the need for the development of efficient parallel automatic mesh generators which can operate directly from the geometric representations housed in CAD systems. Section 13.3 presents an algorithm for parallel three-dimensional automatic mesh generation. Section 13.4 outlines a set of local mesh modification procedures for the effective refinement and coarsening of meshes and discusses their parallel operation.

Given a set of parallel procedures for generation and control of the mesh, the remaining ingredient is the parallel adaptive solver. Section 13.5 outlines the structure

---

[1] Scientific Computation Research Center, Rensselaer Polytechnic Institute, Troy, NY
12180–3590, USA

*Parallel Solution Methods in Computational Mechanics* edited by M. Papadrakakis
© 1997 John Wiley & Sons Ltd

of such a solver for compressible flow problems and discusses specific aspects of the effectiveness of parallel adaptive calculations.

## 13.2   PARALLEL CONTROL OF EVOLVING MESHES

Central to parallel automated adaptive analysis are tools to control the mesh, and its distribution among the processors, as meshes are generated and analyzed. These tools must maintain load balance as the mesh evolves during the computations in such a manner that the interprocessor communications are kept as small as possible. It is also critical that these procedures operate in parallel and scale as the problem size grows so they do not become the bottleneck in the parallel computation process.

The tools required to support parallel automated adaptive analysis include:

1.  data structures and operators to support the model representations employed
2.  interprocessor communication control mechanisms
3.  mechanisms to effectively move portions of the discrete models generated to various processors so load balance can be maintained
4.  techniques to partition the mesh among the processors so the load is balanced and communications are minimized
5.  techniques to up-date the mesh partitions to regain load balance which was lost due to mesh modifications

The minimum data structures needed for an automated adaptive analysis are the problem definition, in terms of a geometric model and analysis attributes, and the mesh, which the discrete representation used by the analysis procedures. A boundary-based representation is used for the problem definition and the mesh, as well as to support the partition model used to control the parallel processing procedures.

### *GEOMETRY-BASED MESH DATA STRUCTURE*

The classic mesh data structure of node point coordinates and element connectivities can not support the needs of automated adaptive analysis. Richer structures are required to support adaptive mesh enrichment and to provide the links to the original domain definition needed by critical functions, including ensuring the automatic mesh generator has produced a valid discretization of the domain. A number of alternative mesh data structures have been proposed for various forms of mesh adaptation. All of them can be effectively captured by a general structure based on a hierarchy of topological entities.

The goal of an analysis is to solve a set of partial differential equations over the domain of interest, $_g\Omega$. Generalized numerical analysis procedures employ a discretized version of the domain in terms of a mesh. Since the mesh domain, $\overline{_m\Omega}$ may not be identical to the original geometric domain, $\overline{_g\Omega}$, and/or various procedures, such as automatic mesh generation, adaptive mesh refinement and element stiffness integration, need to understand the relationship of the mesh to the geometric model, it is critical to employ a representational scheme which maintains the relationships between these two models. Although a number of schemes are possible for defining a geometric domain (Requicha and Voelcker 1983), the most advantageous for the current purposes are boundary-based schemes in which the geometric domain to be analyzed is represented as

$$\overline{_g\Omega}\left(g\{_gS\}, g\{_gT\}\right) \tag{13.1}$$

where $g\{_gS\}$ represents the information defining the shape of the entities which define the domain and $g\{_gT\}$ represents the topological types and adjacencies[2] of the entities which define the domain. In addition to being unique, the use of topological entities and their adjacencies provides a convenient abstraction for defining the relationship of different models of the same domain. Boundary representations also allow the convenient specification of the analysis attributes of material properties, loads, boundary conditions and initial conditions (Shephard 1988, Shephard and Finnigan 1989). An additional advantage is the fact that current computer aided design systems support a boundary representation of the domains defined within them. This allows the effective combination of these packages with automatic mesh generation. A final advantage of recent boundary representations are their ability to properly represent the non-manifold geometric domains commonly used for analysis processes (Weiler 1988, Gursoz et al. 1990).

Since individual volume elements will be limited to simple regions, bounded by simply connected faces, consideration of the topological entities can focus on the basic 0 to $d$ dimensional topological entities, which for the three-dimensional case ($d=3$) are:

$$v\{_vT\} = \{v\{_vT^0\}, v\{_vT^1\}, v\{_vT^2\}, v\{_vT^3\}\} \tag{13.2}$$

where $v\{_vT^d\}$, $d = 0, 1, 2, 3$ are respectively the set of vertices, edges, faces and regions defining the domain.

Critical to the understanding of the relationship of the mesh with the geometric domain is the concept of classification of a derived model to its parent model (Schroeder and Shephard 1990, Schroeder and Shephard 1991).

*Definition: Mesh Classification Against the Geometric Domain — The unique association of a topological mesh entity of dimension $d_i$, $_mT_i^{d_i}$, to a topological geometric domain entity of dimension $d_j$, $_gT_j^{d_j}$, where $d_i \leq d_j$, is termed classification and is denoted*

$$_mT_i^{d_i} \sqsubset {}_gT_j^{d_j} \tag{13.3}$$

*where the classification symbol, $\sqsubset$, indicates that the left hand entity, or set, is classified on the right hand entity.*

Multiple $_mT_i^{d_i}$ can be classified on a $_gT_j^{d_j}$. Mesh entities are always classified with respect to the lowest order object entity possible. Classification of the mesh against the geometric domain is central to (i) ensuring that the automatic mesh generator has created a valid mesh (Schroeder and Shephard 1990, Schroeder and Shephard 1991), (ii) transferring analysis attribute information to the mesh (Shephard and Finnigan 1989), (iii) supporting h-type mesh enrichments, and (iv) integrating to the exact geometry as needed by higher order elements.

In addition to the mesh representation, it is often desirable to consider other derived representations of the domain. The one of importance to the parallel adaptive analysis is the processor representation, $_p\overline{\Omega}$, which is an intermediate representation between that of the mesh and the geometric domain.

---

[2] In the context of a domain representation, adjacencies are the relationships among topological entities which bound each other. For example, the edges that bound a face, is a commonly used topological adjacency.

The adjacencies of various order mesh topological entities and their classification with respect to the higher order models are used to support a great number of the operations required by a parallel automated adaptive analysis. Therefore, it is important that they can be quickly determined. Clearly, if the adjacencies of each order entity against all other entities were stored, all possible adjacency information would be readily available. This approach would be highly wasteful with respect to the amount of data storage required. On the other hand, storing only a minimal number of adjacencies could require extensive searches and sorts to determine other specific adjacencies. An examination of the specific adjacencies used by the various algorithmic operations provides guidance as to the minimum number of adjacencies needed. For example references (Biswas and Strawn 1993, Connell and Holmes 1994, Golias and Tsiboukis 1994, Kallinderis and Vijayan 1993) define adjacencies used in specific finite volume and finite element procedures. Since the procedures considered here support any form of adaptive analysis on conforming unstructured meshes[3], all adjacencies are either stored, or can be quickly determined through a set of local traversals and sorts which are not a function of the mesh size. One set of relationships that can effectively meet these requirements is to maintain adjacencies between entities one order apart. Fig. (13.1) graphically depicts this set of relationships as well as the classification with respect to the geometric domain representation. For more information on the use and implementation of alternative topologically-based mesh data structures see (Beall and Shephard 1995).

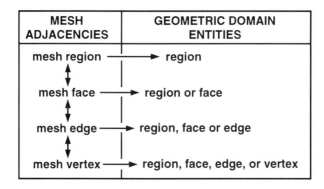

Figure 13.1  Mesh topological adjacencies and classification information

## PARTITION COMMUNICATION AND MESH MIGRATION

Adaptive unstructured meshes on distributed memory computers require data structures which provide efficient queries for various entity and processor adjacency information, as well as fast updates for changes in the mesh. These requirements are met by the distributed mesh environment Parallel Mesh Database (PMDB).

---

[3] A conforming mesh is one where all mesh entities exactly match. For example, a situation where the mesh edge bounding one mesh face has two mesh edges from another mesh face lying exactly on top of it is not allowed. Although possible to extend the procedures presented here to support those situations, they will not be considered in the present document.

## Background on Parallel Mesh Structures

Early distributed memory parallel finite element methods involved static meshes and used the data parallel SIMD constructs. The generation of data parallel communication references of unstructured meshes requires a runtime system such as the PARTI primitives (Saltz *et al.* 1990) which computes needed references prior to entering a loop where the actual computations are done. If the distribution of the mesh changes, then all the references have to be recomputed. Since limited analysis can be done at the level of references only, the data parallel compilers are not well suited for handling the dynamically changing mesh data structures of adaptive applications.

Williams' Distributed Irregular Mesh Environment (DIME) (Williams 1990, Williams 1993) uses a hash table to implement *voxel databases* (Williams 1992) which store a global key associated with an entity. This key is the geometric centroid of the entity. The coordinates of the centroid are converted to integer hash table index by dividing it with a user supplied tolerance. When elements are migrated in DIME, new voxel entries are packaged into a message passed from processor to processor in a ring until each has seen the message. Each processor takes the voxel entry and checks if a match is found in the hash table. If found, the entity is shared and the off-processor address is stored. Since the new voxels are passed in a ring of all processors, the update procedure has a fixed cost dependent on the number of processors.

Vidwans et al. (Vidwans *et al.* 1994) present a procedure to migrate face adjacent tetrahedra between *sender-receiver-disjoint* processors. The sender-receiver-disjoint requirement necessitates processors be paired as either a sender or a receiver, which is part of a divide and conquer dynamic load balancing algorithm. Since a face can be shared by no more than two processors and a processor migrates to its face-adjacent processor, the shared face identification is readily available. Hence global identifications are not needed. A disadvantage of sender-receiver disjoint migration is that elements cannot be piped by a receiver processor to other processors in the same cycle of migration. This can lead to memory problems whereby a receiving processor obtains a large number of elements and has to store them before it can pass them onto other processors.

The Tiling system developed by Devine (Devine 1994) supports hp-adaptive analysis and provides migration routines for regularly structured two dimensional rectangular element meshes which can be hierarchically refined. Each tiling element stores pointers to the four neighboring elements, with partition boundary elements pointing to a ghost-element data which acts as a buffer during communication. The elements are given unique ids which are maintained in a balanced AVL tree (Sedgewick 1990) for efficient insertion and deletion during migration.

## Distributed Mesh Model

The distributed mesh is viewed analogous to the modeling of geometric objects. Fig. (13.2) shows the hierarchical classification of the global mesh entities $_mT_i^d$, the processor model entities $_pT_i^{d'}$ and geometric model entities $_gT_i^{d''}$. Given the set of mesh entities $\{_mT\}$, a partitioning at the $d_m$ dimension level divides the mesh into $n_p$ parts, $_pT_{p_k}^{d_m}$, each of which is assigned to a processor with id $p_k = 0, \ldots, n_p - 1$. As a result of partitioning, some of the entities with dimension $d < d_m$ will be *shared* by more than one processor. The $d_m$-dimensional entity will be held by only one processor. Hence in general, partitioning with $d_m > 0$ defines a one-to-many relation from a mesh entity $_mT_i^d$ to its uses $_m^kT_i^d$ where $k \leq min(\Delta(_mT_i^d), n_p)$. Here $\Delta$ defines the degree of an entity, i.e. given the dimension $d$ of an entity, $\Delta$ is the number of $d + 1$ dimensional entities which use it.

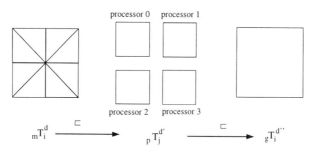

Figure 13.2 The relationship between the mesh
model, processor model and the geometric model

One of the processors holding a given entity $_mT_i^d$ is designated as the owner of that entity. We distinguish the owned entities as $(p_o, a_o)$, which defines a one-to-one and onto mapping of global mesh entities onto the owned distributed mesh entities, Since this mapping can be inverted, the pair $(p_o, a_o)$ can serve as a global key of a distributed entity.

The uses of shared entities are mapped onto the owner entity by a many-to-one relation, $\Phi : (p_k, a_k) \longmapsto (p_o, a_o)$. Fig. (13.2) shows the relationship between the geometric model entities $_gT_i^{d''}$, the global mesh entities $_mT_i^d$ and the processor model. Given the uses $(p_k, a_k)$ of an entity distributed over processors $p_k$, an agreement can be reached among these processors on whether they hold the identical entity by computing the ownership using the function $\Phi$.

The PMDB data structures were designed to provide the full set of adjacencies. A partition boundary entity can get all the links of an entity on other processors. Each partition boundary entity stores all the uses on other processors as a linked list (Fig. (13.3)). The bold edges and vertices indicate the owners of the shared entities. This ownership information can be used in the implementation of the owner computes rule, for example, during link updates in mesh migration or scalar product computation in an iterative linear solver.

Since each processor stores the uses $(p_k, a_k)$ on all the processors that hold a shared entity, the ownership can be computed as a function of these uses. Since the pair $(p_o, a_o)$ *is the global key*, there is no need to generate and store a separate key. On a processor, at the level of entities, the sets of entities that are on the partition boundary or adjacent to a specific processor are organized in doubly linked lists which provide constant insertion and deletion.

## Mesh Migration

Analogous to the owner computes rule, the mesh migration procedure of PMDB uses an *owner updates rule* to collect and update any changes to the links on partition boundaries after moving entities among processors. The migration of a set of mesh entities from a given processor to destination processors proceeds in three stages. Firstly, sender

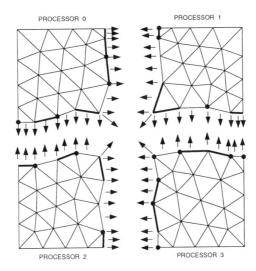

Figure 13.3 PMDB inter processor links and entity ownership

processors migrate the mesh entities to receiver processors. Secondly, the senders and receiver processors report the deletions or new addresses of migrated mesh entities to owner processors. In the last stage, the owner processors inform the affected processors about the updates in links. The processing which is done in the first stage is proportional to the number of mesh entities being migrated, whereas in the second and third stages, it is proportional to union of boundary of the migrated mesh entities. The migration procedure is outlined in Fig. (13.4). An explanation of the procedure and demonstration of it scalability is given in references (Ozturan 1995, Shephard *et al.* 1995).

## LOAD BALANCING OF ADAPTIVELY EVOLVING MESHES

The evolving nature of an adaptive discretization introduces load imbalance into the solution process. Therefore, it is critical that the load be dynamically rebalanced as the adaptive calculation proceeds.

*Recursive Bisection* (RB) is a popular partitioning technique. However, common RB implementations start with the entire mesh on a single processor and partition from there. This is not advantageous for adaptive calculations on a distributed mesh because (i) the time required to gather the distributed mesh together on a single processor, and (ii) the fact that after the mesh has been adapted, it may have grown to the point that it can not fit on a single processor. These problems can be alleviated if the mesh remains distributed during the repartitioning process. The next subsection outlines a parallel implementation of *Inertial Recursive Bisection* that operates on a distributed mesh.

*Iterative Local Migration* techniques have the potential for dynamically balancing adaptive meshes. These techniques exchange load between neighboring processors to improve the load balance and/or decrease the communication volume. *Cyclic pairwise exchange* (Hammond 1991) pairs processors connected by a hardware link and exchanges the nodes of the mesh to improve the communication. Leiss/Reddy (Leiss and Reddy 1989) use the hardware link as the neighborhood to transfer work from heavily loaded to less loaded processors. The `Tiling` system (Devine 1994) extends the Leiss/Reddy algorithm to the case where the neighborhood is defined by the connectivity of the split domains. The algorithm of Lohner and Ramamurti (Löhner and Ramamurti 1993)

**procedure** mesh_migrate $(P_s, \{_mT_{s_i}^{d_i}\}, P_r, \{_mT_{r_i}^{d_i}\})$
**input:** $P_s$: destination processors
$\{_mT_{s_i}^{d_i}\}$ : sets of regions to be migrated
**output:** $P_r$: source processors
$\{_mT_{r_i}^{d_i}\}$: sets of regions received
**begin**
/* 1. senders and receivers to owners */
1    Pack the mesh $\{_mT_{s_i}^{d_i}\}$ to be sent
2    Find the migrated boundary
3    Delete migrated internal entities
4    Pack the owners' uses corresponding to migrated boundary
5    Send packed submeshes and uses to $P_{s_i}$
6    Receive packed submeshes and uses from $P_{r_i}$
7    Unpack the submeshes to get $\{_mT_{r_i}^{d_i}\}$

/* 2. senders and receivers to owners */
8    Establish usage of both sent and received migrated boundary entities
9    Pack local uses of migrated boundary and owners uses to be sent to owner processors
     $P_o$
10   Send packed local and owner uses to owner processors
11   Receive packed uses from senders and receivers

/* 3 owners to affected */
12   Owners update use lists by inserting/deleting received local uses into/from use lists
     pointed to by owner uses and generate new ownerships
13   Pack updated uses list of entities to be sent to affected processors $P_a$
14   Send updated use lists and ownership to owner processors
15   Receive updated uses list and ownership from owner processors
16   $P_a$ update use lists and ownership
17   Delete unused sent migrated boundary entities
     **end**

Figure 13.4 Mesh Migration Algorithm

exchanges elements between subdomains according to a deficit difference function which reflects the imbalance between an element and its neighbors. Vidwans et al. (Vidwans *et al.* 1994) uses a divide and conquer approach to pair processors and uses connectivity as well as coordinate information to decide which elements to migrate. Subsection 2 outlines an iterative load balancing procedure based on the Leiss/Reddy heuristic of requesting load from the most heavily loaded neighbor. The performance of this procedure is compared with repartitioning by the parallel distributed inertia recursive bisection algorithm.

## Dynamic Balancing Based on Repartitioning

The current dynamic balancing based on repartitioning relies on the Inertial Recursive Bisection (IRB) method (Löhner and Ramamurti 1993) which bisects a set of entities by considering the median of the set of corresponding centroids with respect to the inertial coordinate system for the set of entities to be bisected. The main assumption for performing repartitioning in parallel is that the entities are distributed. It is also assumed that there is no reason for the number of entities to be uniform across processors. The result of this repartitioning will be an equal number of entities per processor. The key algorithm in IRB is the determination of the median for a given set of doubles (referred to as "keys") (Sedgewick 1990). With respect to this algorithm, the "keys" are the first coordinates, in the inertial frame, of the entities to be bisected. The method used here is to sort the "keys" and then pick the entry at the middle of the sorted list. In this case, efficiently performing IRB in parallel can be reduced to efficiently sorting in parallel (JaJa 1992). A parallel sample sort algorithm (Blelloch *et al.* 1991) is well suited to large data sets, and is used here to efficiently support IRB on distributed meshes (de Cougny *et al.* 1996, Shephard *et al.* 1995).

Given a set of $n$ "keys" distributed on $p$ processors ($n >> p$), a sample sort algorithm consists of three main steps (de Cougny and Shephard 1995):

1.  $p$-$1$ splitters (or pivots) are chosen among the $n$ "keys"
2.  Each key is routed to the processor corresponding to the bucket the "key" is in
3.  Keys are sorted within each bucket (no communication)

The goal of step 1 is to split the "keys" into $p$ parts (buckets) as evenly and efficiently as possible. The $p$-$1$ splitters which are implicitly sorted (say with respect to increasing value) are labeled from *1* to *p-1*. All distributed "keys" below splitter *1* belong to bucket *0*, all distributed "keys" between splitter $i$ ($0 < i < p$-$1$) and splitter $i+1$ belong to bucket $i$, and all distributed "keys" above splitter $p$-$1$ belong to bucket $p$-$1$. Processor $i$ ($0 \leq i < p$) is responsible for the bucket labeled $i$. In step 2, assuming the $p$-$1$ splitters have been found and broadcasted to all processors, any distributed "key" can tell in which bucket it belongs and is rerouted to the processor that is responsible for that bucket. At this point, any processor has knowledge of all "keys" that belong to the bucket it has been assigned. Step 3 can be performed using any efficient sequential sorting algorithm, like quicksort (Sedgewick 1990). Parallel efficiency is maximal when the sizes of the buckets are near equal. A sampling method is used to obtain "good" splitters (Blelloch *et al.* 1991, de Cougny *et al.* 1996, Shephard *et al.* 1995).

The following pseudo-code shows the process of parallel repartitioning using IRB for entities already distributed on processors. A statement of the form **for** ( $i = 0$ ; $i < n$ ; $i++$ ) { ... } indicates a loop that is executed as long as the loop variable $i$, initially set to 0 and incremented by 1 upon completion of each pass ($i++$), has a value less than $n$. Each processor executes the following pseudo-code (MIMD):

1.  Associate each entity with a "key" structure consisting of (i) the coordinates of the entity's centroid with respect to the current inertial coordinate system, (ii) the processor the actual entity is stored on, (iii) a pointer to the entity, and (iv) the destination processor for the entity
2.  **for** ( $step = 0$ ; $step < \log_2 p$ ; $step$ ++ ) {

    a.  Split the $p$ processors into $2^{step}$ processor sets
    b.  Balance the load such that each processor has approximately the same number of keys
    c.  Get center of gravity, the principal axes of inertia, and transform the keys
    d.  Get $p'$ −1 splitters among the keys

e.   Using the position with respect to the splitters to determine in which bucket (processor) each key goes
f.   Sort the keys (no communication)
g.   Using the position with respect to the median, determine in which bucket (processor) each key goes
h.   Free the processor sets

}

3.   The destination processor is set to the processor the key is currently in
4.   Reroute all keys to the originating processors
5.   Migrate entities according to the destination processor stored at the key level

Steps 2.b through 2.g are done independently on each processor set. Once all keys have been sorted in the processor set (at the end of step 2.f), the median (key that splits the set of keys into two subsets of same cardinality) is easily obtained. Fig. (13.5) is a graphical depiction of steps 2.b through 2.g in the case when $p'$ equals four. At each step, the array of keys (distributed across the processors in the set) is represented by a horizontal line which is cut to show how it is currently distributed. The symbol $<$ indicates that the keys in the array are not sorted if above the processor cutter, it also indicates that any key in the left processor's array is smaller than any key in the right processor's array. If there is no such symbol, the keys are not sorted yet.

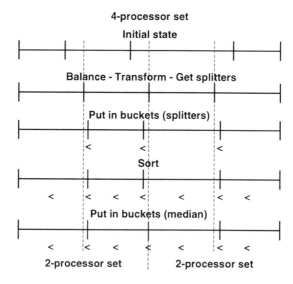

Figure 13.5  Graphical description of the repartitioning algorithm (4–processor set)

Fig. (13.6) shows a randomly distributed mesh (approximately 35,000 elements) and the resulting dynamically repartitioned mesh for eight processors. Fig. (13.7) shows timings (wall-clock seconds on IBM sp-2) for that particular mesh on 2, 4, 8, and 16 processors. The processor assignment timing corresponds to steps 1 to 4 (decision making). The migration timing corresponds to step 5. It should be noted that a randomized mesh as the initial state is a worst-case scenario for the migration part of the repartitioning procedure. Past four processors, the time spent decreases as the number of processors increases, which is a good indication of scalability. It is conjectured that the "abnormal" speed with two processors is due to the fact that (i) the only processor set ever used is the full set of processors and (ii) there is some performance degradation when more than one processor set is defined.

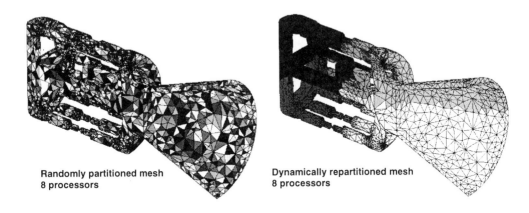

Randomly partitioned mesh
8 processors

Dynamically repartitioned mesh
8 processors

Figure 13.6 Dynamic repartitioning on a randomly distributed mesh

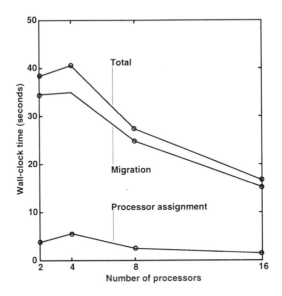

Figure 13.7 Timings for dynamic repartitioning

## Dynamic Balancing based on Iterative Load Migration

The Tiling system which uses the Leiss/Reddy approach calculates the load averages utilizing the immediate neighborhood. To incorporate more global information and to direct load transfers, we view the processor requests for load from heavily loaded processors as forming a forest of trees. Fig. (13.8(a)) shows an example of requests that can be formed. Given this hierarchical arrangement of processors as the nodes of trees, we balance the trees as shown in Fig. (13.8(b)) and iteratively repeat the process until the load distribution converges to optimal load balance within a user supplied tolerance. The full algorithm is given in Fig. (13.9). The procedure details are given as follows.

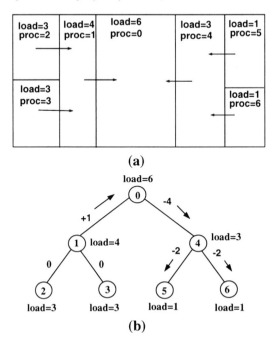

Figure 13.8 Iterative load migration load balancing example: load request (a) load migration on the tree (b)

**procedure** tree_load_balance($tol_{load}, max_{iter}$)
**in** $tol_{load}$imbalance load tolerance
**in** $max_{iter}$ : maximum number of iterations
**begin**
1    iter = 0
2    **while** (max. load difference > $tol_{load}$ ) **and**
     (iter < $max_{iter}$) **do**
3       iter = iter + 1
4       Compute neighboring load differences
5       Request load from neighbor processor having
        largest load difference (creates processor trees)
6       Linearize processor trees
7       Compute amounts of load migration
8       Select and migrate load
9    **endwhile**
**end**

Figure 13.9 Iterative load migration balancing procedure

The steps of balancing the forest of trees are repeated until convergence is achieved.

In step 4, load differences are computed by each processor sending its load value to its neighbors and receiving load values from its neighbors. Step 5 invokes the Leiss/Reddy load request process. Since each processor can receive requests from multiple processors, but can only request from a single processor, a forest of trees is formed.

In step 6, the trees are linearized for efficient scan operations. We linearize using depth-first-links (Kruskal *et al.* 1990, Szymanski and Minczuk 1978) which use between $|T|$ and $2(|T| - 1)$ number of links. Step 7 computes the amounts of load migrations on the tree using logarithmic scan operations on the linearized tree. Let $load\_mig_i$ denote amount of load that will be migrated into or out of a tree node $i$ which represents a processor. Let also $T_i$ denote the subtree with node $i$ as the root of the subtree and $load(T_i)$ be the sum of loads of nodes in this subtree. The amount of load migration is then calculated as

$$load\_mig_i = load(T_i) - avg\_load(T) * |T_i|$$

with $avg\_load(T) = load(T)/|T|$ representing the average load on the tree when balanced. Given $load\_mig_i$, the direction of load migrations can be found as

$$
\begin{aligned}
load\_mig_i \ &= \ 0, \ \text{do nothing with parent,} \\
&< \ 0, \ \text{get load from parent,} \\
&> \ 0, \ \text{send load to parent.}
\end{aligned}
$$

Having calculated the directions of load migration, step 8 migrates the elements on the partition boundary in a slice-by-slice manner until $load\_mig_i$ has been transferred.

## Examples

**Test 1:** A `patella` mesh of 8497 tetrahedrons is refined in the center area to 15329 tetrahedrons. Iterative load migration run on 16 processor converged in 3 iterations for this example. Fig. (13.10), plots the times taken for the iterative load migration procedure, a parallel recursive moment of inertia bisection routine and the parallel inertia repartitioner. The parallel recursive moment of inertia bisection starts with the whole mesh on one processor and recursively splits it in parallel. The parallel inertia repartitioner outperforms the other two strategies. There are various reasons why the parallel inertia repartitioner outperforms the iterative load migration. The performance of the iterative load migration is greatly affected by the distances between the heavily loaded and underloaded processors. For example, suppose there is refinement at the corner of a model on one of the processors. To propagate the excess load to the rest of the processors by local neighborhood transfers requires a number of steps needed at least the size of the diameter of partition graph. For other types of distributions in which the distance between the lightly loaded to heavily loaded processors is small and there is a high frequency of load imbalances, the load balancer will have better performance. The repartitioner bypasses the effects of distance by directly sending load from heavily loaded to lightly loaded processors. On an architecture such as the IBM-SP2, in which communication cost is independent of the distance between the processors and hence

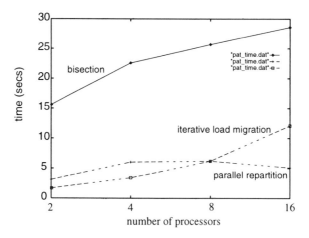

Figure 13.10 Time taken for iterative load
migration, parallel repartitioning and bisection

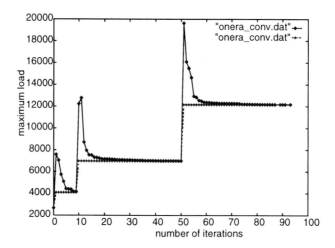

Figure 13.11 The convergence history for iterative
load migration during three stages of refinement

the same between any pair of processors, the repartitioner will be advantageous since
it directly sends the load to its final destination.

**Test 2:** In this test, various statistics are reported for the adaptively refined
onera-m6 wing mesh during an adaptive CFD analysis on 32 processors. The initial
mesh has 85567 tetrahedrons. Three stages of adaptive refinements are performed during
which the number of tetrahedrons increase to 131000, 223501 and finally to 388837. Fig.
(13.11) shows the convergence history of the iterative load migration. In all cases of
load balancing after refinement, the imbalance reduces to less than 4% during the first
8 iterations. One need not run the tree_balance to full convergence. It can be
stopped when a reasonable imbalance is achieved.

Table (13.1) shows the execution time comparisons between iterative load migration and the parallel inertia repartitioner. In all cases, moment of inertia outperforms iterative load balancing for the reasons explained in Test 1.

Table 13.1 Execution times (in seconds) for
iterative load migration and inertia repartitioner

| refinement | 1st | | 2nd | | 3rd | |
|---|---|---|---|---|---|---|
| percent imbalance | 1.7 | 0 | 3.8 | 0 | 1.9 | 0 |
| iterative load migration (sec) | 73 | 85 | 127 | 210 | 189 | 283 |
| inertia repartition (sec) | | 21 | | 48 | | 128 |

Finally, Table (13.2) shows partition quality comparisons between iterative load balancing and the moment of inertia partitioner. The percentage of maximum number and the total number of cut faces are given for both iterative load balance and the inertia repartitioner.

Table 13.2 Maximum and total percentage of cut faces
for iterative load migration and inertia repartitioner

| refinement | 1st | | 2nd | | 3rd | |
|---|---|---|---|---|---|---|
| percent cuts | max | total | max | total | max | total |
| iterative load migration | 24 | 10 | 26 | 11 | 25 | 10 |
| inertia repartition | 26 | 7 | 25 | 7 | 19 | 6 |

## 13.3 PARALLEL AUTOMATIC MESH GENERATION

The development of automatic mesh generation techniques for complex three-dimensional configurations has been an active area of research for over a decade (George 1991, Shephard and Weatherill 1991). The introduction of these mesh generation procedures has removed a major bottleneck in the application of finite element and finite volume analysis techniques. The introduction of scalable parallel computers is allowing the solution of ever larger models. It is now common to see meshes of several million elements. As mesh sizes become this large, the process of mesh generating on a serial computer becomes problematic both in terms of time and storage. Therefore, parallel mesh generation procedures that operate on the same computer, and using similar structures, as the parallel analysis procedures must be developed

With recent advances in the efficiency of automatic mesh generators which create well over two million elements per hour on a workstation (Weatherill and Hassan 1994), one may question the need for the parallel generation of meshes. One answer is that as the problem size grows, the solution process on parallel computers will continue to scale by the addition of more processors. However, mesh generation on a single processor will not scale, therefore becoming the computational bottleneck. A second, more critical, reason for parallel mesh generation is the shortage of memory on a sequential machine when dealing with very large meshes. On a parallel machine, the memory problem is addressed by distributing the mesh over a number of processors, each of which stores its own portion of the mesh.

Efficient parallel algorithms require a balance of work load among the processors while maintaining interprocessor communication at a minimum. Key to determining and distributing the work load and controlling communications is knowledge of the structure of the calculations and communications. Parallel mesh generation is difficult to effectively control since the only structure known at the start of the process is that of the geometric model which has no discernible relationship to the work load needed to generate the mesh. On the other hand, the more useful structure to discern work load and control communications is the mesh which is only fully known at the end of the process. The lack of initial structure and ability to accurately predict work load during the meshing process underlies the selection of algorithmic procedures in the parallel mesh generation procedure presented here. In particular, the procedure employs an octree decomposition of the domain to control the meshing process. The octree structure supports the distribution or redistribution of computational effort to processors.

## BACKGROUND AND MESHING APPROACH

Löhner et al (Löhner *et al.* 1992) have parallelized a two-dimensional advancing front procedure. The approach taken is to subdivide (partition) the domain (with the help of a background grid) and distribute the sub-domains to different processors for triangulation. The interior of subdomains are meshed independently. Then, the inter-subdomain regions are meshed using a coloring technique to avoid conflicts. Finally, the "corners" between more than two processors are meshed following the same basic strategy. A "one master-many slaves" paradigm has been chosen to drive the parallel procedures. This approach has been extended to three dimensions with some modifications (Shostko and Löhner 1995). A load balancing phase follows the initial domain splitting (at the background grid level). The interface gridding incorporates mechanisms (i) to avoid degradation of performance by using fine grain parallelism and (ii) to reduce the number of processors when there is too much communication overhead. Results show scalability of the method.

Saxena and Perruchio (Saxena and Perucchio 1992) describe a parallel Recursive Spatial Decomposition (RSD) scheme which discretizes the model into a set of octree cells. Interior and boundary cells are meshed by either templates or element extraction (removal) in parallel at the octant level meshes. This requires no communication between octants. The main difficulty for this meshing approach is to guarantee that a boundary octant can always be meshed regardless of the complexity of the model. Robust loop building algorithms which include possible tree refinement to resolve invalid configurations are in general difficult to parallelize (Shephard and Georges 1991). Parallel results have been simulated on a sequential machine.

The parallel mesh generator presented here meshes three-dimensional non-manifold objects following the hierarchy of topological entities. That is, the model edges are meshed first, the model faces are meshed second, and the model regions are meshed last. The current discussion focuses on the octree-based region meshing procedure.

Fig. (13.12) graphically depicts the basics of the present mesh generator. The first step in meshing a model region is to develop a variable level octree which reflects the mesh control information and is consistent with the triangulation on the boundary of the model region. Octants containing mesh entities classified on the boundary of the model region to be meshed are constructed to be approximately of the same size as the mesh entities they contain. A one level difference on octants sharing one or more edges is enforced during this process to control smoothness of the mesh gradations. Once the octree is generated, the octants are classified as *interior*, *outside*, or *boundary*. Those classified as *outside* receive no further consideration. Some *interior* octants are reclassified *boundary* if they are too close to mesh entities classified on the boundary of the model region (*boundary-interior*). The purpose of this reclassification is to avoid the complexities caused when *interior* octant mesh entities are too close to the boundary and may lead to the creation of poorly shaped elements in that neighborhood. *Interior* octants are meshed using templates. Face removal procedures are then used to connect the boundary triangulation to the interior octants. Fig. (13.13) graphically describes a face removal in a two-dimensional setting.

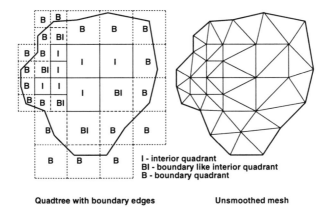

Quadtree with boundary edges          Unsmoothed mesh

Figure 13.12 Graphical depiction of the basics of the presented mesh generator

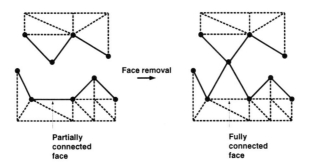

Figure 13.13 Face removal (2–D setting)

## SEQUENTIAL REGION MESHING

The starting point for the region meshing process is a completely triangulated surface. The surface triangulation must satisfy the conditions of topological compatibility and geometric similarity (Schroeder and Shephard 1991) with respect to the model faces. The region meshing process consists of the three steps of (i) generation of the underlying octree, (ii) template meshing of interior octants, and (iii) face removal to connect the given surface triangulation to the *interior* octants. These steps are outlined below and explained in more detail in (de Cougny *et al.* 1994, de Cougny *et al.* 1996).

Mesh faces to which tetrahedral elements will eventually be connected are referred to as partially connected faces. They are basically missing one connected tetrahedron in the manifold case, and one or two in non-manifold situations. Initially, the mesh faces classified on the model boundary are the partially connected mesh faces. Once templates have been applied, the interior mesh faces connected to exactly one tetrahedron are also partially connected mesh faces. The current set of partially connected mesh faces constitutes the front. During face removal, tetrahedra are connected to these faces, therefore eliminating them. Any face created in the construction of a tetrahedra, referred to as a new face, is a partially connected face until it is eliminated. The face removal process is complete when there are no partially connected mesh faces remaining.

## Underlying Octree

The octree is built over the given surface mesh to (i) help in localizing the mesh entities of interest, and (ii) provide support for the use of fast octant meshing templates. Proper localization is achieved by having each terminal octant reference any partially connected mesh face which is either totally or partially inside its volume. This information is used to efficiently guarantee the correctness of the face removal technique. The octree building process can be decomposed into: (i) root octant building, (ii) octree building, (iii) level adjustment, (iv) assignment of partially connected mesh faces to terminal octants, and (v) terminal octant classification.

The root octant must contain the given surface mesh within it. The terminal octants are constructed to be approximately the same size as any partially connected mesh face associated with them to ensure appropriate element sizes and gradations. This is done by visiting each mesh vertex in the initial surface mesh, computing the average size of the connected mesh edges, and refining the octree until any terminal octant around that vertex is at a level corresponding to that size. To ensure a smooth gradation between octant levels, no more than one level of difference is allowed between terminal octants that share an octant edge.

Once the tree is completed, partially connected mesh faces are assigned to terminal octants. Given a mesh face, terminal octants that should know about it can be separated into two groups: (i) those are in the path of each bounding mesh edge and (ii) those whose octant edges are in the path of the mesh face.

Any terminal octant which knows at least one partially connected mesh face is classified *boundary*. Terminal octants classified *boundary* separate *interior* terminal octants from *outside* terminal octants. Given the set of *boundary* octants for a valid surface triangulation, the process of determining the *interior* terminal octants is straightforward (de Cougny *et al.* 1994, de Cougny *et al.* 1996). After the basic octant classification process, *interior* terminal octants can exist which have boundary entities arbitrarily close to surface triangles in *boundary* octants. Since poorly shaped elements can result when these entities are too close, some *interior* terminal octants are reclassified as *boundary* based on closeness to a partially connected mesh face.

## Template Meshing of Interior Octants

Terminal octants classified *interior* are meshed using (i) meshing templates or (ii) fast meshing procedures when a template is not available. Examination of the number of templates required for all cases and the distribution of template usage indicates that octants with eight, nine, thirteen, and seventeen vertices cover over 90% of the interior octants. All the eight, nine, thirteen, and seventeen vertex octant configurations can be meshed by six templates (Fig. (13.14)) with the correct rotations applied. The remaining interior octants are then quickly meshed using a fast procedure which accounts for the fact that the octant is a rectangular prism. One very fast option is to create an interior vertex and to create the correct connections to it (Yerry and Shephard 1984).

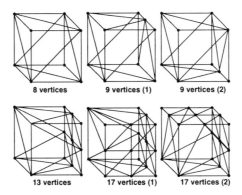

8 vertices     9 vertices (1)     9 vertices (2)

13 vertices     17 vertices (1)     17 vertices (2)

Figure 13.14 Terminal octant meshing templates available: one eight vertex case, two nine vertex cases, one thirteen vertex case, and two seventeen vertex cases

## Face Removal

Removal of a partially connected mesh face consists of connecting it to a mesh vertex. Since the volume to be meshed is the space between the given surface triangulation and the interior octree, the vertex used is usually an existing one. However, in some situations, it is desirable to create a new vertex. The choice of the target vertex (existing or new) must be such that the created element is of good quality and its creation does not lead to poor (in terms of shape) subsequent face removals in that neighborhood. The following pseudo-code indicates how the target vertex is selected for a given partially connected mesh face to be removed. See references (de Cougny *et al.* 1994, de Cougny *et al.* 1996) for an explaination of this process.

1. Collect set of potential target vertices from tree neighborhood,
2. Reorder target vertices with respect to decreasing shape measure (for the element to be created),
3. Consider next vertex in list as a target,
4. If the new element intersects any existing mesh entity, go to 3,
5. If the new element is too close to existing mesh entities, go to 3,
6. If no target vertex selected (yet), create one on the perpendicular to the face at centroid.

## PARALLEL CONSTRUCTS REQUIRED

### Octree and Mesh Data Structures

The two main data structures are the mesh and octree data structures. The mesh data structure (sequential) and parallel mesh data base (PMDB) both described above are used here to support the presented mesh generator. The octree data structure is on top of the mesh data structure. During the course of removing partially connected faces, processors will have to gather tree neighborhoods. Any processor must be able to gather the terminal octants making up a given tree neighborhood (if they are all on processor) or be told that at least one of them is off-processor. In the latter case, the processor will not attempt to perform the given face removal since neighboring information is not currently there. This type of information is easily available when each processor has full knowledge of the basic octree in terms of structure and processor assignment. This is the approach currently implemented. Although the size of the tree is small compared to that of the mesh, it does not scale indefinitely. Any terminal octant stores links to on-processor partially connected mesh faces and off-processor partially connected mesh faces totally or partially within its volume.

Techniques that maintain only portions of the tree on individual processors while providing tree neighboring information efficiently are currently under investigation. It is of interest to be able to retrieve tree neighboring information without having to communicate. This is to avoid having to involve all processors when requesting some tree neighboring information, which can degrade perfomance. In general, this requires processors to store sub-trees which are as complete as possible. The tree partitioner must produce tree partitions such that: (i) the load is balanced and (ii) tree neighboring information can be obtained without communicating. Since these two requirements may be conflicting, it may be necessary to duplicate small tree portions.

### Multiple Octant Migration

When the mesh generation process comes to a point when no face removal can be applied (face removals are not applied when needed tree neighborhoods are not fully on processor), the tree and associated mesh is repartitioned. The migration of octants is key to repartitioning once decisions concerning new destinations of terminal octants (classified *boundary*) have been made. Multiple octant migration itself relies on the multiple migration of partially connected mesh faces and/or mesh regions (described above). Note that multiple mesh region migration is also used in the final repartitioning at the region level once the mesh has been fully generated.

Any processor can send any number of terminal octants to another processor. When a terminal octant is migrated from one processor to another, the partially connected mesh faces not connected to any mesh region (these are the mesh faces remaining from the given surface triangulation) owned by the octant and/or the mesh regions that are bounded by at least one partially connected mesh face owned by the octant are migrated as well. An octant owns a mesh entity when it knows about it (has it within its volume) and has its centroid within its volume. Fig. (13.15) shows a two-dimensional example of the mesh regions to be migrated within an octant.

### Dynamic Repartitioning

Dynamic repartitioning enables redistribution of the load among processors as evenly as possible at key stages of the mesh generation process. These key stages are:

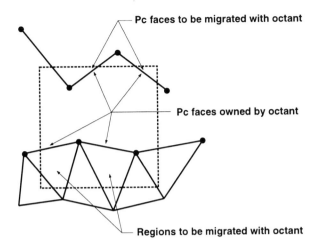

Figure 13.15 Octant migration

1. at the beginning of template meshing,
2. at the beginning of each face removal step, and
3. at completion of the mesh generation process.

Repartitioning for stages 1 and 2 is done at the terminal octant level (1 with respect to terminal octants classified *interior* and 2 with respect to terminal octants classified *boundary*). Repartitioning for stage 3 is performed at the mesh region level. The strategy is identical for both cases, only the process of migrating differs. The methods used here are geometry-based dynamic balancing (repartitioning) procedures which are described in section 1.

## PARALLEL REGION MESHING

### Template Meshing of Interior Octants

Once all terminal octants have been properly classified, those classified *interior* are partitioned. The parallel application of templates is a straight forward process in which there is no communication required during the process of creating the octant level meshes. The application of templates to octants sharing the same octant face implicitly lead to the same octant face triangulation. The finite elements generated in these octants are loaded into the processor mesh data structure. The interprocessor communication required at the end of this step is for the updating of interprocessor mesh entity links for mesh entities created on the boundaries of interior octants which are on processor boundaries. The cost for the application of templates is small compared to the cost of performing face removals. Therefore, the parallel efficiency of parallel region meshing is dictated primarily by the face removal part only.

### Face Removal

Parallel face removal is an iterative process where each iteration consists of three steps:

1. Tree repartitioning at the terminal octant (classified *boundary*) level,
2. Face removal step, and
3. Reclassification of terminal octants from *boundary* to *meaningless*

The goal of step 1 is to make sure that all processors will have an equal amount of work to perform during step 2. It is difficult to predict how much work or, more precisely, how many face removals (step 2) any processor will perform and the total amount of effort for a particular face removal. However, a terminal octant classified *boundary* is a good unit of work load since the set of all terminal octants classified *boundary* approximately corresponds to the domain still to be meshed. The difficulty of performing face removals in parallel resides in the fact that any face removal requires the knowledge of tree neighborhoods. If, at any point during the face removal, a tree neighborhood is not fully on-processor, the face removal is aborted and the next mesh face is considered for removal. The process of performing face removals and repartitioning the tree continues until there are no more partially connected mesh faces in the mesh. After a few iterations, the efficiency of the face removal stage can be low because information required to perform face removals is almost always off-processor. When more than half of the processors have an efficiency below some given threshold (25%), the processor set is reduced (by half). Since migration of terminal octants only deals with those classified *boundary* and only worries about mesh regions bounded by partially connected mesh faces, it is very likely that the final mesh will be scattered across processors with no real structure. It is therefore necessary to repartition in parallel the distributed mesh using IRB at the mesh region level with the original full set of processors. Fig. (13.16) shows the whole process of parallel face removal on four processors. The first 8 pictures display the currently partially connected mesh faces after the terminal octants classified *boundary* have been repartitioned. Note that iterations 1, 2, 3, and 4 use all four processors, iterations 5, 6, and 7 use two processors, and iteration 8 uses one processor. The final picture displays the final three-dimensional repartitioned mesh on four processors.

Table (13.3) shows the four processor speed-up for the *connecting rod* model and the final repartitioned mesh is shown in Fig. (13.17). Table (13.4) shows the eight processors speed-up for the *mechanical part* model and the final repartitioned mesh is shown in Fig. (13.18). The number of mesh regions created indicated in the captions corresponds to parallel face removal only and does not include template meshing. Face removal speed-up indicates speed-up for step 2 of the parallel face removal procedure. Total speed-up indicates speed-up for all steps (1, 2, and 3). In that case, the first repartitioning (iteration 1) is not counted since it can be considered an initial partitioning step. Note that the time taken to perform the first repartitioning depends on the size of the problem and not the number of processors. The speed-up is by definition set to 1.0 for the run with the smallest number of processors. The results show good speed-ups as long as the size of the problem is adequate with the number of processors on hand.

## 13.4  PARALLEL MESH ENRICHMENT

Since the mesh enrichment is an integral part of an adaptive solver, it must be run in parallel as well in order to not become a bottleneck. A variety of approaches are possible to perform the required mesh enrichments. Currently the local mesh modification procedures of edge-based refinement, edge-based derefinement and local mesh optimization, as described in (de Cougny and Shephard 1995), are employed. These procedures have

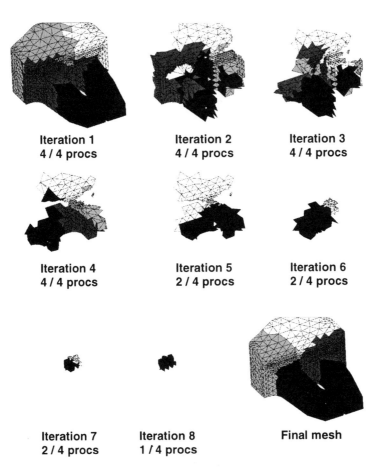

Figure 13.16  Successive face removal iterations and final repartitioned mesh for *chicklet*

Table 13.3  Face removal statistics for *connecting rod* (35,000 mesh regions created by face removals — 70,000 total)

| Procs | 1 | 2 | 4 |
|---|---|---|---|
| Iterations | 1 | 5 | 7 |
| Face removal speed-up | 1.0 | 1.9 | 3.3 |
| Total speedup | 1.0 | 1.8 | 2.9 |

been developed to interact directly with the geometric representations and deal with all the complexities introduced by curved geometries. The basic steps in this process are:

1. Derefinement using edge collapsing,
2. Grid optimization to improve element shapes,
3. Refinement using full set of subdivision patterns without consideration for stability,
4. Refinement vertex snapping (to the model boundary), and
5. Grid optimization to improve element shapes.

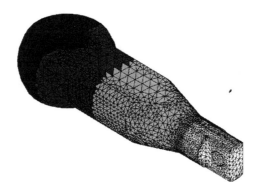

Figure 13.17  Final repartitioned mesh for *connecting rod* (4 processors)

Table 13.4  Face removal statistics for *mechanical part* (125,000
mesh regions created by face removals — 240,000 total)

| Procs | 2 | 4 | 8 |
|---|---|---|---|
| Iterations | 4 | 7 | 11 |
| Face removal speed-up | 1.0 | 2.0 | 3.4 |
| Total speedup | 1.0 | 1.9 | 3.2 |

Figure 13.18  Final repartitioned mesh for *mechanical part* (8 processors)

The next three subsections discuss the parallelization of these steps.

## DEREFINEMENT

If a mesh edge $_mT_1^1$ is marked for derefinement, it is attempted to be collapsed. If the affected polyhedron $pol(_mT_1^0)$ (made up of all connected mesh regions) is on processor $p_i$, the edge collapsing is performed on $p_i$. If $pol(_mT_1^0)$ is not fully on $p_i$, the missing mesh regions are requested from the appropriate processors. When all processors are done traversing their lists of mesh edges, the processors that have received requests send (migrate) the requested mesh regions. In Fig. (13.19), processor $p_0$ requests mesh regions from processors $(p_1, p_2, p_3)$ and the requested mesh regions are migrated. If there is conflict, the processor with lowest $p_i$ has priority. On the next iteration, it is the processor with highest $p_i$ that will have priority. This switching is done to prevent too much load imbalance at completion. The process of traversing the list of mesh edges and sending/receiving requests continues until all marked mesh edges have been collapsed (more exactly, have been attempted to be collapsed). Because mesh regions are migrated, it is possible that the processors are not well balanced after the derefinement step. The triangulation is therefore submitted to a load balancing step (at the region level) before going further. For a triangulation of 85,000 elements where 50% of the mesh edges are derefined (the resulting triangulation has approximately 46,000 elements), the speed-ups are approximately 1.8, 3.3, and 5.4 for 2, 4, and 8 processors, respectively.

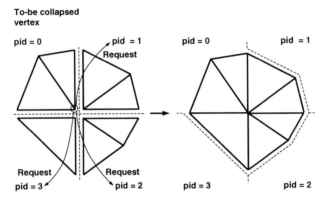

Figure 13.19 Mesh migration to support parallel distributed derefinement

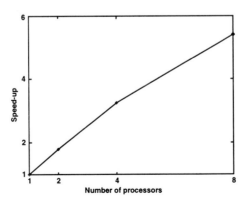

Figure 13.20 Speed-ups for derefinement (85,000 elements — 50% edges derefined)

## *TRIANGULATION OPTIMIZATION*

Assuming the current triangulation is partitioned, each processor $p_i$ ($0 \leq i < n_p$) optimizes its own partition ($_pT_{p_i}$). As processor $p_i$ pops a mesh region from its queue $Qu_i$, two situations may occur:

1.  All polyhedra to be considered by the mesh optimization tools are on processor $p_i$, in which case, the proper local retriangulation tool can be applied (if needed) and any poorly shaped new mesh region is pushed in $Qu_i$
2.  At least one polyhedron is not fully on $p_i$, in which case, $p_i$ requests any needed mesh region that is not on $p_i$ and push back the mesh region in $Qu_i$

The mesh region popping process continues until $Qu_i$ is empty or stuck (does not change). Clearly, requests concerning mesh regions that have already been deleted are cancelled. After a synchronization step, all processors examine the requests they have received and send (migrate) the appropriate mesh regions to the appropriate processors. If a mesh region is requested by several processors, the processor with lowest $p_i$ has priority and will be granted the mesh region. On the next iteration, it is the processor with highest $p_i$ that will have priority. This switching is done to prevent too much load imbalance at completion. Each processor $p_i$ adds to its queue $Qu_i$ any new mesh region of unacceptable shape that it has received and restarts popping mesh regions. The combined process of emptying the queue and migrating requested mesh regions terminates when all queues $Qu_i$ ($0 \leq i < n_p$) are empty or stuck and there is no mesh region to migrate. Because mesh regions are migrated, it is possible that the processors are not well balanced after the optimization step. The triangulation is therefore submitted to a load balancing step (at the region level) before going further. For a triangulation of 85,000 elements, the speed-ups are approximately 1.7, 3.1, and 4.8 for 2, 4, and 8 processors.

## *REFINEMENT*

Refinement relies on the marking (for refinement) of mesh edges. Any mesh region is given a weight proprotional to the number of marked bounding edges. Prior to applying the parallel refinement scheme, the triangulation is load-balanced based on that weight. Any mesh face on some partition boundary with at least one marked mesh edge is triangulated using two-dimensional subdivision patterns. Once all mesh faces on the partition boundary are subdivided, links for all new mesh entities are updated. Then, each processor can apply three-dimensional templates on any mesh region with at least one marked edge without any communication (de Cougny and Shephard 1995). Once all appropriate mesh regions have been subdivided, the refinement vertices which are classified on the model boundary need to be snapped to the corresponding model entity. Since snapping makes use of the local retriangulation tools, the technique to parallelize that process is similar to the one used to parallelize the derefinement and optimization steps. For a triangulation of 36,000 elements when 20% of the mesh edges are refined (resulting triangulation has 88,000 elements), speed-ups are approximately 1.9, 3.4, and 5.8 for 2, 4, and 8 processors, respectivley.

# 13.5   PARALLEL ADAPTIVE ANALYSIS

## *STRUCTURE AND FINITE ELEMENT FORMULATION*

Although the most computationally intensive operations in an adaptive analysis are of the same type as those of a fixed mesh analysis, an adaptive analysis must use more general structures which effectively account for the evolution of the discretization. The structure of a parallel adaptive analysis procedure follows directly from the procedures used for the parallel control of evolving meshes presented in the previous sections.

Two main processing phases naturally emerge in the finite element method, the "form phase", where the local finite element arrays at the sub-domain level are generated, and the "solve phase", where the global problem is solved. Parallel implementation of the form phase is straightforward.

On the other hand, the efficient scalable realization of the solve phase is a non trivial task. Current Multiple Instruction/Multiple Data (MIMD) computers tie together independent processors using a high speed switch under a message passing paradigm. The resulting system incorporates relatively powerful individual processors with large local memories. The communication bandwidth between processors remains well below that of the individual processor to memory bandwidth resulting in significant cost for interprocessor communication compared to local computation. This type of architecture has significant impact on the design of parallel algorithms for the solution of large linear systems. Such algorithms must amortize any communication costs over large amounts of simultaneous parallel computation. Additionally, the large local data space is still only a fraction of the global memory space and data cannot be highly duplicated over multiple processors if full advantage is to be taken of the available memory. Under these constraints, Krylov space based iterative solvers and domain decomposition techniques become attractive approaches.

The procedure developed employs a stabilized finite element formulation which is valid for general steady and unsteady compressible flow problems. The linear algebra is solved by means of a scalable implementation of the standard and matrix–free GMRES algorithms. Simple techniques are used for estimating regions of high error with the purpose of driving the adaptive procedures.

Details of the finite element formulation of our parallel adaptive code for compressible flow problems are discussed in (Bottasso and Shephard 1995). Here we simply review its basic capabilities.

The program is able to deal with steady and unsteady problems, in fixed and rotating frames, and it has been extensively used for the simulation of helicopter rotors in hover and forward flight. The formulation is based on the Time–Discontinuous Galerkin Least–Squares finite element method (Shakib 1988, Shakib *et al.* 1991). The TDG/LS is developed starting from the symmetric form of the Euler equations expressed in terms of the entropy variables and it is based upon the simultaneous discretization of the space–time computational domain. A least–squares operator and a discontinuity capturing term are added to the formulation for improving stability without sacrificing accuracy. This highly sophisticated approach to the solution of the compressible Euler equations is characterized by a very limited amount of numerical diffusion, allowing a nice resolution of vortical structures and sharp shocks in transonic regions.

Two different three dimensional space–time finite elements have been implemented in our code. The first is based on a constant in time interpolation, and, having low order of time accuracy but good stability properties, it is well suited for solving steady problems using a local time stepping strategy. The second makes use of linear–in–time basis functions and, exhibiting a higher order temporal accuracy, is well suited for addressing unsteady problems. In these cases, moving boundaries are handled by means of the space–time deformed element technique (Bottasso 1995, Tezduyar *et al.* 1992).

Discretization of the weak form implied by the TDG/LS method leads to a non–linear system, which is solved iteratively using a quasi–Newton approach. At each Newton iteration, a non–symmetric linear system of equations is solved using the GMRES algorithm. We have developed scalable parallel implementations of the preconditioned GMRES algorithm and of its matrix–free version (Johan *et al.* 1992, Johan 1992). This latter algorithm approximates the matrix–vector products with a finite difference stencil. This avoids the storage of the tangent matrix, thus realizing a substantial savings of computer memory at the cost of additional on–processor computations. Preconditioning is achieved by means of a nodal block–diagonal scaling transformation.

Error indication is currently limited to the evaluation of the norm of the gradient of the flow variables, or on estimates of the second derivatives of the solution due to (Löhner 1987). The edge values of the error indicator are computed by averaging the corresponding two nodal values. These edgewise error indicator values are then used for driving the mesh adaptation procedure. Appropriate thresholds are supplied for the error values, so that the edge is refined if the error is higher than the maximum threshold, while the edge is collapsed if the error is less than the minimum threshold.

## *EFFECTIVENESS OF PARALLEL ADAPTIVE ANALYSIS*

In the following we address some of the key questions related to the evaluation of the performance of a parallel adaptive code. To this aim, the discussion will be concentrated on a classical problem in CFD, namely the problem of the Onera M6 wing in transonic flight. This wing has been studied experimentally by Schmitt and Charpin (Schmitt and Charpin 1982). The wing is characterized by an aspect ratio of 3.8, a leading edge sweep angle of $30^o$, and a taper ratio of 0.56. The airfoil section is an Onera D symmetric section with 10% maximum thickness-to-cord ratio.

We consider a steady flow problem characterized by an angle of attack $\alpha = 3.06^o$ and a value of $M = 0.8395$ for the freestream Mach number. In such conditions, the flow pattern around the wing is characterized by a double–lambda shock on the upper surface of the wing with two triple points.

As a first step, we analyize the parallel performance of the code on a fixed mesh. To this end, we consider one single mesh and we measure the speed-ups attained by the program varying the number of processing nodes.

The simulation was performed using a mesh consisting of 128,172 tetrahedra, using the matrix–free GMRES algorithm with reduced integration of the interior elements and full integration of the boundary elements. A local time stepping strategy was employed with one single Newton iteration per time step, using a CFL condition of 5 in the first 20 time steps and a CFL equal to 10 for another 80 time steps, attaining a drop in the residual of three orders of magnitude. The mesh was partitioned using a parallel implementation of the IRB algorithm. The analysis was run on 4, 8, 16, 32, 64 and 128 processors of an IBM SP–2 and the results are presented in Fig. (13.21) in terms of the inverse of the wall clock time versus the number of processing nodes. The highly linear behavior of the parallel algorithm shows the excellent characteristics of scalability of the code.

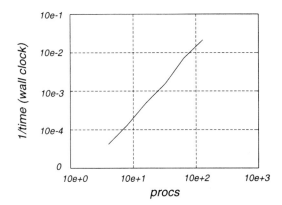

Figure 13.21  Parallel efficiency evaluated at fixed mesh for the Onera M6 wing in transonic flight. 128,172 tetrahedra, IRB partitions.

We then consider some aspects of the performance of the code, when mesh adaptivity is performed for more accurately resolving the features of the flow. An initial coarse mesh of 85,567 tetrahedra was partitioned with the IRB algorithm on 32 processing nodes and the analysis was carried out to convergence as previously explained. The results obtained were then used for computing an error indicator based on density and Mach number, which was employed for performing a first level of refinement, bringing the mesh to 131,000 tetrahedra. The solution was projected on the new vertices using a simple edge interpolation technique, and the analysis was then performed on the refined mesh for 80 time steps at a CFL number of 10. Similarly, two levels of refinement followed by subsequent analysis were performed, obtaining an intermediate 223,499 tetrahedron mesh and a final 388,837 tetrahedron mesh.

Fig. (13.22) shows the density isocontour plots on the upper surface of the wing corresponding to the initial and the final mesh discretizations. Note that the forward shock is barely visible in the results obtained with the initial coarse mesh, the aft shock presents significant smearing and the lambda shock located at the tip of the wing is not resolved. As expected, considerable improvement in the resolution of the shocks can be observed when mesh adaptation is employed.

Fig. (13.23) shows the initial and final meshes. Once again, elements assigned to the same subdomains are denoted by the same grey level. For the final mesh, the partitions shown are those obtained with the iterative load balancing algorithm.

We consider the evolution during the analysis of two fundamental parameters: (i) the surface–to–volume ratio for the subdomains, and (ii) the number of neighbors of each subdomain. The first of these two parameters essentially dominates the volume of communication in terms of the size of the messages to exchange, while the second parameter dominates the number of messages that each processor must send and receive.

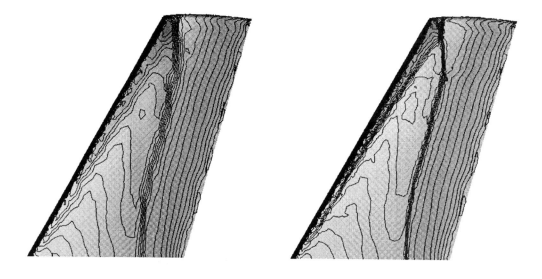

Figure 13.22  Onera M6 wing in transonic flight, $\alpha = 3.06^{o}$,
$M = 0.8395$. Density isocontour plots for the initial and final meshes.

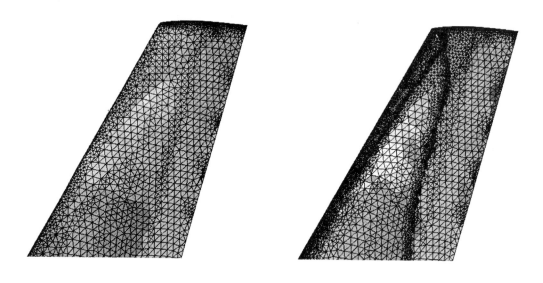

Figure 13.23  Onera M6 wing in transonic flight, $\alpha = 3.06^{o}$, $M = 0.8395$.
Initial and final meshes.  Grey levels indicate processor assignment.

Two distinct runs were made, the only difference between them being the repartitioning strategy adopted. In both cases, all the stages of the analysis —initial IRB partitioning, flow solution, error sensing, adaptation and load balancing— were performed automatically in parallel on 32 processing nodes, i.e. without ever leaving the parallel environment. The load balancing algorithm was activated three times during the adaptation of each of the meshes, after the refinement, after the snapping of the newly generated vertices to the curved boundaries of the model and after the local retriangulation[4]. At every call, the algorithm was requested to perform only approximately eight migration iterations, yielding a maximum out of balance number of elements per processing node equal to one at the end of each refinement level. This strategy allows better efficiency of the various stages of the adaptive algorithm that can then operate on balanced or nearly balanced meshes. This "incremental" rebalancing capability represents a nice advantage of the iterative load balancing scheme over other algorithms. The parallel repartitioning algorithm was instead activated just once at the end of each adaptive step.

The meshes obtained during the two previously mentioned parallel adaptive simulations of the Onera M6 wing were analyzed for gathering data on the overall performance of the analysis. Fig. (13.24) reports plots of the boundary faces and neighbor statistics. The quantities plotted are defined as:

*(i)* Surface–to–volume measures:

$$S_{\max} = \max_i \left(\text{Boundary Faces}_i / \text{Faces}_i\right), .$$

$$S_{\text{glob}} = \text{Boundary Faces}/\text{Faces}.$$

*(ii)* Neighbor measures:

$$N_{\max} = \max_i \left(\text{Neighbors}_i / (\text{Procs} - 1)\right),$$

$$N_{\text{avrg}} = \left(\sum_i \text{Neighbors}_i / (\text{Procs} - 1)\right)/\text{Procs}.$$

All these quantities are reported in Fig. (13.24) versus the number of tetrahedra in the mesh at a certain adaptive level normalized by the number of tetrahedra in the initial mesh. The solid line represents the values of the parameters obtained for the parallel adaptive analysis where the iterative load migration procedures were employed. The dashed line corresponds to the parallel adaptive analysis where the refined meshes were repartitioned after each adaptive step using the parallel IRB algorithm.

From the analysis of the first two plots at the top of Fig. (13.24), it is clear that the iterative load migration procedure effectively controls the surface–to–volume ratios, which in fact remain constant and fairly similar to the ones obtained with the IRB partitioning for the whole simulation. On the other hand, the second two plots of the same figure show that the number of neighbors of each subdomain tends to increase with the number of adaptive steps performed. A more detailed analysis shows that in

---

[4] We remark that in the current implementation, snapping can also cause load imbalance since it makes use of local triangulation.

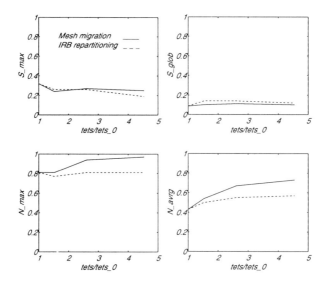

Figure 13.24 Boundary faces and neighbor statistics for the parallel–adaptive
analysis of the Onera M6 wing in transonic flight using the iterative
load migration (mesh migration) and IRB rebalancing schemes.

general each subdomain is connected by a significant amount of mesh entities (vertices, faces, edges) only with a reduced number of neighbors, while it shares a very limited number of mesh entities with the other neighbors. We are currently investigating ways of removing such small contact area interconnections, in order to achieve better control on the number of neighbors.

The different partition statistics provided by the two rebalancing algorithms and shown in the previous figure clearly have an impact on the performance of the flow solver. For example, the ratio of the wall clock timings for the flow solutions performed on the final adapted mesh was found to be 0.83, in favor of the repartitioning algorithm.

The two approaches were also compared in terms of relative wall clock timing cost. The repartitioning algorithm outperformed the iterative load migration scheme at each adaptive step. The ratio of the iterative migration wall clock time to the rebalancing wall clock time was found to be 4.07 at the first level (131,000 tetrahedron mesh), 4.41 at the second (223,499 tetrahedron mesh) and 2.21 at the third (388,837 tetrahedron mesh).

These preliminary test results seem to indicate that the iterative load migration scheme tends to be more computationally expensive than the parallel IRB algorithm, and at the same time does not yield the same quality of the partitions, at least with the currently implemented heuristics. A more complete analysis of the relative merits of the two approaches is however required for assessing their real merits. In particular, adaptive unsteady problems must be analyzed, where the adaptation is activated more frequently and usually the change in the mesh from one adaptation to the successive is not as relevant as in the steady problem considered here. Preliminary results seem to indicate that in such cases the iterative migration strategy could outperform the parallel repartitioning.

## 13.6  CLOSING REMARKS

This paper has presented progress made to date on the development of parallel automated adaptive analysis procedures for unstructured meshes which operate on distributed memory MIMD computers. The procedures presented allow for the reliable analysis, through the use of automated adaptive analysis, of large problems which can only be supported by the computational power of parallel computers. Specific emphasis was placed on the techniques needed to effectively support evolving meshes such that computational load balance was maintained throughout the simulation process.

## 13.7  ACKNOWLEDGMENTS

The authors would like to acknowledge the support of NASA Ames Research Center under grants NAG 2–832 and NCC 2–9000, the Army Research Office under grant DAAH04–93–G–0003, and the National Science Foundation under grant DMS-93–18184.

We would also like to acknowledge the input of several others that contributed to the material presented in this document. They are Saikat Dey, Pascal Frey, Rao Garimella, Marcel Georges, Ramamoorthy Ravichandran and Wes Turner.

## 13.8  REFERENCES

Beall, M. W. and Shephard, M. S. 1995. Mesh data structures for advanced finite element computations. 19-1995, Scientific Computation Research Center, Rensselaer Polytechnic Institute, Troy, NY 12180-3590. submitted to Int. J. Num. Meth. Engng.

Biswas, R. and Strawn, R. 1993. A new procedure for dynamic adaptation of three-dimensional unstructured grids. In *31st Aero. Sci. Meet.*, 1993.

Blelloch, G., Leiserson, C., Maggs, B., Plaxton, C., Smith, S. and Zagha, M. 1991. A comparison of sorting algorithms for the connection machine cm-2. *ACM*, pages 3–16.

Bottasso, C. L. and Shephard, M. S. 1995. A parallel adaptive finite element flow solver for rotary wing aerodynamics. In *12th AIAA CFD Conference*, San Diego, CA.

Bottasso, C. L. 1995. On the computation of the boundary integral of space–time deforming finite elements. *Comm. Num. Meth. Eng.* under review.

Connell, S. D. and Holmes, D. G. 1994. 3-dimensional unstructured adaptive multigrid scheme for the euler equations. *AIAA J.*, 32:1626–1632.

de Cougny, H. L. and Shephard, M. S. 1995. Parallel mesh adaptation by local mesh modification. 21-1995, Scientific Computation Research Center, Rensselaer Polytechnic Institute, Troy, NY 12180-3590. submitted.

de Cougny, H. L., Shephard, M. S. and Ozturan, C. 1994. Parallel three-dimensional mesh generation. *Computing Systems in Engineering*, 5(4-6):311–323.

de Cougny, H. L., Shephard, M. S. and Ozturan, C. 1996. Parallel three-dimensional mesh generation on distributed memory MIMD computers. *Engineering with Computers*, 12(2):94–106.

Devine, K. M. 1994. *An adaptive HP-finite element method with dynamic load balancing for the solution of hyperbolic conservation laws on massively parallel computers.* PhD thesis, Computer Science Dept., Rensselaer Polytechnic Institute, Troy, New York.

George, P. L. 1991. *Automatic Mesh Generation.* John Wiley and Sons, Ltd, Chichester.

Golias, N. and Tsiboukis, T. 1994. An approach to refining three-dimensional tetrahedral meshes based on delaunay transformations. *Int. J. Numer. Meth. Engng.*, 37:793–812.

Gursoz, E. L., Choi, Y. and Prinz, F. B. 1990. Vertex-based representation of non-manifold boundaries. In Wozny, M. J., Turner, J. U. and Priess, K, editors, *Geometric Modeling Product Engineering*, pages 107–130. North Holland.

Hammond, S. W. 1991. *Mapping Unstructured Grid Computations to Massively Parallel Computers*. PhD thesis, Computer Science Dept., Rensselaer Polytechnic Institue, Troy,.

JaJa, J. 1992. *An introduction to Parallel Algorithms*. Addison Wesley, Reading Mass.

Johan, Z., Hughes, T. J. R., Mathur, K. K. and Johnsson, S. L. 1992. A data parallel finite element method for computational fluid dynamics on the connection machine system. *Comp. Meth. Appl. Mech. Engng.*, 99:113–134.

Johan, Z. 1992. *Data Parallel Finite Element Techniques for Large-Scale Computational Fluid Dynamics*. PhD thesis, Stanford University.

Kallinderis, Y. and Vijayan, P. 1993. Adaptive refinement-coarsening scheme for three-dimensional unstructured meshes. *AIAA J.*, 31(8):1440–1447.

Kruskal, C. P., Rudolph, L. and Snir, M. 1990. Efficient parallel algorithms for graph problems. *Algorithmica*, 5:43–64.

Leiss, E. and Reddy, H. 1989. Distributed load balancing: Design and performance analysis. Vol. 5, W. M. Keck Research Computation Laboratory.

Löhner, R. and Ramamurti, R. 1993. A parallelizable load balancing algorithm. In *Proc. of the AIAA 31st Aerospace Sciences Meeting and Exhibit*, 1993.

Löhner, R., Camberos, J. and Merriam, M. 1992. Parallel unstructured grid generation. *Comp. Meth. Appl. Mech. Engng.*, 95:343–357.

Löhner, R. 1987. An adaptive finite element scheme for transient problems in cfd. *Comp. Meth. Appl. Mech. Engng.*, 61:323–338.

Ozturan, C. 1995. *Distributed Environment and Load Balancing for Adaptive Unstructured Meshes*. PhD thesis, Rensselaer Polytechnic Institute, Troy, NY.

Requicha, A. A. G. and Voelcker, H. B. 1983. Solid modeling: Current status and research directions. *IEEE Computer Graphics and Applications*, 3(7):25–37.

Saltz, J., Crowley, R., Mirchandaney, R. and Berryman, H. 1990. Run-time scheduling and execution of loops on message passing machines. *Journal of Parallel and Distributed Computing*, 8(2):303–312.

Saxena, M. and Perucchio, R. 1992. Parallel fem algorithms based on recursive spatial decompositions - I. automatic mesh generation. *Computers and Structures*, 45:817–831.

Schmitt, V. and Charpin, F. 1982. Pressure distributions on the onera m6 wing at transonic mach numbers. R-702, AGARD.

Schroeder, W. J. and Shephard, M. S. 1990. A combined octree/Delaunay method for fully automatic 3-D mesh generation. *Int. J. Numer. Meth. Engng.*, 29:37–55.

Schroeder, W. J. and Shephard, M. S. 1991. On rigorous conditions for automatically generated finite element meshes. In Turner, J., Pegna, J. and Wozny, M, editors, *Product Modeling for Computer-Aided Design and Manufacturing*, pages 267–281. North Holland.

Sedgewick, R. 1990. *Algorithms in C*. Addison-Wesley Publishing Company.

Shakib, F., Hughes, T. J. R. and Johan, Z. 1991. A new finite element formulation for computational fluid dynamics: X. The compressible Euler and Navier Stokes equations. *Comp. Meth. Appl. Mech. Engng.*, 89:141–219.

Shakib, F. 1988. *Finite Element Analysis of the Compressible Euler and Navier-Stokes Equations.* PhD thesis, Stanford University.

Shephard, M. S. and Finnigan, P. M. 1989. Toward automatic model generation. In Noor, A. K. and Oden, J. T, editors, *State-of-the-Art Surveys on Computational Mechanics*, pages 335–366. ASME.

Shephard, M. S. and Georges, M. K. 1991. Automatic three-dimensional mesh generation by the Finite Octree technique. *Int. J. Numer. Meth. Engng.*, 32(4):709–749.

Shephard, M. S. and Weatherill, N. P, editors. 1991. *Int. J. Numer. Meth. Engng.*, volume 32. Wiley-Interscience, Chichester, England.

Shephard, M. S., Flaherty, J. E., de Cougny, H. L., Ozturan, C., Bottasso, C. L. and Beall, M. W. 1995. Parallel automated adaptive procedures for unstructured meshes. In *Parallel Computing in CFD*, volume R-807, pages 6.1–6.49. AGARD, Neuilly-Sur-Seine, France, 1995.

Shephard, M. S. 1988. The specification of physical attribute information for engineering analysis. *Engineering with Computers*, 4:145–155.

Shostko, A. and Löhner, R. 1995. Three-dimensional parallel unstructured grid generation. *Int. J. Numer. Meth. Engng.*, 38:905–925.

Szymanski, B. K. and Minczuk, A. 1978. A representation of a distribution power network graph. *Archiwum Elektrotechniki*, 27(2):367–380.

Tezduyar, T. E., Behr, M., Mittal, S. and Liou, J. 1992. A new strategy for finite element computations involving moving boundaries and interfaces - the deforming-spatial-domain/space time procedure: I. the concept and preliminary tests. *Comp. Meth. Appl. Mech. Engng.*, 94:339–351.

Vidwans, A., Kallinderis, Y. and Venkatakrishnan. 1994. Parallel dynamic load-balancing algorithm for three-dimensional adaptive unstructured grids. *AIAA Journal*, 32(3):497–505.

Weatherill, N. P. and Hassan, O. 1994. Efficient three-dimensional delaunay triangulation with automatic point creation and imposed boundary constraints. *Int. J. Numer. Meth. Engng.*, 37:2005–2039.

Weiler, K. J. 1988. The radial-edge structure: A topological representation for non-manifold geometric boundary representations. In Wozny, M. J., McLaughlin, H. W. and Encarnacao, J. L, editors, *Geometric Modeling for CAD Applications*, pages 3–36. North Holland.

Williams, R. 1990. *DIME: Distributed Irregular Mesh Environment.* Supercomputing Facility, California Institute of Technology.

Williams, R. D. 1992. Voxel databases: A paradigm for parallelism with spatial structure. *Concurrency*, 4:619–636.

Williams, R. D. 1993. Dime++: A parallel language for indirect addressing. CCSF-34, Caltech Concurrent Supercomputing Facilities, Pasadena.

Yerry, M. A. and Shephard, M. S. 1984. Automatic three-dimensional mesh generation by the modified-octree technique. *Int. J. Numer. Meth. Engng.*, 20:1965–1990.

# Index

*Index compiled by Geoffrey C. Jones*